Catastrophe Theory and its Applications

Catastrophe Theory and its Applications

Tim Poston BATTELLE, GENEVA
and Ian Stewart UNIVERSITY OF WARWICK

Pitman

London · San Francisco · Melbourne

PITMAN PUBLISHING LIMITED
39 Parker Street, London WC2B 5PB

FEARON–PITMAN PUBLISHERS INC.
6 Davis Drive, Belmont, California 94002, USA

Associated Companies
Copp Clark Ltd, Toronto
Pitman Publishing New Zealand Ltd, Wellington
Pitman Publishing Pty Ltd, Melbourne

First published 1978

AMS Subject Classifications: (main) 58C25, 58E05
(subsidiary) 70C10, 70K20, 73B30,
73C20, 73H05, 78A10

Filmset in Northern Ireland at the Universities Press (Belfast) Ltd.
Printed by photolithography and bound in Great Britain at The Pitman
Press, Bath

ISBN 0 273 01029 8

to Christopher Zeeman
at whose feet we sit
on whose shoulders we stand

Preface

Since the first rumours in the mid-1960's of René Thom's *Stabilité Structurelle et Morphogénèse*, which finally appeared in 1972, there has been a rapid growth of interest in the subject now known as *catastrophe theory*. Thom suggested using the topological theory of dynamical systems, originated by Poincaré, to model discontinuous changes in natural phenomena, with special emphasis on biology; and he pointed out the importance in this context of *structural stability*, or insensitivity to small perturbations. He further remarked that in one context this requirement implied that the system could be described, locally, by one of seven standard forms, the *elementary catastrophes*.

As well as great interest, Thom's ideas have generated great confusion and, more recently, controversy. Early claims for the theory's universality (partly misstatements based on confusion between elementary and non-elementary catastrophes, and partly overstatements attributable to the 'youthful enthusiasm' for a new subject) have been repeated too often without proper qualification. In some circles, too, the belief has arisen that catastrophe theory is 'purely qualitative', with a split between those who think this is a good thing and those who do not. The wide range of precursors to the theory in many fields (itself a product of the typicality that we shall account for in Chapter 7) has led some to infer that the theory contains no new material at all. Speculative extensions of the theory to realms where its applicability is not guaranteed by the appropriate mathematical formalism have been misinterpreted as definitive applications; and disputes arising in these areas have contaminated other fields where the problems are quite different. These misunderstandings may largely be traced to the unfamiliar mathematical language in which the theory is couched, and the tendency of mathematicians to emphasize aspects of the theory that are not always in sympathy with the practical requirements of the working scientist. Thus Turing, charged that computers only operated *deterministically*, replied that was how he was asked to design them. The same goes for *qualitatively* with topologists and catastrophe descriptions, except that they asked themselves. You want numbers, we have numbers; but, simply, most topologists *don't* want numbers, they want qualities – though these sometimes acquire a fearsomely algebraic, even numerical, expression. The problems have been exacerbated by the lack of suitable source material between the extremes of hard-core topology and soft-centre popularization.

Our first aim in this book is to explain the mathematical ideas involved, in terms which will be accessible to the practising scientist who is familiar with calculus in several variables and a little linear algebra. This places catastrophe theory in its rightful position as an extension of, or a development within, the calculus (rather than a radical new departure, or a replacement for current methods, as is sometimes thought to be the case). It also makes clear the limitations of the theory. Unless one understands in some detail the precise mathematical hypotheses involved, and the way in which they lead to the conclusions, one will fail to have an adequate feel for what the theory can or cannot be made to do. It has been said more than once that it is possible to apply Thom's theorem without understanding the mathematics behind it: we disagree. In fact we disagree with the implication that it is Thom's *theorem* that should be applied: analysis of the most solid and successful applications shows that the methods and concepts that lie behind the theorem are often of greater importance than the result itself.

Our second aim is to explode the myth that catastrophe theory is purely qualitative. We achieve this by the direct method of surveying some of its quantitative applications. We concentrate on the physical sciences, where the existing mathematical theory leads naturally to problems that fall within the domain of catastrophe theoretic methods, and where these methods may be used as mathematical tools to provide quantitative information that may be tested by experiment. We place considerable emphasis on the computational aspects of the subject and the explicit calculations that may be performed using it, illustrating these both by mathematical examples and in applications. Catastrophe theory methods have a clearly defined, though not universal rôle to play in the physical sciences, and it is important that controversy over the less well established applications should not be permitted to obscure this fact.

The mathematical chapters which form the first half of the book do no more, in principle, than to expound the theory as it now stands; but the novelty of our approach is to use the mathematics of the working scientist to motivate the style of thinking involved and the results obtained. We do not give the rigorous proofs of the more powerful theorems (where the deepest and newest mathematics lies) but we do give in a new way the geometric heart of these arguments, which explains (better than a strictly formal treatment) why the results are true. Independently of any question of applications, the mathematical theorems of catastrophe theory form an essential contribution to an important and natural problem: the study of singularities of families of smooth functions. Our treatment of them here may also prove useful as introductory motivation for those who wish to study the mathematics in its full depth.

One result that we do prove explicitly (using only elementary calculus) is the important *Splitting Lemma* whereby the number of variables in a problem may be (often drastically) reduced. To some scientists this result has appeared as the most surprising of the theory: its essentially classical nature deserves to be better known, as does the result itself.

The second half of the book is on applications. We not only discuss the more established and familiar of these, but include some very recent work that is less well known, and a certain amount of material that has not seen print before. In the latter category are the stability analysis of an idealized oil-rig in Chapter 10; the sections on mirages and sonic booms in Chapter 11; the quantitative exploitation of the *non-local* bifurcation set of the elliptic umbilic catastrophe in Chapter 12 in connection with fluid flow; parts of Chapter 13 on elasticity, especially the treatment of a double-cusp catastrophe in a buckling plate; much of Chapter 14 on thermodynamics; the bee theory and the new *constraint catastrophes* of Chapter 16, along with other material on the formation of biochemical and ecological frontiers. Chapter 15 is also new, and is due entirely to Bob Gilmore and Lorenzo Narducci. The reader who has had previous contact with the catastrophe literature will find much that he has not seen before, as well as some of the old favourites. Indeed, the current explosive growth of catastrophe theory is well illustrated by the fact that if this book had gone to press in Summer 1976, roughly half of our material on applications could not have been included. The time pressures on the writing (which we hope are not too visible via consequent errors) were only partly exerted by the publisher: the subject itself was breathing down our necks.

A detailed list of the applications discussed will be found on the contents pages and will not be given here; but two points should be made. The first is that we have attempted, as far as possible, to consult research workers in the fields of application, to check that our exposition is in line with current thinking in those subjects. This of course does not imply that the experts agree with our theorizing. But it helps to avoid the effect, too apparent in some of the literature, whereby the physics encountered in applications is that of the nineteenth century. When arguing that physics relies too heavily on the mathematics of the nineteenth century – as in some areas we feel it does – it seems best to avoid committing the same error in reverse. Not being physicists, naval architects, biologists, engineers ourselves, it is too much to hope that we have entirely succeeded in this attempt, but we have at least tried.

The second point is that many of the applications, notably in Chapters 11, 12 and 13, are to systems governed by *partial* differential equations. This is so despite the fact that the obvious application is only to a special class of *ordinary* differential equations, a fact that is often advanced as an objection. It happens this way because mathematics does not always respect the tidy categories into which it is habitually placed. The bifurcations describable by (elementary) catastrophe theory occur commonly in a much wider class of equations than in the special class (gradient ordinary differential equations) in which it is most obvious that they are the *only* ones that can stably occur. Rigorous studies of partial differential equations can often – though certainly not always – lead to elementary catastrophes. Any study of a mathematical problem may transfer it to a different area from that in which it was posed.

The penultimate chapter gives a brief survey of some of the uses that have been made of catastrophe-inspired models in the social and behavioural sciences. By relating these to our earlier work we are able to offer reasoned and (we hope) constructive criticism of these types of model. The chapter also serves to exhibit the wide spectrum across which attempts to apply catastrophe theory have been made. We hope that a non-partisan discussion of this controversial area may help to set it in perspective. We have therefore concluded with one simple example where both elementary catastrophe modelling, and the implicit equilibrium hypotheses of traditional verbal economics ('the invisible hand of the market') fail in spectacular fashion.

We do not consider the resolution of this particular controversy to be of essential importance to the development of catastrophe theory *in general*; any more than the arguments about astrology or biorhythms or general relativity can affect the status of spherical trigonometry, Fourier analysis, or differential geometry. Our own views on the likely future development of catastrophe theory are given briefly in the final chapter, and may be summarized as follows. In the immediate future only the physical sciences will see solid benefits, because of their selection of 'simple' systems, and more recently of those with *disorganized* complexity, which can be 'statistically simple'. The organized complexity of biology offers the best hope for the medium term, but here it is the whole of dynamical systems theory that will be required (with catastrophe theory a small but essential component). The organized complexity of social systems is unlikely to be well understood until after we have come to grips with biological systems. The only important question to be resolved at the moment is whether catastrophe theory is worth pursuing at all. We feel that, if only for the immediate returns to be gained in physics, the answer must be 'yes': whether we may thereby be taking a small step towards understanding the more spectacular problems of human existence is a question that may reasonably be left to future generations.

A book of this type, cutting across traditional subject lines, would not have been possible without the generous assistance of experts in the fields involved, which we acknowledge with gratitude. Christopher Zeeman alone is almost too numerous to mention; it is questionable whether, without his pioneering efforts and enthusiasm, the subject would have advanced to the stage at which sufficiently many people were even aware of its existence, let alone embarking on criticism of it. Both of the authors of this book made their initial contacts with catastrophe theory by way of his lectures; and we hope he will consider it a compliment if we say that his teaching was so successful that we not only began to understand a tiny part of Thom's theories, and Zeeman's efforts to develop them, but on occasions found ourselves in disagreement with both! To him this book is respectfully dedicated.

Every contact with René Thom shed a new and often mysterious light on the beauties of mathematics and the sciences. Ken Ashton supplied us with his ecological data. Ruth Bellairs told us of her biological experiments and corrected our use of biological language. Michael Berry allowed us to borrow his extensive knowledge of optics, along with many of his photographs; he and Malcolm Mackley were similarly generous regarding fluid dynamics, and Malcolm Mackley went to considerable lengths to supply us with pre-publication copies of the beautiful photographs of fluid flow that adorn Chapter 11, each of which represents many hours of work on his part. As experimental evidence for a solid, technological application of the theory, these photographs are essential to the message of the book. Bill Supple, Giles Hunt, Michael Thompson and Michael Sewell kept us informed of developments in engineering and admonished our ignorance. Edgar Ascher and Bob Gilmore instructed us in thermodynamics. Colin Renfrew, Alan Wilson, Robert Magnus and David Pitt allowed us to quote from unpublished work; Bob Gilmore essentially wrote Chapter 15 for us. Our original plan to collaborate with Ted Woodcock, who was to have provided several chapters on biology, was prevented by the pressures of time: however, he is with us in spirit and, more physically, represented by some of his computer graphics. The exigencies of space forbid the further listing of several dozen people who made important contributions, either to the content of the book or to the morale of its authors.

I. N. S. & T. P.
March 1977
Coventry and Geneva

The work of Tim Poston was supported by the Fonds National Suisse de la Recherche Scientifique (grant no. 2.461-0.75) with additional support by Battelle Institute, Ohio (grant no. 333-207).

Contents

1 Smooth and Sudden Changes

Classical physics (from Newton to General Relativity) is essentially the theory of various kinds of smooth behaviour; above all the awe-inspiring fall of the planets around the sun: unresting, unhasting and utterly regular. Even the wobbles that have dethroned Earth's rotation as the standard clock happen smoothly. No coherent and mathematical description of celestial mechanics can allow, say, a huge comet falling into the solar system, parting the Red Sea as it passes Earth, and *then* losing most of its kinetic energy and settling down into an almost perfectly circular orbit as the planet Venus (a widely held pseudoscientific theory). Planets interact much too evenly for that.

1 Catastrophes

Other things, however, jump. Water suddenly boils. Ice melts. Earths and moons quake. Buildings fall. The back of a camel is stable, we are told, under a load of N straws, but breaks suddenly under a load of $N+1$. Stock markets collapse.

These are *sudden* changes caused by *smooth* alterations in the situation: an analogous astronomical event would be the Sun's steady motion around the galaxy causing the Earth to switch (instantly or in a matter of days) to an orbit ten million miles wider, when some critical position was reached. Such changes are far more awkward for prediction and analysis than the stars in their courses, and the sciences (from physics to economics) are still gathering together the analytical techniques to handle such jumping behaviour.

Now there are many kinds of jump phenomena. There are forces that build up until friction can no longer hold them: the roar of an earthquake, and the rustle of rhubarb growing, are made by the movements when friction gives way. There is a critical population density below which certain creatures grow up as grasshoppers, above which as locusts: this is why locusts, when they *do* occur, do so in a huge swarm. A cell suddenly changes its reproductive rhythm and doubles and redoubles, cancerously. A man has a vision on the road to Tarsus.

Many of these still defy analysis: many have been analysed, with a tremendous variety of mathematical methods. We shall be concerned in this book with one particular mathematical context which covers a broad range of such phenomena in a coherent manner. The techniques involved were developed by the French mathematician René Thom and became widely known through his book *Stabilité Structurelle et Morphogénèse* [1] in which he proposed them as a foundation for biology. The sudden changes involved were christened by Thom *catastrophes*, to convey the feeling of abrupt or dramatic change: the word's overtones of disaster are, for most applications, misleading. The subject has since become known as *catastrophe theory*, a phrase which is open to a variety of interpretations depending on the scope accorded it.

These techniques apply most directly (but far from exclusively) to systems that through varying situations seek at each moment to minimize some function (e.g. energy) or maximize one (e.g. entropy). We shall clarify in Chapter 3 what this means mathematically. For the present a good picture is that of a ball rolling around a landscape and 'seeking' through the agency of gravitation to settle in some position which, if not the lowest possible, is at least lower than any other nearby. (But meanwhile the landscape itself is changing.) The particular geometrical forms that arise in this setting have become known, following Thom, as *elementary* catastrophes, in the sense of fundamental entities (like chemical elements) and their use as expounded in this book is thus 'elementary catastrophe theory' (a phrase misinterpreted by Sussman and Zahler [1a] to resemble 'elementary arithmetic'), though it is deep both mathematically and scientifically. For some systems more complicated phenomena can occur (we give an easily explained example in Chapter 17 Section 7), whose onset Thom [1] classes collectively as *generalized* catastrophes. Their theory is by no means so complete.

Physical intuition is important for the understanding of catastrophe theory. In this chapter we shall describe three simple physical systems exhibiting typical catastrophic behaviour, having the advantage that (unlike earthquakes or stock markets) they are simple enough to build, and small enough to carry around. In addition they may be used for elementary experiments. They are well adapted to analysis, although we shall not analyse them at this stage, and will be used repeatedly as examples. The reader will find his intuition very much assisted if he actually makes them (for which reason we give some practical indications as to their construction) and plays with them. No description can compete with direct experience. But it must be emphasized that these machines bear a similar relation to catastrophe theory as do the toys known as 'Newton's cradle' and 'the simple pendulum' to Newtonian mechanics.

2   The Zeeman Catastrophe Machine

We begin with the first machine invented. E. C. Zeeman, of the University of Warwick, devised it in 1969: after three weeks of

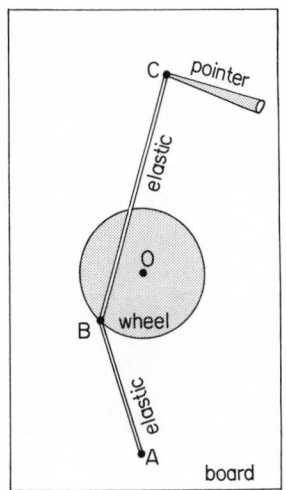

Fig. 1.1

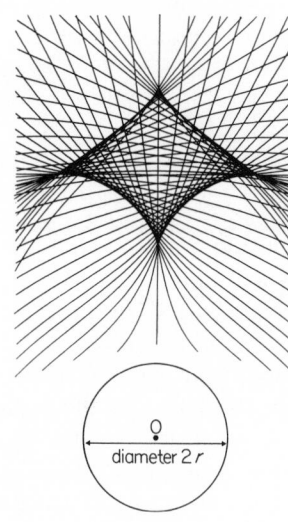

Fig. 1.2

experimentation with rubber bands and paperclips he refined it to the version we describe. The first appearance in print was Zeeman [2]: other references include Poston and Woodcock [3] and Dubois and Dufour [4].

It consists of a wheel (Fig. 1.1) mounted flat against a board, able to turn freely, and not too heavy: too much friction resisting the movement or inertia prolonging it obscure the behaviour we wish to study. To one point (B) on its edge are attached two lengths of elastic. One of these has its other end fixed to the board at point (A), far enough from the hub (O) of the wheel to keep the elastic BA always tight. The second has its other end (C) attached to a pointer, to be held in the hand. (The position of C can thus be controlled from a little distance without obscuring it.) Dimensions which work well in practice are a wheel of radius 3 cm, OA of length 12 cm, and each piece of elastic of unstretched length 6 cm.

Regardless of the radius r of the wheel, the unstretched lengths a and b of the elastic BA and BC, and the distance OA (as long as this is more than $a + r$) the machine will show qualitatively the behaviour to be described below. This is a part of the property of 'structural stability' which we discuss later: changes in the parameters make no essential qualitative difference, in a sense to be made precise at the relevant time. However, we shall be analysing the machine for one particular set of numbers, with the aid of computer graphics; so we now give detailed instructions for a machine whose behaviour has exactly the geometry drawn by the computer.

Photocopy Fig. 1.2. An enlarged or reduced photograph is perfectly acceptable, since scale does not affect the behaviour, or its subsequent analysis. Mount the result on board or heavy card, and attach a wheel at point O. Attach a stiff wire to the wheel, perpendicular to the plane of the board, at radius r from O. (It may be convenient to make the wheel itself a little larger, since only the position of the wire matters, and combine the mounting of the wire with that of the wheel as in Fig. 1.3.) Fix another stiff wire perpendicular to the board at A. File a groove round each wire, both at the same distance above the board, and higher than any central raised point of the wheel. Take a piece of good quality rubber cord (not a cut rubber band or sewing elastic: better the square section cord sold for catapults† or model aeroplanes), somewhat longer than four times the diameter $2r$ shown for the wheel in your copy of Fig. 1.2. Attach the middle to the wire at A, binding it with cotton to form a loop round the groove (Fig. 1.4(a)). Mark the point whose distance along the unstretched elastic is $2r$, holding the elastic doubled and straight: bind on each side of it to form a tight loop around the groove in wire B (Fig. 1.4(b)). Bind the point whose distance is $2r$ further along to a pointer (Fig. 1.4(c)). Care will be needed in making AB exactly $2r$ long: it may help to bind the

† In the USA: slingshots.

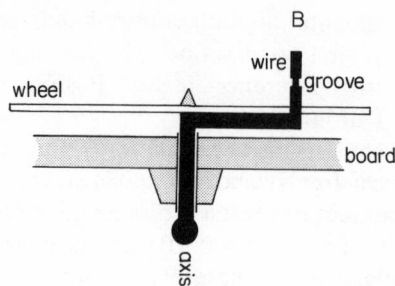

Fig. 1.3

doubled end to a loop, attach point B, and only then slip the loop over wire A.

Now experiment with the machine. You will find that if C is held equidistant with A from the board over a point on the board outside the four-pointed region ✧, only one position can be occupied by the wheel under the influence of the elastic alone. If you push it to some other position and release it, it jumps back again. This one position will depend on that of C, but a smooth change in C will lead to a smooth change in the position of the wheel.

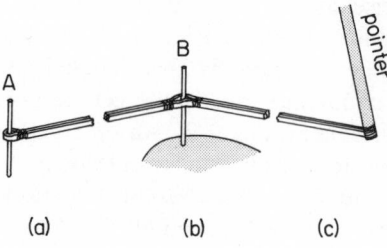

Fig. 1.4

(a)   (b)   (c)

When C is inside ✧, however, two such positions are possible. By entering ✧ smoothly from one side, the wheel moves smoothly to one of them; entering on the other side and taking C to the same point carries the wheel to the other.

Only if you can now decide by pure thought what happens if you enter from the left by the upper/lower edge and leave on the right by the upper/lower edge of ✧ (four possibilities altogether) do you *not* need to make this machine to understand it properly.

If time is of the essence, a few minutes will make a qualitatively accurate version of the machine, using stiff card for the board, drawing pins† as wheel axle and wires, and elastic bands (all found in any office). Link the bands over pin B as shown in Fig. 1.5: if you fasten them down to the wheel they get tangled when it rotates. The position and exact shape of ✧ will change a little, but can be found experimentally. (How?)

Fig. 1.5

† In the USA: thumbtacks.

3 Gravitational Catastrophe Machines

Photocopy Fig. 1.6 (again scale does not matter) and back it with light card, about postcard thickness. Cut round the figure accurately (a knife or razor blade is best) and cut another piece of card into a ring a few centimetres wide whose outer edge is identical to that of the first. Make six triangular beams of equal length, about one

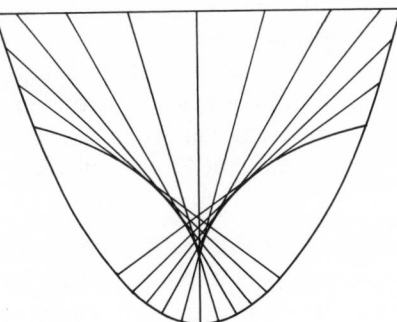

Fig. 1.6

quarter that of the axis of the parabola, as in Fig. 1.7. Glue them to points near the edge of the parabola, evenly spaced, with one at each corner; and to the corresponding positions on the ring, so that when laid on its face, the solid card has its boundary directly below that of the ring. A small heavy magnet behind the solid card will grip a light piece of metal in front (Fig. 1.8) and can be slid to any desired position while retaining a good grip.

Since most of the mass of the assembled device is in the magnet, we may take the centre of gravity of the whole to be the position of the magnet. When the machine balances steadily on edge, the centre of gravity must be vertically above the point of contact. If the machine rests on a level plane, the plane must be a tangent to the edge, so the centre of gravity lies on the corresponding *normal* (the line through the point of contact perpendicular to the tangent). The straight lines in Fig. 1.6 are some of these normals.

Experiments with the machine, or geometric thought along the above lines, will answer the following questions.

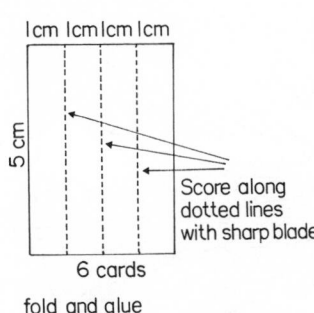

Score along dotted lines with sharp blade

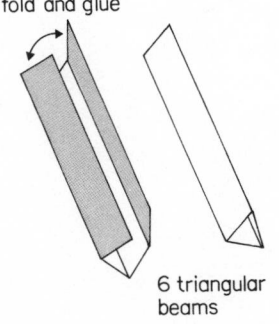

fold and glue

6 triangular beams

Fig. 1.7

(a) What, if any, positions of the magnet give the machine N possible angles at which it can balance (where $N = 0, 1, 2, 3, \ldots$ and there is a new question for each choice of N)?

(b) Putting the magnet *anywhere* on the normal at a point P places the centre of gravity vertically above P, so the machine can balance at P. However, for some positions of the magnet on this normal the machine will return to a point of balance after a small wobble (that is, the equilibrium is *stable*), for others, it will topple over like an egg that

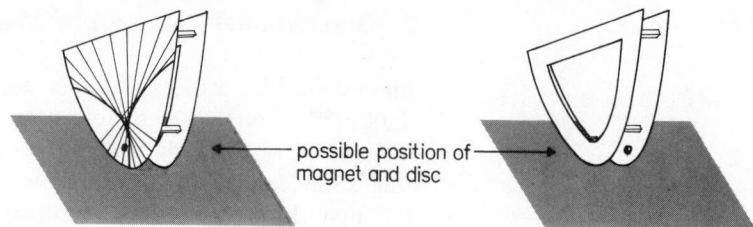

possible position of magnet and disc

Fig. 1.8

has been stood on end (the equilibrium is *unstable*). What distinguishes the two?

(c) When does a small change in the position of the magnet leave the machine sitting in a position which is also slightly different, and when does it make the machine roll right over (a dramatic 'catastrophic change' which is unmistakable in a practical experiment)?

Now repeat the construction, with the parabola replaced by an ellipse. Fig. 1.6 is replaced by Fig. 1.9. Answer the same three questions in this case.

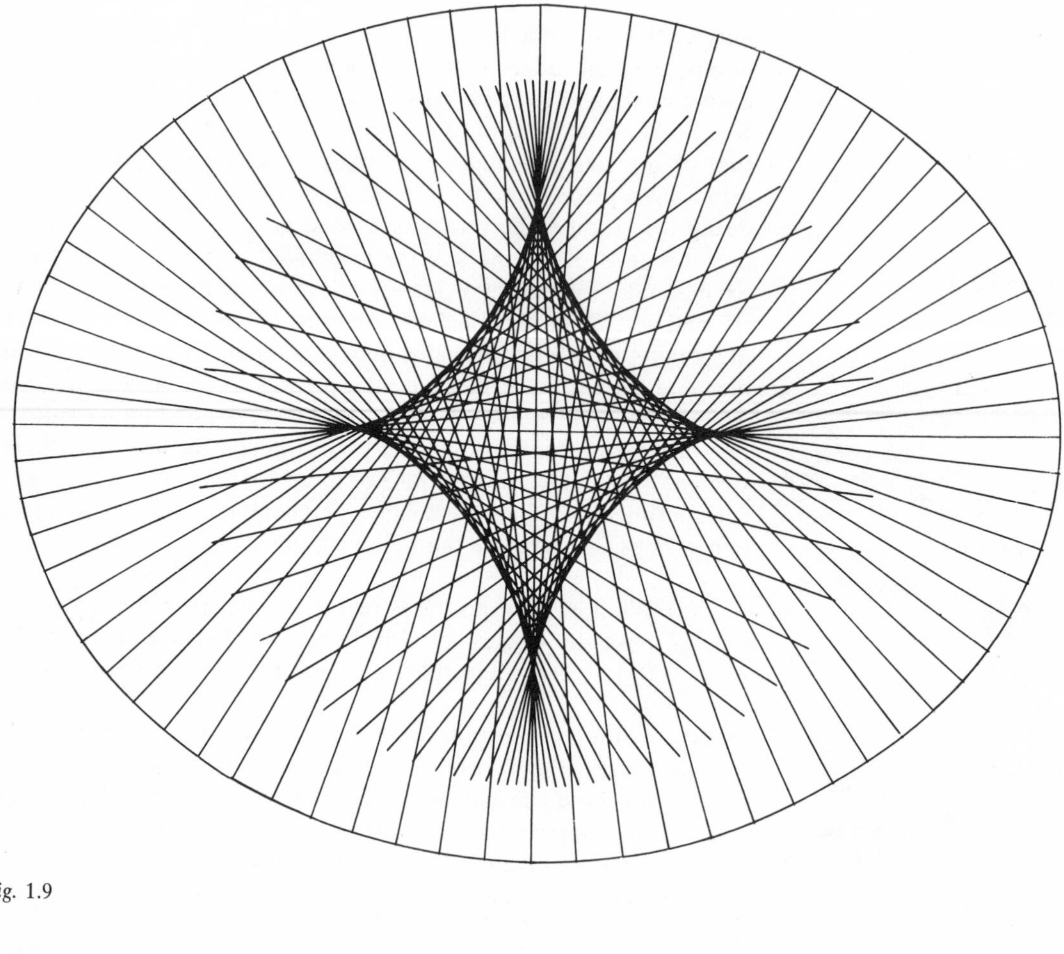

Fig. 1.9

These machines are not as artificial as they may seem. Both of them turn out to correspond closely (Chapter 10) to larger scale phenomena in the behaviour of ships.

4 Catastrophe Theory

The complicated behaviour of the above machines shows that even simple problems in classical statics conceal many subtleties. A deeper analysis reveals that there are some underlying regularities in the mathematical structure which permit routine calculations of how such systems behave, based on the traditional applied mathematician's use of Taylor series approximations. But these techniques also conceal many subtleties. The main mathematical thrust of this book is to develop a proper understanding of the geometric and algebraic methods used to handle Taylor series *properly*. Once developed, the methods provide powerful tools for tackling a wide range of problems, going far beyond simple statics of artificial machines, and opening up perspectives to which the traditional use of Taylor expansions as a source of approximations, to be justified *post hoc* by experiment, is blind.

Catastrophe theory is not a single thread of ideas; it resembles more closely a web, with innumerable interconnected strands; these include physical intuition and experiment, geometry, algebra, calculus, topology, singularity theory and many others. This web is itself connected to and embedded in a larger web: the theory of dynamical systems. A proper perspective on the theory involves some appreciation of *all* of these strands and the way they combine. The elementary catastrophes of René Thom are but one strand, though an important one. That they only come in seven basically different shapes is an intriguing fact, but it is not the only significant feature to be dealt with. It is not Thom's *theorem*, but Thom's *theory*, that is the important thing: the assemblage of mathematical and physical ideas that lie behind the list of elementary catastrophes and make it work.

2 Multidimensional Geometry

A proper understanding of catastrophe theory involves a feeling for the geometry of space of many dimensions, backed up by suitable algebraic and analytic techniques. This permits a geometric approach to the calculus of several variables: an important viewpoint which can motivate and simplify calculations by relating them to geometric insights.

The first few sections of this chapter review essential linear algebra, presenting the geometric view that is sometimes missed in treatments of 'matrix theory'. (For a more detailed geometric account, with proofs and many more pictures, see Dodson and Poston [5], which also develops the rigorous geometry of calculus in several variables.) We then take our first, classical steps in catastrophe theory. The most widely publicized feature of the theory has been the classification theorem mentioned above and discussed in Chapter 7: up to suitable changes of coordinates, a small number of standard forms are 'typical' for many phenomena. Coordinate changes thus play a key role in the theory. Here we show *linear* coordinate changes in action, reducing *polynomial* functions to a few standard expressions. This is both a key example of the kind of 'classification' the theory achieves in a far more glorious context, and a vital ingredient in what we do later. Much of the subsequent material aims to reduce other problems to those we solve in this chapter.

1 Set-theoretic Notation

It is convenient to make use of some elementary notions from set theory. A *set* is a collection of objects (of arbitrary nature) and these objects are called the *elements, members* or *points* of the set. The notation

$x \in X$

means that x is a member of the set X. Usually the sets we shall consider will be sets of points in a multidimensional space. A set with no members, we call *empty*.

A set X is a *subset* of another set Y if every element of X is an element of Y. We write

$X \subseteq Y$

for this, and say also that X is *contained* in Y, or that Y *contains* X.

The set of all elements x for which a particular property or condition $P(x)$ holds is denoted

$$\{x \mid P(x)\}.$$

The *union* of two sets X and Y is defined to be

$$X \cup Y = \{x \mid x \in X \text{ or } x \in Y\},$$

and the *intersection* is

$$X \cap Y = \{x \mid x \in X \text{ and } x \in Y\}.$$

For arbitrary elements x, y we may introduce the *ordered pair* (x, y) with the property that $(x, y) = (u, v)$ if and only if $x = u$ and $y = v$. The *Cartesian product* of two sets X and Y is then defined to be

$$X \times Y = \{(x, y) \mid x \in X \text{ and } y \in Y\}.$$

As standard notation we use $\mathbb{R}$ to denote the set of real numbers.

Cartesian products may be visualized geometrically. For example, if $\mathbb{R}^2 = \mathbb{R} \times \mathbb{R}$ and

$$X = \{(x, y) \in \mathbb{R}^2 \mid x^2 + y^2 = 1\}$$
$$Z = \{z \in \mathbb{R} \mid 1 \leqslant z \leqslant 2\}$$

then $X \times Z$ may be thought of as a set of points (x, y, z) in three-dimensional space $\mathbb{R}^3$ looking like the surface of a cylinder, as in Fig. 2.1.

There is a more general notion of an ordered *n-tuple*

$$(x_1, x_2, \ldots, x_n)$$

which may be thought of as belonging to a repeated Cartesian product

$$X_1 \times X_2 \times \cdots \times X_n.$$

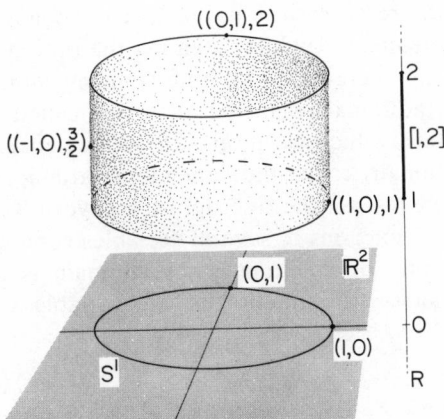

Fig. 2.1

One of the most important concepts for our purposes is that of a function. If X and Y are sets, then a *function f* with *domain X* and *codomain Y* is a rule† which associates to each $x \in X$ a *unique* element $f(x) \in Y$. Functions are also called *maps* or *mappings*. We write

$$f : X \rightarrow Y$$

and read this as 'f is a function from X to Y'. We say that f maps X to Y, and x to $f(x)$. When discussing the effect of f on *elements* we use a different type of arrow, thus $x \mapsto f(x)$. The *image* $f(x)$ of f is the subset

$$\{f(x) \mid x \in X\}$$

of Y. The image $f(A)$ of $A \subseteq X$ *under f* is

$$\{f(a) \mid a \in A\}.$$

For example the function

$$\sin : \mathbb{R} \rightarrow \mathbb{R}$$

$$x \mapsto \sin(x)$$

maps x to $\sin x$. The function log which maps x to $\log x$ cannot be defined unless x is positive. So

$$\log : \{x \in \mathbb{R} \mid x > 0\} \rightarrow \mathbb{R}$$

$$x \mapsto \log x$$

is a function whose domain is the set of positive real numbers.

There is a certain freedom of choice as regards the codomain. For example, sin may also be thought of as having codomain

$$\{x \in \mathbb{R} \mid -1 \leqslant x \leqslant 1\} = \sin(\mathbb{R}),$$

or indeed any set containing this. It is customary to choose any codomain which is convenient.

Traditional texts usually define functions by phrases like:

'The function $f(x) = x^2$'.

In our context this will be interpreted as 'the function $f : \mathbb{R} \rightarrow \mathbb{R}$ which maps $x \in \mathbb{R}$ to $x^2 \in \mathbb{R}$'. This allows us to use the slightly imprecise traditional language whenever it is clear what the precise meaning is, while retaining the option of being more pedantic if we can thereby avoid confusion in ambiguous cases. When using the traditional language x is often called the (independent) *variable*, a term which we retain for convenience. (The value $y = f(x)$ is traditionally called the dependent variable, which term we avoid, though we reserve the right to use the word 'variable' with reference to y.)

Functions of several variables come under the same heading if we think of functions whose domain is a Cartesian product. For instance, the function of two variables

$$f(x, y) = x^2 + y^2$$

†A more rigorous definition can be given, but is not required for our purposes. *See* Stewart and Tall [6].

may be viewed as a function f with domain $\mathbb{R} \times \mathbb{R}$ and codomain $\mathbb{R}$, mapping $(x, y) \in \mathbb{R} \times \mathbb{R}$ to $x^2 + y^2$. A function of n variables is just a function

$$f : X_1 \times \cdots \times X_n \rightarrow Y$$

which maps $(x_1, \ldots, x_n) \in X_1 \times \cdots \times X_n$ to $f(x_1, \ldots, x_n) \in Y$.

If $f : A \rightarrow B$ and $g : C \rightarrow D$ are functions, and if $f(a) \in C$ for all $a \in A$, we define the *composition* $g \circ f$ of f and g by

$$g \circ f(a) = g(f(a)).$$

Then $g \circ f$ is a function $A \rightarrow D$. In particular we can make this definition when $B = C$.

If $f : A \rightarrow B$ and $g : B \rightarrow A$ are such that

$$g(f(a)) = a$$
$$f(g(b)) = b$$

for all $a \in A$, $b \in B$, then we say that g is the *inverse function* to f and write

$$g = f^{-1}.$$

(Note that many of the traditional 'inverse functions' such as $\sin^{-1}$ are either not functions in our strict sense, being 'multivalued', or must be defined on carefully chosen domains.) Even when f has no inverse we use the notation

$$f^{-1}(Y)$$

to denote the set of all $a \in A$ such that $f(a) \in Y$, for a subset Y of B.

If $X \subseteq A$, we define the *restriction* of f to X to be the function

$$f|_X : X \rightarrow B$$

for which

$$f|_X(a) = f(a).$$

It differs from f only in being defined on a smaller domain X. (For further treatment of these concepts and especially the non-traditional notation and terminology, *see* Stewart and Tall [6].)

2   Euclidean Space

High dimensional spaces are studied by using a generalized kind of coordinate geometry. For any integer $n > 0$ we define *n-dimensional Euclidean space* to be

$$\mathbb{R}^n = \{(x_1, \ldots, x_n) \mid x_i \in \mathbb{R} \text{ for } i = 1, \ldots, n\} \qquad (2.1)$$
$$= \mathbb{R} \times \cdots \times \mathbb{R}.$$

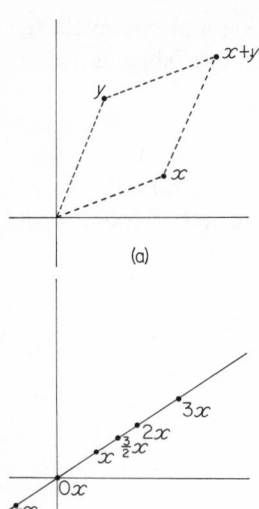

(a)

(b)

Fig. 2.2

It is convenient to use the abbreviated notation

$$x = (x_1, \ldots, x_n)$$

and refer to the x_i as the (*i*th) *components* or *coordinates* of x.

If $x, y \in \mathbb{R}^n$ and $\lambda \in \mathbb{R}$ we define *addition* and *multiplication by a scalar* λ by

$$x + y = (x_1 + y_1, \ldots, x_n + y_n)$$
$$\lambda x = (\lambda x_1, \ldots, \lambda x_n),$$

and we write 0 for $(0, \ldots, 0)$. These operations give $\mathbb{R}^n$ the structure of a *real vector space*. When we have this structure in mind, its points are called *vectors*.

The operations have a geometric interpretation in $\mathbb{R}^2$ or $\mathbb{R}^3$. The addition rule corresponds to the 'parallelogram law' (Fig. 2.2(a)), and scalar multiplication to a change of scale (and of direction for a negative scalar) as in Fig. 2.2(b). By analogy we can imagine similar geometric operations on $\mathbb{R}^n$: this gives a vivid language but needs algebraic verification that expected properties carry over. In particular when $x \neq 0$, the set of all λx ($\lambda \in \mathbb{R}$) is called the *straight line through* 0 and x; and 0 is called the *origin*.

A *subspace* of $\mathbb{R}^n$ is a subset W with the properties

$$\text{If } x, y \in W \text{ then } x + y \in W \tag{2.2}$$

$$\text{If } x, y \in W \text{ and } \lambda \in \mathbb{R}, \text{ then } \lambda x \in W. \tag{2.3}$$

To see what this means geometrically, consider $\mathbb{R}^3$. Condition (2.3) says that if any $x \in W$ then so does the line through x and 0; then condition (2.2) says that given any two points in W, the vertex of the corresponding parallelogram lies in W. We have several cases to consider.

(a) $W = \{0\}$. This certainly is a subspace.

 If $W \neq \{0\}$ we can find $x \in W$ with $x \neq 0$. Then the line through x and 0 also lies in W. This line may be the whole of W:

(b) $W = \{\lambda x \mid \lambda \in \mathbb{R}\}$, a line through 0.

 If not, there exists $y \in W$ not lying on the line $\{\lambda x\}$. Then the line through y and 0 is in W, and also the vertices of all parallelograms whose sides lie along these two lines, namely, points $\lambda x + \mu y$ for $\lambda, \mu \in \mathbb{R}$. Clearly (Fig. 2.3) these are the points in the plane through 0, x and y. If this is the whole of W,

(c) $W = \{\lambda x + \mu y \mid \lambda, \mu \in \mathbb{R}\}$, a plane through 0.

 Finally, W may contain another point z not on this plane. A picture like Fig. 2.3, but with three lines and using parallelepipeds, shows that $W = \{\lambda x + \mu y + \nu z \mid \lambda, \mu, \nu \in \mathbb{R}\} = \mathbb{R}^3$. Hence the last possibility is:

(d) $W = \mathbb{R}^3$.

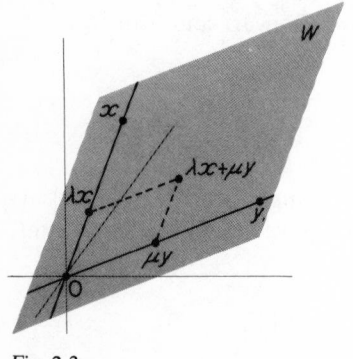

Fig. 2.3

Thus subspaces of $\mathbb{R}^3$ are either points, lines, planes or $\mathbb{R}^3$ itself (solids) which contain the origin. Intuitively speaking, they are 'flat' subsets containing the origin; and at least for $\mathbb{R}^3$ they may be classified as 0-dimensional (the origin alone), 1-dimensional (lines), 2-dimensional (planes) and 3-dimensional ($\mathbb{R}^3$). To generalize these ideas and make them precise we introduce some algebra.

A set of points $\{v^1, v^2, \ldots, v^r\} \in \mathbb{R}^n$ is *linearly dependent* if there exist scalars $\lambda_1, \lambda_2, \ldots, \lambda_r$, not all zero, such that

$$\lambda_1 v^1 + \cdots + \lambda_r v^r = 0. \tag{2.4}$$

If no such equation holds (or in other words if such an equation implies that all scalars $\lambda_i = 0$) then the set is *linearly independent*.

Geometrically, two points are linearly independent if neither lies on the line through 0 to the other; three points are linearly independent if none lies on the plane through 0 and the other two; and so on.

A set of points $v^1, \ldots, v^r$ is said to *span* a subspace W if every element of W can be written as a *linear combination*

$$\lambda_1 v^1 + \cdots + \lambda_r v^r \qquad (\lambda_1, \ldots, \lambda_r \in \mathbb{R}),$$

and if all such linear combinations lie in W (or equivalently each v^i does). A *basis* for a subspace W is a linearly independent set of elements which spans W. The *dimension* of W is the number of elements in a basis: an important theorem states that this is independent of the basis chosen. As a convention, $\{0\}$ has dimension 0. We write

$$\dim W$$

to denote the dimension of W.

In some contexts we have to refer to *infinite-dimensional* spaces, namely, those for which no finite set is a basis. (For instance, the vector space of polynomials in x: any finite list has a highest degree, say k, so that x^{k+1} cannot be a linear combination of polynomials in the list.) But we will usually be able to avoid most of the technical complications that can result from this.

It can be proved that a subspace W of $\mathbb{R}^n$ must have dimension in the range

$$0 \leqslant \dim W \leqslant n,$$

and that any dimension in this range occurs for suitably chosen W. Usually the choice of W is not unique, but if $\dim W = 0$ then $W = \{0\}$, and if $\dim W = n$ then $W = \mathbb{R}^n$. The difference $n - \dim W$ is called the *codimension* of W in $\mathbb{R}^n$. A *cobasis* for W in $\mathbb{R}^n$ is a set of vectors $v^1, \ldots, v^r$ which, together with a basis for W, yield a basis for $\mathbb{R}^n$. Necessarily r then equals the codimension of W. Note that a cobasis is not uniquely determined by W, but involves arbitrary choices.

3 Linear Transformations

A *linear transformation* or *linear map* from $\mathbb{R}^n$ to $\mathbb{R}^m$ is a function $f: \mathbb{R}^n \to \mathbb{R}^m$ with the properties

$$f(x + y) = f(x) + f(y)$$
$$f(\lambda x) = \lambda f(x)$$

for all $x, y \in \mathbb{R}^n$, $\lambda \in \mathbb{R}$. To find the general form of a linear transformation we take bases $u^1, \ldots, u^n$ for $\mathbb{R}^n$ and $v^1, \ldots, v^m$ for $\mathbb{R}^m$. Then for each i, $f(u^i)$ is some element of $\mathbb{R}^m$, so there must exist scalars λ_{ji} for which

$$f(u^i) = \lambda_{1i} v^1 + \cdots + \lambda_{mi} v^m$$

$$= \sum_j \lambda_{ji} v^j. \tag{2.5}$$

Each element on $\mathbb{R}^n$ is uniquely expressible in the form

$$x = \sum_i \mu_i u^i \qquad (\mu_i \in \mathbb{R}),$$

so that

$$f(x) = f\left(\sum_i \mu_i u^i\right) = \sum_i \mu_i f(u^i) = \sum_{i,j} \mu_i \lambda_{ji} v^j.$$

Thus every linear transformation is of this form. It is easy to verify the converse: everything of this form is a linear transformation, no matter what values the scalars λ_{ji} take. (Physicists please note: we never sum over repeated indices without a $\sum$ to say so.)

It is usually most convenient to take for $\mathbb{R}^n$ the *standard basis*

$$u^1 = (1, 0, 0, \ldots, 0)$$
$$u^2 = (0, 1, 0, \ldots, 0)$$
$$\cdots$$
$$u^n = (0, 0, 0, \ldots, 1),$$

with a similar choice of v^j, so that

$$x = \sum_i \mu_i u^i = (\mu_1, \mu_2, \ldots, \mu_n).$$

Then the μ_i are the coordinates of x, which we usually write x_i. Making this notation change, we have

$$f(x_1, \ldots, x_n) = (\lambda_{11} x_1 + \cdots + \lambda_{1n} x_n, \ldots, \lambda_{m1} x_1 + \cdots + \lambda_{mn} x_n).$$

What, geometrically, is a linear map? To see this we take the easiest case, maps $f: \mathbb{R}^2 \to \mathbb{R}^2$. Suppose that (Fig. 2.4)

$$f(1, 0) = (\alpha, \beta)$$
$$f(0, 1) = (\gamma, \delta).$$

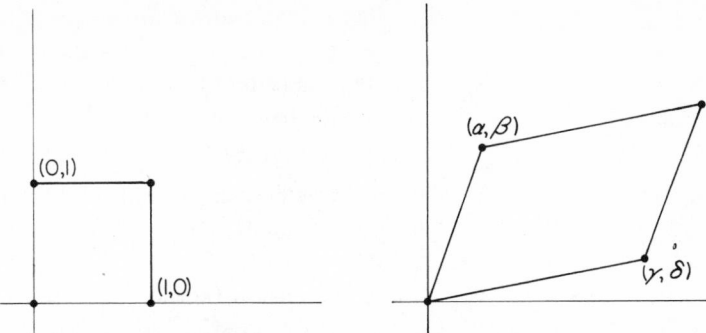

Fig. 2.4

Then, for example $f(1, 1) = (\alpha + \gamma, \beta + \delta)$. The effect of f is to distort the plane in a manner which preserves straight lines through the origin, sending squares into parallelograms (Fig. 2.5(a)). This, at least, is the case when (α, β) and (γ, δ) are linearly independent. If they are linearly dependent (but not both 0) then f maps $\mathbb{R}^2$ to a line, squashing it flat (Fig. 2.5(b)). Should $\alpha = \beta = \gamma = \delta = 0$, then f maps the whole of $\mathbb{R}^2$ to the origin (Fig. 2.5(c)), squashing still further.

Similar things occur for maps $\mathbb{R}^3 \to \mathbb{R}^3$, $\mathbb{R}^2 \to \mathbb{R}^3$, $\mathbb{R}^3 \to \mathbb{R}^2$; and so we view any linear transformation (even in higher dimensions) in geometric terms as a distortion which preserves straight lines through 0 and maps (multidimensional) cubes to (multidimensional) parallelepipeds.

Returning to the maps $f : \mathbb{R}^2 \to \mathbb{R}^2$, notice that when $\mathbb{R}^2$ is mapped to a line, every point on that line is the image of a whole line of points; when f maps to 0, this is the image of the whole plane; but when f maps $\mathbb{R}^2$ to itself, each point is the image of a unique point. Roughly, the greater the amount of squashing required (in terms of

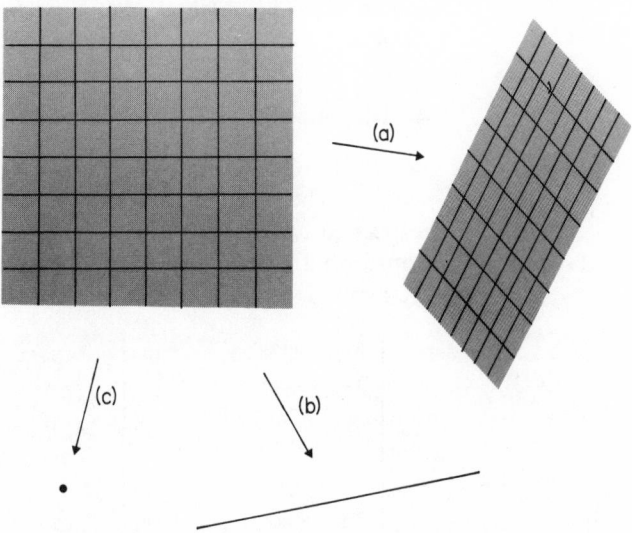

Fig. 2.5

dimension), the more things get squashed to a point (in the same sense). To make this observation respectable, define the *rank* of f to be the dimension of the image $f(\mathbb{R}^n) = \{f(x) \mid x \in \mathbb{R}^n\}$. The *nullity* of f is the dimension of its *kernel*

$$\{x \in \mathbb{R}^n \mid f(x) = 0\}.$$

It may now be proved that the rank of f plus the nullity of f equals n. An equivalent statement, and often more revealing geometrically, is that

codimension $(\mathrm{kernel}\, (f)) = \mathrm{rank}\, (f)$.

That is, the number of directions *not* squashed flat by f is the number of directions in its image.

We say that f is *non-singular* or *invertible* if there is an *inverse* linear map $g : \mathbb{R}^m \rightarrow \mathbb{R}^n$ such that $f(g(x)) = x$ for all $x \in \mathbb{R}^m$ and $g(f(x)) = x$ for all $x \in \mathbb{R}^n$. This is the case if and only if $m = n$ and the rank of f is also n, or equivalently, its nullity is 0. Thus the rank may be used to see how far f is from being invertible, or to what extent it squashes points together: how *singular* it is.

We can use these ideas to classify linear maps $\mathbb{R}^n \rightarrow \mathbb{R}$ up to change of coordinates. If $f : \mathbb{R}^n \rightarrow \mathbb{R}$ is linear and non-zero, then its rank is 1 and its nullity is $n - 1$. Taking a basis $v^1, \ldots, v^{n-1}$ for the kernel, and extending to a basis for $\mathbb{R}^n$ by adding in v^n such that $f(v^n) = 1$, we find that f is specified by

$$f(v^i) = 0 \quad 1 \leqslant i \leqslant n-1,$$

$$f(v^n) = 1.$$

If g is another linear map $\mathbb{R}^n \rightarrow \mathbb{R}$, the same goes for g relative to a basis $u^1, \ldots, u^n$; hence a change of coordinates from v's to u's makes f and g look the same. Thus all non-zero linear maps $\mathbb{R}^n \rightarrow \mathbb{R}$ look the same up to change of coordinates, all zero maps *are* the same: end of classification.

The same ideas show that if f and g are maps $\mathbb{R}^n \rightarrow \mathbb{R}$ with the *same* kernel, then they are scalar multiples of each other.

4 Matrices

We turn from the geometric to the computational aspects of the subject. The scalars λ_{ji} which define a linear transformation with respect to two bases (as in (2.5)) are said to form the *matrix* of the transformation relative to these bases, and are usually written in a rectangular array

$$\begin{bmatrix} \lambda_{11} & \lambda_{12} & \cdots & \lambda_{1n} \\ \lambda_{21} & \lambda_{22} & \cdots & \lambda_{2n} \\ \cdot & & \cdots & \\ \cdot & & \cdots & \\ \cdot & & \cdots & \\ \lambda_{m1} & \lambda_{m2} & \cdots & \lambda_{mn} \end{bmatrix} = (\lambda_{ji})_{j \leqslant m, i \leqslant n}$$

This is an $m \times n$ matrix, with m *rows* and n *columns*. Matrices are extremely convenient for making calculations with linear transformations. Given linear transformations $f : \mathbb{R}^n \to \mathbb{R}^m$, $g : \mathbb{R}^n \to \mathbb{R}^m$, and $h : \mathbb{R}^m \to \mathbb{R}^p$, we define scalar multiplication, addition, and the product by

$$(\alpha f)(x) = \alpha f(x)$$
$$(f + g)(x) = f(x) + g(x)$$
$$h \circ f(x) = h(f(x))$$

for $x \in \mathbb{R}^n$, $\alpha \in \mathbb{R}$. If we pick bases $u^1, \ldots, u^n$ for $\mathbb{R}^n$, $v^1, \ldots, v^m$ for $\mathbb{R}^m$, and $w^1, \ldots, w^p$ for $\mathbb{R}^p$ then the corresponding operations on the matrices (λ_{ji}), (μ_{ji}), (ν_{kj}) are as follows. Scalar multiplication and addition correspond to the formulae

$$\alpha(\lambda_{ji}) = (\alpha \lambda_{ji}) \qquad (\alpha \in \mathbb{R})$$
$$(\lambda_{ji}) + (\mu_{ji}) = (\lambda_{ji} + \mu_{ji}).$$

More interesting is matrix multiplication, corresponding to composition of linear transformations. The matrices for f, h are determined by

$$f(u^i) = \sum_j \lambda_{ij} v^j$$

$$h(v^j) = \sum_k \mu_{jk} w^k.$$

Then

$$h \circ f(u^i) = h\left(\sum_j \lambda_{ji} v^j\right)$$

$$= \sum_j \lambda_{ji} h(v^j)$$

$$= \sum_j \lambda_{ji} \sum_k \nu_{kj} w^k$$

$$= \sum_k \left(\sum_j \nu_{kj} \lambda_{ji}\right) w^k.$$

If $(\rho_{ki})_{k \leqslant p, i \leqslant n}$ is the matrix of fh with respect to the given bases, it follows that

$$\rho_{ki} = \sum_j \nu_{kj} \lambda_{ji}$$

Hence we are led to define a product of two matrices of sizes $m \times n$, $p \times m$, using the rule

$$(\nu_{kj})(\lambda_{ji}) = \left(\sum_j \nu_{kj} \lambda_{ji}\right).$$

The *zero matrix* 0_{mn} is the $m \times n$ matrix with all entries 0. The *identity matrix* 1_n is the $n \times n$ matrix whose only entries are 1's on the main diagonal,

$$
1_n = \begin{bmatrix}
1 & 0 & 0 & \cdots & 0 \\
0 & 1 & 0 & \cdots & 0 \\
0 & 0 & 1 & \cdots & 0 \\
\cdot & \cdot & \cdot & \cdots & \cdot \\
\cdot & \cdot & \cdot & \cdots & \cdot \\
\cdot & \cdot & \cdot & \cdots & \cdot \\
0 & 0 & 0 & \cdots & 1
\end{bmatrix}.
$$

This corresponds to the identity transformation $f(u) = u$. From the many computational techniques using matrices, we shall here single out the one which computes the rank of a linear transformation, by the process of *row reduction* to *echelon form*. This we now explain.

We permit three different types of *row operation* on a given matrix:

> T_{ij}: interchange rows i and j
> λR_i: multiply row i by the scalar $\lambda \neq 0$
> $R_i + R_j$: replace row i by the sum of rows i and j, leaving row j unchanged.

(The symbolic notation is used only for the purpose of the illustration which follows.)

By using a sequence of row operations, every matrix can be put into *echelon form*, which looks like

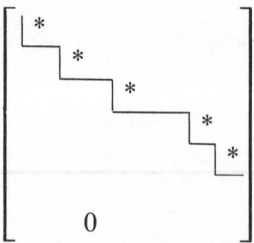

where each * is non-zero, and everything below the 'staircase' is zero. (The 'treads' of the staircase need not have equal length.) In lieu of a proof, we give a typical example. To shorten the process, notice that the sequence of operations λR_j, $R_i + R_j$, $\lambda^{-1} R_j$ adds λ times row j to row i, and leaves all other rows unchanged. It is natural to denote this by $R_i + \lambda R_j$.

Take the matrix

$$
\begin{bmatrix}
0 & 1 & 2 & 1 \\
1 & 3 & 2 & 6 \\
2 & 1 & 4 & 3 \\
-3 & -5 & -8 & -10
\end{bmatrix}.
$$

Perform the following sequence of row operations.

$$T_{12}: \begin{bmatrix} 1 & 3 & 2 & 6 \\ 0 & 1 & 2 & 1 \\ 2 & 1 & 4 & 3 \\ -3 & -5 & -8 & -10 \end{bmatrix},$$

$$R_3 - 2R_1: \begin{bmatrix} 1 & 3 & 2 & 6 \\ 0 & 1 & 2 & 1 \\ 0 & -5 & 0 & -9 \\ -3 & -5 & -8 & -10 \end{bmatrix},$$

$$R_4 + 3R_1: \begin{bmatrix} 1 & 3 & 2 & 6 \\ 0 & 1 & 2 & 1 \\ 0 & -5 & 0 & -9 \\ 0 & 4 & -2 & 8 \end{bmatrix}.$$

At this point we can repeat the process on the 3×3 block in the lower right-hand corner, since the first row and column are in the desired form. Thus:

$$R_3 + 5R_2: \begin{bmatrix} 1 & 3 & 2 & 6 \\ 0 & 1 & 2 & 1 \\ 0 & 0 & 10 & -4 \\ 0 & 4 & -2 & 8 \end{bmatrix},$$

$$R_4 - 4R_2: \begin{bmatrix} 1 & 3 & 2 & 6 \\ 0 & 1 & 2 & 1 \\ 0 & 0 & 10 & -4 \\ 0 & 0 & -10 & 4 \end{bmatrix}.$$

Now we can concentrate on the 2×2 block:

$$R_4 + R_3: \begin{bmatrix} 1 & 3 & 2 & 6 \\ 0 & 1 & 2 & 1 \\ 0 & 0 & 10 & -4 \\ 0 & 0 & 0 & 0 \end{bmatrix}.$$

and the process is complete: the rank of the matrix (or the corresponding linear transformation) is 3.

The proof that this process does give the rank is not too hard: roughly speaking, each row operation leaves the rank unchanged, and the rank of a matrix in echelon form is easily seen to be the number of non-zero rows.

A few more pieces of matrix algebra will be found useful later. If M is an n by n matrix, then its *inverse* M^{-1} (if it exists) is a matrix with the property

$$MM^{-1} = 1_n = M^{-1}M.$$

If M has an inverse it is said to be *non-singular;* if not then it is *singular.* A matrix is non-singular if and only if the corresponding linear transformation is; and an $n \times n$ matrix is non-singular if and only if its rank is n.

Another test for non-singularity, more useful in theory than in practice, is a numerical invariant called the *determinant.* This is denoted by $\det(M)$ and satisfies the equation

$$\det(MN) = \det(M)\det(N).$$

We shall not define it here. Geometrically it corresponds to 'what M does to volume'. For full details, see Dodson and Poston [5]. A square matrix is non-singular if and only if its determinant is non-zero. (This is a useful theoretical result, but it is cumbersome to compute determinants in specific cases, and row reduction is better.)

The *transpose* M^T of a matrix (λ_{ij}) is defined by

$$(M^\mathsf{T})_{ij} = \lambda_{ji}.$$

That is, the transpose is obtained by reflection in the diagonal. We have

$$\det(M^\mathsf{T}) = \det(M).$$

5   Quadratic Forms

A *quadratic form* in n variables $x_1, \ldots, x_n$ is an expression

$$q(x) = \sum_{ij} \lambda_{ij} x_i x_j.$$

In matrix language we introduce the *row vector*

$$x = [x_1 \cdots x_n]$$

whose transpose is the *column vector*

$$x^\mathsf{T} = \begin{bmatrix} x_1 \\ \cdot \\ \cdot \\ \cdot \\ x_n \end{bmatrix}.$$

We call $\Lambda = (\lambda_{ij})$ the matrix of the quadratic form; and then we have

$$q(x) = x\Lambda x^\mathsf{T}.$$

If we replace Λ by $\frac{1}{2}(\Lambda + \Lambda^\mathsf{T}) = M$, the quadratic form remains unchanged since $x_i x_j = x_j x_i$, and now M is *symmetric*, in the sense that $M = M^\mathsf{T}$. Hence every quadratic form can be written as

$$q(x) = xMx^\mathsf{T}$$

for symmetric M, and we shall normally assume this is the case.

Every quadratic form in n variables can be reduced, by a non-singular linear transformation of the variables, to the shape

$$d_1 y_1^2 + \cdots + d_n y_n^2. \tag{2.6}$$

We sketch how this goes, since we shall base a more sophisticated proof (the Morse Lemma, Chapter 4) on it later. The first step is to reach a form with $\lambda_{11} \neq 0$. If any diagonal term $\lambda_{ii} \neq 0$ then the linear transformation which interchanges x_1 and x_i achieves this. If all diagonal terms are zero then some off-diagonal term λ_{ij} $(i \neq j)$ is non-zero, and by symmetry $\lambda_{ji} \neq 0$. So the form is

$$q(x) = 2\lambda_{ij} x_i x_j + \text{terms involving other variables.}$$

Putting

$$y_i = \tfrac{1}{2}(x_i + x_j), \qquad y_j = \tfrac{1}{2}(x_i - x_j)$$

and leaving all other x_k unchanged ($y_k = x_k$), we obtain a term

$$2\lambda_{ij} y_i^2 - 2\lambda_{ij} y_j^2$$

with non-zero diagonal coefficients, and the previous argument applies.

Having made this preparatory step, we can revert to the old notation and assume $\lambda_{11} \neq 0$. Now

$$q(x) = \lambda_{11}\left(\sum_{ij} \mu_{ij} x_i x_j \right)$$

where $\mu_{ij} = \lambda_{ij}/\lambda_{11}$. The terms in which x_1 occurs are

$$x_1^2 + 2 \sum_{j=2}^n \mu_{1j} x_1 x_j = \left(x_1 + \sum_{j=2}^n \mu_{1j} x_j \right)^2 - \left(\sum_{j=2}^n \mu_{1j} x_j \right)^2.$$

Put

$$y_1 = x_1 + \sum_{j=2}^n \mu_{1j} x_j$$

$$y_i = x_i \qquad (i > 1),$$

a non-singular linear transformation. The form becomes

$$q(y) = \lambda_{11} y_1^2 + \tilde{q}(y_2, \ldots, y_n)$$

where $\tilde{q}$ is a quadratic form in variables $y_2, \ldots, y_n$ *only*. Repeating the process sufficiently many times leads to the desired result.

We can now go a little further. Replacing x_i by

$$z_i = x_i/\sqrt{|d_i|} \qquad (d_i \neq 0)$$

leads to an expression of similar type, but with only the coefficients $0, 1, -1$. Changing the order to get 1's first, then -1's, and discarding 0's, we obtain the result that *any quadratic form in n variables can, by a non-singular real linear transformation of the variables, be put in the form*

$$z_1^2 + z_2^2 + \cdots + z_r^2 - z_{r+1}^2 - \cdots - z_s^2$$

where $s \leqslant n$. The number s is called the *rank* of the quadratic form and can be shown to equal the rank of its matrix; hence does not depend on the linear transformation chosen. It is computable by reduction to echelon form. *Sylvester's Law of Inertia* says that r is also independent of the choice of linear transformation. The number

$$2r - s = r - (s - r) = \sum(\text{coefficients})$$

is called the *signature* of the quadratic form $q(x)$. Up to a non-singular linear transformation, any quadratic form is uniquely determined by its rank and signature.

This is a slightly more sophisticated instance of a central theme of this book: reduction to simple form by changing coordinates. A quadratic form $\mathbb{R}^n \rightarrow \mathbb{R}$ takes $\frac{1}{2}n(n+1)$ numbers to specify (55 when $n = 10$). But up to change of coordinates it can be specified far more simply by n numbers, each of which is ± 1 or 0, namely the coefficients of the above expression; or indeed by the two numbers r and s.

We can give a vivid geometrical interpretation of the classification by rank and signature in the case of real two-variable quadratic forms

$$q(x, y) = ax^2 + 2bxy + cy^2.$$

The different possibilities are:

(i) $u^2 + v^2$ (rank 2, signature 2)
(ii) $u^2 - v^2$ (rank 2, signature 0)
(iii) $-u^2 - v^2$ (rank 2, signature -2)
(iv) u^2 (rank 1, signature 1)
(v) $-u^2$ (rank 1, signature -1)
(vi) 0 (rank 0, signature 0).

Now case (iv) occurs precisely when q is a perfect square, say $(px + vy)^2$. But the equation $q = 0$ can be solved by the usual formula for quadratics, to yield

$$x = \left[\frac{-b \pm \sqrt{(b^2 - ac)}}{a} \right] y.$$

But the solutions of $(px + vy)^2 = 0$ are given by

$$x = -\frac{v}{p} y,$$

and the only way these solutions can agree is if $b^2 - ac = 0$, so that the two solutions reduce to one, repeated twice. On the other hand, if $b^2 - ac = 0$, then

$$\left[x\sqrt{a} + \frac{b}{\sqrt{a}} y \right]^2 = ax^2 + 2bxy + \frac{b^2}{a} y^2 = q(x, y).$$

Thus $q(x)$ can be reduced to u^2 exactly when $b^2 = ac$ and when a and c are *positive*, and it looks like $-u^2$ when $b^2 = ac$ and a and c

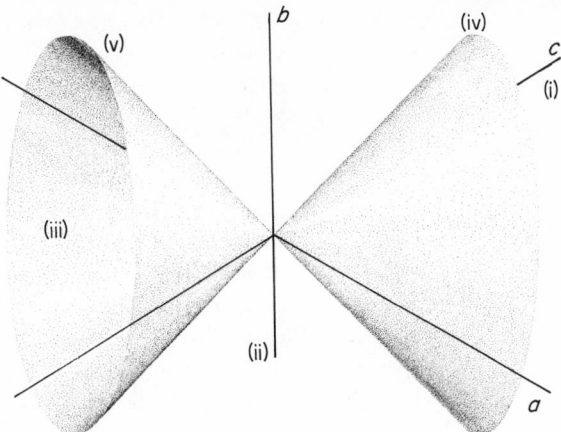

Fig. 2.6

are *negative*. Case (vi) of course has $a = b = c = 0$, so $b^2 = ac$ trivially. Thus cases (iv), (v), (vi) are distinguished by the condition $b^2 = ac$, and among them we distinguish further by the sign $(+, 0, -)$ of a (or c).

If we let (a, b, c) be the coordinates of a point in $\mathbb{R}^3$ and plot the set S of such points for which $b^2 = ac$, we see that it is a double cone, with vertex the origin, on which both the a- and c-axes lie; further it separates the remaining three cases as shown in Fig. 2.6. (The method by which it is seen to be a double cone will be elucidated in Section 7 of this chapter.) Thus q is of type (ii) when $b^2 > ac$, the 'outside' of the cone; type (i) when $b^2 < ac$ and a is positive, the inside of the positive half of the cone; type (iii) when $b^2 < ac$ and a is negative, the inside of the negative half of the cone.

In view of these results, we shall call a quadratic form *degenerate* if its rank r is less than the number n of variables. The difference $n - r$ is the *corank* and measures the number of independent directions in which it is degenerate, as is particularly clear in the diagonalized form. Degeneracy happens if and only if the determinant of its matrix is zero. Thus the matrix of the two-variable form above is

$$\begin{bmatrix} a & b \\ b & c \end{bmatrix}$$

and its determinant is $ac - b^2$. This determinant is known as the *discriminant* of the quadratic form. We call S, defined as above, the *discriminant cone*.

6 Two-variable Cubic Forms

Now we take up the analogous problem of classifying *cubic* forms, the solution to which we shall require in Chapter 7, as a key step in the classification theorem. Since we shall there reduce the problem to the study of forms in only two variables, we analyse only this restricted case here.

A cubic form in two variables x, y is an expression

$$c(x, y) = \alpha x^3 + \beta x^2 y + \gamma xy^2 + \delta y^3$$

with real coefficients $\alpha, \beta, \gamma, \delta$. We let K be the four-dimensional space of all such forms, which can be identified with $\mathbb{R}^4$ by taking the coefficients $(\alpha, \beta, \gamma, \delta)$ as coordinates.

In Section 5 we found six types of quadratic function (four up to change of sign). Analogously there are five types of cubic: the question of sign changes makes no difference this time, since the change of coordinates $(u, v) = (-x, -y)$ will automatically change the sign. The types are well described by their *root structures*: the set of points

$$R_A = \{(x, y) : \alpha x^3 + \beta x^2 y + \gamma xy^2 + \delta y^3 = 0\}$$

where $A = (\alpha, \beta, \gamma, \delta)$. Now if $\lambda \in \mathbb{R}$ and $(x, y) \in R_A$, then it is clear that $(\lambda x, \lambda y) \in R_A$ as well. So for any A, R_A consists of straight lines through the origin.

To find out which arrangements of straight lines can occur we find their intersection with the line $x = 1$. (There is a minor problem here, in that the y-axis does not intersect $x = 1$ at all; but we take care of this case separately. Those happy with projective geometry can introduce 'points at infinity' to similar ends.) Putting $x = 1$ gives a cubic equation in y,

$$P_A(y) = \alpha + \beta y + \gamma y^2 + \delta y^3 = 0.$$

This has at most three solutions y_1, y_2, y_3, unless $A = (0, 0, 0, 0)$; and hence there are at most three lines of roots. To deal with the y-axis, note that it has equation $x \leftharpoondown 0$. Now $x = 0$ is a solution if and only if $\delta = 0$, and then the cubic $P_A(y)$ becomes a quadratic, with at most two solutions. Hence again we find at most three root lines, as in Fig. 2.7. Thus the possibilities for R_A are:

(i) Three distinct root lines, as in Figs 2.7 and 2.8(a).
(ii) A single real line, as in Fig. 2.8(b).
(iii) Three lines, two of which 'coincide', as in Fig. 2.8(c).
(iv) Three coincident lines, as in Fig. 2.8(d).
(v) The whole plane, with $A = (0, 0, 0, 0)$.

Clearly no change of coordinates will alter the number of root lines which a cubic function has. (Notice that it is important to distinguish between 'the function P on the plane defined in (x, y)-coordinates by $x^2 y$' and 'the polynomial $X^2 Y$'. If we define new coordinates $u = \frac{1}{2}(x + y)$, $v = \frac{1}{2}(x - y)$, then the function has the new expression $(u + v)^2(u - v)$ in terms of the new coordinates. Thus a single polynomial *function* can have many polynomial *expressions*.)

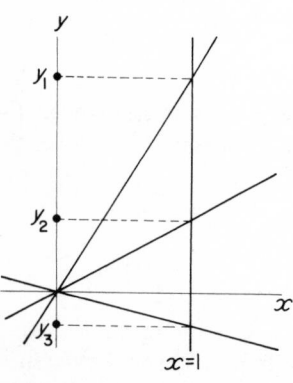

Fig. 2.7

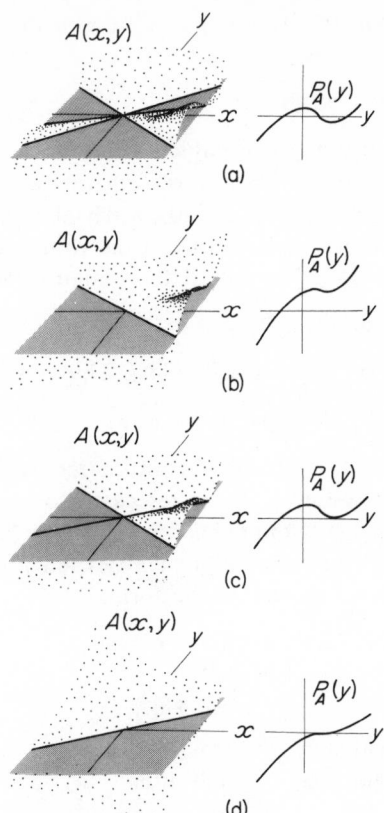

Fig. 2.8

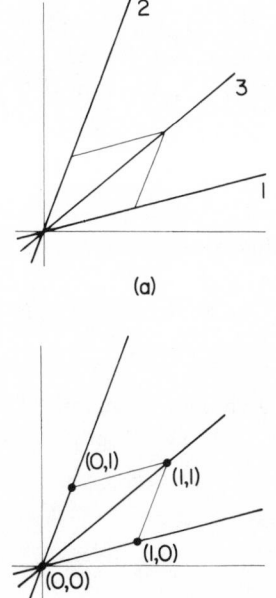

(a)

(b)

Fig. 2.9

Also, no smooth change can alter case (iv), where the gradient vanishes entirely along a single root line, into case (ii) where it vanishes only at $(0, 0)$. Hence the root structure separates K into five classes.

Remarkably, whenever two cubic functions f_1 and f_2 belong to the same class, we can find a *linear* change of coordinates by which f_2 has the same expression in the new coordinates as f_1 had in the old ones. So 'up to change of coordinates' they are the same: more carefully, in a very strong sense they have the same shape.

We begin with case (i), three distinct lines of roots. We can always draw a parallelogram as in Fig. 2.9(a), with sides along two of the root lines (christened 1 and 2) and diagonal along root line 3. Choose new (u, v)-coordinates with the u-axis along line 1, the v-axis along line 2, and scales that make the sides of the parallelogram have unit length (as in Fig. 2.9(b)). Then in these coordinates the equations of the root lines are

line 1: $v = 0$,
line 2: $u = 0$,
line 3: $u - v = 0$.

Thus whenever u, v or $u - v$ vanishes, so does f. It follows that f is a multiple of $uv(u - v)$, and by a further change of scale we can put f in the form

$$f(u, v) = uv(u - v).$$

Hence any two cubics with three real distinct root-lines can both be put in the same form $uv(u - v)$.

Unless $A = (0, 0, 0, 0)$, a trivial case which we henceforth ignore, we can use at least one of the root lines as above to give us a linear factor of f. Namely, if the equation of the line is

$$lx + my = 0$$

then $lx + my$ must divide f. Thus f may be expressed as

$$(lx + my)(ax^2 + bxy + cy^2)$$

even before changing coordinates. Using the results of Section 5 we can now find coordinates (u, v) to simplify the quadratic term: the linear factor then becomes $Lx + My$ for certain L, M. The quadratic factor can have its sign absorbed into the linear one, so we can express f in one of the forms

(a) $(Lu + Mv)(u^2 - v^2)$
(b) $(Lu + Mv)u^2$
(c) $(Lu + Mv)(u^2 + v^2)$.

We will show that case (a) corresponds either to Fig. 2.9 or to case (b), and then we shall discuss (b) and (c). Now (a) factorizes further as

$$(Lu + Mv)(u - v)(u + v).$$

Unless $Lu + Mv$ is a scalar multiple of $u + v$ or of $u - v$, we are in the case discussed above. If it is a scalar multiple then either $L = M = c$, or $L = -M = c$. Writing $U = (u \pm v)c^{-1/3}$, $V = vc^{-1/3}$, we find U^2 divides f and we are in case (b).

In case (b), if $M \neq 0$ we take new coordinates

$$U = Lu + Mv, \qquad V = u$$

which gives f the expression UV^2. If $M = 0$ this is not a legitimate coordinate change, but then we can set

$$U = L^{1/3}u, \qquad V = v,$$

to give f the form U^3.

In case (c), if either L or M vanishes then a change of scale and relabelling of axes will put f in the form $(U^2 + V^2)V$. If not, we rotate the axes by setting

$$w = \frac{Lv - Mu}{\sqrt{L^2 + M^2}}, \qquad z = \frac{Lu + Mv}{\sqrt{L^2 + M^2}}$$

which, after some algebra, leads to

$$f(w, z) = \sqrt{(L^2 + M^2)}(w^2 + z^2)z$$

and a change of scale gives this the form $(U^2 + V^2)V$.

Thus by a suitable linear change of coordinates we can reduce any non-zero homogeneous cubic to one of the expressions

$$(U - V)UV, \quad (U^2 + V^2)V, \quad U^2V, \quad U^3.$$

These correspond to root structures of types (i), (ii), (iii), (iv) respectively, and hence are geometrically distinct. Thus we have exactly four different classes of two-variable cubic, plus the zero cubic. For neatness we make a further coordinate change in $(U - V)UV$, as follows:

$$x = \tfrac{1}{2}(U + V), \qquad y = \tfrac{1}{2}(U - V),$$

which transforms it to $x^2y - y^3$. For uniformity of notation we change U to x and V to y in the other cases, yielding the following list of standard cubics:

 (i) $x^2y - y^3$
 (ii) $x^2y + y^3$
 (iii) x^2y
 (iv) x^3.

Next we consider how the four types are arranged in the 4-dimensional space K of cubic forms. In other words, we seek the analogue of Fig. 2.6. Since K is 4-dimensional we must resort to some trick to get the geometry down on two dimensions of paper.

The trick is to ignore scale. Multiplying a quadratic by a non-zero positive constant c does not change its type, since the change can be undone by a change of variable to $c^{1/2}x$, $c^{1/2}y$. Similarly for homogeneous cubics we can compensate by a cube-root change of scale. Now Fig. 2.6 is fully described by taking the unit sphere

$$\{(a, b, c) : a^2 + b^2 + c^2 = 1\}$$

around the origin of I, marking it as in Fig. 2.10, and taking the cone on this picture with vertex the origin. The two circles on this sphere are the quadratic forms (a, b, c) which lie on the sphere and have $b^2 = ac$. Similarly we can describe the distribution of cubics of the various types in K by first finding what happens in the 3-dimensional sphere

$$\{(\alpha, \beta, \gamma, \delta) : \alpha^2 + \beta^2 + \gamma^2 + \delta^2 = 1\}$$

and then taking the cone on this formed by the straight lines radiating from $(0, 0, 0, 0)$.

Unfortunately there is no accessible 3-sphere in the physical world (apart, perhaps, from the entire universe, or rather a spacelike section of it) so we must introduce some distortion to bring the interesting part of the geometry within reach, without making any drastic alterations to it. The same problem occurs in Fig. 2.10, with the perennial cartographer's nightmare of representing the surface of a sphere in the plane; it is solved in the same way, by taking a

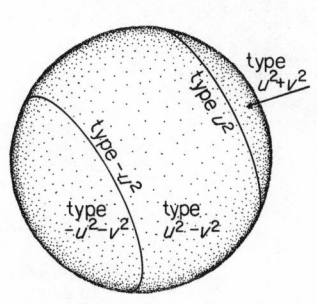

Fig. 2.10

'projection' which introduces serious distortions only far away from the region of interest. Thus, if we call the point $(0, 1, 0)$ at the top the North Pole, and the antipodal point the South Pole, we must smoothly stretch some parts of the sphere and contract others to move the circles down near the South Pole. Then we can flatten out the relevant piece of spherical surface, near the South Pole, without destroying the configuration of two circles. The analogous operation on the 3-sphere of cubics flattens it out into ordinary 3-dimensional space, and leads to the beautiful surface shown (carved in wood) in Plate 1. This may be described as the result of taking a three-cusped hypocycloid (Fig. 2.11), rotating it round a circle, and joining up with a one-third twist.

We can identify the position of the various types of cubic relative to this surface. In Plate 1 the cubics of type (i) are inside the surface; type (ii) outside; type (iii) on the smooth part of the surface; and type (iv) on its cusped edge. This geometry is in agreement with a

Plate 1. Zeeman's umbilic bracelet, carved in hardrock maple by T.P.

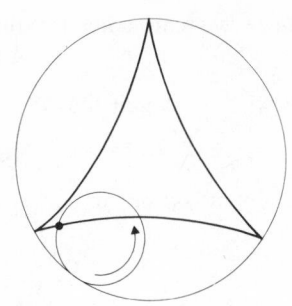

Fig. 2.11

number of plausible inferences from Fig. 2.8. Thus types (i) and (ii) are separated by type (iii) (together with type (iv)). Small changes in type (iii) can be made to yield either of types (i) or (ii). Type (iv) has all other types arbitrarily close to it.

Zeeman [7] who first analysed this shape, christened it the *umbilic bracelet:* 'umbilic' from the catastrophes derived from the types of cubic involved, and 'bracelet' from his wife's jewellery activities. His paper gives a full proof that Plate 1 is correct, but this is not necessary here. The picture helps to *explain* some of the assertions of Chapter 7, but these are best *proved* by more powerful methods. The geometry of the umbilic bracelet as a whole first becomes involved in an essential way in the analysis of the 8-dimensional catastrophes known as *double cusps.*

7 Polynomial Geometry

A *polynomial* in variables $x_1, \ldots, x_n$ is an expression of the form

$$p(x) = \sum \lambda_{i_1 \ldots i_n} x_1^{i_1} \cdots x_n^{i_n}$$

where the i_j are positive integers (or 0), the $\lambda_{i_1 \ldots i_n}$ are real, and the number of terms is finite. Its *degree* is the largest value of $i_1 + \cdots + i_n$ for which $\lambda_{i_1 \ldots i_n} \neq 0$. If only one coefficient is not 0, $p(x)$ is a *monomial.* An *algebraic set* in $\mathbb{R}^n$ is a set of points of the form

$$\{x \in \mathbb{R}^n \mid p_1(x) = p_2(x) = \cdots = p_k(x) = 0\}$$

where $p_1, \ldots, p_k$ are polynomials in $x_1, \ldots, x_n$. For example the *unit sphere*

$$S^{n-1} = \{x \in \mathbb{R}^n \mid x_1^2 + \cdots + x_n^2 = 1\}$$

is an algebraic set. A *semi-algebraic set* is one specified by polynomial *inequalities* $q_i(x) \rho 0$, where ρ is any of $=, <, \leqslant, \geqslant, >$. Such sets tend to arise naturally, for the following reason. Suppose $f : \mathbb{R}^n \to \mathbb{R}^m$ is a function defined by polynomials:

$$f(x) = (q_1(x), \ldots, q_m(x))$$

with each q_i a polynomial in $x_1, \ldots, x_n$. Let A be an algebraic set in $\mathbb{R}^n$. Then $f(A)$ need not be algebraic (take $m = n = 1$, $f(x) = x^2$), but it is always *semi*-algebraic.

We are less concerned here with the mathematical properties of algebraic and semi-algebraic sets (the historical starting point for the very deep subject known as algebraic geometry) than with obtaining a good grasp of their geometry, especially for $n \geqslant 4$ when it is no longer feasible to draw pictures (without using special tricks). The most useful method is to 'stack' cross-sections: rather than dealing with the general case we shall apply the method to some examples. Many further instances will arise as we proceed.

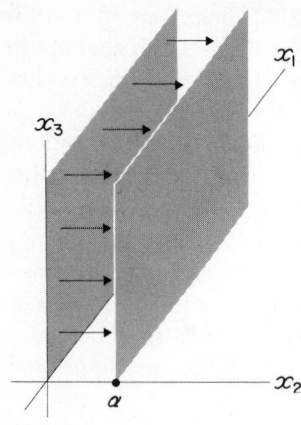

Fig. 2.12

The first two examples are in $\mathbb{R}^3$ where we can compare the sections with the actual object.

(1) Let $n = 3$, $p(x) = x_1^3 + x_1 x_2 + x_3$. We wish to sketch the set

$$S = \{x \in \mathbb{R}^3 \mid p(x) = 0\}.$$

Fix a value of x_2, say $x_2 = \alpha$. Then a point (x_1, α, x_3) lies in S if and only if

$$x_1^3 + \alpha x_1 + x_3 = 0,$$

that is,

$$x_3 = -x_1^3 - \alpha x_1. \qquad (2.7)$$

We can think of (x_1, x_3) as a system of coordinates on the plane $x_2 = \alpha$, by projecting the (x_1, x_3)-plane according to the function

$$(x_1, x_3) \rightarrow (x_1, \alpha, x_3)$$

as in Fig. 2.12. Then for each choice of α, the equation (2.7) defines a curve in (x_1, x_3)-space. The shape of this cubic curve depends mainly on the sign of α, and Fig. 2.13 shows three typical cases. We 'stack' these curves by allowing α to vary (Fig. 2.14(a)), which makes it clear that S has the shape of a pleated surface as shown more smoothly in Fig. 2.14(b). We shall encounter this particular surface many times in catastrophe theory applications.

(2) Let $n = 3$, $p(x) = x_2^2 - x_1 x_3$. We wish to sketch the set

$$T = \{x \in \mathbb{R}^3 \mid p(x) = 0\}.$$

(Up to changes of notation, this is the set of points (a, b, c) with $b^2 - ac = 0$ referred to in Section 5, where it was claimed to be a cone.)

This time it is convenient to change axes first. We put

$$u = (x_1 + x_3)/\sqrt{2}$$
$$v = (x_1 - x_3)/\sqrt{2}$$
$$w = \sqrt{2} x_2.$$

Fig. 2.13

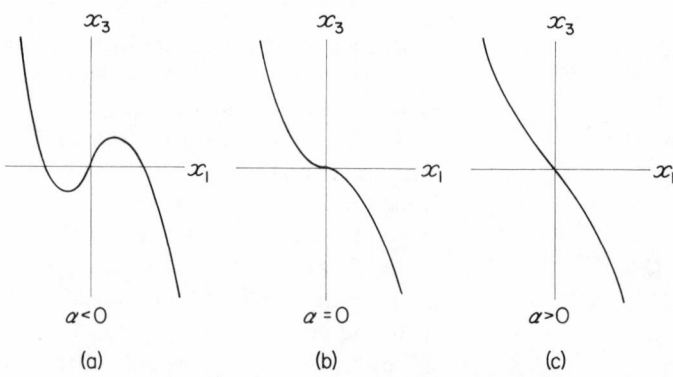

$\alpha < 0$	$\alpha = 0$	$\alpha > 0$
(a)	(b)	(c)

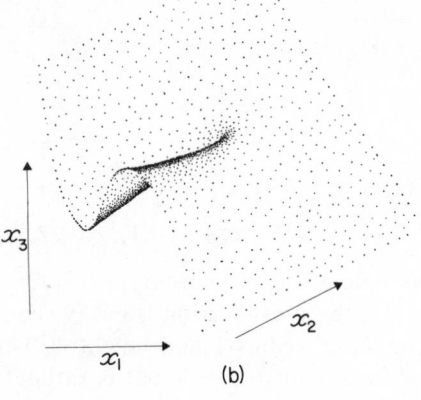

Fig. 2.14

Then the equation becomes

$$u^2 - v^2 = w^2,$$

or

$$v^2 + w^2 = u^2.$$

Each section $u = \alpha$ clearly gives a circle of radius $|\alpha|$, and since this grows linearly with α, stacking the sections yields a double cone as in Fig. 2.15. Now the transformation of coordinates we used corresponds to rotating the (x_1, x_3)-plane through 45° and stretching everything by a factor $\sqrt{2}$ in the x_2-direction. So the axis of the cone is the u-axis, which in original coordinates is given by

$$0 = v = w,$$
$$x_2 = 0, \qquad x_1 = x_3.$$

Hence the double cone lies as shown in Fig. 2.16. Note that because of the stretching in the x_2-direction, the cone is elliptical rather than circular in cross-section.

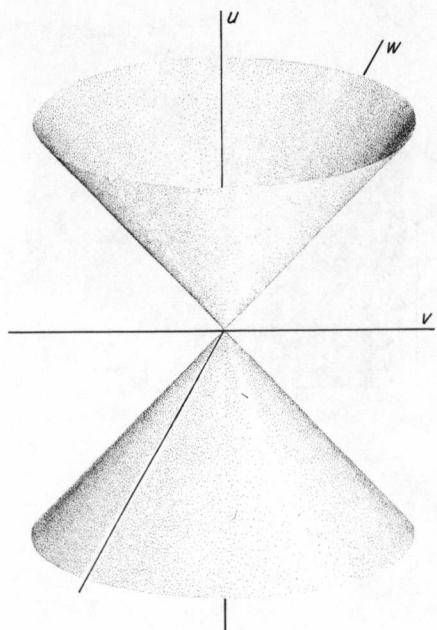

Fig. 2.15

(3) Let $n = 4$,

$$U = \{x \in \mathbb{R}^4 \mid x_1^2 + x_2^2 = 1, |x_3| \leqslant 2, |x_4| \leqslant 1\},$$

this time a semi-algebraic set (why?). For a fixed value $x_4 = \alpha$, with $|\alpha| \leqslant 1$, the cross-section is easily seen to be the curved surface of a cylinder of radius 1 and height 4. This cross-section does not vary with α provided $|\alpha| \leqslant 1$, but is empty for other α. Hence U is a kind of '4-dimensional cylinder' of height 2, whose base is the afore-mentioned cylinder of height 4.

(4) Let $n = 4$,

$$V = \{x \in \mathbb{R}^4 \mid x_1^2 + x_2^2 + x_3^2 - x_4^2 = 0\},$$

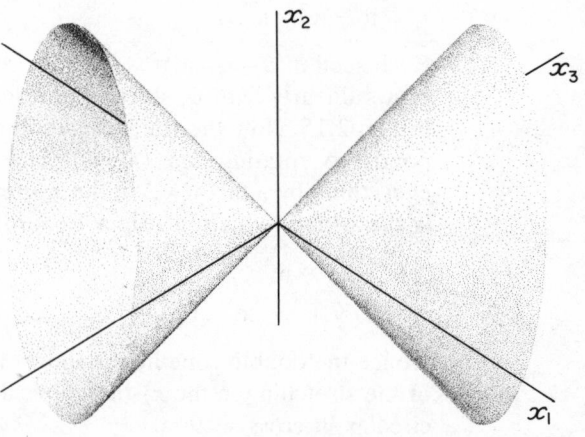

Fig. 2.16

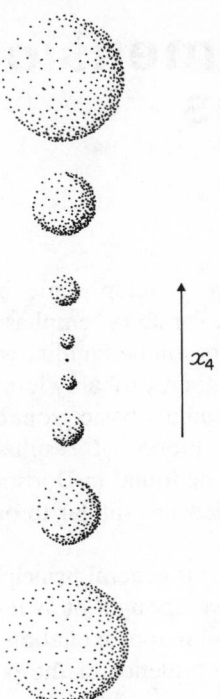

x_4

Fig. 2.17

a set familiar to relativists! For fixed $x_4 = \alpha$ the cross-section is given by

$$x_1^2 + x_2^2 + x_3^2 = \alpha^2,$$

so is a sphere of radius $|\alpha|$. Since the radius varies in proportion to α, the set V is a kind of 'double cone' on this sphere, analogous to example (2) above. The spheres stack along the x_4-direction as indicated schematically in Fig. 2.17.

It takes plenty of practice to visualize sets in four or more dimensions, and the above examples are a good deal simpler than many that we have to deal with later; however further elaboration at this point would probably be unhelpful.

3 Multidimensional Calculus

In this chapter we develop some basic principles of differential calculus of several variables, emphasizing the geometric viewpoint. Again, most readers will be familiar with much of the subject matter in one form or another, probably less geometrical than that which is given here. We assume a basic grounding in calculus of functions of a single variable. Proofs of results stated here will usually be omitted: they may be found in Dodson and Poston [5] or Spivak [8] (whose points of view are similar to ours) or in many other standard texts.

The most significant general principle to emerge from this chapter is that the derivative, thought of as the best *linear* approximation to a given function, is an approximation worth making. Later chapters give ample further evidence of the power of this principle.

1 Distance in Euclidean Space

As well as a linear structure, $\mathbb{R}^n$ has a metric structure. For $x, y \in \mathbb{R}^n$ define the *norm*

$$\|x\| = \sqrt{(x_1^2 + \cdots + x_n^2)},$$

and define the *distance* between x and y to be

$$\|x - y\|.$$

This takes its inspiration from Pythagoras' theorem. It has the basic properties required of a distance, namely

$$\|x\| \geq 0, \quad \text{and} \quad \|x\| = 0 \quad \text{if and only if} \quad x = 0$$

$$\|x - y\| = \|y - x\|$$

$$\|x - y\| + \|y - z\| \geq \|x - z\|.$$

By using the norm we can extend to higher dimensions notions such as limits or continuity, replacing the usual absolute value $|x|$ in the one-dimensional case by the norm. We assume that the reader has encountered the relevant definitions in the one-dimensional case: as an example of the kind of extension contemplated, we define a limit. (For more detail, and a geometric interpretation, *see* Dodson and Poston [5].) Let

$$f : \mathbb{R}^n \to \mathbb{R}^m$$

be a function. We say that $f(x)$ *tends to a limit* λ *as* x *tends to* a if, given any $\varepsilon > 0$, there exists $\delta > 0$ such that whenever $x \neq a$ and $\|x - a\| < \delta$, it follows that $\|f(x) - \lambda\| < \varepsilon$. That is, if we pick x near enough to a (but not equal to it) then $f(x)$ can be made as close as we wish to λ. We write

$$\lim_{x \to a} f(x) = \lambda.$$

Then f is *continuous at* a if $\lim_{x \to a} f(x) = f(a)$; and *continuous* if this holds for all $a \in \mathbb{R}^n$.

Continuity will be less important for us than differentiability, which is where we make much use of the norm: the definition is more subtle, and does not arise automatically from the single-variable case.

There is a similar notion of limit for sequences (a_n) as $n \to \infty$. An infinite *series*

$$\sum_{n=1}^{\infty} a_n$$

is *convergent* if

$$\lim_{N \to \infty} \sum_{n=1}^{N} a_n$$

exists, and otherwise is *divergent*.

If $x \in \mathbb{R}^n$ and $r > 0$, then the *open disc* of radius r, centre x, is defined to be

$$\{y \in \mathbb{R}^n \mid \|x - y\| < r\}.$$

A subset X of $\mathbb{R}^n$ is *open* if for each $x \in X$ there is some open disc centre x lying inside X. A subset Y is a *neighbourhood* of x if x lies in some open set X which is contained in Y. A property holds *locally* at x (or *around* x) if it is true for all y in some neighbourhood of x.

2 The Derivative as Tangent

For this section we concentrate on the following question: what is a *sensible* and *useful* definition of the derivative of a function $f : \mathbb{R}^n \to \mathbb{R}^m$?

Recall that for a function $f : \mathbb{R} \to \mathbb{R}$ we define the derivative to be the function Df (or alternatively and more traditionally df/dx) for which

$$Df|_x = \lim_{h \to 0} \frac{f(x+h) - f(x)}{h} \tag{3.1}$$

provided this limit exists. (We here write $Df|_x$ instead of $Df(x)$, since this leads to clearer notation later.) Now this definition does not carry over to higher dimensions without some changes. To begin with, x and h would have to come from $\mathbb{R}^n$, whereas $f(x)$ and $f(x+h)$ lie in $\mathbb{R}^m$, so the right-hand side of (3.1) would tell us to divide an element of $\mathbb{R}^n$ by an element of $\mathbb{R}^m$, which cannot be done in any useful way. Even if we somehow get round this, the limit will not exist even in cases where the derivative ought to: everything depends on the direction from which h approaches 0. For example, let

$$f : \mathbb{R}^2 \to \mathbb{R}$$

$$f(x) = f(x_1, x_2) = x_1 + 2x_2.$$

Then

$$\frac{f((x_1, x_2) + (0, h)) - f(x_1, x_2)}{h} = 2 \quad \text{for all } h,$$

$$\frac{f((x_1, x_2) + (h, 0)) - f(x_1, x_2)}{h} = 1 \quad \text{for all } h.$$

This means that the limit would have to be both 1 and 2, a manifest absurdity.

Instead of trying to generalize the formula (3.1), we go back to fundamentals. Geometrically, the value of Df at a point x expresses the slope of the tangent line to the graph of f at the point $(x, f(x))$, as illustrated in Fig. 3.1. It is helpful to translate the origin of coordinates to the point $(x, f(x))$, temporarily, because then the tangent line is the graph of a *linear* function $\lambda : \mathbb{R} \to \mathbb{R}$ since the tangent line passes through the origin. In fact the tangent line is the *best linear approximation* to f. We can measure how good an approximation it is by comparing

$$f(x+h) \quad \text{and} \quad f(x) + \lambda(h)$$

for small h (see Fig. 3.1). It turns out (and the motivation is Fig. 3.3 which we discuss at the appropriate place) that what we need is for

$$|f(x+h) - f(x) - \lambda(h)|$$

to tend to 0 *faster* than h does; that is to say, the error in approximating $f(x+h)$ by $f(x) + \lambda(h)$ should be 'of smaller order than h'. More precisely,

$$\frac{|f(x+h) - f(x) - \lambda(h)|}{|h|} \to 0$$

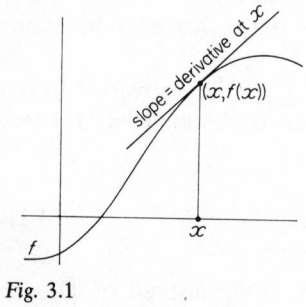

Fig. 3.1

as $h \to 0$.

Next, consider a function $f : \mathbb{R}^2 \to \mathbb{R}$. We can draw the graph of this in $\mathbb{R}^3$, and it will resemble a 'landscape' as in Fig. 3.2. If this graph is sufficiently 'smooth' at $(x, f(x))$ there will be a tangent plane there. This is again (after a shift of origin) the graph of a *linear* map

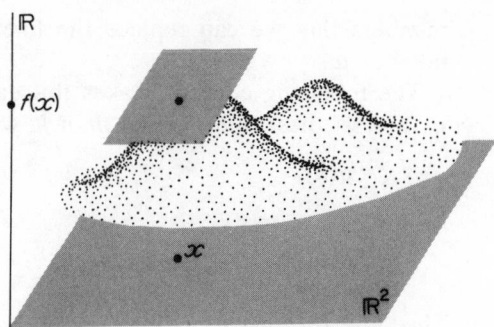

Fig. 3.2

$\lambda:\mathbb{R}^2\to\mathbb{R}$, and its tangency is expressed by the fact that

$$\|f(x+h)-f(x)-\lambda(h)\|$$

tends to 0 faster than h; that is,

$$\lim_{h\to 0}\left(\frac{\|f(x+h)-f(x)-\lambda(h)\|}{\|h\|}\right)=0. \qquad (3.2)$$

It should be clear how to generalize this formulation to higher dimensions: seek a best approximating linear map, corresponding geometrically to a tangent hyperplane, and use (3.2) as a definition of 'best approximating'. Thus we say that a function $f:\mathbb{R}^n\to\mathbb{R}^m$ is *differentiable* at $x\in\mathbb{R}^n$ if there exists a linear map $\lambda:\mathbb{R}^n\to\mathbb{R}^m$ such that

$$\lim_{h\to 0}\left(\frac{\|f(x+h)-f(x)-\lambda(h)\|}{\|h\|}\right)=0. \qquad (3.3)$$

The linear map λ is unique if it exists (we postpone the proof for a moment). We call it the *derivative* of f at x, and write

$$Df\big|_x=\lambda.$$

First we show how to reinterpret the usual derivative of a function in these terms. Let $f:\mathbb{R}\to\mathbb{R}$ have derivative $f':\mathbb{R}\to\mathbb{R}$, and, for fixed x, put

$$f'(x)=\lambda.$$

We claim that the map $h\to\lambda h$ is the derivative at x, in the new sense. For

$$\frac{|f(x+h)-f(x)-\lambda h|}{|h|}=\left|\frac{f(x+h)-f(x)}{h}-\lambda\right|$$

As $h\to 0$, $(f(x+h)-f(x))/h\to\lambda$, so the whole expression $\to 0$ as claimed.

In other words: every linear map $\mathbb{R}\to\mathbb{R}$ is of the form $x\mapsto\mu x$ for $\mu\in\mathbb{R}$. The value of μ giving the derivative at x (in the new sense) is the *value* of f' at x (in the old sense). It is only because of the coincidence that linear maps $\mathbb{R}\to\mathbb{R}$ correspond naturally to real numbers that we can replace the linear map $x \mapsto \mu x$ by the real number μ.

The following concept makes this easier to handle. We say that $g : \mathbb{R}^n \to \mathbb{R}^m$ has *smaller order than* $.h$, written

$$g = o(h),$$

if

$$\|g(h)\|/\|h\| \to 0 \quad \text{as} \quad h \to 0.$$

Then the formula (3.3) in the definition of the derivative can be written in the equivalent form

$$f(x+h) - f(x) - \lambda(h) = o(h). \tag{3.4}$$

Think of this as follows:

$$\underbrace{(f(x+h) - f(x))}_{\substack{\text{how } f \text{ varies} \\ \text{near } x}} - \underbrace{\lambda(h)}_{\substack{\text{how a linear} \\ \text{map varies} \\ \text{near } x}} = \underbrace{o(h)}_{\substack{\text{closer than} \\ \text{something} \\ \text{linear in } h.}}$$

how good an approximation is λ?

If p, q are functions such that $p = o(h)$, $q = o(h)$, and if $\alpha, \beta \in \mathbb{R}$, then $\alpha p + \beta q = o(h)$. Using this, we can prove uniqueness of the derivative. For suppose that, in addition to (3.4),

$$f(x+h) - f(x) - \mu(h) = o(h).$$

Subtracting from (3.4),

$$\lambda(h) - \mu(h) = o(h).$$

But it is easy to see that a linear map which is $o(h)$ must be zero. For suppose θ is linear, $\theta = o(h)$. Then for any h,

$$\|\theta(th)\|/\|th\| \to 0 \quad \text{as} \quad t \to 0,$$

that is,

$$|t| \|\theta(h)\|/|t| \|h\| \to 0,$$

so

$$\|\theta(h)\|/\|h\| \to 0.$$

But h is fixed here, so $\theta(h) = 0$; therefore $\theta = 0$.

Now $\lambda - \mu$ is linear and $o(h)$, so $\lambda - \mu = 0$, and $\lambda = \mu$. This proves uniqueness.

Geometrically it is clear (by similar triangles) that the difference between two distinct hyperplanes cannot be $o(h)$. Fig. 3.3 illustrates this.

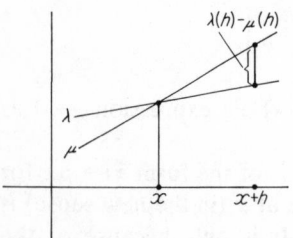

Fig. 3.3

Now we say that $f:\mathbb{R}^n \to \mathbb{R}^m$ is *differentiable* if it is differentiable at each $x \in \mathbb{R}^n$. Its *derivative* is then the function Df whose value $Df|_x$ at x is the linear map defined above. If we write

$$L(\mathbb{R}^n, \mathbb{R}^m)$$

for the set of all linear maps $\mathbb{R}^n \to \mathbb{R}^m$, then we have

$$Df:\mathbb{R}^n \to L(\mathbb{R}^n, \mathbb{R}^m).$$

This makes Df quite a complicated object. When $n = m = 1$ we can identify $L(\mathbb{R}, \mathbb{R})$ with $\mathbb{R}$, which is why then $Df:\mathbb{R} \to \mathbb{R}$. But note that in general Df is *not* a map $\mathbb{R}^n \to \mathbb{R}^m$. It is its *value* $Df|_x$ at any x which is a map $\mathbb{R}^n \to \mathbb{R}^m$, and is linear to boot.

The derivative obeys a number of standard rules, of which the following are mentioned without proof. Proofs are given in Spivak [8].

(a) If $f:\mathbb{R}^n \to \mathbb{R}^m$ is constant then $Df|_x = 0$.

(b) If $f:\mathbb{R}^n \to \mathbb{R}^m$ is linear, then $Df|_x = f$ for all x.

(c) If $f:\mathbb{R}^n \to \mathbb{R}^m$ is given by $f(x) = (f_1(x), \ldots, f_m(x))$, then f is differentiable if and only if each f_i is, and then

$$Df|_x = (Df_1|_x, \ldots, Df_m|_x).$$

(d) If $f, g:\mathbb{R}^n \to \mathbb{R}^m$ and $\lambda \in \mathbb{R}$, then

$$D(f+g)|_x = Df|_x + Dg|_x,$$

$$D(\lambda f)|_x = \lambda Df|_x.$$

(e) If $f:\mathbb{R}^n \to \mathbb{R}^m$ is differentiable at x, and $g:\mathbb{R}^m \to \mathbb{R}^p$ is differentiable at $f(x)$, then $g \circ f:\mathbb{R}^n \to \mathbb{R}^p$ (defined by $g \circ f(x) \doteq g(f(x))$) is differentiable at x, and

$$Dg \circ f|_x = Dg|_{f(x)} \circ Df|_x.$$

This formula is the *chain rule*, often referred to in elementary mathematics as 'differentiating a function of a function'.

There are also rules for differentiating products or quotients of functions $\mathbb{R}^n \to \mathbb{R}$, analogous to the usual ones for $\mathbb{R} \to \mathbb{R}$.

3 Contours

A function $f:\mathbb{R}^n \to \mathbb{R}$ can sometimes be usefully visualized in terms of its *contours*, the sets

$$\{x \in \mathbb{R}^n \mid f(x) = c\}$$

for all possible choices of the constant $c \in \mathbb{R}$. These contours (or in practice a suitable selection of them) convey information about the graph of f in exactly the way that contours of a geographical map convey information about the height of the land: the graph is

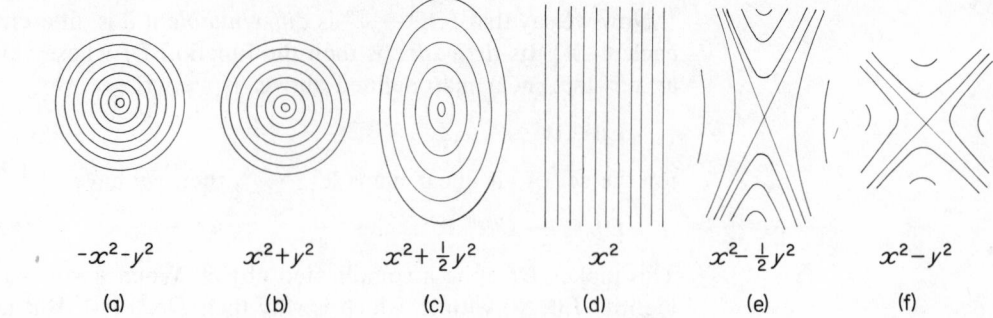

$-x^2-y^2$ x^2+y^2 $x^2+\frac{1}{2}y^2$ x^2 $x^2-\frac{1}{2}y^2$ x^2-y^2

Fig. 3.4   (a)   (b)   (c)   (d)   (e)   (f)

obtained by 'stacking' the contours, as described in Section 7 of Chapter 2. Fig. 3.4 shows contours of six functions $\mathbb{R}^2 \to \mathbb{R}$ which we shall encounter again in Chapter 11.

If f is linear and non-zero, its contours are parallel, straight, and of dimension $n-1$. For smooth f, the contours of Df at a point are the best approximation by straight, evenly spaced contours to the contours of f, around that point (Fig. 3.5).

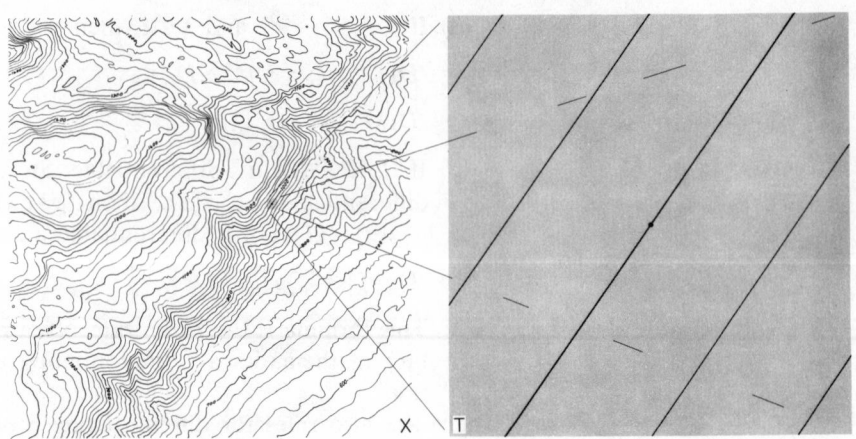

Fig. 3.5

4 Partial Derivatives

If $f:\mathbb{R}^n \to \mathbb{R}$ and $x \in \mathbb{R}^n$, then the limit

$$\lim_{h \to 0} \left(\frac{f(x_1, \ldots, x_i + h, \ldots, x_n) - f(x_1, \ldots, x_n)}{h} \right),$$

if it exists, is called the ith *partial derivative* of f at x, denoted

$$D_i f \big|_x$$

or

$$\left.\frac{\partial f}{\partial x_i}\right|_x \quad \text{or} \quad \frac{\partial f}{\partial x_i}(x) = \frac{\partial f}{\partial x_i}(x_1, \ldots, x_i, \ldots, x_n)$$

in more traditional notation (in which x_i has two different meanings!). It is equal to the ordinary derivative of the function $g : \mathbb{R} \to \mathbb{R}$ defined by

$$g(y) = f(x_1, \ldots, x_{i-1}, y, x_{i+1}, \ldots, x_n)$$

at the point $y = x_i$. Here the other variables x_j, for $j \neq i$, are to be treated as constants. It follows that $D_i f|_x$ is the slope of the tangent line to the graph of f in the x_i-direction, which is the line in which the tangent hyperplane to the graph of f meets the coordinate plane $x_j = \text{constant}$ ($j \neq i$). *See* Fig. 3.6.

The importance of partial derivatives is that they give an explicit expression for the derivative in standard coordinates. Namely, if $f : \mathbb{R}^n \to \mathbb{R}^m$ is differentiable at x, then $D_i f_j|_x$ exists for all i, j, and $Df|_x$ has the matrix

$$[D_i f_j|_x] \quad \text{or} \quad \left[\left.\frac{\partial f_j}{\partial x_i}\right|_x\right]$$

in standard coordinates for $\mathbb{R}^n, \mathbb{R}^m$. Here $f(x) = (f_1(x), \ldots, f_m(x))$ as usual.

For example, let $f : \mathbb{R}^2 \to \mathbb{R}^3$, where

$$f(x, y) = (x^2 + y^2, e^x, y).$$

Then

$$f_1(x, y) = x^2 + y^2 \qquad f_2(x, y) = e^x \qquad f_3(x, y) = y,$$

and

$$\frac{\partial f_1}{\partial x} = 2x \qquad\qquad \frac{\partial f_2}{\partial x} = e^x \qquad\qquad \frac{\partial f_3}{\partial x} = 0$$

$$\frac{\partial f_1}{\partial y} = 2y \qquad\qquad \frac{\partial f_2}{\partial y} = 0 \qquad\qquad \frac{\partial f_3}{\partial y} = 1.$$

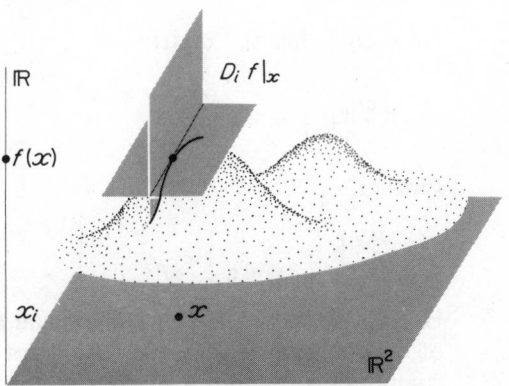

Fig. 3.6

Consequently

$$Df\big|_{(x,y)} = \begin{bmatrix} 2x & 2y \\ e^x & 0 \\ 0 & 1 \end{bmatrix}.$$

It may be proved that $Df\big|_x$ exists provided all $\partial f/\partial x_i$ exist and are continuous (though the converse need not be true). Such a function f is called *continuously differentiable*.

5 Higher Derivatives

Suppose that $f:\mathbb{R}^n \to \mathbb{R}^m$ is differentiable, with derivative

$$Df:\mathbb{R}^n \to L(\mathbb{R}^n, \mathbb{R}^m),$$

where $L(\mathbb{R}^n, \mathbb{R}^m)$ is the space of all linear maps from $\mathbb{R}^n$ to $\mathbb{R}^m$. Then each element of $L(\mathbb{R}^n, \mathbb{R}^m)$ is specified by an $n \times m$ matrix, and by taking the entries of this matrix in a fixed order we may identify $L(\mathbb{R}^n, \mathbb{R}^m)$ with $\mathbb{R}^{mn}$. Thus we may think of Df as a map from $\mathbb{R}^n$ to $\mathbb{R}^{mn}$. It may or may not be differentiable. If it is, then $D(Df)\big|_x$ exists for all $x \in \mathbb{R}^n$, and we say that f is *twice differentiable*, and write

$$D^2 f\big|_x = D(Df)\big|_x.$$

Then for each $x \in \mathbb{R}^n$, the *second derivative* $D^2 f\big|_x$ is a linear map from $\mathbb{R}^n$ to $L(\mathbb{R}^n, \mathbb{R}^m)$. This somewhat complicated idea may be interpreted in terms of coordinates. Let $x = (x_1, \ldots, x_n) \in \mathbb{R}^n$, and let $f(x) = (f_1(x), \ldots, f_m(x)) \in \mathbb{R}^m$. Then we have the matrix

$$Df\big|_x = \left[\frac{\partial f_j}{\partial x_i}\bigg|_x \right]$$

So Df is given in natural coordinates by $Df = g$, where

$$g(x) = [g_{ji}(x)], \qquad g_{ji}(x) = \frac{\partial f_j}{\partial x_i}\bigg|_x$$

Now $Dg\big|_x$ has the matrix

$$\left[\frac{\partial g_{ji}}{\partial x_k}\bigg|_x \right]$$

Further,

$$\frac{\partial g_{ji}}{\partial x_k}\bigg|_x = \frac{\partial}{\partial x_k}\frac{\partial f_j}{\partial x_i}\bigg|_x = \frac{\partial^2 f_j}{\partial x_k\, \partial x_i}\bigg|_x.$$

Hence the matrix of $D^2 f$ contains all the second partial derivatives of the coordinate functions f_j. It is not surprising that it is relatively complicated to define!

In the case of most interest to us, namely maps $\mathbb{R}^n \to \mathbb{R}$, we may think of the second derivative as having the *Hessian* matrix

$$\left[\frac{\partial^2 f}{\partial x_i\, \partial x_j} \Big|_x \right].$$

On the whole, we shall not require third derivatives explicitly. However, we shall often make the assumption that the functions we are interested in are *smooth*; that is, they possess derivatives of arbitrary order. In coordinates this means that their partial derivatives of arbitrary order exist (and are continuous).

If partial derivatives up to order r exist and are continuous we say that f is *r-fold differentiable*, or *of class* C^r. Smooth functions are of class C^∞. Occasionally we shall relax the condition of differentiability to *piecewise* differentiability, where the function may be decomposed into a finite number of functions, each defined on some region and differentiable there.

6 Taylor Series

A common mathematical technique is to take a smooth function $f : \mathbb{R} \to \mathbb{R}$ and expand it in a *Taylor series*

$$f(x_0 + x) = a_0 + a_1 x + a_2 x^2 + a_3 x^3 + \cdots \tag{3.5}$$

about a point x_0. Traditionally this series representation has been considered useful only if the series converges in some neighbourhood U of x_0, and its sum is equal to $f(x_0 + x)$ in U. In this case f is said to be *analytic* at x_0. The series may then be differentiated term by term in a (perhaps smaller) neighbourhood V of x_0, and it follows that the coefficients a_r are given by

$$a_r = \frac{1}{r!} D^r f \big|_{x_0}.$$

Even if f is not analytic, this formula may be used to define a formal Taylor series of f, but of course we no longer have any guarantee of convergence, or, if the series converges, we cannot be sure that it converges to f.

A function such as $\sin(x)$ is analytic, and its Taylor series at the origin is well known to be

$$\sin(x) = x - \frac{x^3}{3!} + \frac{x^5}{5!} - \frac{x^7}{7!} + \cdots .$$

Fig. 3.7 shows the graphs of the polynomial functions obtained by truncating this series after n terms, for $n \leqslant 43$. It illustrates neatly the *way* in which the Taylor series converges. Even with a large number of terms the approximation is very bad too far from the origin; on the other hand near the origin the approximation is very

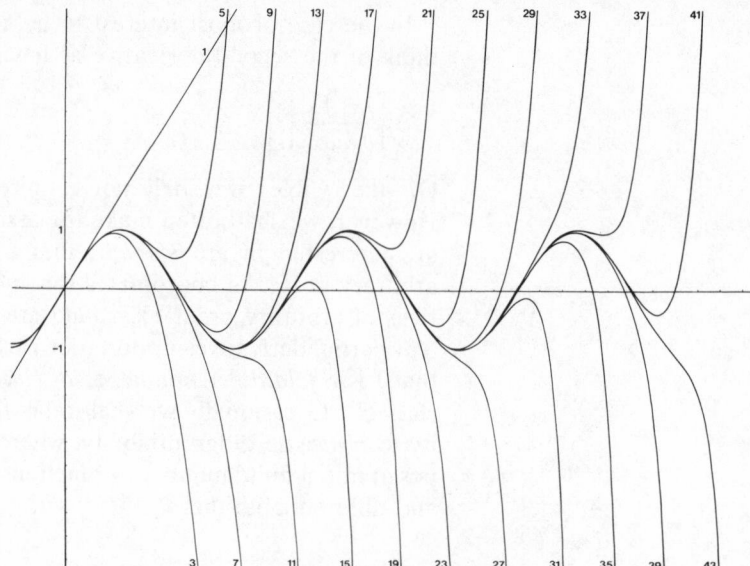

rapid. As the number of terms increases, therefore, both the degree of approximation and the range over which it is valid improve.

It is not always appreciated that for smooth functions the Taylor series may not converge, or may converge but to the wrong sum. For example, let

$$f(x) = \begin{cases} 0 & \text{if } x = 0 \\ e^{-1/x^2} & \text{if } x \neq 0 \end{cases}$$

(*see* Fig. 3.8). One verifies easily that for all n

$$\lim_{x \to 0} D^n f \big|_x = 0$$

and it follows that f is smooth (in other words, patching in the value 0 at 0 does no harm because e^{-1/x^2} is very flat near the origin). Now

$$D^n f \big|_0 = 0$$

for all n, and the Taylor series for f about 0 is

$$0 + 0x + 0x^2 + 0x^3 + \cdots.$$

This certainly converges – but not to $f(x)$.

A theorem of Borel (*see* Bröcker and Lander [9] p. 32) shows that the situation can be worse than this. Namely, given any sequence of real numbers whatsoever, say $b_0, b_1, b_2, \ldots$, then there exists a smooth function $f : \mathbb{R} \to \mathbb{R}$ such that its Taylor series at 0 is

$$b_0 + b_1 x + b_2 x^2 + \cdots.$$

It is easy to choose b_r to make this series diverge for all $x \neq 0$; for example $b_r = r!$ will do this.

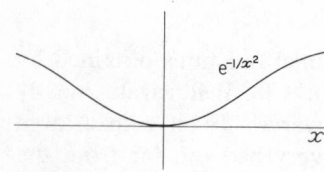

Fig. 3.8

Normally, in applied mathematics or physics, the Taylor series has been used to approximate f, and it is not surprising that *analyticity* has been emphasized. But in fact analyticity is neither necessary nor sufficient for such approximations to be valid, in terms of the use to which they are put. Analyticity has been over-emphasized, and the vanishing of the remainder term

$$R_k = f(x_0, x) - (a_0 + a_1 x + \cdots + a_k x^k)$$

as k tends to ∞ has figured too prominently, thus – in an elegant phrase of Zeeman's – allowing the tail of the Taylor series to wag the dog.

For example, the Taylor series $0 + 0x + 0x^2 + \cdots$ mentioned above is, despite its failure to converge to $f(x)$, an excellent qualitative approximation near the origin. It captures precisely the fact that f is *very flat* at that point. What it does *not* capture is that the origin is a local minimum of f; for $-f$ has the same Taylor series but a local maximum.

For any smooth function f we define its *Taylor series* at the origin to be the *formal* power series

$$\sum_{r=0} \frac{1}{r!} D^r f \big|_0 \cdot x^r.$$

Truncating this series at degree k gives the *k-jet*

$$j^k f(x) = \sum_{r=0}^{k} \frac{1}{r!} D^r f \big|_0 \cdot x^r.$$

(In the rigorous theory this term applies to a slightly more abstract object, defined in coordinate-free terms, but which reduces to this on taking coordinates. This aids rigour in the game of musical coordinates that the theory involves, but since we are omitting the deeper proofs we may forego this precision.) These truncated series represent perfectly good polynomial functions

$$j^k f : \mathbb{R} \to \mathbb{R}$$

whether or not the Taylor series converges. Further, they capture certain important information about f. Say that a function $g : \mathbb{R} \to \mathbb{R}$ has *order* k at the origin if

$$0 = g(0) = Dg \big|_0 = \cdots = D^{k-1} g \big|_0.$$

(By convention a constant has order 0.) The expression $O(k)$, as distinct from $o(h)$, will mean 'some function of order k' throughout. Then $j^k f$ is the unique polynomial of degree $\leq k$ such that $f - j^k f$ has order $k+1$. In other words, *the rth derivatives at 0 of f and $j^k f$ agree for $r = 0, 1, 2, \ldots, k$*. In this way the Taylor series, and its truncations, become a formal device to exhibit information on the derivatives of f, and hence its shape, near the origin. When (occasionally) we wish to examine f around a point $y \neq 0$, we shall write

$j_y f$ for the *k-jet of f at* y, alias the *k*-jet in the above sense of the function $x \mapsto f(y + x)$.

Note that for a polynomial its *order* is the lowest power of the variable appearing, whereas its *degree* is the highest power.

For a smooth function $f: \mathbb{R}^n \to \mathbb{R}^m$ there is a generalization of the Taylor series. In coordinate-free terms it is given by

$$f(x_0) + Df|_{x_0}(x) + \cdots + \frac{1}{r!} D^r f|_{x_0}(x, x, \ldots, x) + \cdots$$

where there are r x's in the rth term. For analytic f the series converges to $f(x_0 + x)$ in some neighbourhood.

We now switch to coordinates and for simplicity (it being the only case we need) set $m = 1$, so that $f: \mathbb{R}^n \to \mathbb{R}$. Then

$$x_0 = (x_{01}, \ldots, x_{0n})$$

$$x = (x_1, \ldots, x_n).$$

The rth term of the Taylor series becomes

$$\frac{1}{r!} \sum \frac{\partial^r f}{\partial x_{i_1} \cdots \partial x_{i_r}}\bigg|_{x_0} x_{i_1} \cdots x_{i_r}$$

where the sum is over all sequences $i_1, \ldots, i_r$ with $0 \le i_j \le n$ for each j. For example if $f: \mathbb{R}^2 \to \mathbb{R}$, $x_1 = x$, $x_2 = y$, $x_0 = (0, 0)$, then the initial terms of the Taylor series are

$$f(0, 0) + \left(\frac{\partial f}{\partial x}\bigg|_0 x + \frac{\partial f}{\partial y}\bigg|_0 y \right)$$

$$+ \frac{1}{2} \left(\frac{\partial^2 f}{\partial x^2}\bigg|_0 x^2 + \frac{\partial^2 f}{\partial x\, \partial y}\bigg|_0 xy + \frac{\partial^2 f}{\partial y\, \partial x}\bigg|_0 yx + \frac{\partial^2 f}{\partial y^2}\bigg|_0 y^2 \right) + \cdots.$$

Jets $j_{x_0}^k f$ and $j^k f = j_0^k f$ are obtained by truncating as before, and their partial derivatives up to given order agree with those of f. The 'tail' $f - j^k f$ of the Taylor series has its first k derivatives 0 at the origin. We shall refer so often to this expression that we abbreviate its name to *Tayl*.

7 Truncated Algebra

In dealing with jets we often need to perform 'truncated' algebraic operations. Given any polynomial $p(x)$, where $x = (x_1, \ldots, x_n)$, we define its *truncation at degree k*, denoted

$$\overline{p(x)}^k$$

to be the polynomial formed by taking only those terms of $p(x)$ whose degree is k or less. For example,

$$\overline{3x - 2y + 7xy - 9x^3 + 43x^7 y^2}^2 = 3x - 2y + 7xy,$$

$$\overline{3x - 27 + 7xy - 9x^3 + 43x^7 y^2}^3 = 3x - 2y + 7xy - 9x^3.$$

In fact, we have

$$\overline{p(x)}^k = j^k p(x),$$

but the different notation emphasizes the more algebraic point of view.

Fix the value of k, and let p and q be arbitrary polynomials. Then clearly

$$\overline{p+q}^k = \bar{p}^k + \bar{q}^k.$$

For products, it is a little less straightforward:

$$\overline{pq}^k = \overline{\bar{p}^k\bar{q}^k}^k$$

since terms of degree higher than k may occur as products of two terms of lower degree. However, it follows from these equations that in working to a fixed degree k, we may truncate and perform algebraic operations in any convenient order without altering the result. In fact it is easy by inspection to spot terms in a product that will turn out to have too high a degree, and ignore them.

We can also *substitute* a truncated polynomial in another one, provided we truncate the result at the same point. Thus, truncating at degree 2, if $p(x) = 1 - 2x + 3x^2$ and $q(x) = 3x + x^2$, we have

$$\overline{p(q(x))}^2 = \overline{1 - 2(3x + x^2) + 3(3x + x^2)^2}^2 = 1 - 6x + 25x^2.$$

In terms of this truncated notation, we can express concisely how k-jets behave relative to sums, products and composition (the jet analogue of the chain rule), as follows:

$$j^k(f+g) = j^k f + j^k g,$$

$$j^k(fg) = \overline{j^k f \cdot j^k g}^k,$$

$$j^k(f \circ g) = \overline{j^k f \circ j^k g}^k.$$

8 The Inverse Function Theorem

In the sequel we shall make considerable use of coordinate-changing techniques, and it is important to restrict the type of change allowed so that important information is not destroyed. We shall insist that the changes be *smooth* and *reversible* (and that the reverse change is also smooth). The technical term is *diffeomorphism*, and since this is found throughout the literature we shall occasionally make use of it.

Everything that we have done above for maps $f : \mathbb{R}^n \to \mathbb{R}^m$ can be reworked for maps $f : U \to \mathbb{R}^m$ where U is an open set in $\mathbb{R}^n$. By doing this we can work locally and avoid certain global problems. Let U, V be open sets in $\mathbb{R}^n$. Let $f: U \to \mathbb{R}^n$ and suppose that $f(U) = V$. Then f is a diffeomorphism provided:

(a) f is smooth,
(b) f has an inverse function $g: V \to \mathbb{R}^n$ such that $f \circ g = 1_V$, $g \circ f = 1_U$,
(c) g is smooth.

For example if U is the set of real numbers greater than 0, and

$$f(x) = x^2,$$

then we can set

$$g(x) = \sqrt{x}$$

and $V = U$. However, if we replace U by $\mathbb{R}$, then f is no longer a diffeomorphism because no inverse can be defined on the whole of $\mathbb{R}$. More subtle is the case $U = V = \mathbb{R}$,

$$f(x) = x^3.$$

Now we *can* find an inverse

$$g(x) = \sqrt[3]{x},$$

but g is not smooth: specifically $Dg|_0$ does not exist.

The geometric effect of a diffeomorphism is to bend the coordinates in a smooth way, as shown in Fig. 3.9. Hence if $h: U \to \mathbb{R}^m$ is smooth, and $f: U \to U$ is a diffeomorphism, the composite map $h \circ f$ is obtained from h by a smooth local change of coordinates (where in this case 'local' means inside U).

More generally, a *local diffeomorphism* at a point x is a map whose restriction to some open set containing x is a diffeomorphism onto its image. There is a simple and useful test for this property, known as the Inverse Function Theorem.

Theorem 3.1   (*Inverse Function Theorem*). *Let $f: U \to \mathbb{R}^m$ be smooth, and let $x \in U$. If the linear map $Df|_x$ is non-singular, then f is a local diffeomorphism at x.*

Thus we can test smooth maps for being local diffeomorphisms by computing the derivative. Note that $Df|_x$ is non-singular if and only

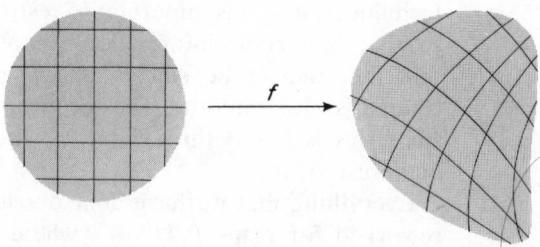

Fig. 3.9

if the *Jacobian determinant*

$$Jf|_x = \det Df|_x \neq 0.$$

9   The Implicit Function Theorem

A very similar result to the Inverse Function Theorem is the Implicit Function Theorem. This says that if the straight approximation (tangent line, plane, hyperplane or whatever) to the set S of solutions of an equation

$$f(x, y) = 0$$

is the graph of a *function*

$$y = y(x),$$

then so, locally, is the set of solutions itself. This is true for any finite number of dimensions for $x = (x_1, \ldots, x_n)$, $y = (y_1, \ldots, y_m)$. For example, in Fig. 3.10(a), we have

$$f(x, y) = x^2 + y^2 - 1$$

so the set S is a circle. The tangent at P is the graph of a function, hence so is S around P. Specifically, S is locally the graph of

$$y = +\sqrt{(1 - x^2)}.$$

The tangent lines at $(1, 0)$ and $(-1, 0)$ are not graphs; and neither is S around these points.

However, when the conditions are not fulfilled, S *may* locally be a graph. If

$$f(x, y) = x - y^3$$

then the set S is as shown in Fig. 3.10(b) and is the graph of $y = \sqrt[3]{x}$. But this, as the tangent is vertical, is not the graph of a *differentiable* function.

More specifically, if f is smooth and the tangent is a graph, then the function $y(x)$ which the Implicit Function Theorem tells us exists, is also smooth around the relevant point.

In more formal terms, we can state:

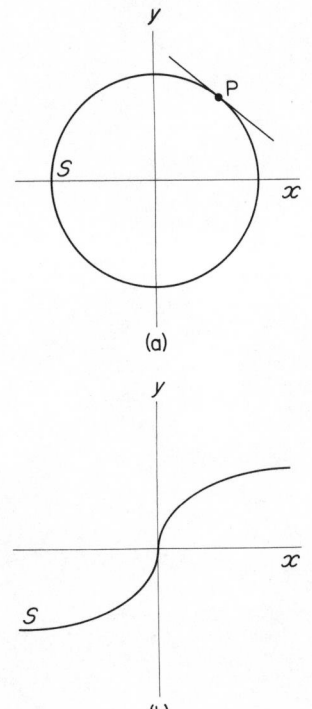

(a)

(b)

Fig. 3.10

Theorem 3.2 *(Implicit Function Theorem)*. *If $f : \mathbb{R}^m \times \mathbb{R}^n \to \mathbb{R}^p$ is smooth, and if*

$$\{(x, y) \mid Df|_{(x,y)} = 0\}$$

is the graph of a function $y = y(x)$, then

$$\{(x, y) \mid f(x, y) = 0\}$$

is locally the graph of a smooth function $y = \bar{y}(x)$.

This theorem is the justification of implicit differentiation, and is crucial in treating any kind of constrained system (where $f(x, y) = 0$ represents the constraint). It is of much the same depth as the Inverse Function Theorem, and each is a simple deduction from the other: which to take as a starting point is a matter of taste.

4 Critical Points and Transversality

In both physical and mathematical applications, an important feature of a smooth real-valued function is the occurrence of 'critical points' at which its derivative vanishes. For example, the commonest types of critical point possessed by a smooth function are its (local) maxima and minima. But even here we encounter difficulties, such as 'points of inflexion', and a closer study separates each of these three kinds into a whole range of types. For two or more variables the problem is compounded by the wide range of geometric phenomena that become possible.

A major mathematical source of catastrophe theory is the classification of types of critical points. In this chapter we make a start on the problem by proving the Morse Lemma, which classifies 'nice' critical points for any number of variables; and extending the classification to 'nasty' critical points in the case of a single variable. These two mathematical shoots nourish us by what they reach, and later develop into a network of roots which penetrates all the mathematical earth, and is firm enough to let us reach towards the stars.

By generalizing the proof of the Morse Lemma we obtain the important Splitting Lemma, which under certain circumstances permits a reduction in the number of variables in a problem. We have already remarked in the Preface that this result seems to some scientists to be the most surprising of the theory; yet its proof involves none of the deeper machinery developed for catastrophe theory, being no more than an extended exercise in elementary calculus.

Morse critical points have an important stability property, which may be stated intuitively as 'preservation of type under small perturbations'. This property can be seen algebraically, but it becomes clearer if reformulated as a geometric transversality property. This leads into a general discussion of transversality, which is discussed first for vector spaces, and generalized to smooth manifolds. We end by further exploring the notion of 'codimension' introduced in Chapter 2 and its relation to transversality.

The type of classification of critical points obtained here motivates and illuminates much of our later work. In particular the importance of Taylor expansions becomes apparent. The computational side of catastrophe theory, which we explain in Chapter 8, is designed to handle these problems. Some examples given in this chapter show that the behaviour of Taylor expansions becomes much more complicated for functions of two or more variables – indeed, there are several striking surprises, and the problem is far from being a straightforward extension of the single-variable case. Understanding these problems requires more than the algebraic formalism of Taylor series, and is largely unrelated to convergence questions: the key piece of insight is a geometric formulation exploiting transversality arguments.

1 Critical Points

Let $f:\mathbb{R}^n \to \mathbb{R}$ be a smooth function. A point $u \in \mathbb{R}^n$ is a *critical point* of f if

$$Df|_u = 0,$$

or in coordinates, if

$$\frac{\partial f}{\partial x_1}\bigg|_u = \cdots = \frac{\partial f}{\partial x_n}\bigg|_u = 0.$$

The value $f(u)$ at a critical point u is called a *critical value* of f.

Critical points occur when the graph of f has a horizontal tangent. If $n = 1$ these are traditionally classified as *(local) maxima, minima and points of inflexion* (Fig. 4.1). It will soon become apparent that a more delicate classification is required.

For $n = 2$, when $f:\mathbb{R}^2 \to \mathbb{R}$, there are more possibilities. The most common are *(local) maxima, minima and saddles*. Examples are (respectively)

$$f(x, y) = -x^2 - y^2, \quad x^2 + y^2, \quad x^2 - y^2$$

at the origin (Fig. 4.2). However, there are a wide variety of more complicated types, a few of which are shown in Fig. 4.3, for functions

$$f(x, y) = x^3 - 3xy^2, \quad x^2, \quad x^2 y^2,$$

known respectively as the *monkey-saddle, pig-trough and crossed pig-trough*. Of these, the monkey-saddle is only mildly bad, in that

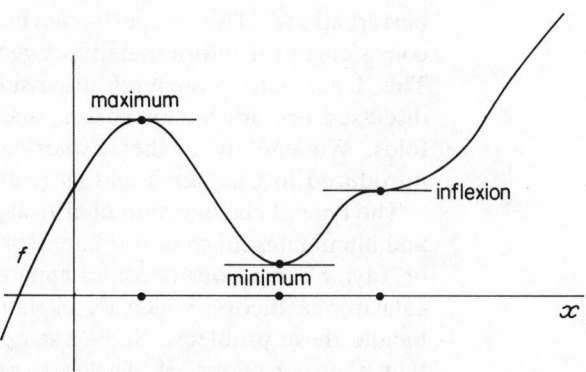

Fig. 4.1

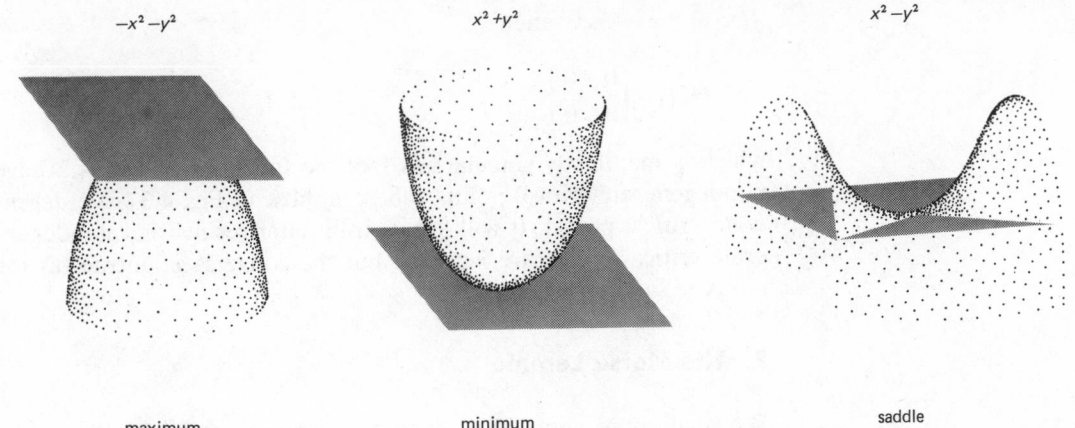

$-x^2-y^2$ x^2+y^2 x^2-y^2

maximum   minimum   saddle

Fig. 4.2

the critical point at the origin is *isolated:* sufficiently close to it there exist no other critical points. In the other two the origin is not an isolated critical point: it lies on (respectively) one or two lines of critical points. Non-isolated critical points are especially nasty, but in a strong sense extremely uncommon (*see* Chapter 8 Section 7), so for many purposes may be ignored.

The most important distinction is the following. We say that f has a *nondegenerate* critical point at u if $Df|_u = 0$ and if $D^2f|_u$ is a nondegenerate quadratic form (that is, its rank is equal to the number of variables n). Equivalently, the *Hessian matrix*

$$Hf|_u = \left[\frac{\partial^2 f}{\partial x_i \, \partial x_j} \Big|_u \right]$$

is non-singular, so the *Hessian determinant*

$$\det (Hf|_u) \neq 0.$$

For example, if $f(x, y) = x^2 + y^2$ then

$$Hf|_0 = \begin{bmatrix} 2 & 0 \\ 0 & 2 \end{bmatrix}$$

which is non-singular (of determinant 4). On the other hand if

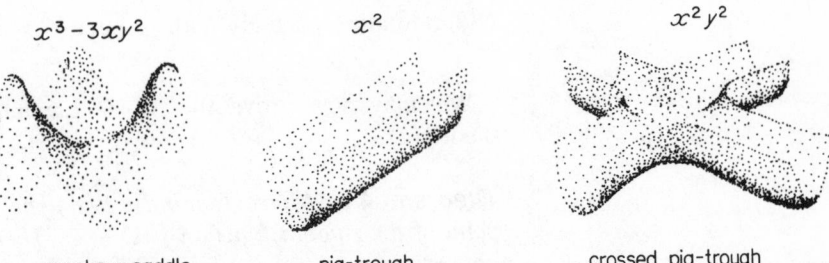

x^3-3xy^2 x^2 x^2y^2

Fig. 4.3   monkey – saddle   pig-trough   crossed pig-trough

$$f(x, y) = x^3 - 3xy^2 \text{ then}$$

$$Hf|_0 = \begin{bmatrix} 0 & 0 \\ 0 & 0 \end{bmatrix}$$

which is manifestly singular! In fact the functions of Fig. 4.2 have nondegenerate critical points, whereas those of Fig. 4.3 have *degenerate* critical points. It will follow from later results that nondegenerate critical points are isolated, but the converse is untrue (as the monkey-saddle shows).

2 The Morse Lemma

We shall now show that near a nondegenerate critical point a function f may be put into a simple standard form by changing its coordinates. Since this material is not a standard part of most scientists' equipment, we give detailed proofs.

Lemma 4.1 Let $f : \mathbb{R}^n \to \mathbb{R}$ be smooth in a neighbourhood of 0, with $f(0) = 0$. Then in a possibly smaller neighbourhood, there exist functions $g_i : \mathbb{R}^n \to \mathbb{R}$ such that

$$f = \sum_{i=1}^{n} x_i g_i,$$

where each g_i is smooth, and $g_i(0) = \dfrac{\partial f}{\partial x_i}\Big|_0$.

Proof We have

$$f(x_1, \ldots, x_n) = \int_0^1 \frac{\mathrm{d}}{\mathrm{d}t}(f(tx_1, \ldots, tx_n))\,\mathrm{d}t$$

$$= \int_0^1 \sum_{i=1}^{n} \frac{\partial f}{\partial x_i}\Big|_{(tx_1, \ldots, tx_n)} \cdot x_i \,\mathrm{d}t.$$

Hence we define

$$g_i(x_1, \ldots, x_n) = \int_0^1 \frac{\partial f}{\partial x_i}\Big|_{(tx_1, \ldots, tx_n)} \mathrm{d}t.$$

Differentiating partially with respect to x_i shows that $g_i(0) = \dfrac{\partial f}{\partial x_i}\Big|_0$.

We may now prove the *Morse Lemma* (though for us it is a theorem!).

Theorem 4.2 (Morse Lemma). Let u be a nondegenerate critical point of the smooth function $f : \mathbb{R}^n \to \mathbb{R}$. Then there is a local coordinate system $(y_1, \ldots, y_n)$ in a neighbourhood U of u, with $y_i(u) = 0$ for *all i, such that*

$$f = f(u) - y_1^2 - \cdots - y_l^2 + y_{l+1}^2 + \cdots + y_n^2$$

for all $u \in U$.

Proof We can translate the origin to u, and hence assume $u = 0$, and $f(u) = f(0) = 0$. By Lemma 4.1 we may then write

$$f(x) = \sum_{j=1}^{n} x_j g_j(x)$$

in some neighbourhood of 0. Since 0 is a critical point, we have

$$g_j(0) = \frac{\partial f}{\partial x_j}\bigg|_0 = 0.$$

Hence, using Lemma 4.1 again, there exist smooth functions h_{ij} such that

$$g_j(x) = \sum_{i=1}^{n} x_i h_{ij}(x),$$

and we can write

$$f(x) = \sum_{i,j=1}^{n} x_i x_j h_{ij}(x). \tag{4.1}$$

If we replace h_{ij} by

$$\hat{h}_{ij} = \tfrac{1}{2}(h_{ij} + h_{ji})$$

this equation still holds, and further $\hat{h}_{ij} = \hat{h}_{ji}$.

Partially differentiating (4.1) twice, we see that

$$\frac{\partial^2 f}{\partial x_i \, \partial x_j}\bigg|_0 = 2\hat{h}_{ij}(0),$$

and hence the matrix

$$[\hat{h}_{ij}(0)] = \left[\frac{1}{2}\frac{\partial^2 f}{\partial x_i \, \partial x_j}\bigg|_0\right]$$

is non-singular since 0 is a nondegenerate critical point.

Suppose inductively that there exist local coordinates $u_1, \ldots, u_n$ in a neighbourhood U_1 of 0 such that

$$f = \pm u_1^2 \pm \cdots \pm u_{r-1}^2 + \sum_{i,j \geq r} u_i u_j H_{ij}(u_1, \ldots, u_n),$$

where $H_{ij} = H_{ji}$. By a linear change in the final r coordinates (as in the reduction of quadratic forms in Section 5 of Chapter 2) we may assume that $H_{rr}(0) \neq 0$. Let

$$g(u_1, \ldots, u_n) = \sqrt{|H_{rr}(u_1, \ldots, u_n)|}.$$

By the Inverse Function Theorem this is smooth in some neighbourhood U_2 of 0, contained in U_1. (This step is the main reason why the Morse Lemma only holds *locally* in general.) We change coordinates to $v_1, \ldots, v_n$ defined by

$$v_i = u_i \qquad (i \neq r),$$

$$v_r = g(u_1, \ldots, u_n)\left(u_r + \sum_{i>r} \frac{u_i H_{ir}(u_1, \ldots, u_n)}{H_{rr}(u_1, \ldots, u_n)}\right)$$

which, again using the Inverse Function Theorem, is a local diffeomorphism. Now

$$f(v_1, \ldots, v_n) = \pm v_1^2 \pm \cdots \pm v_r^2 + \sum_{i,j \geqslant r+1} v_i v_j H'_{ij}(v_1, \ldots, v_n),$$

a formula like that for the u_i, but with r replaced by $r+1$. Hence, by induction, we obtain the conclusions of the theorem.

The proof of this should be compared with the diagonalization procedure for quadratic forms.

A function of the form

$$z_1^2 + z_2^2 + \cdots + z_{n-l}^2 - z_{n-l+1}^2 - \cdots - z_n^2$$

is called a *Morse l-saddle*. Hence the Morse Lemma implies that every nondegenerate critical point may be transformed by a diffeomorphism (smooth reversible coordinate change) to a Morse l-saddle, for some l. When $l = n$ we have a maximum, when $l = 0$ a minimum. (Some authors use $n - l$ in place of l, but for applications it is more convenient to have '0-saddle' mean always 'minimum', independently of n, as we can sometimes discard most of the variables without even knowing how many there are. Minima, and l-saddles for small l, are far more often significant than maxima (n-saddles) in applications.)

Since a Morse saddle clearly is an isolated critical point, and since smooth coordinate changes leave isolated critical points isolated, it follows that a nondegenerate critical point is always isolated.

The number l is an invariant of the topological type of the critical point, in the following sense: a smooth reversible coordinate change does not alter l.

At a non-Morse critical point, the Hessian is degenerate. We can measure *how* degenerate by computing its corank (Chapter 2 Section 5), the 'number of independent directions in which it is degenerate'. This number is independent of smooth invertible coordinate changes and comes to prominence in Chapters 7 and 8.

3 Functions of a Single Variable

We now study degenerate critical points in the simplest case of a function $f : \mathbb{R} \to \mathbb{R}$. Assuming (as we always can by a translation of coordinates) that f has a critical point at the origin and that $f(0) = 0$, we therefore also have $Df|_0 = 0$. The critical point is nondegenerate if and only if $D^2f|_0 \neq 0$. By the Morse Lemma there then exists a smooth local change of coordinates under which f becomes

$$\pm x^2,$$

where the sign is that of $D^2f|_0$.

If, however, $D^2f|_0 = 0$, we may obtain a finer classification by taking more terms of the Taylor series for f. However, this classification will say nothing about functions such as e^{-1/x^2} for which the whole Taylor series is zero.

Lemma 4.3 *Let $q:\mathbb{R} \to \mathbb{R}$ be a smooth function such that*

$$q(0) = Dq|_0 = \cdots = D^k q|_0 = 0.$$

Then in some neighbourhood of 0, there exists a smooth function l such that

$$q(x) = x^k l(x),$$

and $l(0) = 0$.

Proof We use induction on k. When $k = 0$ Lemma 4.1 applies. For $k \neq 0$ we use the same lemma to show that

$$q(x) = x l_1(x)$$

where l_1 is smooth. Differentiating this relation m times we get

$$D^m q|_x = x D^m l_1|_x + m D^{m-1} l_1|_x.$$

Putting $x = 0$ we deduce that

$$l_1(0) = Dl_1|_0 = \cdots = D^{k-1} l_1|_0 = 0.$$

Inductively,

$$l_1(x) = x^{k-1} l(x)$$

where l is smooth (in some neighbourhood of 0) and $l(0) = 0$. Then

$$q(x) = x^k l(x)$$

as claimed.

Following our major theme, of using coordinate changes to reduce functions to simple forms, amenable to classification, we obtain:

Theorem 4.4 *Let $f:\mathbb{R} \to \mathbb{R}$ be a smooth function, such that*

$$f(0) = Df|_0 = \cdots = D^{k-1} f|_0,$$

but

$$D^k f|_0 \neq 0.$$

Then there exists a smooth local change of coordinates under which f takes the form

$$x^k \quad (k \text{ odd})$$

$$\pm x^k \quad (k \text{ even})$$

and in the latter case the sign is that of $Df|_0$.

Proof   Taking the k-jet we have

$$f(x) = \frac{1}{k!} D^k f|_0 \cdot x^k + q(x)$$

where q is a function of order $k+1$. (This is true independently of convergence questions, as noted earlier.) Let

$$\frac{1}{k!} D^k f|_0 = \alpha \neq 0.$$

By Lemma 4.3

$$q(x) = x^k l(x)$$

for smooth l, in a neighbourhood of 0. Hence

$$f(x) = x^k (\alpha + l(x))$$
$$= \pm x^k |\alpha + l(x)|$$

where the sign is that of α, provided x lies in a neighbourhood U of 0 small enough to make $|l(x)| < |\alpha|$. If x lies in U then $|\alpha + l(x)| \neq 0$. Define

$$g(x) = x |\alpha + l(x)|^{1/k},$$

where the kth root is the unique positive one. We claim that $g : \mathbb{R} \to \mathbb{R}$ is a local diffeomorphism. To see this we invoke the Inverse Function Theorem. Clearly g is smooth. Now

$$Dg|_0 = \left[x \cdot \frac{1}{k} |\alpha + l(x)|^{(1/k)-1} Dl|_x + |\alpha + l(x)| \right]_0$$

$$= |\alpha| \neq 0,$$

so by the Inverse Function Theorem g is a local diffeomorphism. Now

$$f(x) = \pm g(x)^k,$$

so the change of coordinates $y = g(x)$ puts f into the form $\pm y^k$. Thus f transforms by a smooth local change of coordinates to the function $\pm x^k$. When k is odd we may replace x by $-x$ to make the sign positive.

Theorems 4.2 and 4.4 motivate an important notion. Two smooth functions $f, g : \mathbb{R}^n \to \mathbb{R}$ are said to be *equivalent* around 0 if there is a local diffeomorphism $y : \mathbb{R}^n \to \mathbb{R}^n$ around 0 and a constant γ such that, around 0,

$$g(x) = f(y(x)) + \gamma.$$

(Then y is a smooth reversible local change of coordinates, and the *shear term* γ adjusts the *value* of the function at 0, taking care of the various translations of the origin used above.) Then Theorem 4.2 says that near a nondegenerate critical point a function is equivalent to one of the Morse standard forms; and Theorem 4.4 that a function $f:\mathbb{R} \to \mathbb{R}$ with non-zero Taylor series is equivalent to $\pm x^k$ for some k. (In the mathematical literature this relation is called *right equivalence*, to distinguish it from other relations that occur, but in this book we do not need any of the others, and drop the adjective 'right'.)

It is fairly easy to check that $\pm x^k$ and $\pm x^l$ are equivalent if and only if $k = l$ and (for k, l even) the signs are the same; for whether or not a particular derivative vanishes is unaffected by such changes. This means that, say, the points of inflexion of x^3 and x^5 at the origin are different types of critical point in this classification, even though superficially they resemble each other. Likewise, the x^2 minimum is different from an x^4 minimum. These differences are important when we come to consider stability properties of critical points, which we do later in this chapter.

Intuitively, what we have proved is that *the behaviour of a critical point of a function of one variable is determined by the first non-zero jet (if it exists)*. It seems a widespread myth that similar results hold for two or more variables: actually this is not the case, and one of the more mathematical aims of catastrophe theory is to find out what really happens. Examples in the next section will explode the myth: the more delicate task of replacing it by suitable true statements will be left to Chapter 8.

The local nature of Theorem 4.4 is essential. The function

$$f(x) = x^2 - x^4$$

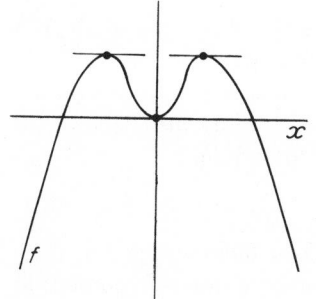

Fig. 4.4

has a local minimum of type x^2 at the origin, but further away acquires two local maxima (Fig. 4.4) and hence cannot be *globally* equivalent to x^2. The question 'how local is local' can be answered here by examining the proof of Theorem 4.4: the answer is 'until any other critical point is encountered'. For several variables the prescription is more complicated.

4 Functions of Several Variables

The previous section showed that a function of one variable may be replaced by the first non-zero term of its Taylor series, near to a critical point, without introducing any qualitative changes. This

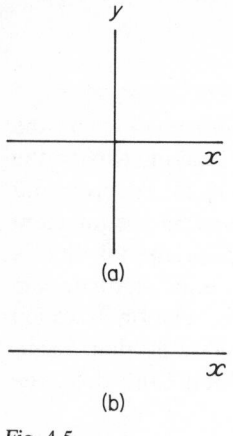

(a)

(b)

Fig. 4.5

happy circumstance does not extend to functions of several variables. For example, let

$$f(x, y) = x^2 y + y^{2001}.$$

The first non-vanishing jet is $x^2 y$. But this is not equivalent to f under local coordinate changes, since the solutions of the equation

$$x^2 y = 0$$

are given by the lines $x = 0$, $y = 0$ (Fig. 4.5(a)), whereas the solutions of

$$f(x, y) = 0$$

are given by

$$y(x^2 + y^{2000}) = 0,$$

that is,

$$y = 0,$$

as in Fig. 4.5(b). These two solution sets are clearly qualitatively different.

It follows that f is not equivalent to any jet $j^k f$ for $k \leq 2000$. The source of difficulty is not the large size of y^{2001}, for a similar analysis holds if we take instead the function

$$x^2 y + 10^{-1000000} y^{2001}.$$

But it *is* true (though we cannot explain why until Chapter 8) that if we take any function of the form

$$g(x, y) = x^2 y + y^{2001} + (\text{order } 2002)$$

then f and g *are* equivalent. So, for some functions, *some* jet, but not necessarily the first non-vanishing one, may provide a good qualitative approximation. Whether or not this is the case depends only on the form of this jet, and therefore does *not* depend on whether or not the function is analytic. The function

$$x^2 y + y^{2001} + e^{-1/x^2} \cdot e^{-1/y^2}$$

is adequately represented by its 2001-jet, for small enough x and y.

The problem is not simply that $x^2 y$ is a degenerate polynomial: to emphasize its subtle nature, we quote an example which will be discussed in full in Chapter 8, Section 13. The function

$$f(x, y) = \tfrac{3}{2} x^2 + x^3 - 3xy^2$$

has the following curious property. If $g(x, y)$ is a function whose 3-jet is equal to f, then g may not be equivalent to f. But if the 4-jet of g is equal to f, then g *is* equivalent to f. Thus we must examine terms of degree 4 in the jet of g, *even though these do not occur in f*; but terms of degree 5 or higher are irrelevant.

If a function has a jet which can be guaranteed to describe its qualitative behaviour, in the above sense, it is called *determinate* and the jet is called *sufficient*. When we are neglecting the technical distinction between jets and functions, we may call jets determinate too. In Chapter 8 we explain in considerable detail how to test for determinacy, along with many other important computational techniques. These techniques are extracted from the mathematical literature, and based on the papers of Mather [10–15] which contain full proofs. Chapter 8 explains, using less highbrow mathematics, why the results take the form that they do; but does not prove them.

5 The Splitting Lemma

A more glorified version of the Morse Lemma allows us to tidy up, somewhat, a degenerate critical point, by 'splitting' the function into a Morse piece on one set of variables and a degenerate piece on a different set, whose number is equal to the corank. This is an important and powerful result, basic to the theory, but need not have been as recent as it is. Its proof can be wholly elementary, and does not require any of the more sophisticated theorems of catastrophe theory. Formally stated, it is:

Theorem **4.5** *Let* $f : \mathbb{R}^n \to \mathbb{R}$ *be a smooth function, whose Hessian at* 0 *has rank* r *(and corank* $n - r$*). Then* f *is equivalent, around* 0*, to a function of the form*

$$\pm x_1^2 \pm \cdots \pm x_r^2 + \hat{f}(x_{r+1}, \ldots, x_n)$$

where

$$\hat{f} : \mathbb{R}^{n-r} \to \mathbb{R}$$

is smooth.

Proof By a linear change of coordinates $u = u(x)$ we can transform the Hessian of f at 0 into the form

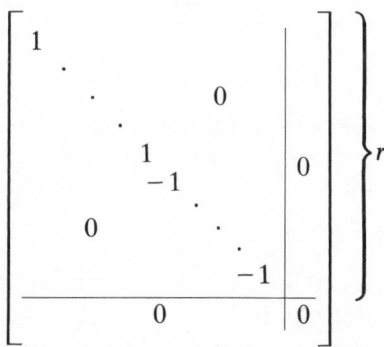

The Implicit Function Theorem now allows us to express the set

$$\left\{ u \,\middle|\, \frac{\partial f}{\partial u_1} = \cdots = \frac{\partial f}{\partial u_r} = 0 \right\}$$

(locally) as the graph

$$\{(g_1(u_{r+1}, \ldots, u_n), \ldots, g_r(u_{r+1}, \ldots, u_n), u_{r+1}, \ldots, u_n)\}$$

of a smooth function

$$g : \mathbb{R}^{n-r} \to \mathbb{R}^r.$$

We use g to turn this graph into the $(u_{r+1}, \ldots, u_n)$-axis, by a map ϕ, easily seen to be a local diffeomorphism, defined by

$$\phi(u_1, \ldots, u_n) = (u_1 - g_1(u_{r+1}, \ldots, u_n), \ldots$$

$$\ldots, u_r - g_r(u_{r+1}, \ldots, u_n), u_{r+1}, \ldots, u_n).$$

(This step is crucial to the argument, as an initial tidying step to get the correct kind of dependence on $u_{r+1}, \ldots, u_n$.)

Let $F = f \circ \phi$. Locally, each function

$$F_{(u_{r+1}, \ldots, u_n)} : \mathbb{R}^r \to \mathbb{R}$$

$$(u_1, \ldots, u_r) \mapsto F(u_1, \ldots, u_r, u_{r+1}, \ldots, u_n)$$

has a Morse critical point at the origin of $\mathbb{R}^r$, though not necessarily taking the value 0 at that point. We write

$$\hat{f}(u_{r+1}, \ldots, u_n) = F(0, \ldots, 0, u_{r+1}, \ldots, u_n).$$

Now the argument appearing in the first part of the proof of the Morse Lemma may be used (after generalizing Lemma 4.1 to the case where f vanishes along a multi-axis) to write

$$F(u) = \hat{f}(u_{r+1}, \ldots, u_n) + \sum_{k,m \leqslant r} u_k u_m h_{km}^{(u_{r+1}, \ldots, u_n)}(u_1, \ldots, u_r),$$

where for each choice of $u_{r+1}, \ldots, u_n$ the function

$$h_{km}^{(u_{r+1}, \ldots, u_n)} : \mathbb{R}^r \to \mathbb{R}$$

is smooth. The remainder of the proof of the Morse Lemma, applied in a $(u_{r+1}, \ldots, u_n)$-dependent fashion to this expression, reduces F to the form

$$\hat{f}(u_{r+1}, \ldots, u_n) \pm v_1^2 \pm \cdots \pm v_r^2$$

and proves the theorem. In essence, the whole *process* of reduction to standard form used to prove the Morse Lemma depends smoothly on $u_{r+1}, \ldots, u_n$, once the initial tidying has been done.

This theorem says, in a strong sense, that the behaviour of a function near a degenerate critical point can be found by studying a function involving a number of variables equal to the *corank* of the Hessian. Thus, say, a critical point of a function of 2001 variables, of corank 3, requires us to study only a function of three variables. This reduction to a small number of variables is what makes the Splitting Lemma so useful, and so surprising. The corank of the Hessian is also the *corank of the function* at the critical point concerned.

In Chapter 6 we shall obtain more extensive versions of the Morse and Splitting Lemmas for *families* of functions – a reformulation which actually involves very little work – and give a formal proof of the most general version.

6 Structural Stability

We next consider the effect of a small perturbation on a critical point. Speaking intuitively, we say that a function p is *small* if all of its partial derivatives are small for x in a fixed neighbourhood of 0. Suppose we take a function f with a critical point at 0, and perturb it a little by adding a small enough function p: what happens?

If the derivative of p at the origin is non-zero, 0 will no longer be a critical point at all. Indeed, f may no longer *have* a critical point at all. But to begin our investigation of the simplest case, we shall make life easy by assuming that $Dp\,|_0 = 0$.

Suppose that f is *Morse*: the critical point at 0 is nondegenerate. Then the determinant of the Hessian

$$\det Hf\,|_0 \neq 0.$$

If now p is small enough, we must have

$$\det H(f+p)\,|_0 \neq 0,$$

for the determinant of the Hessian varies continuously. Thus $f+p$ *is also Morse*. It is not hard to strengthen the argument to show that the type of saddle is the same for f and for $f+p$: the critical points are both l-saddles for the same l, and hence are equivalent.

We say that f is *structurally stable* if, for all sufficiently small smooth functions p, the critical points of f and $f+p$ have the same type; or in other words if f and $f+p$ are equivalent, after a suitable translation of the origin. The above discussion shows that near a Morse critical point any function is structurally stable. We shall generalize this notion to the most important case for catastrophe theory, families of smooth functions, in Chapter 6.

Now relax the condition $Dp\,|_0 = 0$. A typical example of what happens is if we take $f(x) = x^2$, $p = 2\varepsilon x$ (ε a small constant). Then

$$f(x) + p(x) = x^2 + 2\varepsilon x = (x+\varepsilon)^2 - \varepsilon^2$$

which has a Morse critical point at $x = -\varepsilon$. Thus the critical point is moved (in a way which depends smoothly on ε) but still does not change its type.

For degenerate critical points, the picture is quite different. If we perturb x^3 by adding a term εx, then for positive ε we find *no* critical points at all, and for negative ε we find *two*: one Morse

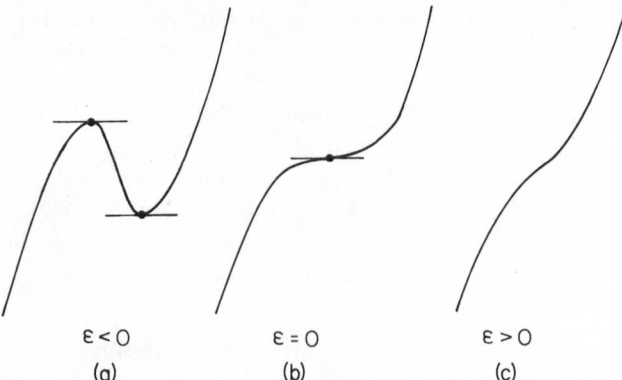

Fig. 4.6

maximum and one Morse minimum (Fig. 4.6). Thus x^3 is structurally unstable as a critical point. Similarly x^4, perturbed by a term
εx^2, either turns into a single Morse minimum, or two minima and
one maximum (Fig. 4.7). The higher the degree n, the worse x^n
behaves in this respect: perturbing x^5 can lead to *four* critical points,
two maxima and two minima; and this is true no matter how small a
perturbation we allow (Fig. 4.8).

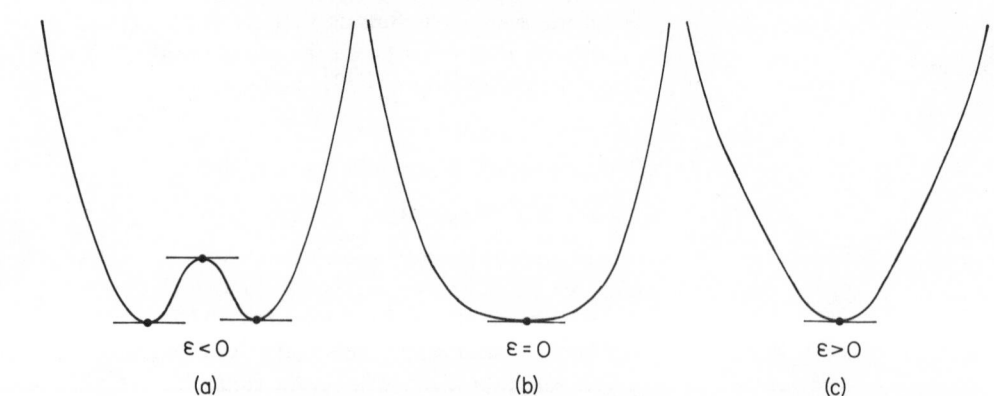

Fig. 4.7

It can be proved (*see* Milnor [16]) that a critical point is structurally stable if and only if it is nondegenerate; hence every degenerate
critical point is structurally unstable. Structural stability is also
typical, in a sense we shall explore. However, we shall see in
Chapter 7 that when we look at *families* of smooth functions the
situation is quite different.

Notice how the analysis of small perturbations allows us to make
a clear distinction between x^3 and x^5, despite their superficial
similarity: under perturbations they split into different numbers of
critical points. This is a qualitative difference, and clearly can be
important in applications.

Our next task is to explain how the typicality of Morse functions
may be seen geometrically. The next two sections set up the relevant
notions; the third section gives the geometric reasons for structural
stability.

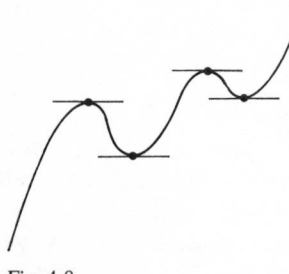

Fig. 4.8

7 Manifolds

A *manifold* is just a higher-dimensional analogue of a smooth curve or surface, and we shall not require any of the deep mathematical properties of manifolds, just the concept itself. A *smooth submanifold* of $\mathbb{R}^n$ is a subset M with the following properties:

(a) locally it looks like a small piece of $\mathbb{R}^m$;
(b) it is embedded in $\mathbb{R}^n$ smoothly, that is, has a unique tangent hyperplane at each point.

The number m is called the *dimension* of the manifold. A one-dimensional manifold is a curve, a two-dimensional manifold a surface. Fig. 4.9 shows some typical manifolds in $\mathbb{R}^3$, all of dimension 2. Fig. 4.10 shows some sets which are not manifolds because they have the wrong local structure; Fig. 4.11 sets which are not smoothly embedded.

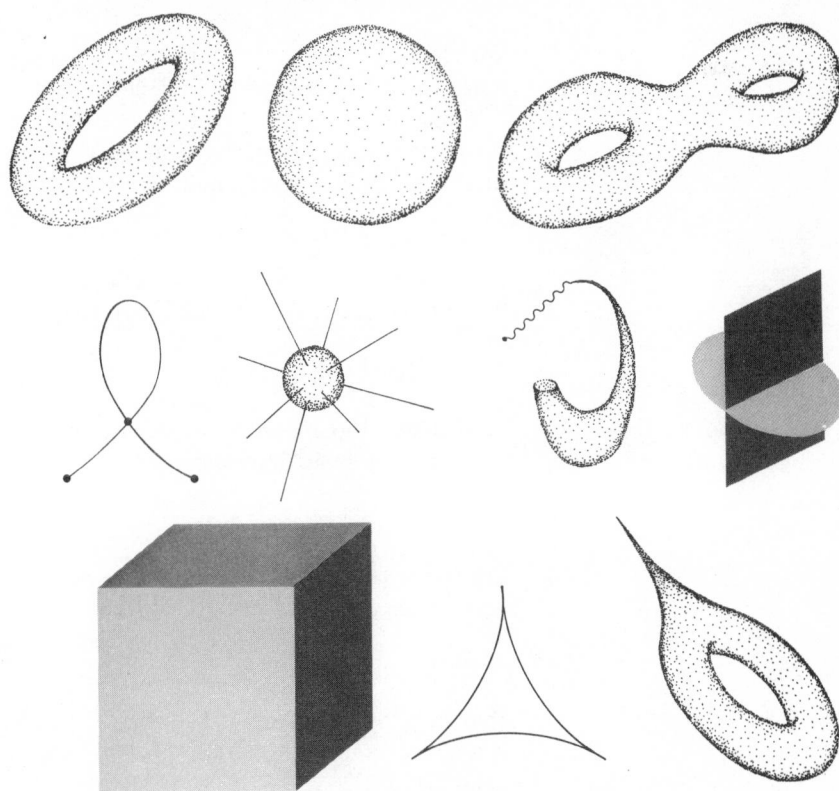

Fig. 4.9

Fig. 4.10

Fig. 4.11

If we define (a) in a more technical manner, we have to introduce the notion of a *chart* or *local coordinate system* at a point $x \in M$. This is a local diffeomorphism $f : U \to M$ where U is an open set in $\mathbb{R}^n$ and $f(U)$ is a neighbourhood of x in M. The standard coordinates on $\mathbb{R}^n$ induce (local) coordinates on M as shown in Fig. 4.12.

In physics or engineering texts these are often called *curvilinear* or *generalized coordinates*. The word *parameter* includes 'coordinate' but we sometimes *parametrize* objects less well behaved than manifolds by *singular* maps (not local diffeomorphisms).

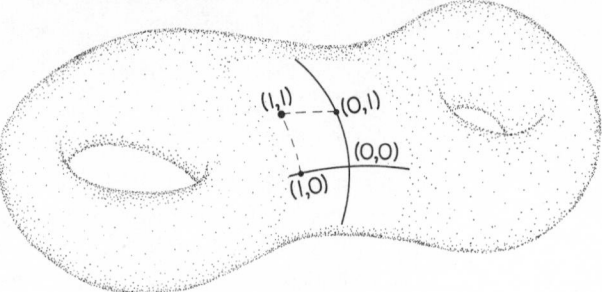

Fig. 4.12

8 Transversality

We shall eventually define transversality for manifolds, but to begin with we restrict attention to vector spaces.

Two subspaces U, V of $\mathbb{R}^n$ are *transverse* if they meet in a subspace whose dimension is as small as possible. If dim $U = s$ and dim $V = t$ then this minimal dimension is

$$\max(0, s + t - n).$$

For example, two planes in $\mathbb{R}^3$ are transverse if they meet in a line (or, equivalently, do not coincide), for

$$\max(0, 2 + 2 - 3) = 1.$$

A 4-dimensional and a 6-dimensional subspace of $\mathbb{R}^7$ are transverse if they meet in a subspace of dimension

$$\max(0, 4 + 6 - 7) = 3.$$

On the other hand two 1-dimensional subspaces of $\mathbb{R}^3$ are transverse if they meet in a space of dimension

$$\max(0, 1 + 1 - 3) = 0,$$

that is, a point (namely the origin). Equivalently (exercise for the reader) U and V are transverse if and only if U and V together span $\mathbb{R}^n$.

Next we generalize to *affine* subspaces: vector subspaces translated away from the origin. Specifically, an affine subspace of $\mathbb{R}^n$ is a subset of the form

$$X = V + a = \{v + a \mid v \in V\}$$

where a is a fixed element of $\mathbb{R}^n$ and V is a subspace (Fig. 4.13). The dimension of X is defined to be that of V.

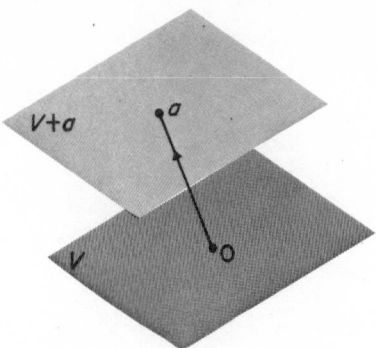

Fig. 4.13

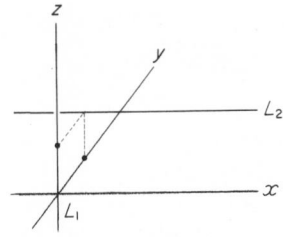

Fig. 4.14

The extra point to take care of now is that affine subspaces need not meet at all (e.g. two skew lines in $\mathbb{R}^3$, such as

$$L_1 = \{x, y, z) \mid x = y = 0\}$$
$$L_2 = \{(x, y, z) \mid y = z = 1\}$$

cannot meet, for any point (x, y, z) in their intersection has $y = 0$ and $y = 1$, *see* Fig. 4.14.) Indeed, this is the typical way for two affine lines to meet in $\mathbb{R}^3$: that is, not to meet at all.

Let X and Y be affine subspaces of $\mathbb{R}^n$ of dimensions s, t respectively. They meet *transversely* if either

(a) their intersection $X \cap Y$ is empty, or
(b) $s + t \geq n$ and dim $X \cap Y = s + t - n$.

Fig. 4.15 shows some typical intersections in $\mathbb{R}^3$, stating which are transverse.

Note that transversality depends on the dimension of the surrounding space: for example Figs 4.15 (e), (h) and (j) would be transverse intersections in $\mathbb{R}^2$.

Now we move on to manifolds. Two submanifolds of $\mathbb{R}^n$ meet transversely at a given point provided either they do not meet, or their tangent hyperplanes meet transversely. Unlike those of affine subspaces, these tangents may coincide for manifolds which meet in isolated points. For example the curves (1-manifolds) $y = 0$ and $y = x^3$ in $\mathbb{R}^2$ are not transverse at 0, even though this is an isolated point of intersection, since both have the x-axis as tangent (Fig. 4.16).

Fig. 4.17 shows corresponding pictures to Fig. 4.15 for manifolds, together with a more subtle one, Fig. 4.17 (m). This is not transverse since the two tangent planes coincide, even though it looks a little like Fig. 4.17 (i).

Transversality is often called *general position* (because 'nothing special happens'). It is *typical* in the following sense (which can be made precise): two manifolds taken at random are infinitely unlikely to intersect non-transversely (in the same way that a real number chosen at random is infinitely unlikely to be *exactly* equal to π). The

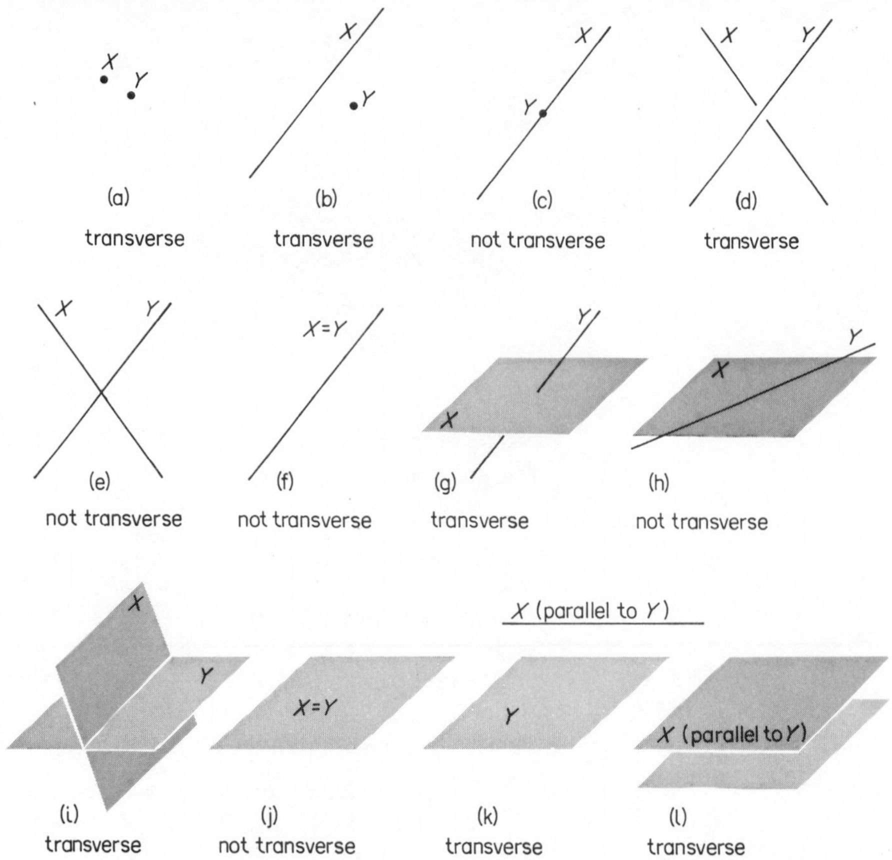

(a) transverse

(b) transverse

(c) not transverse

(d) transverse

(e) not transverse

(f) not transverse

(g) transverse

(h) not transverse

(i) transverse

(j) not transverse

(k) transverse

(l) transverse

Fig. 4.15

precise form of this typicality statement was first discovered and proved by Thom, and as the Thom transversality theorem it has become a powerful technical tool. But in an intuitive form it has long been an implicit assumption of much physics. We will have the language to express this more clearly in Section 11.

Near a point of transverse intersection the two manifolds are closely approximated by their tangents. This implies that the local nature of the intersection depends only on the dimensions of the manifolds: for a pair of spaces of given dimensions, one transverse intersection looks much like any other. For there is essentially only one way to obtain a transverse intersection of vector subspaces U, V; namely, pick a basis $u^1, \ldots, u^p$ for $U \cap V$; extend to a basis for U by adding $u^{p+1}, \ldots, u^s$, and to a basis of V by adding $v^{p+1}, \ldots, v^t$. Then

$$u^1, \ldots, u^s, v^{p+1}, \ldots, v^t$$

is a basis for $\mathbb{R}^n$. Hence, up to a linear transformation, the subspaces may be replaced by

U' spanned by $e^1, \ldots, e^s$

V' spanned by $e^1, \ldots, e^p, e^{s+1}, \ldots, e^n$

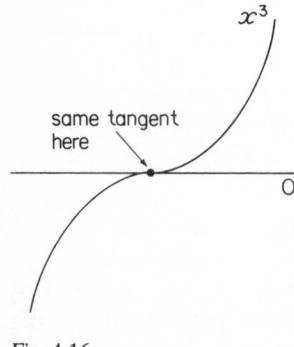

same tangent here

x^3

O

Fig. 4.16

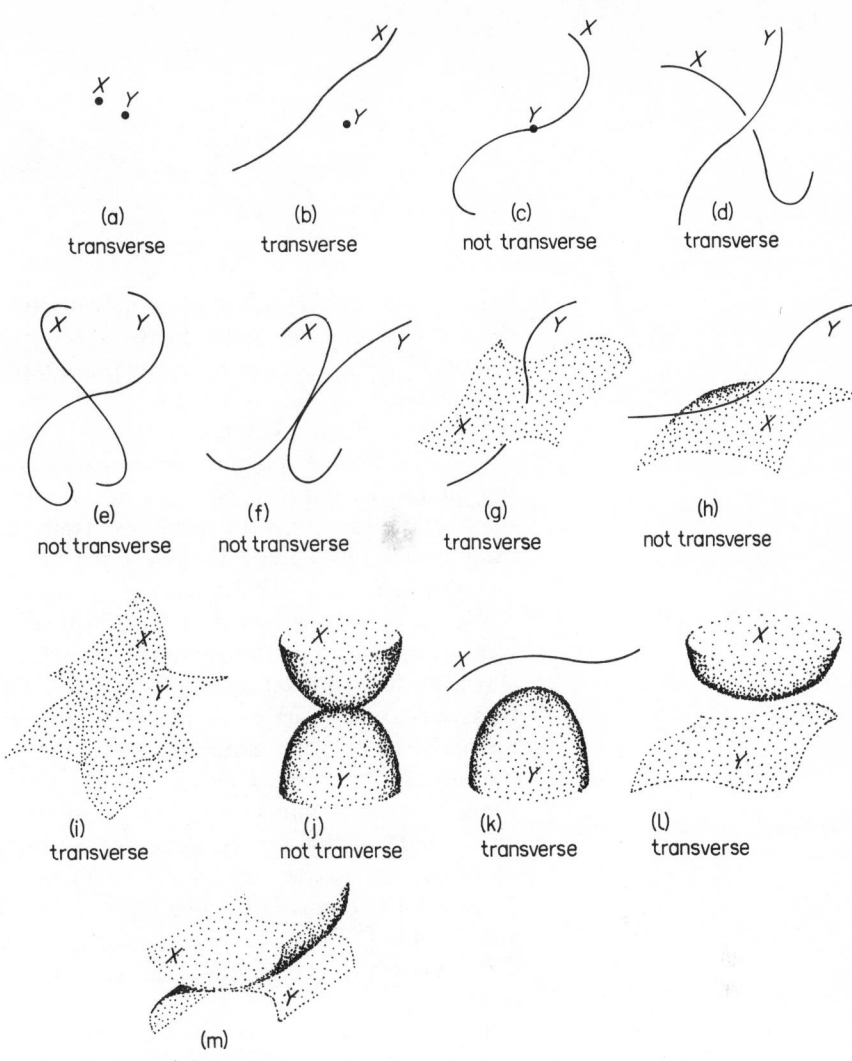

Fig. 4.17

where $e^i = (0, \ldots, 0, 1, 0, \ldots, 0)$ with 1 in the ith place. This configuration is unique, and every transverse intersection of an s-manifold with a t-manifold in $\mathbb{R}^n$ looks, locally, just like it. Fig. 4.18 illustrates this for 2-manifolds in $\mathbb{R}^3$.

Note that the reduction to standard form required two choices, which could have been made in various ways. In language introduced in Chapter 2 Section 2, these were the choices of cobases for $U \cap V$ in U and in V.

9 Transversality and Stability

It is intuitively reasonable, in terms of the figures we have drawn, to suppose that if we wobble a transverse crossing slightly, we still

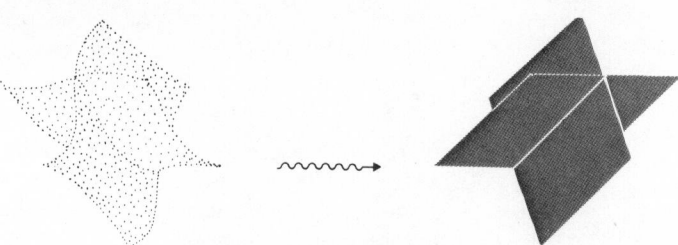

Fig. 4.18

have a transverse crossing. Algebraically this is quite easy to prove By a choice of coordinates in the submanifolds, the transversality condition (the directions in one manifold, plus the directions in the other, equal all the directions in $\mathbb{R}^n$) can be expressed as the requirement that not all of a suitable list of determinants may vanish. These determinants depend continuously on the position of the manifolds, hence if they do not vanish for some position the same is true for all nearby positions. Thus nearby positions also give transverse intersections. Transversality is a *stable* property, maintained under small perturbations.

We do not prove this in detail, since we shall in any case really want to appeal to a stronger result, the Thom isotopy theorem, for which we have not set up the techniques. This says not merely that transversality is a stable *property*, but that transverse crossings are themselves stable. By a suitable change of coordinates in $\mathbb{R}^n$ and in the submanifolds we are looking at, any near enough neighbour of a given transverse intersection can be reduced to *exactly* the same form. (Indeed, this can be done for a whole family of neighbours simultaneously, as we did for the neighbours of a Morse function in Section 6.) The most attractive proof we know is given in Abraham and Robbin [17] which, by elegant coordinate-free notation, makes highly geometric a rigorous proof covering the infinite-dimensional case.

The structural stability of Morse functions can be obtained as a consequence of the Thom isotopy theorem, with some advantages over the more elementary arguments we have given. For the condition that the Hessian be non-singular is exactly the condition that the Jacobian of the mapping

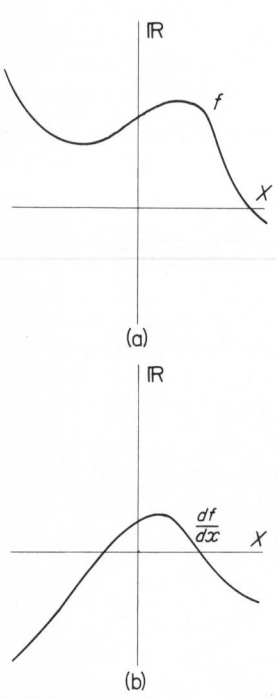

(a)

(b)

Fig. 4.19

$$Df = \left(\frac{\partial f}{\partial x_1} \cdots \frac{\partial f}{\partial x_n} \right) : \mathbb{R}^n \to \mathbb{R}^n$$

be non-singular, which is the condition that the graph of Df meet that of the zero function transversely. (This is easiest to see for $n = 1$, where the condition for Df to cross 0 transversely becomes $\frac{d^2 f}{dx^2} \neq 0$ (Fig. 4.19), which is the Morse condition.)

Thus the stability of Morse functions becomes deducible from that of transverse crossings, and *so does their typicality*. We have seen that the non-Morse functions x^3 and x^4 are not structurally stable.

We now see that their derivatives (Fig. 4.20) have graphs which are atypical in meeting the zero line non-transversely. From the full statement of the Thom transversality theorem, which says that typically a function f has the graphs of Df and of its higher derivatives meeting a particular manifold transversely, we may deduce the typicality of Morse functions. Many proofs of stability and typicality, in more general contexts, work this way.

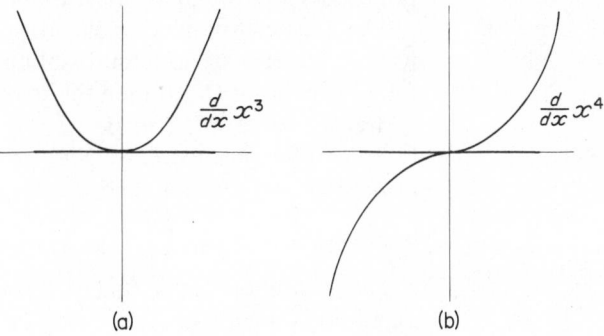

Fig. 4.20   (a)   (b)

Caveat lector   We have been using a subtler fact than 'a typical $f : \mathbb{R}^n \to \mathbb{R}^n$ is transverse to zero', for we have assumed that a typical f has Df transverse to zero: in other words that transversality is typical *among gradient functions*, alias 'exact differentials'. Now gradients are a very special class of functions when $n \geqslant 2$. (For example the map $g : \mathbb{R}^2 \to \mathbb{R}^2$ with $g(x, y) = (y, -x)$ is not a gradient.) On $\mathbb{R}^n$, for $n \geqslant 2$, the gradient functions are those for which a certain differential (the *curl* of 'vector analysis' when $n = 3$) is everywhere zero. To see the delicacy of the problem, consider instead those $f : \mathbb{R}^2 \to \mathbb{R}^2$ whose first derivatives in the x_1-direction vanish everywhere. There is a superficial analogy, in that we are requiring a certain differential operator to vanish. But for this latter class, transversality is *not* typical: explore the geometrical implications of 'constancy along x_1' and you will soon see why.

What is typical depends on the class considered, which is one of the reasons why Thom's transversality theorem, which proves the typicality to which we are appealing, is by no means a trivial or easy result.

Among all continuous curves it is in a strong sense infinitely *atypical* to be differentiable anywhere – yet this whole book is about everywhere differentiable models of the world. Typicality can only be within a chosen class of mathematical models, and the choice must be made – it is not 'given'.

10   Transversality for Mappings

In Chapter 7 we shall require a slightly more elaborate notion of transversality: transversality for a mapping. If C is a *parametrized curve* in the plane, $C:\mathbb{R}\to\mathbb{R}^2$, we can think of the parameter as *time* and the curve as a moving point. Then transversality of this mobile point to a fixed (unparametrized) curve C' requires the following. Whenever $C(t)$ lies on C', then the *velocity* vector of the moving point at t must be linearly independent of the *tangent* vector to C' at $C(t)$. In other words, not only must the track of the point be transversal to C' in the usual sense, but the velocity with which the point passes through C' must be non-zero.

The reason for making the definition this way is to retain the useful property of structural stability. Thus if C' is the y-axis and $C(t)$ is the point $(t^3, 0)$ then the *image* of C is just the x-axis, which is transverse to C'. But perturbing C a little, replacing $(t^3, 0)$ by $(t^3 - \varepsilon t, 0)$ for positive ε, we get a track which crosses C' three times instead of just once, a clear qualitative difference. This can happen because the velocity $\left(\dfrac{d}{dt}t^3\big|_0, 0\right)$ of C as it crosses C' is zero.

The condition for a function $f:\mathbb{R}^m\to\mathbb{R}^n$ to be transverse to a manifold is analogous. The velocity vector is replaced by the space spanned by *all* velocities through $f(x)$ in $\mathbb{R}^n$ that can be obtained by following curves $f(g(t))$ passing through x: and this has to be transverse to the tangent hyperplane of the manifold in $\mathbb{R}^n$. Equivalently, the corresponding *graphs* must be transverse as submanifolds of $\mathbb{R}^m\times\mathbb{R}^n$. (For a more precise statement *see* Golubitsky and Guillemin [18].) This type of transversality is still stable and typical, and the dimensional formula still holds, with care taken for the case $m\geqslant n$. When $f:\mathbb{R}^m\to\mathbb{R}^n$ is transverse to a submanifold C' of $\mathbb{R}^n$, then the intersection X of C' with the image of f has dimension

$$\min(m+r'-n, r')$$

where r' is the dimension of C'.

In the language elaborated in the next section, transversality implies that the codimension of X in the image of f (and consequently that of $f^{-1}(X)$ in $\mathbb{R}^m$) is that of C' in $\mathbb{R}^n$.

11 Codimension

The Implicit Function Theorem (Chapter 3 Section 9) shows that if $f:\mathbb{R}^n\to\mathbb{R}^m$, $m<n$, has $f(x_1,\ldots,x_n)=(c_1,\ldots,c_m)$, and if $Df^{-1}(0)$ is the graph of a function (equivalently, f has rank m) at $(x_1,\ldots,x_n)$, then around $(x_1,\ldots,x_n)$ we can express the set

$$f^{-1}(c)=\{(y_1,\ldots,y_n)\,|\,f_1(y)=c_1,\ldots,f_m(y)=c_m\}$$

as the graph of a smooth function $\mathbb{R}^{n-m}\to\mathbb{R}^m$. In particular $f^{-1}(c)$ can be smoothly parametrized around x by $(y_1,\ldots,y_{n-m})$, and is a smooth submanifold. Conversely, any such submanifold M can be written locally as $f^{-1}(0)$ for a suitable smooth f of maximal rank m

at points of M. (There may be difficulties in doing this for all of M at once. The kind of problem can be seen by considering on which side of a Möbius strip contained in M a function $f:\mathbb{R}^3\to\mathbb{R}$ with $M = f^{-1}(0)$ should be positive! But around any point it is always possible.)

Thus we have a (local) equivalence between $(n - m)$-dimensional submanifolds of $\mathbb{R}^n$ and, in traditional language, 'sets defined by m equations'. It is in this setting that physics most often appeals to the typicality of transversality. If M is defined by m equations, and M' by m', where $m + m' > n$, the combined system of $m + m'$ equations is *overdetermined* and typically has no solutions. If it has solutions, these are unstable unless special conditions hold them in existence – and the details of such special conditions are a necessary part of an adequate theory.

This is just our observation above that if $\dim M + \dim M' < n$, then typically M and M' do not meet at all in $\mathbb{R}^n$. For instance, curves in $\mathbb{R}^3$ are locally 'defined by two equations' each. The intersection of two would satisfy four equations, and typically does not exist.

This leads us to pay special attention to the number $n - \dim M$ of equations in $(x_1, \ldots, x_n)$ locally needed to describe M as a subset of $\mathbb{R}^n$ (as distinct from the number $\dim M$ required to *parametrize* it). This number is the *codimension* of M in $\mathbb{R}^n$, and corresponds precisely to the codimension in $\mathbb{R}^n$ (as defined in Chapter 2 Section 2) of each tangent vector space to M. It can often be useful independently of the dimension.

For example, let X be the island of Great Britain, omitting offshore islands. As represented on the map, X is 2-dimensional and the border between England and Scotland is 1-dimensional, hence has codimension 1. Now the border is not really 1-dimensional. You cross it just as surely if you fly (within the boundaries of X) from London to Glasgow as if you walk, though you cross it a mile or so *higher*. If we add the vertical dimension to those shown on the map, 'England' becomes a 3-dimensional region extending from the centre of the Earth to the fringes of the atmosphere (depending on the current state of international law) and so does Scotland. The border becomes 2-dimensional, but its *codimension* remains 1. If we add time to the picture, the countries become 4-dimensional, the border 3-dimensional, and again the codimension is 1.

This the codimension of the border is better defined than its dimension, and it is the codimension that determines many of its properties. For example, a typical point in Great Britain will *either* be in England + Wales *or* in Scotland (whatever its height and date); a typical curve will meet the border in isolated points, if at all, independently of whatever we are thinking of the 2-dimensional map, the 3-dimensional space, or the 4-dimensional space-time. It is possible usefully to say that the codimension of something is, say, c ('usefully' in the sense that we can make correct deductions from it) even in a context where its dimension may depend on more information than the context specifies.

In some applications the dimensions involved become infinite. The codimension of M may still be thought of as the 'number of equations defining M' but *not* any longer as the difference between two dimensions, now both infinite. As long as the codimension remains finite, all of its useful properties survive.

5 Machines Revisited

In this chapter we shall analyse the behaviour of the catastrophe machines of Chapter 1 and show how it relates to critical point theory and the behaviour of degenerate critical points under perturbation. This motivates the formulation of a general mathematical problem, of which Thom's theorem and its supporting cast of mathematical techniques may be considered the solution. Thus the simple mechanical systems studied here become archetypal examples for the general theory, maintaining the link between theory and practice and strengthening the rôle of physical intuition.

1 The Zeeman Machine

We shall analyse the Zeeman machine for the particular measurements indicated in Chapter 1. It should become clear that varying these measurements will not make serious qualitative differences to the behaviour – a fact which contains the germ of an important idea, to be unfolded later.

The first step is to locate the position of the cusp point P. By symmetry it lies on the axis, and we are thus led to consider Fig. 5.1. We will take distance to be measured in units which make the diameter of the disc equal to 1, the lengths of unstretched elastic equal to 1, and the distance OA equal to 2.

It is clear, again by symmetry, that as the point B runs along the axis there will always be an equilibrium position $\theta = 0$. The point P is where the equilibrium changes from a stable one (local energy minimum) to an unstable one (local maximum), by general principles of statics. We let e, e' be the lengths of the two pieces of elastic when the position of the disc is held at θ, near 0 but not necessarily at it. If λ is the modulus of elasticity of the strings, then the energy (by Hooke's law) is

$$V_s(\theta) = \frac{\lambda}{2}(e-1)^2 + \frac{\lambda}{2}(e'-1)^2.$$

Now

$$e^2 = (2 - \tfrac{1}{2}\cos\theta)^2 + (\tfrac{1}{2}\sin\theta)^2$$

and expanding this in a Taylor series we find it becomes

$$\left(2 - \frac{1}{2}\left(1 - \frac{\theta^2}{2}\right)^2\right)^2 + (\tfrac{1}{2}\theta)^2 + O(4)$$

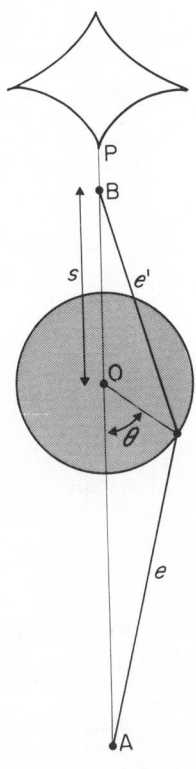

Fig. 5.1

where $O(4)$ denotes a function of order 4. Simplifying this we find that

$$e^2 = \tfrac{9}{4} + \theta^2 + O(4)$$

(absorbing terms in θ^4 into the $O(4)$ term). Hence

$$e = \tfrac{3}{2} + \tfrac{1}{3}\,\theta^2 + O(4).$$

Similarly

$$e' = \left(s - \tfrac{1}{2}\right)^2 - \frac{s(2s-1)}{2(2s+1)}\,\theta^2 + O(4)$$

and therefore

$$V_s(\theta) = \frac{\lambda}{4}\left[\frac{1}{4} + \left(s - \tfrac{1}{2}\right)^2 + \theta^2\left(\frac{1}{3} - \frac{s(2s-1)}{2(2s+1)}\right)\right] + O(4).$$

The Morse Lemma implies that in finding the local type of critical point we may neglect the $O(4)$ term whenever the coefficient of θ^2 is non-zero. Now the coefficient of θ^2 is

positive if $\quad \dfrac{1}{3} > \dfrac{s(2s-1)}{2(2s+1)}$,

negative if $\quad \dfrac{1}{3} < \dfrac{s(2s-1)}{2(2s+1)}$.

It follows that the equilibrium changes from a maximum to a minimum where

$$\frac{1}{3} = \frac{s(2s-1)}{2(2s+1)},$$

that is,

$$6s^2 - 7s - 2 = 0.$$

The solutions of this are

$$s = \frac{7 \pm \sqrt{97}}{12},$$

and P clearly corresponds to the positive value of s, namely

$$s = \frac{7 + \sqrt{97}}{12} \sim 1.40.$$

A similar analysis with $\theta = 0$ replaced by $\theta = \pi$ leads to the position of the top cusp P′, for which $s = \dfrac{27 + \sqrt{489}}{20} \sim 2.46$ (*see* Poston and Woodcock [3]). The two side cusps may also be located, but the analysis is more complicated.

Notice the role played by the stability of Morse functions in ensuring that *most* of the time the wheel does not jump. If V_{ab}, the

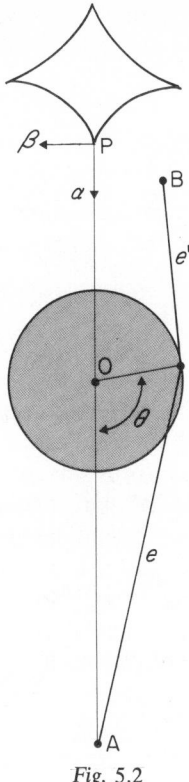

Fig. 5.2

energy function corresponding to the free end having position (a, b), has a Morse minimum at θ, then for (a', b') near enough to (a, b) the function $V_{a'b'}$ has a Morse minimum θ' near θ. Indeed in this neighbourhood θ' can be expressed as a smooth function of (a, b), by the way that Morse functions reduce to standard form. Thus, near the equilibrium point, the wheel moves smoothly with (a, b). However, as we approach a non-Morse point, 'near' becomes more restrictive. It is the geometry around such special points, where the Morse property and the consequent smooth movement of the wheel break down, that will concern us most.

Next we analyse in detail the behaviour near the point P. We have already seen that the term θ^2 in the energy vanishes at P; and by symmetry there is no θ^3 term, so we must look to at least θ^4. Now we work from Fig. 5.2, and let the free end B be at the point (α, β) relative to the coordinates shown (drawn in the opposite direction to the usual one for algebraic convenience later). The formula for e is as before, except that now we work to $O(5)$ and retain the terms in θ^4; whereas

$$e'^2 = (s + \tfrac{1}{2} \cos \theta - \alpha)^2 + (\tfrac{1}{2} \sin \theta + \beta)^2.$$

If we work out $V_{\alpha\beta}(\theta)$, the corresponding energy, to order 4, the result is of the form

$$V_{\alpha\beta}(\theta) = (a_0 + a_1\beta\theta + a_2\alpha\theta^2 + a_3\beta\theta^3 + a_4\theta^4) + O(5).$$

Here $a_0, \ldots, a_4$ are certain constants, whose exact value is immaterial; they are approximately

$$a_0 \sim 0.54$$
$$a_1 \sim 0.24$$
$$a_2 \sim 0.16$$
$$a_3 \sim -0.09$$
$$a_4 \sim 0.045$$

(values supplied by E. C. Zeeman, who is responsible for the whole of this analysis).

Note at this stage that at P, where $\alpha = \beta = 0$, we have a function of the form

$$\mu\theta^4 + O(5) \qquad (\mu > 0)$$

and hence a degenerate critical point, equivalent by Theorem 4.4 to that of θ^4. Thus for qualitative results at P we can neglect the $O(5)$ term. Much deeper theorems, discussed but not proved in Chapter 8, show that the same is true *near* P. Thus we may simplify the expression by omitting $O(5)$. Further simplifications are to take units for elasticity to make $\lambda a_4 = \tfrac{1}{4}$; to eliminate the cubic term by a change of variable

$$x = \theta + \frac{\beta a_3}{4a_4},$$

and to define suitable scalar multiples a of α and b of β, which numerically turn out to be approximately $1.8\,\alpha$ and $1.3\,\beta$. This leads to the energy in the form

$$V_{ab}(x) = \tfrac{1}{4}x^4 + \tfrac{1}{2}ax^2 + bx + c$$

for a constant c. Since we are interested only in critical points of V we may take $c = 0$ (or translate the origin of the energy values) without loss, and we then have

$$V_{ab}(x) = \tfrac{1}{4}x^4 + \tfrac{1}{2}ax^2 + bx \qquad (5.1)$$

This formula defines what we shall later call the *cusp catastrophe*.

The next step, at which we actually get useful information, is the analysis of the critical points of V. Before we perform this analysis (which is now routine) it is worth remarking in what respects the above work differs from a 'classical applied mathematical' treatment. The difference is that the neglect of higher order terms, $O(4)$ in finding the position of P, $O(5)$ in obtaining (5.1), is justified rigorously by appealing to the Morse Lemma, and to the deep mathematics behind Thom's theorem (even though we have not given proofs of the validity of this). The classical applied mathematician has a cavalier approach to such approximations, coupled with a good 'nose' for what is or is not likely to work: a good applied mathematician would not mistakenly truncate V at $O(4)$, which gives nonsense; but he would not have any coherent explanation why $O(5)$ makes no difference. Catastrophe theory, in contrast, comes equipped with rigorous tests which continue to be applicable when the number of variables is too large for 'experience' to be a safe guide. This is an important point which most popularizations of the subject miss entirely.

2   The Canonical Cusp Catastrophe

As a separate subsection we analyse the critical point structure of the functions $V_{ab}(x)$. We shall return to the Zeeman machine once this has been done.

For fixed (a, b) the critical points of (5.1) are given by

$$0 = \frac{\mathrm{d}}{\mathrm{d}x}\,(\tfrac{1}{4}x^4 + \tfrac{1}{2}ax^2 + bx) = x^3 + ax + b. \qquad (5.2)$$

(The fractions in the coefficients are chosen to obtain this simple form.) This is a cubic in x, and so must have at least one and at most three real roots. The nature of the roots depends on the values of a and b: specifically on the *discriminant*

$$D = 4a^3 + 27b^2$$

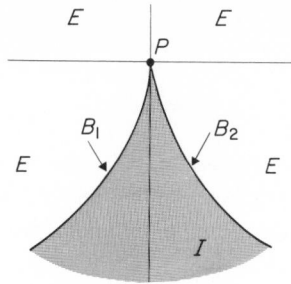

Fig. 5.3

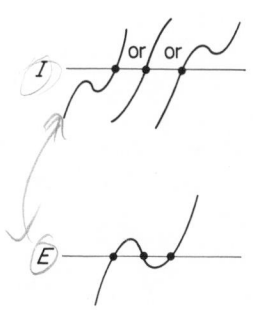

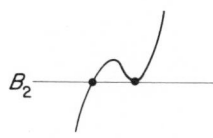

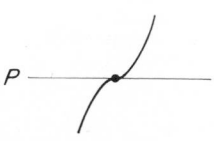

Fig. 5.4

of the cubic equation. It is well known (*see* Salmon [19] p. 183) that if $D < 0$ there are three distinct real roots, and if $D > 0$ there is one real root and a conjugate pair of complex roots. If $D = 0$ there are in a sense three real roots, but some of them coincide: in fact if $D = 0$ but $a \neq 0$ or $b \neq 0$ two of the real roots are equal, if $D = 0$ and $a = b = 0$ then all three roots are equal.

Geometrically, this means that the nature of the roots, and hence of the equilibria of the catastrophe machine, depends on the position (a, b) of the free end of elastic in the (a, b)-plane C, relative to the curve with equation

$$4a^3 + 27b^2 = 0. \tag{5.3}$$

This is illustrated by the heavy lines in Fig. 5.3.

We divide the (a, b)-plane C into five subsets. The shaded region I 'inside' the curve, the region E 'outside' it, the two branches B_1 and B_2 of the curve, and the origin P. The points (a, b) which lie in I are those for which $4a^3 + 27b^2 < 0$, those in E satisfy $4a^3 + 27b^2 > 0$. Thus:

> if (a, b) lies in E there is one real root;
> if (a, b) lies in I there are three distinct real roots;
> if (a, b) lies in B_1 or B_2 there are three real roots, but two of them coincide: for B_1 the coincidence occurs for the smaller root, for B_2 the larger;
> if (a, b) lies at $P = (0, 0)$, in which case $a = b = 0$, there are three coincident real roots (all equal to 0).

These possibilities are illustrated in Fig. 5.4.

The corresponding potential functions

$$V_{ab}(x) = \tfrac{1}{4}x^4 + \tfrac{1}{2}ax^2 + bx$$

take the forms shown in Fig. 5.5. We see that V_{ab} has one minimum for (a, b) in E, two minima on either side of a maximum for (a, b) in I, one minimum and one point of inflexion for (a, b) in B_1 or B_2, and one minimum for (a, b) at P. Note that in this last case the potential function is $x^4/4$ and so the minimum is more complicated, from a mathematical standpoint, than the other minima: the first three derivatives of V_{ab} are zero in this case, whereas in all other cases only the first derivative is zero. This corresponds to the three coincident roots of the cubic for (a, b) at P. Note also that the difference between the potentials for (a, b) in B_1 or B_2 is that in B_1 the inflexion lies to the left of the minimum, in B_2 to the right.

Dynamically, minima of V correspond to stable equilibria, whereas maxima or inflexions correspond to unstable equilibria. So for control parameters in E there is a single stable equilibrium, for control parameters in I there are two stable equilibria and one unstable.

This somewhat complex behaviour of the potential can be captured geometrically, in a very revealing way, if we draw the

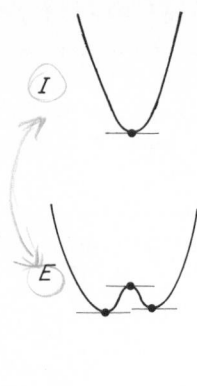

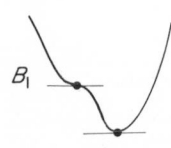

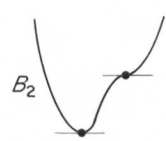

Fig. 5.5

associated *catastrophe manifold* or *equilibrium surface* in (x, a, b)-space. This is the set M of points (x, a, b) satisfying equation (5.2), which we rewrite here:

$$x^3 + ax + b = 0. \tag{5.4}$$

It has the appearance of a folded surface, and is illustrated in Fig. 5.6. (As it happens, we drew this surface (in slightly different coordinates) in Chapter 2, Section 7, and need not repeat the justification here.)

Notice that around most points the surface can locally be thought of as the graph of a function of (a, b), as required by our discussion on p. 77 of how a Morse critical point varies with (a, b). Much of the physical literature attempts to extend this point of view to the places where visibly it will *not* work, by using 'branched functions'. There is no coherent way of doing this around the point P, which in many physical applications is the focus of interest. Conceptual chaos is the usual result. A far clearer picture comes from analysis of the well defined *catastrophe map*

$$\chi : M \to C$$

which projects points on M to the (a, b)-plane C according to

$$(x, a, b) \mapsto (a, b) \qquad (x \in M)$$

in the neighbourhood of the origin.

The catastrophe manifold M is a smooth submanifold of $\mathbb{R}^3$. It is sometimes thought that M is not smooth at the origin, but this is an illusion. The smoothness is immediately visible in a three-dimensional model of the surface. To see it mathematically we

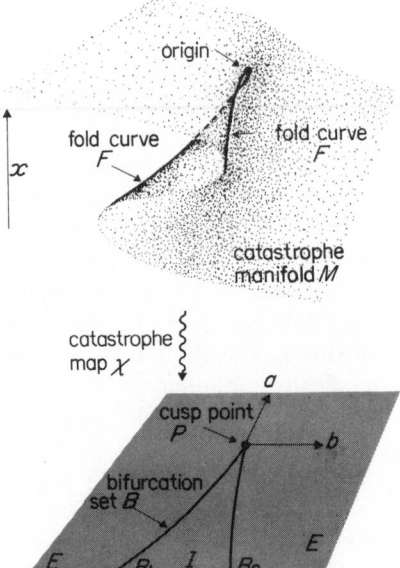

Fig. 5.6

consider a chart for M defined by 'projection' from the (x, a)-plane Y. This is a function

$$\pi : Y \mapsto M$$

such that

$$\pi(x, a) = (x, a, b)$$

where

$$x^3 + ax + b = 0,$$

and eliminating b we obtain

$$\pi(x, a) = (x, a, -x^3 - ax). \tag{5.5}$$

This chart not only provides local coordinates around the origin; it in fact provides coordinates over the whole of M, so a single chart suffices for the whole catastrophe manifold in this case. The Jacobian of $\pi : Y \to \mathbb{R}^3$ takes the form

$$\begin{bmatrix} \dfrac{\partial x}{\partial x} & \dfrac{\partial x}{\partial a} \\[2mm] \dfrac{\partial a}{\partial x} & \dfrac{\partial a}{\partial x} \\[2mm] \dfrac{\partial}{\partial x}(-x^3 - ax) & \dfrac{\partial}{\partial a}(-x^3 - ax) \end{bmatrix}$$

which equals

$$\begin{bmatrix} 1 & 0 \\ 0 & 1 \\ -3x^2 - a & -x \end{bmatrix}$$

which is already in echelon form, and has rank 2 no matter what values a, x take. Hence (5.5) has no critical points and M is smooth.

Intuitively, π corresponds to looking at M horizontally from the side: if this is done with a model it becomes clear that the fold in M is not visible from this direction. A series of computer drawings, taken from Woodcock and Poston [20] demonstrates this clearly in Fig. 5.7.

The points of M at which the surface folds under, and hence the tangent plane is vertical, are precisely the critical points of the catastrophe map $\chi : M \to C$. The easiest way to find these is to solve the simultaneous equations (5.4) and

$$0 = \frac{\partial^2 V}{\partial x^2} = 3x^2 + a. \tag{5.6}$$

This leads to $a = -3x^2$, and substituting in (5.4), $b = 2x^3$. Hence the points of M at which the tangent plane is vertical lie on the curve

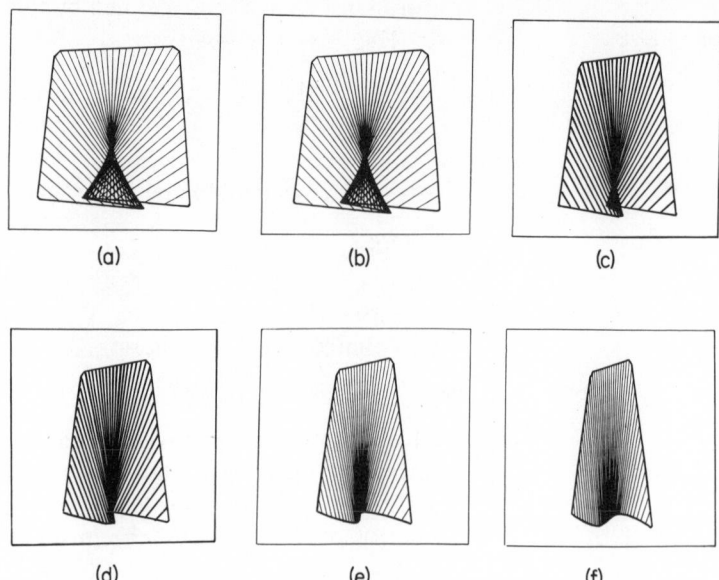

Fig. 5.7

given parametrically by

$$(x, -3x^2, 2x^3)$$

for real x. This is a twisted cubic curve lying on M, which we call the *fold curve* and denote by F. To see F is smooth we consider the parametrization $\phi : \mathbb{R} \to F$ given by

$$\phi(x) = (x, -3x^2, 2x^3).$$

The Jacobian matrix is

$$\begin{bmatrix} 1 \\ -6x \\ 6x^2 \end{bmatrix}$$

which always has rank 1. Hence ϕ is an (obviously smooth) function without critical points, and F is a smooth curve.

Finally we look at the image of F under χ, which is a curve in C obtained by the parametrization

$$(-3x^2, 2x^3).$$

Solving the equations

$$a = -3x^2, \qquad b = 2x^3$$

we find the equation

$$4a^3 + 27b^2 = 0. \tag{5.7}$$

This is a semicubical parabola, and is precisely the curve encountered above in the description of the equilibria. We call it the *bifurcation set* (because two different kinds of function appear when we move (a, b) off it) and denote it by B.

Now we are in a position to give a geometrical interpretation of the nature of the equilibria of the associated dynamical system. For a given pair of parameters (a, b) the equilibria are obtained by solving (5.2). They may therefore be described as the x-coordinates of the points at which a vertical line through (a, b) meets the catastrophe manifold M. It is geometrically obvious that if (a, b) lies in the region E external to the bifurcation set B, there is a unique value of x; because only one 'sheet' of M lies above points in E. However, over a point (a, b) inside I, there are three sheets to the surface, corresponding to the three equilibria. For (a, b) in B_1 the vertical line through (a, b) is tangential to the lower sheet and passes through the upper sheet at a single point: this gives rise to two equal roots of (5.2) and a further root, the equal pair being smaller. For (a, b) in B_2 similar behaviour occurs, except now the tangent is to the upper sheet and the equal pair of roots is the larger. Finally at the cusp point P the vertical line is tangent to the surface M and meets it at a single point, the origin.

Thus almost the entire description of the equilibria is geometrically obvious. To complete the picture it is only necessary to distinguish stable and unstable equilibria: the unstable ones correspond to points on M lying 'inside' the fold curve on the middle of the three sheets; the stable ones lie outside the fold curve.

3 Dynamics of the Zeeman Machine

Actually the section title is misleading: we are really considering the *statics* of the machine – its equilibrium states. A full dynamic description (what happens when we spin the wheel, wiggle the elastic, etc.) is beyond the scope of this book, and belongs to full-blooded dynamical systems theory. However, we shall discuss what is sometimes called the *quasi*-statics: what happens to equilibria when the position (a, b) of the free end of the elastic varies, which experimentally corresponds to a mostly *slow* dynamic behaviour. This works because the machine is heavily damped by friction in the elastic. Indeed in a carefully engineered frictionless machine with a heavy wheel the effects are almost entirely masked by oscillations. (Friction at the axle, conversely, masks them by giving the wheel new equilibria in states where the elastic forces do not balance.)

In terms of the above analysis, we can view the variation of equilibria with (a, b) as follows. Let (a, b) trace out some path in C. Then the observed equilibrium will trace out some path in M, lying above the path in C. Because of the folds in M, this path may have to make a discontinuous jump from one sheet of surface to another. We suppose that the (continuous) physical system makes corresponding jumps, so fast that we may ignore the time taken. The

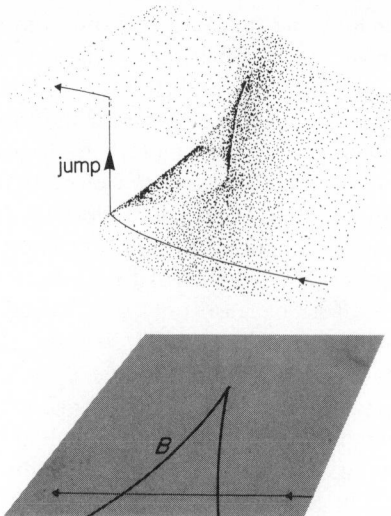

Fig. 5.8

equilibrium analysis will not in itself tell us where the jumps occur, because in principle they are possible whenever there are two or more possible equilibria for the same value of (a, b). The extra information needed is in detail dynamic: instead of invoking the dynamics we hypothesize a convention which approximates the facts, which Thom [1] calls (*perfect*) *delay*: the system only jumps when it has no other choice. That is, the path traced by the equilibria will only switch sheets in the surface when it crosses the edges of folds, and the sheet that it is on disappears. (For fast changes of control this breaks down (Poston [21]).) We will meet another convention in Chapter 14.

Since our analysis so far has been for states near the point P, we shall take the question raised in Chapter 1 for the case where the path enters the region ◇ from the lower left, and emerges at the lower right. The corresponding path in C, together with the path of equilibria predicted by the delay convention, is shown in Fig. 5.8. It will be seen that there is a sudden jump when the path *leaves* the bifurcation set B, but not when it enters it. The sequence of energy curves $V_{ab}(x)$, and the corresponding critical points, are shown in Fig. 5.9: the jump occurs when a minimum and maximum coalesce and annihilate each other.

Such jumps are examples of what Thom and Zeeman call *catastrophic jumps*. If we call a and b the *control variables* (or control *parameters*) and x the *behaviour variable* (or *state* variable), then catastrophic jumps occur when a smooth variation of controls causes a discontinuous change of state.

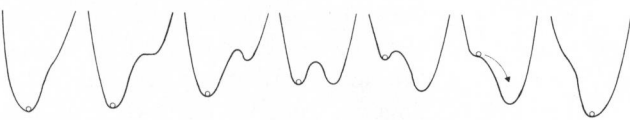

Fig. 5.9

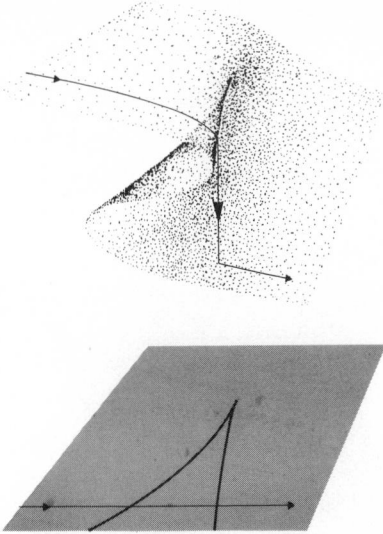

Fig. 5.10

This is the first phenomenon exhibited by the cusp catastrophe, but there are others: all verifiable experimentally on the machine. The second is *hysteresis* (called this by analogy with a similar effect in magnetism which arguably falls within the same framework). Namely, if we reverse the path in the controls, we do not necessarily reverse the path in state space. Fig. 5.10 illustrates this: the delay convention again implies that a jump occurs on leaving the bifurcation set; but since we are travelling the other way this is on the other side.

A third phenomenon is *divergence:* slight differences in the path may (without any jumps) produce large differences in state, even when they start and end at the same control points (Fig. 5.11).

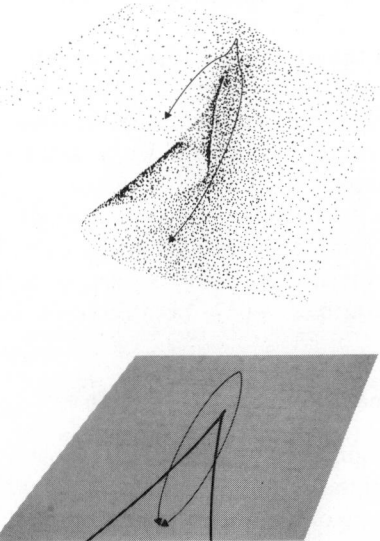

Fig. 5.11

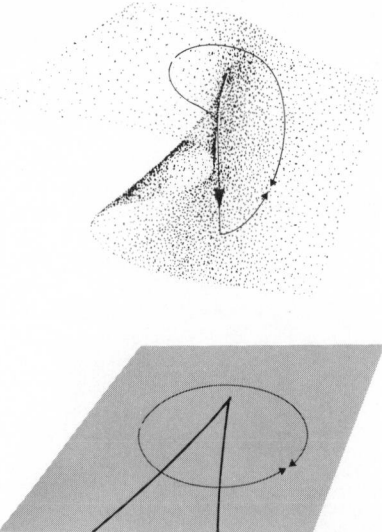

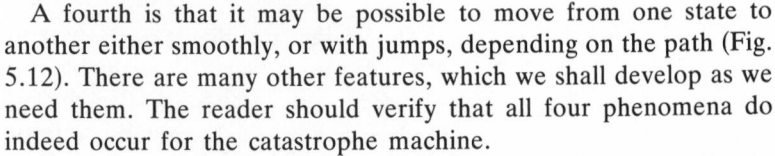

Fig. 5.12

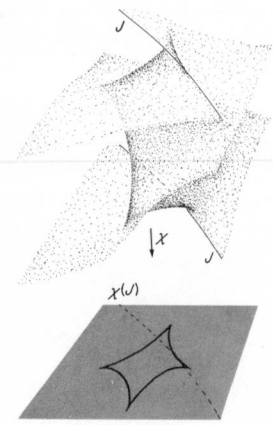

Fig. 5.13

A fourth is that it may be possible to move from one state to another either smoothly, or with jumps, depending on the path (Fig. 5.12). There are many other features, which we shall develop as we need them. The reader should verify that all four phenomena do indeed occur for the catastrophe machine.

So far the analysis has been purely local: what happens near *P*. However, we can draw a similar equilibrium surface for the *global* variations in controls (a, b) over the whole plane *C*. Because the state variable is an angle, hence periodic with period 2π, we have to draw only angles between some θ and $\theta + 2\pi$: then the top of the diagram must be identified with the bottom (so that the apparent edge *J* to the surface at top and bottom does not really exist). The result, it can be shown, is Fig. 5.13. There are actually *four* cusp catastrophe surfaces, one for each cusp point of the region ⟡, fitted together. Two of these are exactly equivalent to the canonical cusp catastrophe analysed above: the other two are *dual cusps* where the potential starts with $-x^4$ instead of x^4. This has the effect of interchanging maxima and minima, hence stable and unstable equilibrium states, but apart from that it has no effect on the geometry.

Using the delay convention in this surface the reader may now deduce the behaviour of the machine for a selection of paths which wind around the various cusp points, without any difficulty: it is doubtful if he would be able to do this without Fig. 5.13 to hand.

4 The Gravitational Machines

We begin by analysing the parabola. Parametrize it by *t*, a point on the parabola being (t, t^2) (Fig. 5.14). The slope at this point is $2t$, so the slope of the normal to the parabola is $-1/2t$, and the normal has

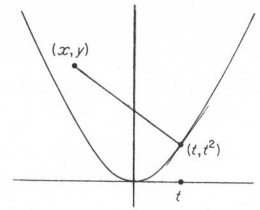

Fig. 5.14

equation

$$x + 2ty = 2t^3 + t.$$

We write this as

$$t^3 + Yt + X = 0 \tag{5.8}$$

where

$$Y = \tfrac{1}{2} - y, \qquad X = -\tfrac{1}{2}x$$

and it takes the same form as equation (5.2) above, with t as state variable and X, Y as controls. Hence the graph of (5.8) in (X, Y, t)-space is the same as the canonical cusp catastrophe manifold.

The *evolute* of the parabola is the *envelope* of its normals; and as the classical textbooks tell us, this is obtained by differentiating the left-hand side of (5.8) with respect to t, and then eliminating t. Differentiating gives

$$3t^2 + Y = 0,$$

and eliminating t we get

$$4Y^3 + 27X^2 = 0.$$

Not only is this answer familiar: the whole *calculation* is familiar: it was done in the previous section when finding the bifurcation set of the catastrophe. This may be explained as follows. For fixed t the equation (5.8) defines a line in (X, Y, t)-space. The union of these lines, for all t, is a ruled surface, and is the usual cusp catastrophe surface. The corresponding catastrophe map projects the surface, hence the lines, down into (X, Y)-space: this is equivalent to thinking of (5.8) as representing a family of lines in (X, Y)-space, as t varies. The envelope of this family is the image in (X, Y)-space of the singularities of the projection χ; *see* Fig. 5.15. (Incidentally, the link with envelopes was one of Thom's starting-points in developing catastrophe theory.)

Now comes the gravity. If the parabola has centre of gravity at (X, Y) and tilts until it is in a position where (t, t^2) rests on the ground, then the potential energy $W_{XY}(t)$ is given by

$$
\begin{aligned}
W_{XY}(t)^2 = V_{XY}(t) &= (x-t)^2 + (y-t^2)^2 \\
&= t^4 + t^2(1-2y) + t(-2x) + x^2 + y^2 \\
&= t^4 + 2Yt^2 + 4Xt + (X^2 + (Y-\tfrac{1}{2})^2).
\end{aligned}
$$

Thus

$$\tfrac{1}{4}V_{XY}(t) = \frac{t^4}{4} + \frac{Y}{2}t^2 + Xt + K$$

where K depends only on X and Y, but not on t.

Now $W_{XY}(t)$ has a critical point if and only if $\tfrac{1}{4}V_{XY}(t)$ has, for the height of the centre of gravity above the floor must be positive, and

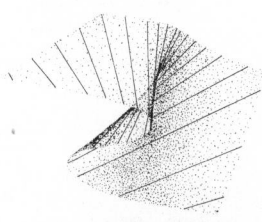

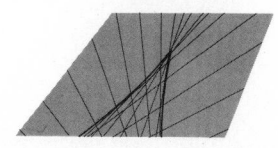

Fig. 5.15

we restrict to this case: then squaring is a diffeomorphism on the positive real numbers and hence preserves critical points. Further, when we differentiate, K vanishes, so it does not affect the critical point analysis. Such a term K is called a *shear term*. This is because its geometric effect is to slide the critical *values* up or down, leaving the equilibrium surface unchanged. The use of 'shear term' in Chapter 4 Section 3 is for the same reason.

Thus we have the same family of functions as in Section 2, and the critical point analysis is the same: this identifies not only the local geometry of the Zeeman machine, the standard cusp, and the parabola machine, but also the families of functions which define the geometry in terms of critical points. In consequence the dynamical behaviour will again follow the cusp catastrophe, as for the Zeeman machine, and the same phenomena will occur. However the *global* behaviour of the parabola lacks the extra three cusps of the Zeeman machine.

A similar calculation applies to the elliptic catastrophe machine, only now the evolute has four cusps, as in Fig. 1.9. The equilibrium surface is now exactly equivalent to the Zeeman machine, and the dynamical analysis again follows the same pattern. In particular the three questions asked have the following answers:

(a) outside the ◇ region there are two possible equilibrium states (one stable, one unstable) whereas inside it there are four (two of each);

(b) if the centre of gravity is below the centre of curvature – which, very classically, is the point where the normal first touches the evolute – at the rest point, the equilibrium is stable; if above, unstable (to see this, approximate the curve locally by its circle of curvature: or check analytically);

(c) catastrophic jumps occur on crossing the ◇ region, but only when folds in the equilibrium surface are encountered (following the delay convention).

It is in fact a theorem that any such smooth closed curve has at least two curvature maxima and at least two curvature minima, so the corresponding gravitational machine (replacing the ellipse by that curve) has at least four critical points in its catastrophe map. Typically these show up as four cusps in the bifurcation set (two ordinary, two 'dual') but in special circumstances there may be an overlap. Thus a smooth involute of a three-cusped hypocycloid has the hypocycloid as the corresponding bifurcation set, with apparently only three cusps; but each corresponds to two cusp points on the catastrophe map.

5   Formulation of a General Problem

The behaviour of all three machines has turned out to be described by essentially the same calculation: critical points of a parametrized family of functions. Thus we are led to consider the general problem of r-parameter families of functions $\mathbb{R}^n \to \mathbb{R}$,

$$f_{u_1, \ldots, u_r}(x_1, \ldots, x_n)$$

or equivalently of functions

$$f : \mathbb{R}^n \times \mathbb{R}^r \to \mathbb{R}.$$

Our aim is in some sense to classify these, up to suitable coordinate changes, in such a way that the critical point structure is not affected qualitatively by the coordinate changes involved. The remarkable theorem of Thom and Arnol'd is that 'almost all' 4-parameter families fall into seven (local) types; this classification extends to 5-parameter families and (with new ideas) far beyond. We explain this in the next chapter but one.

6 Structural Stability

If we consider a single function, only a Morse critical point is structurally stable, and typically only Morse points arise. Chapter 5 shows, however, that in problems involving *families* of functions it may be necessary to deal with unstable, degenerate critical points: indeed these may be the most interesting features.

An extended notion of structural stability for families of functions greatly clarifies these issues. A structurally stable *family* can, typically, include individual functions with degenerate critical points; and roughly speaking, the larger the family the worse the degeneracy may be. The surrounding members of the family exert a kind of 'calming' influence on the degenerate function, formalized in Thom's concept of an unfolding, which we explore in Chapter 8. In this chapter we discuss this type of structural stability and link it to previous examples; the mathematical developments from it appear in Chapters 7 and 8.

1 Equivalence of Families

Recall (Chapter 4, Section 3) that two functions $f, g: \mathbb{R}^n \to \mathbb{R}$ are equivalent around 0 if there is a local diffeomorphism $y: \mathbb{R}^n \to \mathbb{R}^n$ and a constant 'shear term' γ such that

$$g(x) = f(y(x)) + \gamma$$

in a neighbourhood of 0. For families of functions $f, g: \mathbb{R}^n \times \mathbb{R}^r \to \mathbb{R}$ we require an appropriately souped-up notion of equivalence. The diffeomorphism y becomes a *family* of diffeomorphisms $y_s: \mathbb{R}^n \to \mathbb{R}^n$, for $s \in \mathbb{R}^r$, *which vary smoothly with s*; the constant γ becomes a 'family of constants' varying smoothly with s, or what is the same, a smooth function $\mathbb{R}^r \to \mathbb{R}$. Finally, we allow a diffeomorphism $\mathbb{R}^r \to \mathbb{R}^r$ as well: this has no non-trivial counterpart in the case of a single function (the only diffeomorphism on the single point $\mathbb{R}^0$ is the identity!) but is needed for families. Without it, for example, two Zeeman machines, one twice the size of the other, but otherwise identical, would not give equivalent behaviour. The relevant diffeomorphism here just doubles the scale.

Formulating these requirements in a cleaner way leads us to

demand:

(a) a diffeomorphism

$$e : \mathbb{R}^r \to \mathbb{R}^r;$$

(b) a smooth map

$$y : \mathbb{R}^n \times \mathbb{R}^r \to \mathbb{R}^n$$

such that for each $s \in \mathbb{R}^r$ the map

$$y_s : \mathbb{R}^n \to \mathbb{R}^n$$
$$y_s(x) = y(x, s)$$

is a diffeomorphism;

(c) a smooth map

$$\gamma : \mathbb{R}^r \to \mathbb{R}.$$

Then f and g are *equivalent* if there exist e, y, γ, defined in a neighbourhood of 0, such that

$$g(x, s) = f(y_s(x), e(s)) + \gamma(s),$$

for all $(x, s) \in \mathbb{R}^n \times \mathbb{R}^r$ in that neighbourhood.

This relation is important for understanding Thom's theorem, and is needed on several occasions elsewhere. Its geometric meaning is illustrated by Figs 6.1 and 6.2. Thus in Fig. 6.1, which represents $\mathbb{R}^n$ and $\mathbb{R}^r$ schematically by lines, we see that e stretches and bends $\mathbb{R}^r$ smoothly (one would like to say 'diffeomorphs' $\mathbb{R}^r$ but purity of language forbids). For any fixed $s \in \mathbb{R}^r$ the set of points (x, s), for

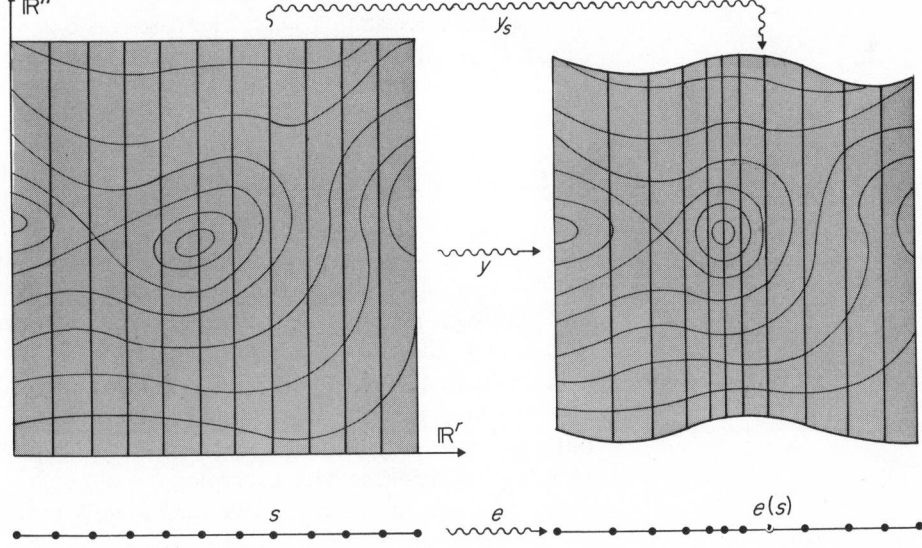

Fig. 6.1

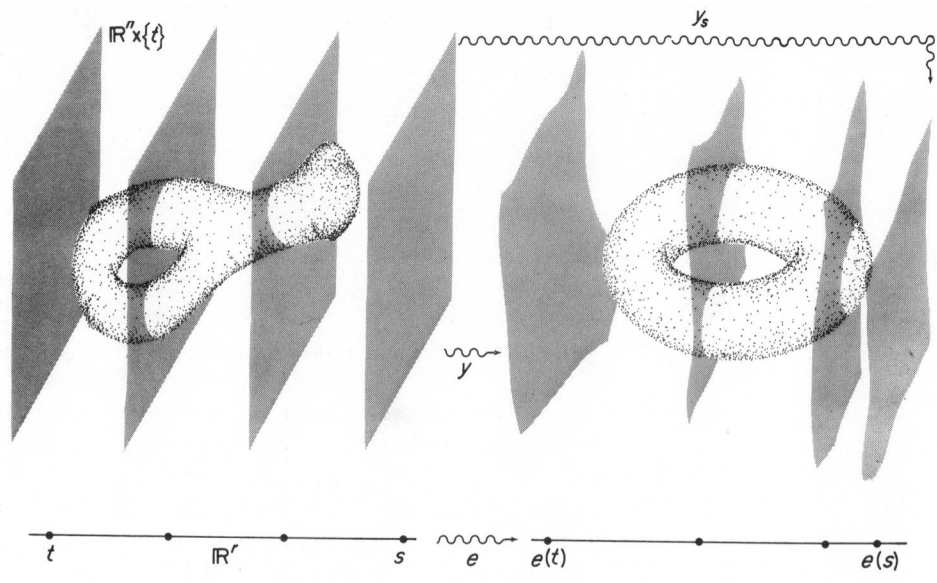

Fig. 6.2

$x \in \mathbb{R}^n$, represented by the vertical line above s, is deformed according to y_s and placed vertically above $e(s)$. Thus the decomposition by vertical lines is preserved, though the lines individually are deformed and moved about along $\mathbb{R}^r$. Finally, we may translate the origin in each vertical line using γ.

The contour lines in Fig. 6.1 are drawn to illustrate the effect of this on $\mathbb{R}^n \times \mathbb{R}^r$: it is deformed smoothly, in a way which preserves the topological features of the projection $\mathbb{R}^n \times \mathbb{R}^r \to \mathbb{R}^r$, $(x, s) \mapsto s$. Thus a contour which is 'multiple valued' above $\mathbb{R}^s$ remains so, and in a qualitatively similar way. The case $n = 2$, $r = 1$ is illustrated in Fig. 6.2, but this time only for one 'contour', drawn as a torus to emphasize the preservation of topological features.

2 Structural Stability of Families

The concept of structural stability now extends to families in a natural way. If $f : \mathbb{R}^n \times \mathbb{R}^r \to \mathbb{R}$ is equivalent, in the above sense, to any family $f + p : \mathbb{R}^n \times \mathbb{R}^r \to \mathbb{R}$, where p is a sufficiently small *family* $\mathbb{R}^n \times \mathbb{R}^r \to \mathbb{R}$, then f is *structurally stable*.

It is usually difficult to *prove* that an r-parameter family, for $r \geqslant 1$, is structurally stable in this sense. The methods of Chapter 8 apply, but they rest on theorems of considerable depth. Here we shall illustrate the concept with examples.

It is in some ways more illuminating to look at *in*stability, because we can more easily what goes wrong, and then say 'stability is when this sort of thing does not happen'. For example, had we analysed the Zeeman machine only for control points (α, β) on the

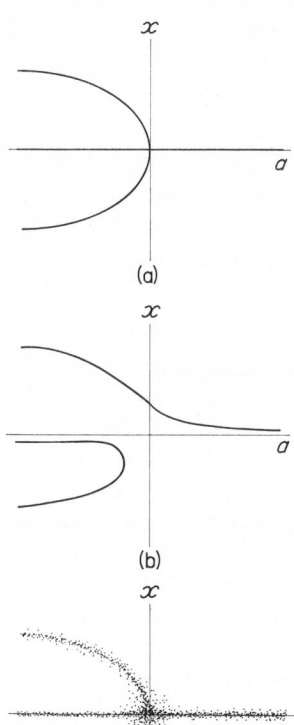

Fig. 6.3

axis $\beta = 0$, we would have been led to the family of functions

$$W_a(x) = \frac{x^4}{4} + \frac{a}{2} x^2. \tag{6.1}$$

The critical points would be given by

$$0 = \frac{d}{dx} W_a(x) = x^3 + ax,$$

that is, by the line $x = 0$ and the parabola $x^2 + a = 0$, as shown in Fig. 6.3(a). This diagram is frequently encountered in books on bifurcation theory. For many purposes it may be a completely adequate description of what happens. However it is structurally unstable and does not always capture the full behaviour. For example, in Fig. 6.3(a), with delay convention, there are no catastrophic jumps.

In fact, let us perturb the family of functions (6.1) by a small term εx, getting

$$\tilde{W}_a(x) = \frac{x^4}{4} + \frac{a}{2} x^2 + \varepsilon x.$$

Now the critical points are given by

$$0 = x^3 + ax + \varepsilon,$$

and the graph (for non-zero small ε) resembles Fig. 6.3(b). The topology of this graph is quite different; for example it is disconnected and has no self-intersection points; and this is true no matter how small ε is.

Of course whether or not this matters depends on the precise question one wishes to answer. Fig. 6.3(b) is a good approximation to Fig. 6.3(a) in other, non-topological senses: for example, both look very like Fig. 6.3(c), which is the sort of graph that one might well be pleased to obtain in an experiment! The point is that one cannot disregard the structural instability at the outset: before it can be seen to be innocuous, one must analyse what effect it has on the properties of interest. The structural instability of the function $x^2 y$, as discussed in Section 4 of Chapter 4, cannot be dismissed by a diagram like Fig. 6.3(c).

On the other hand, the full cusp catastrophe family

$$V_{ab}(x) = \frac{x^4}{4} + \frac{a}{2} x^2 + bx \tag{6.2}$$

is structurally stable. This statement is justified in Chapter 8; but it can be seen in a *very* rough way as follows. Perturbations of order greater than four should not make any qualitative difference, by the argument of Theorem 4.4. Fourth order terms, quadratics and linear terms are accounted for already; cubic terms can be absorbed by a change of coordinates as in the analysis of the Zeeman machine

(Chapter 5 Section 1); and constants do not affect the critical points. To make *this* argument more than an expression of hope is probably impossible, since it fails to take account of all the difficulties: the rigorous proof is quite different and lies deeper.

The structural stability of the cusp catastrophe implies in particular that small errors in building a Zeeman machine will not significantly affect its behaviour. (Experience shows that even quite large errors may be no worse. Sometimes 'local' is less local than expected.)

3   Physical Interpretations of Structural Stability

In scientific work it is customary to insist on the repeatability of experiments: the same experiment should, ideally, give the same result under the same conditions. We say 'ideally', because it is never possible to ensure *exact* repeatability: external factors intrude upon even the most carefully designed experiment. To achieve exactly the same gravitational field, for example, in principle involves fixing the position of every particle of matter in the universe. Thus a more practical view of repeatability is that sufficiently small changes in the conditions under which an experiment is carried out should not significantly affect the result. It follows that the mathematical description of a physical phenomenon should have the same kind of perturbation-insensitivity.

Mathematical formulations of this depend, therefore, on two things:

(a) what kind of perturbations we allow;
(b) what we are prepared to be insensitive to.

In the literature, in a given area, different choices tend to be given different names; but the same name often occurs in different areas with different meanings. And the commonest name of all is *structural stability*. This notion was introduced in the context of differential equations by the Russian mathematicians A. A. Andronov and L. S. Pontryagin in 1937 (as 'coarse' systems), and gave rise to the concept of a structurally stable dynamical system, where for (a) we allow small perturbations of the differential equation involved, and for (b) require *topological* equivalence of the set of solution curves. On the other hand catastrophe theory (a) permits small smooth perturbations of the relevant family of functions, and (b) requires the stronger 'diffeomorphism' equivalence defined above.

Thus there are many different notions of structural stability. Whether in a given application the notion that is mathematically most convenient really corresponds to the physical repeatability of the experiment, depends on the circumstances, and must be investigated separately in each instance. It is a common claim that anything stable enough to be observed repeatedly must be structurally stable, *in whatever sense the claimant feels happiest with at the time.*

A somewhat idealized physical example may help to explain this. A frictionless pendulum hanging in a vacuum performs perfectly regular oscillations. Perturb it by giving a slightly larger initial push, and it will continue to oscillate regularly with almost the same period. In this sense, the system is structurally stable. Perturb instead by letting a little air in, and the oscillations will gradually die away. In this sense it is structurally unstable. But if we now restrict attention to experiments lasting only 50 swings, we cannot detect the change, and it is once more structurally stable. Finally, go back to the original perturbation of the airless system, but look at timekeeping properties over a period of a year (or longer if need be): now we will notice qualitative differences, such as missed appointments. Unstable again!

No mechanical vibrator stably possesses the property of agreeing with a caesium clock.

Thus the notion of stability depends both on the perturbations that we allow, and on the type of equivalence we demand. The wider the class of perturbations, the coarser the useful notion of equivalence is likely to be. The most powerful stability results, and the most interesting, occur when the class of perturbations is broader, or the equivalence finer, than expected.

This said, the 'natural' stability notion for catastrophes seems (at least for the present purposes) to be the one introduced above, preserving as it does most of the striking features of catastrophe geometry and dynamics. It is justifiable to interpret it as physical repeatability *of those features*, subject only to perturbations *of the relevant type*. In the sequel, where we appeal to or assert physical stability, we will discuss its relation to the mathematics.

4 The Morse and Splitting Lemmas for Families

Both the Morse Lemma and the Splitting Lemma extend to families in a strong way. In fact, these extensions may be deduced from the Splitting Lemma by treating the variables parametrizing the family on the same footing as the other variables: for them the initial 'straightening' argument (Theorem 4.5) is unnecessary, so in some ways the proof of the more general theorem is simpler. Here we give a formal proof of the strongest theorem of this nature, without too many frills, the main purpose of which is to demonstrate that even these stronger results are basically elementary, and do not require the deeper theorems of catastrophe theory.

Theorem 6.1 (*Splitting Lemma for families*). *Let* $F:\mathbb{R}^N \times \mathbb{R}^r \to \mathbb{R}$ *be smooth. Denote a point in* $\mathbb{R}^N \times \mathbb{R}^r$ *by* $(x, c) = (x_1, \ldots, x_N, c_1, \ldots, c_r)$. *Suppose that the Hessian*

$$H = \left[\frac{\partial^2 F}{\partial x_i \, \partial x_j} \right]_{1 \leqslant i,j \leqslant N}$$

has corank m at $(x, c) = 0$. Then F is equivalent to a family of the form

$$\tilde{F}(y_1(x, c), \ldots, y_m(x, c), c) \pm y_{m+1}^2 \pm \cdots \pm y_N^2.$$

Proof Find a nondegenerate $(N-m) \times (N-m)$ minor of H and renumber the coordinates to make it

$$\left[\frac{\partial^2 F}{\partial x_i \, \partial x_j} \right]_{1 \leqslant i,j \leqslant N-m}$$

Now define

$$G : \mathbb{R}^N \times \mathbb{R}^r \to \mathbb{R}^{N-m}$$

$$(x, c) \mapsto \left(\frac{\partial F}{\partial x_1}(x, c), \ldots, \frac{\partial F}{\partial x_{N-m}}(x, c) \right).$$

This is of maximal rank $N-m$ on $\mathbb{R}^{N-m} \times \{0\}$, by hypothesis. Hence by the Implicit Function Theorem there is locally a function (defined on a neighbourhood U of 0 in $\mathbb{R}^m \times \mathbb{R}^r$)

$$g : \mathbb{R}^m \times \mathbb{R}^r \to \mathbb{R}^{N-m}$$

such that

$$G(g(x_{N-m+1}, \ldots, x_N, c), x_{N-m+1}, \ldots, x_N, c) = 0,$$
$$g(0) = 0.$$

By continuity, if we put

$$z = (g(x_{N-m+1}, \ldots, x_N, c), x_{N-m+1}, \ldots, x_N, c),$$

the matrix

$$\left[\frac{\partial^2 F}{\partial x_i \, \partial x_j}(z) \right]_{1 \leqslant i,j \leqslant N-m}$$

is nondegenerate for all $c \in U$. Define

$$\phi : \mathbb{R}^N \times \mathbb{R}^r \to \mathbb{R}^N \times \mathbb{R}^r$$

$$(x, c) \mapsto (x_1 + g_1(x_{N-m+1}, \ldots, x_N, c), \ldots$$

$$\ldots, x_{N-m} + g(x_{N-m+1}, \ldots, x_N, c), x_{N-m+1}, \ldots, x_N, c).$$

This is clearly a diffeomorphism which preserves the sets $c =$ constant. Let $F = F \circ \phi$. Then around 0,

$$\frac{\partial F}{\partial x_i}(0, \ldots, 0, x_{N-m+1}, \ldots, x_N, c) = 0, \qquad 1 \leqslant i \leqslant N-m,$$

$$\det \left[\frac{\partial^2 F}{\partial x_i \, \partial x_j}(0, \ldots, 0, x_{N-m+1}, \ldots, x_N, c) \right]_{1 \leqslant i,j \leqslant N-m} \neq 0.$$

Now we may proceed as in the proof of the Morse Lemma, Chapter 4, except that the h_{ij}, H_{ij}, g, etc. (for $1 \leqslant i, j \leqslant N-m$) have $(x_{N-m+1}, \ldots, x_N, c)$ as parameters. For fixed parameter values we only change $x_1, \ldots, x_{N-m}$ in the process. The critical value of each $F|_{\mathbb{R}^{N-m} \times \{(x_{N-m+1}, \ldots, x_N, c)\}}$ is unchanged, and defines the function $F : \mathbb{R}^m \times \mathbb{R}^r \to \mathbb{R}$ in the statement of the theorem.

Corollary 6.2 (*Morse Lemma for families*). *Let $F : \mathbb{R}^N \times \mathbb{R}^r \to \mathbb{R}$ be smooth. Suppose that the Hessian*

$$\left[\frac{\partial^2 F}{\partial x_i \, \partial x_j} \right]_{1 \leq i,j \leq N}$$

is nondegenerate at $(x, c) = 0$. Then F is equivalent to a family of the form

$$\pm y_1^2 \pm \cdots \pm y_N^2.$$

Proof   Put $m = 0$ in Theorem 6.1.

Notice that the parameter $c \in \mathbb{R}^r$ no longer appears in the expression: the function $F_c(y) = F(y, c)$ is Morse, of fixed type, and independent of c. The parameter c has become 'disconnected'.

For a family of functions expressed in the form given in Theorem 6.1, we say that $y_1, \ldots, y_m$ are the *essential* variables, and $y_{m+1}, \ldots, y_N$ the *inessential* variables. This reflects the fact that for many purposes we can neglect the effect of $y_{m+1}, \ldots, y_N$.

5   Catastrophe Geometry

Corollary 6.2 gives us the stability of Morse functions in a particularly strong form. Not only can any small perturbation of a function $f : \mathbb{R}^N \to \mathbb{R}$ with a Morse singularity at $0 \in \mathbb{R}^N$ be 'vanished' around $0 \in \mathbb{R}^N$ by a reparametrization of $\mathbb{R}^N$ and the addition of a constant, restoring the original form; but we can do this *uniformly* for a smooth *family* of perturbations $F : \mathbb{R}^N \times \mathbb{R}^r \to \mathbb{R}$, where $F|_{\mathbb{R}^N \times \{0\}}$ is just f. (The various $F|_{\mathbb{R}^N \times \{c\}}$ are small perturbations of f when c is small – however 'small' is measured – by continuity and smoothness of F.)

Thus if we take a control position (α, β) for the Zeeman machine (Chapter 5 Section 1) for which the elastic energy as a function of θ has only Morse critical points, then (α, β) has a neighbourhood in which variation of the controls has, topologically, no effect. The critical points move as nice functions of the control (this is the point of the Implicit Function Theorem part of the above proof) and up to reparametrization of θ and addition of a control-dependent constant, nothing changes. Around such points, *only*, the catastrophe manifold may be thought of locally as a 'branched function' with several layers to its graph, as may be seen by examining the examples in Chapter 5.

We see that *bifurcation does not happen as long as critical points remain Morse:* the system does not change locally. (Though non-local features may alter, such as which minimum is deepest: see Chapter 11 Section 12.) The collection of critical points can change only by one or more of them becoming non-Morse. Hence the concern in the next chapter with the types of degenerate singularities typically encountered, and the way they are passed through. These 'organize' the changes from one kind of nondegenerate function to another, in ways like that of the cusp catastrophe, analysed in Section 2 of the previous chapter.

7 Thom's Classification Theorem

The aim of this chapter is to give an intuitive, geometrical sketch of the way in which Thom's classification of the elementary catastrophes arises from transversality considerations. The mathematical difficulties which arise in converting the sketched programme into a rigorous proof of the theorem are isolated as a number of simple and relatively plausible assertions which can be taken on trust without interrupting the flow of the argument. They are, as it happens, very difficult to prove, and it is here that the deeper mathematics behind Thom's theorem comes into play. For the mathematically sophisticated reader, Bröcker and Lander [9] or Trotman and Zeeman [22] provide the whole story. Chapter 8 gives additional motivation in this direction. An intermediate step is provided by Lu [28]: more rigorous than us, more geometrical and motivated than any complete treatment. We should observe that Lu's transversality pictures, which include the directions in the domains of the functions in the family examined, correspond more directly than ours to the algebraic transversality conditions in the formal proof. Our approach has the visual advantage of allowing pictures for more cases.

The advantage of such a sketch is that it becomes clear why a theorem like Thom's should be true. The famous list of the seven catastrophes is robbed of some of its mystery. The hypotheses under which Thom's theorem holds are seen as a natural restriction imposed by the underlying idea. It is hoped that a little insight into the way the theorem arises may prove valuable for those who wish to apply it without working through the full mathematical proof, but who are unhappy about taking the result entirely on trust in case this leads to misunderstandings or errors.

In this chapter we concentrate on the following question. In an r-parameter family of functions, which local types do we typically meet? This is one strand in the web that is catastrophe theory. Equally important is the converse: given a function, what does a family which contains it look like, close to the given function? The same mathematical machinery supplies answers to the latter question, which we discuss in the next chapter. The geometry of the catastrophes on Thom's list is dealt with in the chapter after that.

1 Functions and Families of Functions

Let $f : X \to \mathbb{R}$ where X is a manifold (usually $\mathbb{R}^n$). Since Morse critical points (if any critical points exist at all) are typical, it follows that at any point typically $Df \neq 0$, or $Df = 0$ but the Hessian Hf is non-singular; and by the Morse Lemma we can therefore write f locally in one of the forms

$$
\left.
\begin{aligned}
&\text{(a) } f(x_1, \ldots, x_n) = p + x_1 \\
&\text{(b) } f(x_1, \ldots, x_n) = p + x_1^2 + \cdots + x_i^2 - x_{i+1}^2 - \cdots - x_n^2
\end{aligned}
\right\} \qquad (7.1)
$$

which, respectively, represent a non-critical point and a Morse $(n-i)$-saddle.

But what if we have not an individual function, but a whole family of them? In Chapter 5 we considered a number of machines, on each of which it is possible to vary the choice of f among a family of functions. In these examples the variation is caused by mechanical means, for instance by altering the position of the centre of gravity or the tension in a length of elastic. The important mathematical point to notice is that these families may contain functions f which, individually, are non-Morse and hence atypical. However, the families themselves are typical, *as families*. As before, we shall not go into precise formulations of this 'typicality', but will rely on the reader's intuition of the sense in which transversality is typical. We hope to reinforce and sharpen that intuition by the wide range of physical examples in the second half of the book.

2   One-Parameter Families

We have seen how the stability of a Morse function f comes from the stability of transversality, so that f is unchanged 'up to parametrization' by a small perturbation, and its various critical points – maxima, minima and saddles – survive as critical points of the same sort. But if we perturb Fig. 4.18 a lot, as in Fig. 7.1, then elementary calculus shows that en route we must meet a function whose derivative is not transverse to the zero line, as in Fig. 7.1(b).

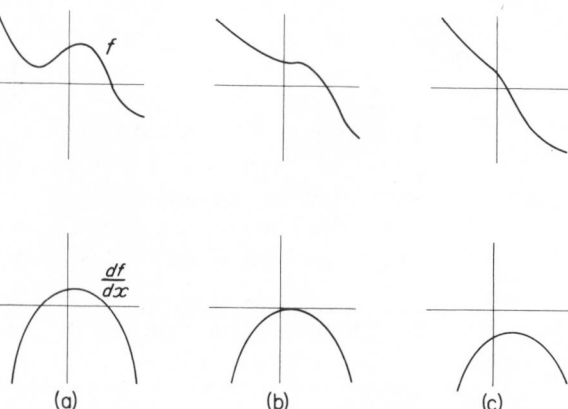

Fig. 7.1   (a)   (b)   (c)

However, we can still use a kind of higher transversality to find the typical way in which the entire sequence of perturbations happens.

We have to look at a function $X \to \mathbb{R}$ which changes according to some extra variable. In applications this may be time, position of centre of gravity, temperature, etc: for definiteness here we call it 'time' to start with. We can write this either as a family of functions $f_t : X \to \mathbb{R}$, for $t \in \mathbb{R}$, or more conveniently as a single smooth function

$$f : X \times \mathbb{R} \to \mathbb{R}$$
$$(x, t) \mapsto f(x, t) = f_t(x).$$
(7.2)

We expect that sometimes the derivative of f_t will not be transverse to zero, as an individual function $X \to \mathbb{R}$, as in Fig. 7.1(b). But by analogy with the previous discussion we can look at the family of maps $Df_t : X \to R$ as a single map

$$\partial_X f : X \times \mathbb{R} \to \mathbb{R}$$
$$(x, t) \mapsto \partial_X f(x, t) = Df_t(x).$$

The X as subscript is to emphasize that we are taking derivatives only in the X direction, not along t. The reason for this is that we are interested in criticality of f_t at a fixed, but arbitrary, time t; and a consideration of the machines in Chapter 5 will verify that the vanishing of the t-derivative does not play this kind of rôle.

Fig. 7.2 shows the graph of a typical $\partial_X f$ for $X = \mathbb{R}$. Notice that there is a particular value $t = c$ for which df_c is not transverse to zero at x_c, but that nonetheless $\partial_X f$ *is* transverse to the zero plane at (x_c, c), and indeed everywhere. It is *this* transversality that is typical, and which we shall exploit.

Let us call the point (x_c, c) by the symbol P. Now the graphs of 0 and $\partial_X f$ are two-dimensional surfaces in $\mathbb{R}^3$; they do not meet (typically) in just isolated points for dimension reasons. How they meet, typically, is in a curve C. Similarly for $X = \mathbb{R}^n$ both 0 and $\partial_X f$ are maps $\mathbb{R}^n \times \mathbb{R} = \mathbb{R}^{n+1} \to \mathbb{R}^n$ and so have $(n+1)$-dimensional

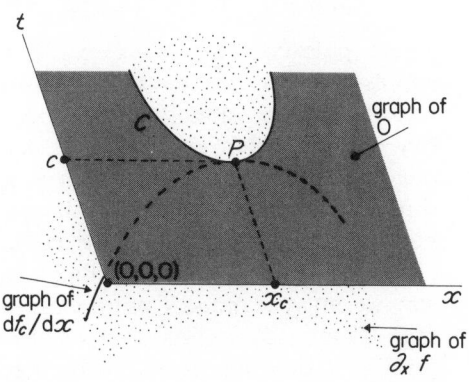

Fig. 7.2

graphs in $\mathbb{R}^{n+1} \times \mathbb{R}^n = \mathbb{R}^{2n+1}$. Typically they intersect in a set of dimension $(n+1)+(n+1)-(2n+1)=1$, which is a curve C; and on C they meet transversely. Thus, typically, we have:

(a) places where $\partial_X f(x, t) \neq 0$, i.e. where f_t is regular;
(b) places where df_t meets zero transversely, so that f_t has a nondegenerate, Morse singularity;
(c) places like P.

The structural stability of cases (a) and (b) as individual functions means that a slight variation in t yields something which is the same up to parametrization. We can in fact reparametrize in a way which depends smoothly on t, which causes t to disappear from the expression for f in a neighbourhood of the relevant point. (We have proved this in detail for case (b) in Corollary 6.2. Case (a) is easier, and left as an easy extension of the unparametrized result.) Around such places, then, $f(x, t)$ (where $x = (x_1, \ldots, x_n)$) can be given the same local expressions as in (7.1) above.

It remains to study case (c). One way to do this is to travel through P along the curve C. (This is not formally necessary in the presence of the full mathematical machinery, but it is helpful in the present pictorial analysis. For an alternative picture *see* Lu [28].) All along C we find critical points: in fact C is precisely the set of points at which the Taylor expansion of f_t in x has no linear term. We have no current interest in the constant term, since we are concerned with the shapes, rather than the values, of the f_t along C. The quadratic term is therefore the first of interest.

In order to bring out certain important points, we shall take $X = \mathbb{R}^2$ in what follows, even though this requires too many dimensions to draw the graphs. We imagine that at time s we are at the point $C(s) = (x_s, y_s, t_s)$ of the curve C. (If, in a particular application, we are thinking of t as time, we must not confuse it with s. For t_s 'turns round' as we pass through P, and so s is a quite distinct 'how far we have travelled in the picture' kind of time.) We can choose s so that we are at the interesting point P when $s = 0$. Now, around (x_s, y_s) the function f_{t_s} can be written

$$f_{t_s}(\tilde{x}, \tilde{y}) = p + a_s \tilde{x}^2 + b_s \tilde{x}\tilde{y} + c_s \tilde{y}^2 + \text{higher order function},$$

where p is constant, $\tilde{x} = x - x_s$, $\tilde{y} = y - y_s$ and a_s, b_s, c_s can be expressed in terms of the second partial derivatives in the usual way. Labelling the possible quadratic expressions in two variables, $ax^2 + bxy + cy^2$, by their coefficients (a, b, c) we see that as s varies the quadratic terms of the Taylor series follow a curve (a_s, b_s, c_s) in the three-dimensional space I of all quadratic expressions. So corresponding to travelling along the curve C in $X \times \mathbb{R}$, we travel along another curve, say $\tilde{C}$, in I. To find what this curve encounters on the way we must examine I more closely for significant features.

In Chapter 2, Section 5 we showed that the type of a quadratic form in I depends on its position relative to a double cone, the

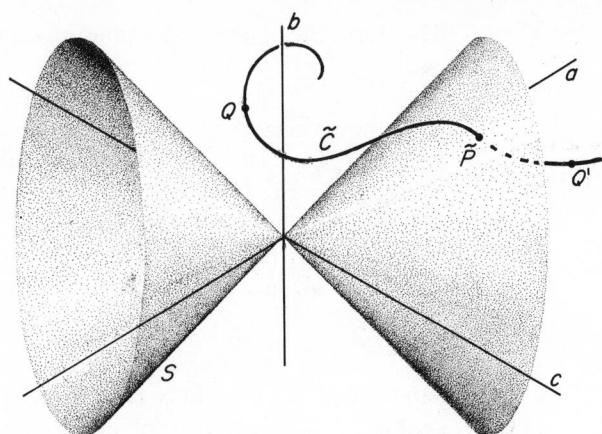

Fig. 7.3

discriminant cone of degenerate quadratic forms. As we move along C the points (a_s, b_s, c_s) trace out a curve $\tilde{C}$ in I, which typically is transverse to the discriminant cone S in Fig. 7.3. We see again that, except at isolated values of s (such as $s = 0$, where $C(s) = P$ in Fig. 7.2 and $\tilde{C}(s) = \tilde{P}$ in Fig. 7.3), the quadratic part of f_{t_s} defined by the second derivative is nondegenerate. Thus, although the first derivative vanishes at all points of C – by definition – away from points like $\tilde{P}$ the quadratic part looks like $u^2 - v^2$ (as at Q) or like $u^2 + v^2$ (as at Q'), or like $-u^2 - v^2$.

Although a typical $\tilde{C}$ may, transversely, meet S, it will not pass through the origin. For a 1-dimensional curve typically meets a 2-dimensional surface in $\mathbb{R}^3$ in a set of dimension $2 + 1 - 3 = 0$, but a curve typically *fails* to pass through any particular point. This means that the form $ax^2 + bxy + cy^2$ cannot be more degenerate than $\pm v^2$, where

$$v = \sqrt{|a|}\, x + \frac{b}{2\sqrt{|a|}}\, y.$$

This is degenerate only 'in the u-direction'. The parametrized Splitting Lemma (Theorem 6.1) now implies that around (x_c, y_c) we can reparametrize (x, y) to (u, v) in a way that varies smoothly with t, to express f in the form

$$f(u, v, t) = \tilde{f}(t, u) \pm v^2$$

where the sign of v^2 depends on which half of the cone $\tilde{P}$ lies in, $\tilde{f}$ being a function whose first and second u-derivatives vanish at $(0, 0)$. The variable v has now been tidied out of the way and can be safely ignored.

If we relax the condition that $X = \mathbb{R}^2$, a similar analysis holds good. If f is a typical 1-parameter family of functions $\mathbb{R}^3 \to \mathbb{R}$ then, around non-Morse points, we can reduce f to the form

$$f(u, v, w; t) = \tilde{f}(u, t) \pm v^2 \pm w^2.$$

This time the space of possible quadratic forms

$$ax^2 + bxy + cy^2 + dyz + ez^2 + fzx$$

is 6-dimensional: we omit the analogues of Figs 7.2 and 7.3. (The case dim $X = 2$ covers the idea perfectly: the general proof requires a rigorous mathematical treatment.) In general, a typical 1-parameter family f of functions $f_t : \mathbb{R}^n \to \mathbb{R}$ has mostly regular and Morse points, plus isolated non-Morse singularities where the Hessian is of corank 1. Around these points it can be given the form

$$f(u_1, \ldots, u_n; t) = \tilde{f}(u_1, t) \pm u_2^2 \pm \cdots \pm u_n^2$$

which again we can study just by looking at $\tilde{f}$. Thus, with no loss of generality, we can restrict attention to a 1-parameter family of functions $\tilde{f}_t$ of a single variable u_1.

For notational convenience put f instead of $\tilde{f}$ and x instead of u_1. We now travel along C looking at the Taylor expansion of the new f_t in more detail, up to order 4:

$$f_{t_s}(\tilde{x}) = k_s + p_s \tilde{x}^2 + q_s \tilde{x}^3 + r_s \tilde{x}^4 + Tayl,$$

k_s being constant and $\tilde{x} = x - x_s$, for each s. The coefficients p_s, q_s, r_s are found by evaluating derivatives. This gives a new curve

$$\tilde{\tilde{C}}(s) = (p_s, q_s, r_s)$$

in a new 3-dimensional space I' with (p, q, r)-coordinates. The quadratic term of the series vanishes for $p = 0$, which is the (q, r)-coordinate plane in I' (Fig. 7.4). The curve $\tilde{\tilde{C}}$ typically meets this plane in isolated points like $\tilde{\tilde{P}}$, where the quadratic term p_s vanishes. Furthermore, typically, when p_s vanishes, q_s does not; for this would mean that $\tilde{\tilde{C}}$ meets the r-axis and typically two 1-dimensional curves in $\mathbb{R}^3$ do not meet at all. Hence, when f_t doesn't look like $p\tilde{x}^2 + Tayl$ (and hence exactly like $\pm u^2$ in a Morse reparametrization) it will look like $q\tilde{x}^3 + Tayl$, $q \neq 0$. By Theorem 4.4 it can then be reduced exactly to the form u^3.

To summarize: a typical 1-parameter family of functions $f : \mathbb{R}^n \to \mathbb{R}$ has non-singular points around which it looks like

$$(u_1, \ldots, u_n; t) \mapsto u_1,$$

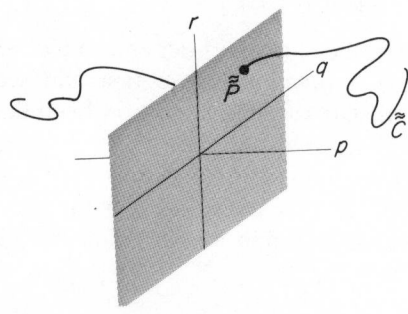

Fig. 7.4

Morse points around which it looks like

$$(u_1, \ldots, u_n; t) \mapsto \pm u_1^2 \pm \cdots \pm u_n^2,$$

and isolated points around which f_t looks like

$$(u_1, \ldots, u_n) \mapsto u_1^3 \pm u_2^2 \pm \cdots \pm u_n^2. \qquad (7.3)$$

We see here the force of the idea of codimension (Chapter 4 Section 11). We can argue equally well in the space of polynomials like $px^2 + cx^3$, like $px^2 + qx^3 + rx^4$, or for that matter like $px^2 + qx^3 + rx^4 + \cdots + sx^k$, that a typical curve will miss the set of polynomials for which $p = q = 0$, because the codimension of this set is in each case 2. Since a non-zero p *or* q is enough to determine the local shape we are seeking, the same transversality arguments apply however many terms in x we allow. The set of k-jets in x which are *not* reducible to $\pm x^2$ or x^3 in a sufficiently small neighbourhood has codimension 2 among the set of all k-jets with no linear part, irrespective of k, for $k \geq 3$. This leads us to the idea that the set of *functions* not everywhere reducible to one of these forms is of codimension 2 in the space of all functions. This notion is useful in seeing what should be true, but less so to date in rigorous proofs of the truths so revealed, owing to technical problems with the (infinite-dimensional) space of 'functions defined near a point', or *germs*, to which the spaces of k-jets may be seen as successive approximations.

Notice also that the set of non-Morse quadratic parts is of dimension 0 in the space $\{px^2 \mid p \in \mathbb{R}\}$ of quadratic forms in one variable, dimension 2 (the cone) in Fig. 7.3, and so on; but of the same *co*dimension, namely 1, in the space of quadratic functions in n variables, whatever n is.

All the reductions to the forms above rest on results that we have proved fully: we only used transversality to *find* the things that we reduced to polynomial form. But we have not yet analysed the t-dependence around (7.3). This leads into the deepest part of the theory. Just as any typical (Df transverse to zero) minimum looks like any other – and in particular like the Morse $u_1^2 + \cdots + u_n^2$ – up to reparametrization, so *any transverse way through any non-Morse function f_t which can be met transversely by a finite-dimensional family, looks like any other.* This is explained further in Chapter 8 Section 7 as the equivalence of 'universal unfoldings'. This is fairly easy to believe for u^3, looking at Figs 7.2 to 7.6; it is true in general, but non-trivial to prove. It means that once we have found *a* transverse way through f_t, we know that it must be the typical way. Now, *one* transverse way through u^3 is given by

$$f_t(u) = u^3 + tu,$$

as can quickly be 'verified' by drawing the analogue of Figs 7.2 and 7.3 for this family. Thus around a point where (7.3) holds there is a reparametrization of t (here, not more than a shift of 0 and perhaps a change in sign) and a reparametrization of $(x_1, \ldots, x_n)$ to $(u_1, \ldots, u_n)$, depending on t, and giving f the expression

$$f(u_1, \ldots, u_n; t) = u_1^3 + tu_1 \pm u_2^2 \pm \cdots \pm u_n^2.$$

It is now easy to see that this family is structurally stable, for Figs 7.1, 7.2 and 7.3 are not altered qualitatively (in particular, transverse crossings remain so) by small perturbations; so we get the same type of point in each case, and the family reduces to the same form after a small perturbation.

The u_1 term of f_t then varies with t as in Fig. 7.5. A local maximum and minimum approach each other, merge at $t = 0$ to a non-Morse singularity, and vanish. If the remaining quadratic terms in $u_2, \ldots, u_n$ are all positive, then the u_1-minimum gives a Morse local minimum for f, and the maximum gives a saddle. Other choices of signs give corresponding types of saddle. Fig. 7.6 shows this for $u_1^3 + tu_1 + u_2^2$, the slopes being indicated by arrows. Two level points, both with $u_2 = 0$, meet and vanish. On the other hand, if f looks like $u_1^3 + tu_1 - u_1^2$, we get the same pictures turned upside down, with maxima and saddles. For $n > 2$ we may have a minimum (0-saddle) meeting a 1-saddle (in the language of Chapter 4 Section 2) or more generally an l-saddle meeting an $(l+1)$-saddle. If $l = n - 1$, this means a saddle meeting a maximum.

These qualitative features could have been deduced already from Fig. 7.2. What is extra to the standard form is that it allows deductions about the algebraic descriptions of various phenomena, in suitable coordinates. This yields interesting predictions: for example, on the intensity of rainbows (*see* Chapter 12). We can already

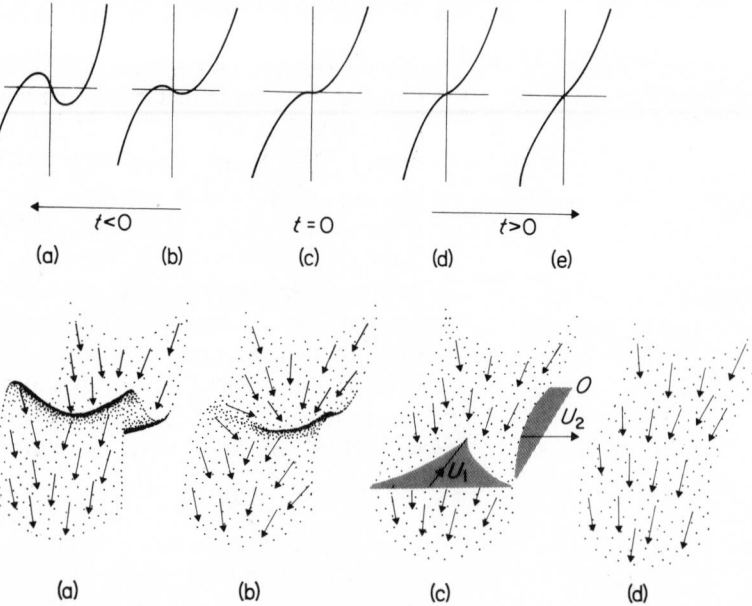

Fig. 7.5

$t<0$ $t=0$ $t>0$

(a)   (b)   (c)   (d)   (e)

Fig. 7.6

(a)   (b)   (c)   (d)

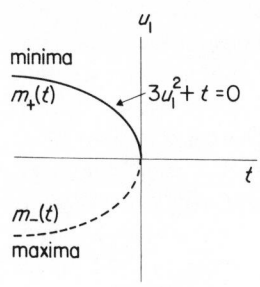

minima

$m_+(t)$

$-3u_1^2 + t = 0$

t

$m_-(t)$

maxima

value of f at critical points

$f_t(m_+(t))$

t

$f_t(m_-(t))$

Fig. 7.7

deduce facts not obvious from Fig. 7.2. The critical points of f_t are given by the solutions $m_+(t)$, $m_-(t)$ of

$$\frac{\mathrm{d}}{\mathrm{d}u_1}(u_1^3 + tu_1) = 0,$$

i.e.

$$m_\pm(t) = \pm\sqrt{(-t/3)}.$$

Hence the critical points merge along the parabola in Fig. 7.7(a). The critical *values*

$$f_t(m_\pm(t)) = (\pm\sqrt{(-t/3)})^3 + t(\pm\sqrt{(-t/3)}) = \pm\frac{4}{3\sqrt{3}}(-t)^{3/2}$$

approach tangentially along a cusp (Fig. 7.7(b).)

These pictures are universal for the typical way that a 1-parameter family of functions passes through a non-Morse singularity. They define the *fold catastrophe*, the simplest of all catastrophes.

3 Non-Transversality and Symmetry

We saw in Chapter 6 that the family

$$W_a(x) = \frac{x^4}{4} + \frac{a}{2}x^2$$

is not structurally stable. It is easy to verify that it is not transverse, in the above sense: that is, the graph of $\partial_x W$ is not transverse to the zero plane. For we have

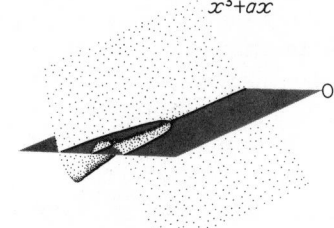

$x^3 + ax$

O

Fig. 7.8

$$\frac{\mathrm{d}}{\mathrm{d}x}W_a(x) = x^3 + ax,$$

and this intersects the zero-plane as shown in Fig. 7.8. (If this picture looks familiar, note that it appears here in a rather different rôle from its usual one. There are various coincidences which occur for catastrophes whose potential involves one variable, which do not extend to two or more variables.) This intersection is clearly not transverse at 0; for example the set in which the surface intersects the plane is the familiar parabola-plus-line whereas for a transverse intersection there should be only a single curve near 0. (More generally: transverse intersections, locally, are manifolds.)

If we perturb W by adding a term bx then the surface shifts up or down by b, and the intersection becomes transverse, as in Fig. 7.9. This makes it at least plausible that for the enlarged family

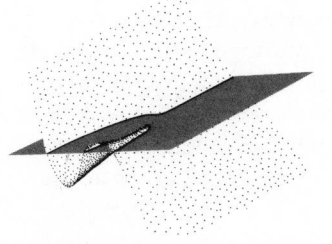

Fig. 7.9

$$V_{ab}(x) = \frac{x^4}{4} + \frac{a}{2}x^2 + bx$$

the intersection has become transverse. The algebraic tests mentioned before will establish this.

Various *special*, that is *a*typical, conditions, sometimes can prevent transversality. For instance, suppose we were only interested in *even* functions of x: those with $f(x) = f(-x)$. It is easy to see that such a function can have no odd powers in its Taylor series. Consequently, if we have a 1-parameter family of such functions, the curve corresponding to $\tilde{C}$ in Fig. 7.4 must lie in the plane $q = 0$, since q is the coefficient of an odd power. *Within this plane* the curve can transversely meet the r-axis. This would be non-transverse, and correspondingly unstable as above, if odd terms could be added, but *given the restriction to even functions* it is transverse and stable. Indeed, specialization of the general techniques to this case show that the family $\frac{1}{4}x^4 + \frac{1}{2}ax^2$, rather than the fold, is the 'universal' local 1-parameter family of functions around $x = 0$.

This is relevant to applications of the theory in two ways. A symmetry condition such as $f(x) = f(-x)$ may be 'given by Nature' to a very good approximation, as in crystallography (see Chapter 14 Sections 15 to 20), and the appropriate analysis then stays within the appropriate space of functions. Thus the assertion 'almost everything is this way' cannot be strengthened to 'everything is this way'. One-parameter families like $\frac{1}{4}x^4 + \frac{1}{2}ax^2$ do occur, stably, in nature, and it is the job of the mathematical scientist to account for the symmetry that stabilizes them – just as the restriction to 'Hamiltonian' systems stabilizes the simple harmonic oscillator (see the pendulum discussion in Chapter 6). On the other hand, engineers and experimentalists often attempt to build systems with particular symmetries, partly because this superficially simplifies the analysis. (For an excellent discussion of this *see* Thompson and Hunt [23].) Consequently, they are often looking for functions that are more singular than is typical, *trying to achieve* the intersection of $\tilde{C}$ with the r-axis of the above example. Typically, they will not succeed perfectly. But the better their technique the more relevant what they are aiming at is to what they achieve, and to the effect of the imperfections by which they fail. (We will see many examples of this in the applications chapters.) One cannot safely use typicality arguments about designed systems without allowing for the atypicality that designers (typically) aim for.

This is well dramatized by a recent paper of Fischer and Marsden [24] which investigates whether mathematical spacetime has a certain stability property often assumed in analyses. Answer: it has, in the absence of certain very special symmetry properties. The set of spacetimes with these symmetries has infinite codimension in the space of all mathematically possible spacetimes, so not merely does a typical spacetime miss it; so does a typical $10^{10^{10}}$-parameter family of spacetimes. In a very strong sense, almost no spacetime has these symmetries, and there is no evidence that physical spacetime has them.

But, thanks to the convenience of these symmetries in analysis, almost no spacetime in a physics text fails to have them.

Whether designing bridges or cosmoi, humans typically think of the atypical. The result is often instability.

4 Two-parameter Families

If we repeat the above analysis for a 2-dimensional (or 2-parameter) family of functions, as occurs in the Zeeman machine, then the set C in Fig. 7.2 is replaced by a set of dimension $(n+2)+(n+2)-(2n+2)=2$: in other words, a surface. We can't draw a picture analogous to Fig. 7.2, as it needs four dimensions: one for x, two for the 2-dimensional parameter $t=(t_1, t_2)$, and one for the values of the various Df_t. But the analogous argument goes through, and we can draw the analogue of Fig. 7.3, namely Fig. 7.10.

This has been drawn to emphasize that $\tilde{C}$ may be quite intricate in the large, with self-intersections or holes; but we are working *locally*, and transversality bars local complications. Now $\tilde{C}$ typically meets the cone S in a curve (rather than a single point as before) but still does not typically pass through O. Thus once more we typically have f_t with a quadratic term degenerate in only one direction. By the Splitting Lemma we can reduce f locally to the form

$$f(u_1, \ldots, u_n; t_1, t_2) = \tilde{f}(u_1; t_1, t_2) \pm u_2^2 \pm \cdots \pm u_n^2.$$

Now we study $\tilde{f}$. The analogue of Fig. 7.4 is Fig. 7.11. This time a typical $\tilde{C}$ *can* meet the r-axis in isolated points like Q. So we have places where the third derivative of f_t with respect to u_1 vanishes, as well as the second, and then

$$\tilde{f}_t(u_1) = k + r u_1^4 + Tayl,$$

k being constant. At such points we can smoothly reparametrize u_1 to get rid of the Tayl, leaving only $\pm u_1^4$.

Before looking at the typical way to go through such a point, we must also discuss the 'cubic' points such as K, where $r \neq 0$. Going

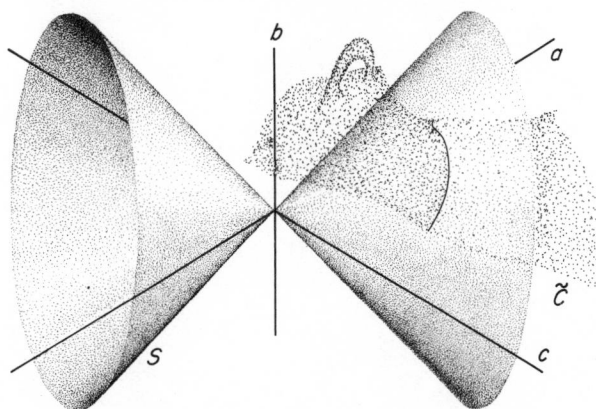

Fig. 7.10

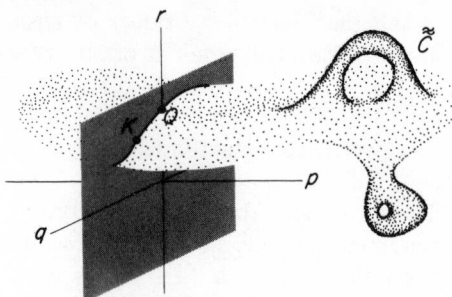

Fig. 7.11

transversely through a cubic function using a 2-parameter family represents a kind of 'overkill': we have already seen that the 1-parameter family of the fold catastrophe is the 'universal' way to do it. But transversality requires that we include such a path, for in general if a curve through a point p in a submanifold of $\mathbb{R}^n$ *can* be non-trivially transverse to M (which of course requires dim $M = n - 1$) then anything of higher dimension will be transverse precisely when it contains such a curve, as in Fig. 7.12. Suppose the point K is given by $t = t_0 = (t_1, t_2)$. In t-space we can find a curve T along which f_t changes transversely through f_{t_0}, in other words a 1-parameter subfamily

$$\hat{f} = \{f_t : t \in T\}$$

which behaves like the fold catastrophe. Now, around T in t-space we pick a family of curves T_λ with $T_0 = T$, varying smoothly with λ, as in Fig. 7.12(c). These give 1-parameter families near to $\hat{f}$. Since $\hat{f}$ is structurally stable, they all look just like $\hat{f}$.

Just as we strengthened (Corollary 6.2) the stability of Morse functions from 'any neighbour looks the same' to 'any family around f, parametrized by $t = (t_1, \ldots, t_n)$, can be made independent of t by

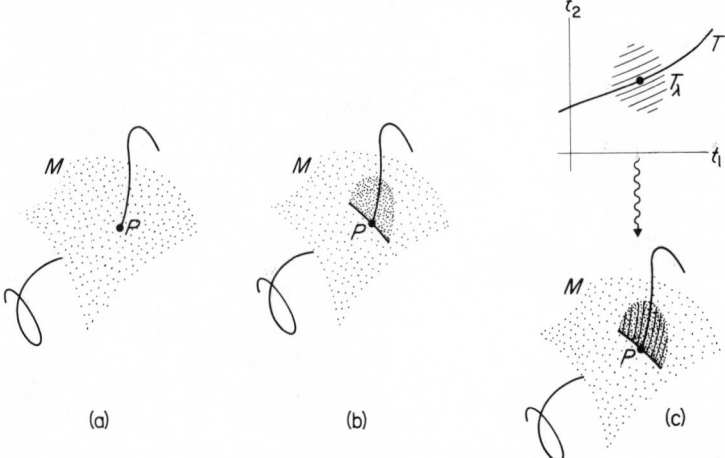

Fig. 7.12

coordinate changes which depend smoothly on t', we can strengthen the stability of $\bar{f}$ from the sameness of neighbours to the constancy of nearby families. This means that if in Fig. 7.1 we pick coordinates $(\bar{t}_1, \bar{t}_2)$ around t_0 making $t_0 = (0, 0)$, T the $\bar{t}_1$-axis, and the T_λ the curves $\bar{t}_2 = \lambda$; then the strengthened stability assertion allows us to pick $\bar{t}_1$ and $\bar{t}_2$ in such a way that, further, changing $\bar{t}_2$ does not alter the form of $f_{\bar{t}_1, \bar{t}_2}$. (Chapter 8 Section 6 gives more details.) This 'disconnects' $\bar{t}_2$ and allows us to write f locally as

$$f(u_1, \ldots, u_n; \bar{t}_1, \bar{t}_2) = u_1^3 + \bar{t}_1 u_1 \pm u_2^2 \pm \cdots \pm u_n^2,$$

so that locally we can ignore $\bar{t}_2$.

The proper proof of this (Bröcker and Lander [9], Trotman and Zeeman [22]), or rather of the general case which the above exemplifies, is in a sense more direct: it does not require choosing T and the T_λ. The above discussion has the advantage that it makes the result seem plausible, without too much appeal to mathematical machinery (which for this case, unlike the stability of Morse functions, goes very deep indeed).

For such a local form of f, the analogue of Fig. 7.7 is obviously Fig. 7.13. Returning to the real meat of the problem, we look at the isolated points like Q in Fig. 7.11. We now need the full 2-dimensionality of $\tilde{C}$, and hence of t, to meet the r-axis transversely. Now *one* transverse way through, and hence as remarked above a universal example, for $r > 0$ is

$$\tilde{f}(u_1; t_1, t_2) = u_1^4 + t_2 u_1^2 + t_1 u_1.$$

Here r has been removed by suitable choice of scale for u_1. If instead $r < 0$, then the typical family around Q looks like the negative of this. These two possibilities are, respectively, the *cusp catastrophe* and the *dual cusp catastrophe*. The analogue of Fig. 7.7 or 7.13 for the cusp is Fig. 7.14. We analysed the geometry of this in Chapter 6. For the dual cusp everything is similar, but we must interchange u_1-minima and u_1-maxima in Fig. 7.14(a), and turn Fig. 7.14(b) upside down.

Transversality in this higher-dimensional case is less easy to 'see', and of course the proper test is algebraic, in the same way that the proper test for being Morse is the non-singularity of the Hessian matrix. (The computational rules for this algebraic test are given in the next chapter.) However, it is worth finding the exact form of Fig. 7.11 for this particular family, to emphasize – if not prove – that it is indeed transverse.

First, notice that Fig. 7.7(a) is exactly the zero-level slice of Fig. 7.11 (given by the horizontal plane drawn in that figure), redrawn for the particular case

$$f(u_1; t) = u_1^3 + t u_1.$$

For if we insert $(u_1; t)$-space as the zero-level

$$\{(u_1, y; t) \mid y = 0\}$$

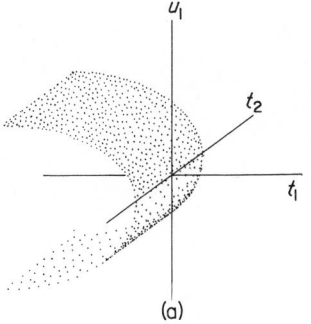

(a)

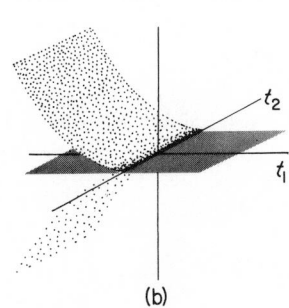

(b)

Fig. 7.13

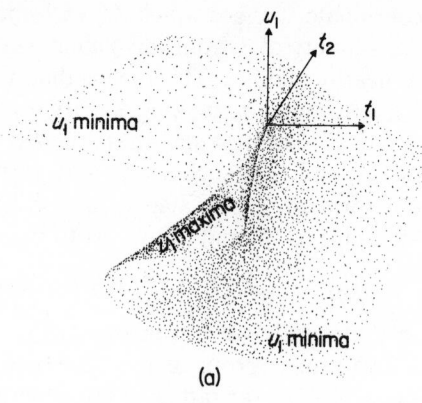

(a)

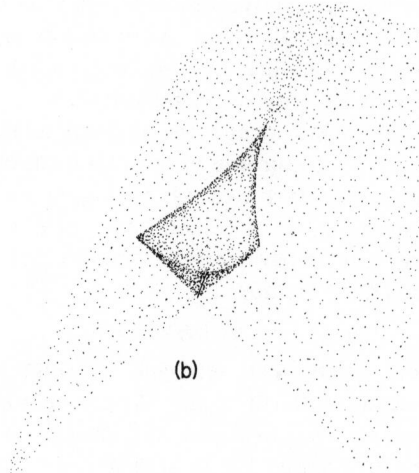

(b)

Fig. 7.14

of the 3-dimensional space of Fig. 2.2, then the curve

$$3u_1^2 + t = \frac{\partial f}{\partial u_1}(u_1; t) = \partial_X f(u_1; t) = 0$$

of Fig. 7.7 is just the intersection C of $(u_1; t)$-space, as the zero-level, with the graph

$$\{(u_1, \partial_X f(u_1; t); t)\}$$

of $\partial_X f$. Analogously, the surface in $(u_1; t_1, t_2)$-space shown in Fig. 7.14(a) is the intersection surface C of

$$\{(u_1, y; t_1, t_2) \,|\, y = 0\}$$

with the graph of

$$4u_1^3 + 2t_2 u_1 + t_1 = \partial_X f(u_1; t_1, t_2)$$

in the 4-dimensional space $\mathbb{R}^2 \times \mathbb{R}^2$ (which we have no room to draw).

Since we shall from now on ignore $u_2, \ldots, u_n$ we shall drop the suffix on u_1 and write just u. Thinking of C as above, we find a very natural choice of coordinates, namely u and t_2. Any choice of (u, t_2) specifies a unique point in C, because the equation of C can be written as

$$t_1 = -4u^3 - 2t_2 u.$$

(This amounts to using the view of C given in Fig. 5.7(f), where the u-axis is horizontal and the t_2-axis vertical.)

At a chosen point $(\bar{u}, \bar{t}_2)$ in C, that is, at the point

$$(u; t_1, t_2) = (\bar{u}; -4\bar{u}^3 - 2\bar{t}_2\bar{u}, \bar{t}_2)$$

$$= (\bar{u}; \bar{t}_1, \bar{t}_2), \quad \text{say},$$

we can look at the 4-jet of $f_{(\bar{t}_1, \bar{t}_2)}$ around the point $\bar{u}$, which takes the form

$$f_{(\bar{t}_1, \bar{t}_2)}(\bar{u} + U) = (\bar{u} + U)^4 + \bar{t}_2(\bar{u} + U)^2 + (-4\bar{u}^2 - 2\bar{t}_2\bar{u})(\bar{u} + U)$$

$$= -(3\bar{u}^4 + \bar{t}_2\bar{u}^2) + (6\bar{u}^2 + \bar{t}_2)U^2 + 4\bar{u}U^3 + U^4.$$

We are not interested in the constant term $-(3\bar{u}^4 + \bar{t}_2\bar{u}^2)$, except for drawing Fig. 7.14(b). As expected (*see* page 102) there is no linear term in U, so we examine the points

$$(p, q, r) = (6\bar{u}^2 + \bar{t}_2, 4\bar{u}, 1)$$

in the space Γ' of coefficients for polynomials

$$px^2 + qx^3 + rx^4$$

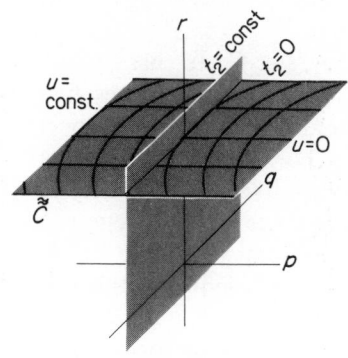

Fig. 7.15

as in Figs 7.4 and 7.11.

Clearly the set of such points is the level plane $r = 1$ of Fig. 7.15, varying with $\bar{u}$ and $\bar{t}_2$ as shown.

This diagram is important in understanding the geometry of the cusp and higher catastrophes, and we shall return to it in Chapter 9. For the present, observe only that $\tilde{C}$ is visibly transverse to the r-axis: both as a set, and in the sense that if we go through $(0, 0)$ in C with non-zero speed, the corresponding movement in Γ' through the r-axis is also at non-zero speed.

We have now carried the subdivision of jets (and thence of functions $\mathbb{R}^n \to \mathbb{R}$) a step further. Among those with no linear part (which is what we are led to by considering the r-parameter set generalizing C of Fig. 7.2, for $t = (t_1, \ldots, t_r)$) the functions whose expansions begin with nondegenerate quadratic terms – and hence are *reducible* to quadratic form – define a set of codimension 0. Those reducible to fold points are a set of codimension 1, those reducible to cusp points a set of codimension 2. 'The Rest' is a set of codimension 3, as we see by combining the arguments around Figs 7.10 and 7.11. We shall decompose it somewhat further in what follows.

Notice again that these codimensions are well defined *independently* of n, although the dimensions of the sets are not (even if we restrict attention, say, to 4-jets). Codimension is clearly here a separate and powerful idea.

5 Three-, Four- and Five-parameter Families

For a family with more than two parameters the arguments in Sections 2 and 3, for degeneracy in at most one direction of the quadratic term, no longer apply. A typical 3-parameter family can stably include a function which is degenerate in two directions at once. However, we do not expect threefold degeneracies until we reach a family with 6 or more parameters, since this demands that we pass transversely through 0 in the 6-dimensional space of quadratic forms

$$ax^2 + bxy + cy^2 + dyz + ez^2 + fzx$$

in three variables. That is, the set of functions $\mathbb{R}^n \to \mathbb{R}$ with *triply* (or more) degenerate critical points, with corank ≥ 3, has codimension 6. Thus for families with up to five parameters we get two classes: the first with exactly one variable u_1 not occurring quadratically; the second with two variables u_1 and u_2 not occurring quadratically. We take these cases in turn, suppressing further reference to those variables u_i which locally we can arrange to appear *only* quadratically, and independently of t, as before: namely, those with $i \geq 2$ in the first case, $i \geq 3$ in the second.

Examine then first a typical $f(x; t_1, t_2, t_3)$. It has a 3-dimensional $\tilde{C}$ in the (p, q, r)-space I', so we will sometimes find isolated points $(\bar{t}_1, \bar{t}_2, \bar{t}_3)$ for which all the derivatives (with respect to x) below fifth order of $f_{(\bar{t}_1, \bar{t}_2, \bar{t}_3)}$ are zero. For these are the points at which $\tilde{C}$ meets the origin of I'. This is just like the possibility of isolated fourth-order points for a 2-parameter family in Section 3. Strictly analogously to that case: if we extend I' to a 4-dimensional space by adding in the quintic coefficient sx^5, this will typically not vanish when the other three p, q, r do. (Count dimensions: a typical 3-dimensional 'hypersurface' misses the origin in 4 dimensions.) Thus we can give $f_{(\bar{t}_1, \bar{t}_2, \bar{t}_3)}$ the local form u^5, and f correspondingly the local form

$$f(u; t_1, t_2, t_3) = u^5 + t_3 u^3 + t_2 u^2 + t_1 u,$$

(putting $u = u_1$ as usual), much as in the case of the fold or cusp.

There is obviously an r-parameter family of this general pattern

$$t_1, \ldots, t_r = \pm(u^{r+2} + t_r u^r + t_{r-1} u^{r-1} + \cdots + t_1 u_1)$$

for every r. The cusp is the simplest which really exhibits the characteristic flavour of these families, so they have collectively been christened the *cuspoids*, though in fact the fold is the first. The 3-dimensional one above is called the *swallowtail*; the 4-dimensional

$$f(u; t_1, t_2, t_3, t_4) = \pm(u^6 + t_4 u^4 + t_3 u^3 + t_2 u^2 + t_1 u)$$

is the *butterfly* (and *dual butterfly*); and the 5-dimensional

$$f(u; t_1, t_2, t_3, t_4, t_5) = u^7 + t_5 u^5 + t_4 u^4 + t_3 u^3 + t_2 u^2 + t_1 u$$

is the *wigwam*. These names are intended purely as mnemonics for the characteristic geometry which the various catastrophes display, and which we discuss in the next chapter.

We now take up the second case, of double degeneracy (corank 2), where $\tilde{C}$ *does* meet the origin. In algebraic terms, this means that the Hessian of some (isolated) $f_{(\bar{t}_1, \bar{t}_2, \bar{t}_3)}(x, y)$ vanishes completely at some point $(\bar{x}, \bar{y})$, so that its matrix becomes

$$\begin{bmatrix} 0 & 0 \\ 0 & 0 \end{bmatrix}$$

(rather than merely being singular, with zero *determinant*. This implies degeneracy, but fails to distinguish between single or multiple.) To find the shape of the function in this case we have to consider terms of higher degree than quadratic, in both x and y simultaneously.

Life is thus more complicated, since we need more powerful machinery to decide how much of the Taylor series to look at. We shall describe the relevant criteria in the next chapter. Their proofs lie deeper than those of the analogues for Morse and single-variable functions proved in Chapter 4, but their application requires only algebra and calculus already covered. The reader should find it a useful exercise to check, using these rules, the relevant statements below (for instance, that no higher order part can change the local shape of $x^3 + y^4$).

Now $f_{(\bar{t}_1, \bar{t}_2, \bar{t}_3)}$ has no linear or quadratic terms in its Taylor series expansion around $(\bar{x}, \bar{y})$. Any nearby point $(x, y; t)$ either lies outside C – in which case f_t has a non-zero linear part when expanded around (x, y) – or is in C but corresponds to a non-zero point in $\tilde{C}$. (This follows since $\tilde{C}$ passes transversely through $(0, 0, 0)$. With both $\tilde{C}$ and I 3-dimensional, this requires that going through $(\bar{x}, \bar{y}; \bar{t})$ with non-zero speed in C entails going through $(0, 0, 0)$ with non-zero speed in I; cf. Chapter 4.) Around, but not at, $(\bar{x}, \bar{y}; \bar{t})$ we can have only one direction which is worse than quadratic, hence only cuspoid points. A typical 3-parameter family in (p, q, r)-space I' will meet the plane $p = 0$ in 2-dimensional sets of cubic (fold) points, the r-axis $p = q = 0$ in 1-dimensional sets of quartic (cusp) points, and the origin $p = q = r = 0$ in *isolated* points distinct from $(\bar{x}, \bar{y}; \bar{t})$. The latter we may ignore by restricting attention to a small enough region around $(\bar{x}, \bar{y}; \bar{t})$. Hence in its immediate vicinity we expect only fold-point surfaces and cusp-point curves. This expectation is borne out by the geometrical analyses of the next chapter.

Taking $(\bar{x}, \bar{y}; \bar{t}_1, \bar{t}_2, \bar{t}_3)$ as origin, and ignoring constants as usual, we have a function whose Taylor series expansion has no linear or quadratic terms, but begins with cubic terms:

$$f_0(x, y) = \alpha x^3 + \beta x^2 y + \gamma x y^2 + \delta y^3 + Tayl.$$

At this point we refer to the geometric classification of cubic forms given in Chapter 2 Section 6. We let K be the space of all cubic forms $(\alpha, \beta, \gamma, \delta)$, so $K = \mathbb{R}^4$. The decomposition of K into types of cubic is given by Zeeman's umbilic bracelet. Thus we must take our typical 3-parameter family of functions f, and look at the cubic term $F = j^3 f_{(0,0,0)}$: this corresponds to a point $(\alpha, \beta, \gamma, \delta)$ in K.

Typically, F will not be the origin of K, so by a change of scale we may suppose that $\alpha^2 + \beta^2 + \gamma^2 + \delta^2 = 1$ and bring it into Plate 1 (*see* page 28). Typically it will miss the surface and the cusp line, so it must be expressible as

$$x^2 y + y^3 \quad \text{or} \quad x^2 y - y^3.$$

It follows (though without an elementary proof of the type which works for the function $x^k + Tayl$) that $f_{(0,0,0)}$ itself can be locally expressed as $u^2 v \pm v^3$ (with the corresponding sign) by a smooth change of coordinates. As an exercise, the reader should check this using the rules in the next chapter.

The family f typically goes through $f_{(0,0,0)}$ transversely, as before, and thus can be expressed in terms of any transverse way through that we care to choose. Convenient forms are

$$f(u_1, u_2; t_1, t_2, t_3) = u_1^2 u_2 \pm u_2^3 + t_3 u_1^2 + t_2 u_2 + t_1 u_1.$$

(We could equally well choose $t_3(u_1^2 + u_2^2)$ for the quadratic term: experiment with the rules in Chapter 8.) The $+$ and $-$ signs give canonical forms for the *hyperbolic umbilic catastrophe* and *elliptic umbilic catastrophe* respectively.

Note that since sign changes can be accomplished by a change of coordinates, these two catastrophes are self-dual like the fold, swallowtail and wigwam, and unlike the cusp and butterfly.

Next, consider a 4-parameter family for which $\tilde{C}$ meets the origin of I. It is squashed by the three dimensions of I, and in fact we should really look at it in the 7-dimensional space $I \times K$ of polynomial functions like

$$ax^2 + bxy + cy^2 + \alpha x^3 + \beta x^2 y + \gamma xy^2 + \delta y^3$$

where $\tilde{C}$ has room to exist as a 4-dimensional 'surface'. As long as not all of a, b, c are zero the quadratic part is degenerate in at most one direction, so as before the f_t will typically be locally of the form

$$\pm x^2 \pm y^2, x^3 \pm y^2, \pm x^4 \pm y^2, x^5 \pm y^2, \pm x^6 \pm y^2,$$

for suitable coordinates; and the ways through them will give regular points, folds, cusps, swallowtails and butterflies as above. But where $\tilde{C}$ *does* meet the 4-dimensional space K of homogeneous cubics given by $a = b = c = 0$, it must typically do so in a curve D: that is how two 4-dimensional things meet in a 7-dimensional space. So we must see where in K this curve goes.

D typically does not go through the origin, so once more we can get our information from the corresponding curve in the unit

3-sphere, and use Plate 1 (*see* p. 28). Typically D can (though it need not) meet the surface of the umbilic bracelet; but it cannot typically meet the cusped curve of perfect cubes (type (iv) cubics). Thus we get whole curves along which f_t is expressible locally as $x^2 y \pm y^3$, with the $\pm$ sign changing at isolated points $\bar{t}$ where $f_{\bar{t}}$ can be written as

$$x^2 y + \text{higher order.}$$

Unlike the case of $x^2 y \pm y^3$, the cubic term $x^2 y$ does not guarantee that we can kill off higher terms, as we saw in Chapter 3 Section 4. Thus we must look at the quartic term. Fortunately, we do not need to describe the quartic analogue of the umbilic bracelet (though this has been done, cf. Poston and Stewart [25] pp. 110–147). For, although the space of quartics is 5-dimensional, only one dimension's worth comes in here, and an algebraic analysis suffices.

In fact, quite generally, let us suppose that

$$f_t(x, y) = x^2 y + \text{homogeneous polynomial of degree } k + O(k+1).$$

Then

$$f_t(x, y) = x^2 y + ax^k + 2xyQ(x, y) + by^k + O(k+1)$$

where Q is homogeneous of degree $k - 2 \geqslant 2$. Write

$$u = x + Q(x, y), \qquad v = y + ax^{k-2}.$$

Then

$$
\begin{aligned}
u^2 v + bv^k &= (x + Q(x, y))^2 (y + ax^{k-2}) + b(y + ax^{k-2})^k \\
&= x^2 y + ax^k + 2xyQ(x, y) + by^k + O(k+1) \\
&= f_t(x, y) + O(k+1).
\end{aligned}
$$

So if $b \neq 0$ then f_t can be written as

$$U^2 V \pm V^k + O(k+1)$$

where $U = |b|^{-1/2k} u$, $V = |b|^{1/k} v$, and the rules show that there is a local smooth change of coordinates to express f_t exactly in the form

$$f_t(x, y) = x^2 y \pm y^k.$$

If b is 0 then we have reduced f_t to the form

$$u^2 v + (\text{homogeneous polynomial of degree } k+1) + O(k+2),$$

and the whole process repeats: we can get it to the form

$$u^2 v + b' v^{k+1} + O(k+2)$$

and remove the order $k+2$ remainder if $b' \neq 0$. If $b' = 0$ we repeat again. Either the process stops somewhere, in which case $f_t(x, y)$ takes the form $x^2 y + y^k$ for *some* k, or else we must have

$$f_t = x^2 y + O(\infty),$$

(an $O(\infty)$ function being one with *all* derivatives zero at the origin), which is *highly* atypical!

Thus the problem reduces to finding which values of k we typically meet as the cubic part of f_t travels along the curve D.

The answer is very much in the same pattern as the cuspoids. The one point on the curve D at which f_t fails to be expressible as $x^2y \pm y^3$ is typically nothing more degenerate than $\pm(x^2y + y^4)$. (Note that we do not have *four* distinct possibilities $\pm x^2y \pm y^4$, as one might at first suppose, since $x^2y - y^4 = -(x^2(-y) + (-y)^4)$.) The typical ways through these are

$$\pm(x^2y + y^4 + t_4x^2 + t_3y^2 + t_2x + t_1y),$$

the canonical forms of the *parabolic umbilic catastrophe* and its dual.

For a 5-parameter family, D becomes 2-dimensional, and thus meets the surface of the umbilic bracelet in curves. At most points of these curves f_t is reducible to $\pm(x^2y + y^4)$: but at isolated points $\bar{t}$ the quartic term vanishes and we typically get $f_{\bar{t}}$ expressible in one of the forms $x^2y \pm y^5$, each being self-dual. These are passed through transversely by the *second hyperbolic umbilic catastrophe*

$$u_1^2u_2 + u_2^5 + t_5u_2^3 + t_4u_2^2 + t_3u_1^2 + t_2u_2 + t_1u_1$$

and the *second elliptic umbilic catastrophe*

$$u_1^2u_2 - u_2^5 + t_5u_2^3 + t_4u_2^2 + t_3u_1^2 + t_2u_2 + t_1u_1.$$

The names suggest, correctly, that there is a sequence of higher umbilics of the form

$$\pm(u_1^2u_2 + u_2^r + t_ru_2^{r-2} + \cdots + t_4u_2^2 + t_3u_1^2 + t_2u_2 + t_1u_1)$$

for even r, the *higher parabolics* and their duals, or

$$u_1^2u_2 \pm u_2^r + t_ru_2^{r-2} + \cdots + t_4u_2^2 + t_3u_1^2 + t_2u_2 + t_1u_1$$

for odd r, the self-dual *higher hyperbolics* and *higher elliptics*. One might call this whole sequence of stable types the *conic* umbilics, gathering ellipses, hyperbolae and parabolae under their usual umbrella.

A 2-dimensional D not only meets the surface of the umbilic bracelet in curves, but also it can typically meet the cusped line of perfect cubes, in isolated points. For isolated $\bar{t}$, in suitable coordinates, we have

$$f_{\bar{t}}(x, y) = x^3 + Tayl.$$

Here again we must look for higher terms: we find that for a 5-parameter family we typically meet only $f_{\bar{t}}$ locally expressible as $\pm(x^3 + y^4)$, passing through it in a way equivalent to

$$\pm(u_1^3 + u_2^4 + t_5u_1u_2^2 + t_4u_2^2 + t_3u_1u_2 + t_2u_2 + t_1u_1),$$

the *symbolic umbilic catastrophe* and its dual.

This begins another family of stable types, but one which peters out rather quickly (*see* Arnol'd [26] and Rand [26a]).

6 Higher Catastrophes

We have listed above many standard local forms that r-parameter families of functions can take, for $r>5$: the higher cuspoids and the higher conic umbilics. But whereas stability is typical for $r\leqslant 5$, the new families that start in higher dimensions break some of the method's teeth. It is easiest to see why in a 2-variable case.

Just as in a 3-parameter family we can stably have isolated $f_t(x,y)$ with no quadratic part, a 7-parameter family can include an $f_t(x,y)$ whose Taylor expansion about some $(\bar{x},\bar{y})$ has neither quadratic nor cubic terms, but starts with quartic terms. By analogy with Sections 2 and 5, we should find out how many types of homogeneous quartic in (x,y) there are, up to change of coordinates as usual.

Unfortunately there are infinitely many.

Take an arbitrary quartic with four distinct real root lines (Fig. 7.16) and try to find coordinates that give it some standard form in the manner of Chapter 2 Section 6. By choosing two of the lines as axes we get x and y as factors, and by choosing the units on them we can put a third line into any desired form, giving say a factor $x-y$; and by multiplying by a scalar we can eliminate awkward fractions. But now we have used up all our freedom in choosing a linear change of coordinates, and we have no control over the fourth root line. There seems to be no way in general to adjust the coordinates any further, say to express the quartic in the form $xy(x-y)(x+y)$.

Not only can we see no way to do it: we can prove that there is none. The gadgetry required is very much in the spirit of 19th century mathematics. Define the *cross-ratio* of the four lines in Fig. 7.16 as the number

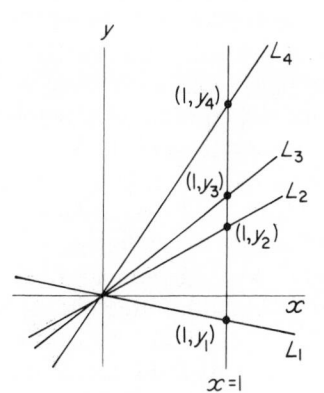

Fig. 7.16

$$(y_1-y_3)(y_2-y_4)/(y_2-y_3)(y_1-y_4)$$

where the y_i are the y-coordinates of the points where the root lines cross the line $x=1$. If one of the lines, say L_4, is the y-axis, we define the cross-ratio to be

$$(y_1-y_3)/(y_2-y_3).$$

An easy exercise (which will be found in any good textbook on projective geometry) is to show that:

(a) any real non-zero c can occur as cross-ratio for a case like Fig. 7.16;
(b) the cross-ratio does not change under any linear change of coordinates, provided the lines are numbered the same way.

We thus have an infinity of sets of four lines through $(0,0)$, and hence of quartics, which cannot be transformed into one another by linear changes of coordinates. (The reader may care to prove that two quartics with four distinct root lines can be transformed into

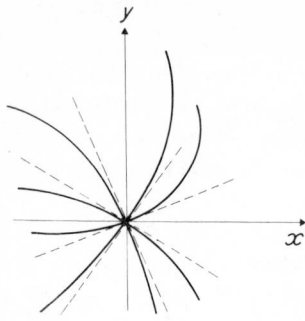

Fig. 7.17

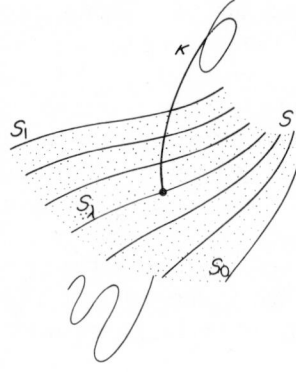

Fig. 7.18

each other up to sign *if and only if* their root structures have the same cross-ratio (with respect to suitable numbering).) It follows that no *differentiable* change of coordinates will work either: it may bend the root lines around (as in Fig. 7.17) but the tangents at $(0, 0)$ change *linearly* by the derivative, and *their* cross-ratio cannot change.

By transversality it is stable for a seven-parameter family to include a point at which the first, second and third derivatives in x and y all vanish; but it is *not* stable to meet a *fixed* type of quartic this way, since quartics of given type form a set of codimension at least 8. In fact, small perturbations can alter the cross-ratio of the root lines of the quartic jet, and as we have seen this cannot be compensated for by a smooth coordinate change.

The argument that transversality, and hence stability, is typical requires that we try only to be transverse to a finite set of things. If a family $\{S_\lambda\}_{\lambda \in \mathbb{R}}$ of curves in 3-space together make up a surface S (Fig. 7.18) then a curve κ passing through S will meet some S_λ non-transversely. (It has to: the dimensions add up to something too small.) Perturb κ, and it will typically become transverse to S_λ (i.e. fail to meet it), but non-transverse to some $S_{\lambda'}$. The problem with 7-dimensional κ meeting sets S of quartics, sliced into layers S_λ by cross-ratio λ, is of higher dimension, but strictly analogous.

6-Parameter families give similar trouble when they meet functions (as they typically can, *see* Section 5) with threefold degeneracies. The space of cubics in x, y, z contains a similarly infinite set of types, the gadget corresponding to cross-ratio being the classical *j-invariant* (which we shall not define, but the infinitude can be established without appeal to this, as done below). Thus only for $r \leqslant 5$ is it *typical* for an r-parameter family of functions to be stable, although even for $r > 5$ it *may* be. Thus the original statement of Thom's theorem ('Among 4-parameter families of functions it is typical to be stable and to have, around any singular point, up to sign and choice of coordinates, one of seven expressions') can be extended in full strength only to the 5-parameter version given in the next section. Higher dimensions are important, and not intractable, but they need application of the theory rather than the theorem.

(For those who have met elementary 'Lie group action' theory: the space of all cubics in x, y, z has dimension 10, and the nondegenerate ones form a 10-dimensional open set. This cannot be the union of finitely many $GL_3(\mathbb{R})$-orbits, since this group has dimension only 9 and its orbits, therefore, dimension $\leqslant 9$. Similarly 4-dimensional $GL_2(\mathbb{R})$ cannot have open orbits in the 5-dimensional space of quartics in x, y. The jet space dimensions grow faster with corank and k than those of the groups, so there are lots of infinite families out there. Indeed Arnol'd's classification [40] covers *all* 'simple' types, plus all infinite families with 1- or 2-dimensional invariants like cross-ratio, met in *any* dimension. *Everything* not included is in a triply infinite family, or worse.)

The set of nondegenerate (no roots equal) quartics in x, y has codimension 7. It breaks up into four connected components: the quartics with all roots real, those with two real and two complex, those with all roots complex and taking only positive values, and those with all roots complex and taking only negative values. Each of these four sets decomposes as a 'parametrized family' of 'leaves', each leaf of codimension 8, and consisting of equivalent quartics under diffeomorphism. The parameter in each case is related to the cross-ratio. By using a more extensive set of coordinate changes, not necessarily smooth at the origin, it is possible to collapse these four families, or *moduli*, into four types, representatives being $x^4 + y^4$, $-x^4 - y^4$, $x^4 - y^4$ and $x^4 - 6x^2y^2 + y^4$. These underlie the *double cusp* catastrophes. For more details *see* Stewart and Poston [25] pp. 110–147, Gibson, Wirthmüller, du Plessis and Looijenga [27] and Lu [28].

7 Thom's Theorem

We can now state Thom's theorem, with as much precision of language as we have so far established, as follows.

Typically an r-parameter family of smooth functions $\mathbb{R}^n \to \mathbb{R}$, for any n and for all $r \leqslant 5$, is structurally stable, and is equivalent (in the sense of Chapter 6) around any point to one of the following forms.

Non-critical

$$u_1.$$

Nondegenerate critical, or Morse

$$u_1^2 + \cdots + u_i^2 - u_{i+1}^2 - \cdots - u_n^2 \qquad (0 \leqslant i \leqslant n).$$

These two types are not catastrophe forms: they involve no change with t. All those that follow *are* catastrophes. We give both the pet names they have acquired, and their symbols in the systematic classification of Arnol'd [26] (on a system too deep to explain here).

(1) *Cuspoid catastrophes:*

the *fold* (A_2)

$$u_1^3 + t_1 u_1 + (M);$$

the *cusp* (A_3)

$$\pm(u_1^4 + t_2 u_2^2 + t_1 u_1) + (M);$$

the *swallowtail* (A_4)

$$u_1^5 + t_3 u_1^3 + t_2 u_2^2 + t_1 u_1 + (M);$$

the *butterfly* (A_5)

$$\pm(u_1^6 + t_4 u_1^4 + t_3 u_1^3 + t_2 u_1^2 + t_1 u_1) + (M);$$

the *wigwam* (A_6)

$$u_1^7 + t_5 u_1^5 + t_4 u_1^4 + t_3 u_1^3 + t_2 u_1^2 + t_1 u_1 + (M).$$

(2) *Umbilic catastrophes:*

the *elliptic umbilic* (D_4^-)

$$u_1^2 u_2 - u_2^3 + t_3 u_1^2 + t_2 u_2 + t_1 u_1 + (N);$$

the *hyperbolic umbilic* (D_4^+)

$$u_1^2 u_2 + u_2^3 + t_3 u_1^2 + t_2 u_2 + t_1 u_1 + (N);$$

the *parabolic umbilic* (D_5)

$$\pm(u_1^2 u_2 + u_2^4 + t_4 u_2^2 + t_3 u_1^2 + t_2 u_2 + t_1 u_1) + (N);$$

the *second elliptic umbilic* (D_6^-)

$$u_1^2 u_2 - u_2^5 + t_5 u_2^3 + t_4 u_2^2 + t_3 u_1^2 + t_2 u_2 + t_1 u_1 + (N);$$

the *second hyperbolic umbilic* (D_6^+)

$$u_1^2 u_2 + u_2^5 + t_5 u_2^3 + t_4 u_2^2 + t_3 u_1^2 + t_2 u_2 + t_1 u_1 + (N);$$

the *symbolic umbilic* (E_6)

$$\pm(u_1^3 + u_2^4 + t_5 u_1 u_2^2 + t_4 u_2^2 + t_3 u_1 u_2 + t_2 u_2 + t_1 u_1) + (N).$$

Here $(u_1, \ldots, u_n) \in \mathbb{R}^n$, $(t_1, \ldots, t_r) \in \mathbb{R}^r$; the symbol (M) indicates a Morse function of the form

$$u_2^2 + \cdots + u_i^2 - u_{i+1}^2 - \cdots - u_n^2 \qquad (1 \leqslant i \leqslant n)$$

which must be added on; and (N) similarly the function

$$u_3^2 + \cdots + u_i^2 - u_{i+1}^2 - \cdots - u_n^2 \qquad (2 \leqslant i \leqslant n).$$

It should be emphasized that there may be fewer than r t-coordinates present in the appropriate form, as in Section 3, the remainder of the t-coordinates being 'dummies'. Thus, for example, the swallowtail may be found as a 4-parameter family with parameters (t_1, t_2, t_3, t_4), the form of f being independent of t_4; or in general as a $p+3$-parameter family in $(t_1, \ldots, t_{p+3})$ where the form of f is independent of $t_4, \ldots, t_{p+3}$. Indeed, a 2001-parameter family may include only Morse points, one for each member, with all 2001 parameters dummies.

The $\pm$ signs indicate 'dual' possibilities, which we have already encountered: those with the negative sign are referred to as the 'dual cusp', 'dual butterfly', etc.

The original *seven elementary catastrophes* of Thom are those in the above list not involving t_5, not distinguishing duals. The reason for not distinguishing duals is that the geometry is not significantly altered, but the interchange of maxima and minima can be important for applications.

The geometry of these families is analysed in Chapter 9.

8 Determinacy and Unfoldings

This chapter is the real heart of the book. The reasoning of Chapter 7, properly refined, shows what beginnings of Taylor series we may typically expect to find (subject to caution as to whether in particular cases we can really expect things to be typical; though an anthology of conventional physics resting on the implicit assumption that the non-transverse does not happen could fill more volumes than many physicists might expect). But if only *whole* Taylor series mattered, what use would that be?

In this chapter we analyse the rules by which one may obtain precise information about functions from finite Taylor expansions, whether 'typical' or not. We do not prove these rules, but we do show geometrically why, if any such rules work, these are the ones to expect. If our exposition is obscure or unconvincing, we apologise: but the results claimed are true. Most are proved in most of Mather [10–15], Levine [29] (which do far more but are not focussed upon catastrophe theory), Bröcker and Lander [9], Gibson [30], Lu [28], Martinet [31], Siersma [32, 33], Trotman and Zeeman [22] and Wassermann [36]. We refer to these collectively as 'the exact sources' with no systematic effort to give original credit for particular results.

For ease of reference, and access for those who find the reasoning difficult, we gather the computational rules and definitions together in Section 12, and give examples of their use in Section 13. These two sections may be read as an independent unit, requiring only language from Chapters 2 to 6, though they may appear somewhat arbitrary and strange without the motivating arguments of previous sections of this chapter. Sections 10 and 11 cover some computational points that follow from the main results; they may be read as pendant either to Sections 1 to 9 or 12.

We do not discuss a growing body of devices for finding, or finding out more about, transformations in particular cases whose existence is guaranteed by general theory. A winnowing of the hard science literature on applications of the theory must in due course generate a monograph on *Computational Methods of Catastrophe Theory* addressed to an audience of non-mathematicians acquainted already with the material treated here; *not* necessarily written by the present authors. While talking of computations, it should be mentioned that Siersma [33], while not highly motivated as an introduction to the rigorous theory, is rich in careful, explicit calculations (though directed at rather pure mathematical questions). For the reader who already knows what kind of thing is being done, it is an excellent treatise on how to do it.

We should remark that the corresponding theory for functions $\mathbb{R}^n \to \mathbb{R}^p$, not just scalar-valued ones, is much more difficult and not in such a satisfactory state. However, if one is interested more in the zero set (as in Fig. 4.5) life becomes rather simpler again; *see* Martinet [34, 35].

The use of the rules in Section 12 could easily be included in any software package that will handle symbolic differentiation. We include as Appendix 1 a program to perform everything but the Taylor expansions themselves (these latter being already standard), due to Olsen, Carter and Rockwood at BYU.

1 Determinacy and Strong Determinacy

Recall that the k-jet $j^k f$ of a smooth function $f \colon \mathbb{R}^n \to \mathbb{R}$ is obtained by taking its Taylor series up to degree k, and that the difference $f - j^k f$ is called a Tayl.

Often the k-jet of f is such that *any* function of the form

$$j^k f + Tayl,$$

that is, $f + g$ for any g of order $k+1$, is locally equivalent by a smooth change of coordinates to $j^k f$. Such an f is called k-*determinate* at 0, and our aim is to find conditions for this to hold. For example, $f(x, y) = x^2 y$ is not k-determinate for any k, as we saw in Chapter 4 Section 4, since $x^2 y + y^{2k+1}$ has a line fewer of zeros than $x^2 y$. In Chapter 4 we showed that a function with non-zero derivative at 0 is 1-determinate there, that a Morse critical point is 2-determinate, and that $f \colon \mathbb{R} \to \mathbb{R}$ is determined by its first non-zero derivative.

Evidently any k-determinate function is k'-determinate for all $k' \geq k$, and we call the *determinacy* of f the lowest k for which f is k-determinate; we denote it by $\sigma(f)$. If, as for $x^2 y$, there is no such finite k, we write $\sigma(f) = \infty$ and say that f is *indeterminate*.

Sometimes we are particularly interested in coordinate changes whose derivative at 0 is the identity. We can illustrate the importance of this condition most clearly in two dimensions. It is nice to know that around the origin $f(x, y) = xy + (x + y)^3$, whose zeros are sketched in Fig. 8.1(a), can be reduced to $x^2 - y^2$ (Fig. 8.1(b)). But if f has arisen in the context of an application we may need to know not merely that it has a saddle at $(0, 0)$, but the directions of its curves of zeros. These directions are preserved (Fig. 8.1(c)) by any local coordinate change with derivative

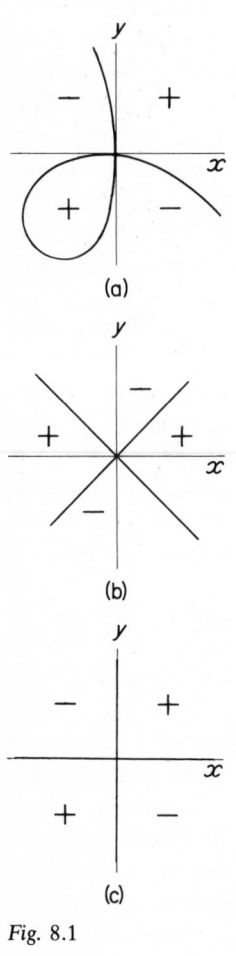

(a)

(b)

(c)

Fig. 8.1

$$\begin{bmatrix} 1 & 0 \\ 0 & 1 \end{bmatrix}$$

at $(0, 0)$. When $f : \mathbb{R}^n \to \mathbb{R}$ is not merely k-determinate, but such that any Tayl of $j^k f$ may be removed by a coordinate change of this restricted type, we call f *strongly k-determinate*.

It is worth repeating that we are *not* assuming analyticity; indeed the k-jet is a useful concept as long as f is just k times differentiable. For convenience of statement, we do assume f smooth.

There are certain vector spaces associated with jets, which we call (collectively) *jet spaces*. These include

$$E_n^k = \{ j^k f \mid \text{all } f : \mathbb{R}^n \to \mathbb{R} \}$$

$$J_n^k = \{ j^k f \mid \text{all } f : \mathbb{R}^n \to \mathbb{R} \text{ with } f(0) = 0 \}$$

$$I_n^k = \{ j^k f \mid \text{all } f : \mathbb{R}^n \to \mathbb{R} \text{ with } f(0) = 0, \ Df|_0 = 0 \}$$

$$M_n^k = \{ j^k f \mid \text{all } f : \mathbb{R}^n \to \mathbb{R} \text{ of order } k \}.$$

Thus E_n^k is the vector space of polynomials of order k; J_n^k those without constant term; I_n^k those without constant or linear term; and M_n^k the homogeneous polynomials of degree k. We have already used in Chapter 7 the geometry of several of these spaces: $I_2^2 = M_2^2$ is the space of homogeneous quadratic polynomials in two variables analysed in Section 5 of Chapter 2, M_2^3 the space of cubics in Section 6 of Chapter 2, while I_1^4 is the space I' of polynomials $px^2 + qx^3 + rx^4$ introduced in Section 2 of Chapter 7, and first drawn as Fig. 7.4.

Our next step is to analyse the geometry of the last space more deeply, to look at simplification of Taylor series in a new way. Unlike that of Section 3 of Chapter 4, this approach does generalize to functions of several variables.

2 One-variable Jet Spaces

The analysis of Chapter 4 Section 3 shows that $px^2 + qx^3 + rx^4$ can be reduced, around 0, by change of coordinates, to

$\pm x^2$ if and only if $p \neq 0$

x^3 if and only if $p = 0, \ q \neq 0$

$\pm x^4$ if and only if $p = q = 0, \ r \neq 0$.

This reduction is *complete*, though local: the function is expressed exactly in the form $\pm x^k$. Now the change of variables involved is not usually polynomial: the proof of the theorem involved taking kth roots, and it is easy to show that *no* polynomial change will reduce, say, $x^2 + x^4$ to x^2, even locally.

But consider the simpler problem of simplifying only the 4-jet. How may we reduce

$$P(x) = px^2 + qx^3 + rx^4$$

to the form

$$\pm x^k + O(5)?$$

A general change of coordinates keeping the origin fixed may be written, using Taylor expansions, as

$$y(x) = (1+\alpha)x + \beta x^2 + \gamma x^3 + O(4).$$

By the Inverse Function Theorem, this is a local diffeomorphism precisely when $\alpha \neq -1$. (Just how local 'local' is depends on β, γ and $O(4)$.) A quick calculation yields

$$P(y(x)) = [(1+\alpha)^2 p]x^2 + [2(1+\alpha)\beta p + (1+\alpha)^3 q]x^3$$
$$+ [(2(1+\alpha)\gamma + \beta^2)p + 3(1+\alpha)^2 \beta q + (1+\alpha)^4 r]x^4 + O(5)$$
$$= p'x^2 + q'x^3 + r'x^4 + O(5),$$

say. Thus the 4-jet of the resulting function is

$$j^4 P(y(x)) = p'x^2 + q'x^3 + r'x^4,$$

where

$$\begin{bmatrix} p' \\ q' \\ r' \end{bmatrix} = \begin{bmatrix} (1+\alpha)^2 & 0 & 0 \\ 2(1+\alpha)\beta & (1+\alpha)^3 & 0 \\ 2(1+\alpha)\gamma + \beta^2 & 3(1+\alpha)^2\beta & (1+\alpha)^4 \end{bmatrix} \begin{bmatrix} p \\ q \\ r \end{bmatrix}. \tag{8.1}$$

Thus the change of variables induces a linear transformation A with the above matrix, mapping I_1^4 to itself. This has determinant $(1+\alpha)^9$ and, as we expect, is invertible whenever the change of variables is invertible around 0. (The analogous result for jet spaces in n variables follows from the generalization of the chain rule to higher derivatives.)

We could deduce information from this matrix for A, but its dependence on α, β and γ is rather complicated. We obtain a clearer picture by looking at a *small* change of coordinates. That is, for the present purposes, we may take α, β, γ to be small enough for us to neglect their powers and products. This reduces the matrix to the form

$$\begin{bmatrix} 1+2\alpha & 0 & 0 \\ 2\beta & 1+3\alpha & 0 \\ 2\gamma & 3\beta & 1+4\alpha \end{bmatrix}. \tag{8.2}$$

Strictly what we have done is to look at the first derivative of A (its linear approximation at $(0, 0, 0)$) as a function of α, β, γ. Thus the matrix (8.2) tells us in what *direction* we move a point (p, q, r) in I_1^4 by starting to change variables in a time-dependent way

$$y_t(x) = (1+t\alpha)x + t\beta x^2 + t\gamma x^3 + O(4),$$

starting at the identity. This gives a remarkable grip on the geometry.

We restrict ourselves now to what might seem the harder problem: *strong k-determinacy*. In the one-variable case, the extra condition that the derivative of the coordinate change be the identity at 0 is just $\alpha = 0$. Then (8.2) becomes

$$\begin{bmatrix} 1 & 0 & 0 \\ 2\beta & 1 & 0 \\ 2\gamma & 3\beta & 1 \end{bmatrix}.$$

What directions of movement in I_1^4 does this indicate? For a point on the r-axis, none at all, since

$$\begin{bmatrix} 1 & 0 & 0 \\ 2\beta & 1 & 0 \\ 2\gamma & 3\beta & 1 \end{bmatrix} \begin{bmatrix} 0 \\ 0 \\ r \end{bmatrix} = \begin{bmatrix} 0 \\ 0 \\ r \end{bmatrix}$$

identically. For a point in the (q, r)-plane,

$$\begin{bmatrix} 1 & 0 & 0 \\ 2\beta & 1 & 0 \\ 2\gamma & 3\beta & 1 \end{bmatrix} \begin{bmatrix} 0 \\ q \\ r \end{bmatrix} = \begin{bmatrix} 0 \\ q \\ r+3\beta q \end{bmatrix}.$$

Thus if the point is not on the r-axis ($q \neq 0$), then a suitable change of x-variable will start to move (p, q) up or down parallel to the r-axis.

For a point off the (q, r)-plane we have

$$\begin{bmatrix} 1 & 0 & 0 \\ 2\beta & 1 & 0 \\ 2\gamma & 3\beta & 1 \end{bmatrix} \begin{bmatrix} p \\ q \\ r \end{bmatrix} = \begin{bmatrix} p \\ q+2\beta p \\ r+2\gamma p+3\beta q \end{bmatrix}.$$

This keeps p fixed, but for suitable β, γ gives any change $(p, q+a, r+b)$ parallel to the (q, r)-plane, by setting

$$\beta = \frac{a}{2p}$$

$$\gamma = \frac{b}{2p} - \frac{3aq}{4p^2}.$$

The directions of motion possible for representative points (p, q, r) are indicated in Fig. 8.2. These directions are easily integrated to give the *whole* motions possible for such points (Fig. 8.3).

Points on the r-axis remain fixed; points in the (q, r)-plane may be moved at will up and down lines parallel to the r-axis; and points off the (q, r)-plane may be moved anywhere within planes parallel to the (q, r)-plane. This may be verified directly from (8.1) for 'non-infinitesimal' changes of variable.

A proof of the general analogue for n variables, where structures like the umbilic bracelet – not just lines and planes – are involved, requires some elementary theory of Lie group actions.

We call the set of points to which (p, q, r) may be moved its *orbit* under the change-of-variable *group*. Gibson [30] makes these concepts precise and emphasizes this aspect of the theory. The group of k-jet transformations arising from local changes of n coordinates will be denoted G_n^k (whether acting on E_n^k, J_n^k or I_n^k). We

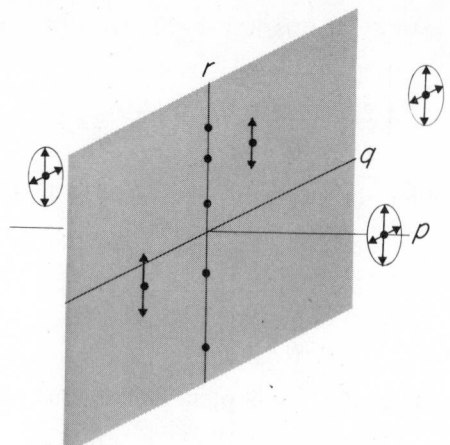

Fig. 8.2

denote by $\tilde{G}_n^k$ the subgroup of k-jet transformations arising from coordinate changes whose linear part at 0 is the identity. Readers not previously exposed to group theory should think of these as sets of matrices describing the ways coordinate changes push k-jets about.

Note, incidentally, that the orbits will not always be flat, as in I_1^k: the cone and umbilic bracelet geometries of I_2^2 and M_2^3 are precisely their subdivisions into G_n^k-orbits (for any $k \geqslant 1$).

In Fig. 8.3, observe that if $p \neq 0$ the orbit of (p, q, r) includes $(p, 0, 0)$; and if $q \neq 0$ the orbit of $(0, q, r)$ includes $(0, q, 0)$. In our original context this says that we can reduce the 4-jets of $px^2 + qx^3 + rx^4$ and $qx^3 + rx^4$ to the forms px^2 and qx^3 respectively, by changes of variable with derivative 1 at 0.

It should be clear that the analogous theory goes through for any I_1^k, $k \geqslant 2$. For a given

$$z = a_s x^s + a_{s+1} x^{s+1} + \cdots + a_k x^k \in I_1^k$$

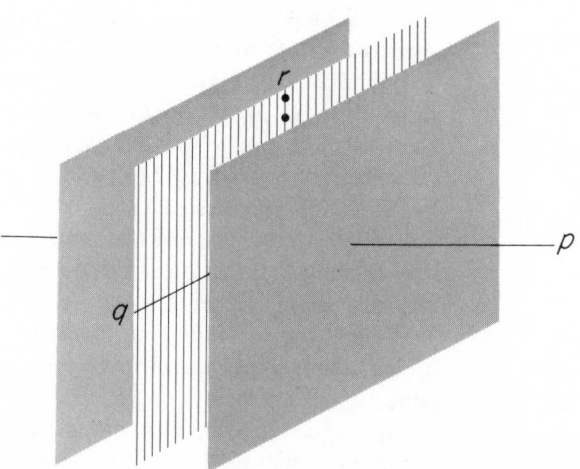

Fig. 8.3

with $a_s \neq 0$, we can specify $\alpha_2, \ldots, \alpha_{k-1}$ such that any change of variable of the form

$$y = x + \alpha_2 x^2 + \cdots + \alpha_{k-1} x^{k-1} + O(k)$$

reduces z to the k-jet $a_s x^s$. It might seem tempting to say 'inductively we can remove the whole Tayl' but there are technical difficulties in constructing a proof this way; not least that after removing all higher polynomial terms we may be left with something like $a_s x^s + \mathrm{e}^{-1/x^2}$. However, the way in which the rules we are leading up to are relevant to removing higher terms gives a very clear intuition of their relevance to removing whole Tayls. In fact Rand [26a] overcomes the technical difficulties. Everything in the rigorous proof has its direct analogue in our discussion.

Now, the spaces of 'directions of possible motion' of points (p, q, r) shown in Fig. 8.2 are exactly the spaces of vectors tangent at (p, q, r) to the orbit of (p, q, r). The removability of higher terms, for example when $p \neq 0$, arose from the following fact: whatever the values of q and r, the q- and r-*directions* were tangent to the orbit of (p, q, r). Thus, *while remaining in the orbit*, we could change the values of q and r arbitrarily (in particular to 0), hence we could find a transformation that would actually make the change. This movement around an orbit by integrating 'infinitesimal' changes of variable has its original counterpart in the use of solutions to differential equations to remove the entire Tayl. It is this condition:

'Whatever the values of the higher terms, the directions parallel to the subspace of higher terms are tangent to the orbit of the jet concerned',

that geometrically underlies the algebraic conditions for removing higher terms.

We must therefore be able to find these spaces of tangent directions systematically, with as little calculation as possible. In the above we found the group of linear transformations (8.1), 'differentiated' that with respect to (α, β, γ) and found the tangent spaces from the result. But by a careful analysis of what the group does, we can remove it from explicit involvement in the calculations, simplifying them considerably.

3 Infinitesimal Changes of Variable

Consider a general time-dependent change of variable, which is the identity at time $t = 0$ and always has the identity as its linear part, of the form

$$y^t(x) = x + tq(x)$$

where

$$q(0) = \frac{dq}{dx}\Big|_0 = 0.$$

(In the above, $q(x)$ was given in partially expanded form as $\beta x^2 + \gamma x^3 + O(4)$.) The function

$$f \circ y^t : x \longrightarrow f(y^t(x))$$

which we hope to steer to a simpler local expression than f, changes with t. How does its $(k+1)$-jet start to move? We find this by differentiating, obtaining a vector in $(k+1)$-jet space,

$$v = \lim_{t \to 0} \frac{j^{k+1}(f \circ y^t) - j^{k+1}f}{t}$$

which we should think of as attached to the point $j^{k+1}f$. The linear nature of Taylor expansion (Chapter 3 Section 5) implies that

$$v = \lim_{t \to 0} \frac{j^{k+1}(f \circ y^t - f)}{t}.$$

If we define $F(x, t) = f(y^t(x)) = (f \circ y^t)(x)$, then

$$v = \lim_{t \to 0} \frac{j^{k+1}(F(x, t) - F(x, 0))}{t}.$$

The components of the vector $j^{k+1}(F(x, t) - F(x, 0))$ are various derivatives with respect to x. Since F is smooth (by construction) we may interchange the order of differentiation, obtaining

$$v = j^{k+1}\left(\lim_{t \to 0} \frac{F(x, t) - F(x, 0)}{t}\right)$$

$$= j^{k+1}\left(\lim_{t \to 0} \frac{f(x + tq(x)) - f(x)}{t}\right).$$

We Taylor expand $f(x + tq(x))$ in t around x. By the definition of the 1-jet,

$$f(x + tq(x)) = f(x) + tq(x)Df|_x + t^2 l(x, t)$$

where $l(x, t)$ is smooth. (As usual, this does not depend on convergence of the Taylor series.) Then the limit inside the brackets equals

$$\lim_{t \to 0} \left(\frac{1}{t}(f(x) + tq(x)Df|_x + t^2 l(x, t) - f(x))\right) = q(x)Df|_x$$

or, in traditional notation,

$$= q(x)\frac{df}{dx}(x).$$

Therefore v is the $(k+1)$-jet of the function $q\dfrac{\mathrm{d}f}{\mathrm{d}x}$,

$$v = j^{k+1}\left(q\frac{\mathrm{d}f}{\mathrm{d}x}\right).$$

[*Technical remark* We have been differentiating, throughout, a motion in the jet space I_1^{k+1}, which is a finite-dimensional real vector space; hence elementary calculus applies. In the infinite-dimensional space $\mathscr{E}$ of all 'germs' – roughly functions defined near 0 – used in the rigorous theory, there is no obvious definition of limits. Contrary to at least one published account of the detailed proof, the fact that $\mathscr{E}$ is a real vector space (of infinite dimension) does *not* give this differentiation an automatic meaning.

Note also that while we have established that all vectors of the form $j^{k+1}\left(q\dfrac{\mathrm{d}f}{\mathrm{d}x}\right)$ are tangent to the orbit, we have not shown that all tangent vectors are of this form. They are, but a proof that for n variables every direction of motion in the orbit can be realized by a motion away from the identity in the Lie group $\tilde{G}_n^{k+1}$ requires appeal to the orbit being an 'embedded' manifold (Mather [14]) or delicate arguments about general, possibly only 'immersed', Lie group orbits, (Magnus [30a]). This point is skated over in some of the exact sources.

The tangent space does consist exactly of vectors of this form, however, and we shall assume this – and its analogue for the larger group G_n^{k+1} – without further comment.]

Thus we can find all tangent vectors at $j^{k+1}f$ to the orbit of $j^{k+1}f$ under $\tilde{G}_1^{k+1}$ by looking at all jets of the form

$$j^{k+1}\left(q\frac{\mathrm{d}f}{\mathrm{d}x}\right), \qquad q(0) = \frac{\mathrm{d}q}{\mathrm{d}x}\bigg|_0 = 0.$$

(Recall that the conditions on q keep the origin fixed and make the change of variable have derivative 1 at the origin.) Applying Section 7 of Chapter 3, these jets may be expressed as

$$\overline{j^{k+1}q \cdot j^{k+1}\left(\frac{\mathrm{d}f}{\mathrm{d}x}\right)}^{\,k+1}.$$

Now the collection of jets $j^{k+1}q$ is the vector space I_1^{k+1}, which may be thought of as the space of all polynomials whose terms have degree between 2 and $k+1$. Therefore we can find a basis for the tangent space by taking a basis $p_1, \ldots, p_r$ for I_1^{k+1} and listing the jets

$$j^{k+1}\left(p_i\frac{\mathrm{d}f}{\mathrm{d}x}\right), \qquad 1 \leqslant i \leqslant r.$$

We may have to discard some of these, reduced by truncation to being linear combinations of others, but we have produced a spanning set and it is easy to extract from it a basis. Sometimes a more natural basis may be found by choosing suitable linear combinations of these. The obvious basis to choose for I_1^{k+1} is $\{x^2, x^3, \ldots, x^{k+1}\}$.

For example, working as above in I_1^4, suppose

$$j^4 f(x) = px^2 + qx^3 + rx^4.$$

Then

$$j^4 \left(\frac{\mathrm{d}f}{\mathrm{d}x}\right)(x) = 2px + 3qx^2 + 4rx^3 + \alpha x^4,$$

where α depends on the fifth derivative of f, which we cannot find from the 4-jet. Choosing x^2, x^3, x^4 as a basis for I_1^4 we have, spanning the tangent space to the orbit of $j^4 f$ under $\tilde{G}_1^4$,

$$P_1 = j^4[x^2(2px + 3qx^2 + 4rx^3 + \alpha x^4)] = 2px^3 + 3qx^4$$
$$P_2 = j^4[x^3(2px + 3qx^2 + 4rx^3 + \alpha x^4)] = 2px^4$$
$$P_3 = j^4[x^4(2px + 3qx^2 + 4rx^3 + \alpha x^4)] = 0,$$

which we notice do not depend on r or α. If $p \neq 0$ we get $x^4 = \frac{1}{2p} P_2$ and $x^3 = \frac{1}{2p} \left(P_1 - \frac{3q}{2p} P_2\right)$, so the tangent space is the plane parallel to the (q, r)-plane. Similarly we get vertical lines (spanned by x^4) as tangent spaces at $(0, q, r)$ for $r \neq 0$, and the zero subspace at $(0, 0, r)$.

We found these tangent spaces in the previous section by a simpler *argument*, but in the above paragraph by a simpler *calculation* – the argument having been done once for all. We need only extend it to several variables. The time-dependent change of variable y^t with derivative the identity at 0 now becomes

$$(y_1^t(x), \ldots, y_n^t(x)) = (x_1 + tq_1(x), \ldots, x_n + tq_n(x))$$

where, as usual, we abbreviate $(x_1, \ldots, x_n)$ to x. The same sequence of steps leads to the general tangent vector to the orbit of $j^{k+1} f$ being

$$v = j^{k+1} \left(\lim_{t \to 0} \frac{f(x_1 + tq_1(x), \ldots, x_n + tq_n(x)) - f(x)}{t}\right)$$

$$= j^{k+1} \left(q_1 \frac{\partial f}{\partial x_1} + \cdots + q_n \frac{\partial f}{\partial x_n}\right)$$

$$= \sum_{i=1}^{n} Q_i \overline{j^{k+1} \left(\frac{\partial f}{\partial x_i}\right)}^{k+1},$$

where Q_i is an arbitrary polynomial in I_n^{k+1}. Since (as with $4rx^3$ and αx^4 above) any terms of order k or $k+1$ in $j^k(\partial f/\partial x_i)$ disappear when we multiply by terms Q_i of order ≥ 2 and truncate above $k+1$, the tangent space to the orbit of $j^{k+1} f$ depends only on the polynomials $j^{k-1}(\partial f/\partial x_i)$, which depend only on $j^k f$.

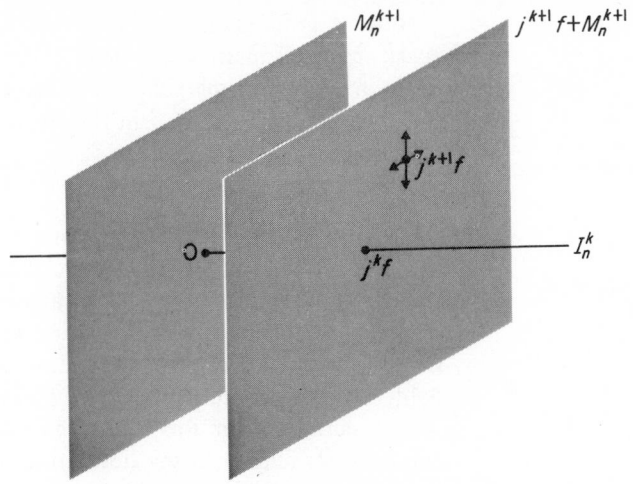

Fig. 8.4

Fig. 8.4 represents the subspace I_n^k of I_n^{k+1} rather schematically, by a line, and the subspace M_n^{k+1} by a plane. The set

$$K = j^{k+1}f + M_n^{k+1} = \{j^{k+1}f + Q \mid Q \text{ homogeneous of degree } k+1\}$$

is an affine (not vector) subspace of I_n^{k+1}, parallel to M_n^{k+1}. *If*, at any point κ in K,

> all directions tangent to K are tangent also to the
> orbit of κ under $\tilde{G}_n^k$, (8.3)

then it is obvious that K is actually contained in the orbit of $j^{k+1}f$. (It is also very easy to prove, given that the orbit is a manifold; and this is an elementary fact about Lie group actions.) If this is so, then the orbit of $j^{k+1}f$ includes the point $j^kf \in I_n^k$, considered as a $(k+1)$-jet (Fig. 8.4) in the obvious way. That is to say, some local transformation with the identity as linear part removes the terms of order $k+1$ from $j^{k+1}f$, reducing it to j^kf.

As we have seen, the presence of a vector in the tangent space to the orbit of $j^{k+1}f$ depends only on j^kf. Thus if (8.3) is true at j^kf, it is indeed true at every point of K. Thus to show that a change of coordinates exists which will perform the reduction, it suffices to show that every direction in M_n^{k+1} is in that tangent space. Algebraically, this is the statement that

$$M_n^{k+1} \subset \left\{ \overline{\sum_{i=1}^n Q_i j^{k-1}\left(\frac{\partial f}{\partial x_i}\right)}^{\,k+1} \;\middle|\; Q_i \in I_n^{k+1} \right\}. \qquad (8.4)$$

It is clearly a necessary condition as well as a sufficient one, for arbitrary terms of degree $k+1$ to be removable, because the latter implies directly that K is contained in the orbit, which implies that the tangent space M_n^{k+1} of K at each point is contained in that of the orbit.

Now, it is a remarkable fact that if condition (8.4) holds, then *every* smooth function $\mathbb{R}^n \to \mathbb{R}$ of order $k+1$ (of which the polynomials in M_n^{k+1} are the simplest examples only) can be expressed in some neighbourhood of 0 in the form

$$q_1 \frac{\partial f}{\partial x_1} + \cdots + q_n \frac{\partial f}{\partial x_n}$$

where the q_i are smooth functions of order ≤ 2. This statement about functions actually follows very easily from the corresponding statement (8.4) about jets, by an algebraic trick called Nakayama's lemma; and this in turn is easy to prove, but the language needed is that of modules over rings (generalizations of vector spaces over fields). We refer the reader to the exact sources. (Incidentally, most treatments for topologists give a needlessly messy proof of Nakayama's lemma, using determinants. *See* Wassermann [36] for the easier proof favoured by algebraists.)

From this it follows that if we work in the jet space I_n^{k+2001}, say, any polynomial whose terms are all of degree $\geq k+1$, $\leq k+2001$, lies in the tangent space

$$\left\{ \sum_{i=1}^{n} \overline{Q_i j^{k+2000}\left(\frac{\partial f}{\partial x_i}\right)}^{k+1} \;\middle|\; Q_i \in I_n^{k+2001} \right\}$$

to the orbit of $j^{k+2001}f$ under the action of $\tilde{G}_n^{k+2001}$ on I_n^{k+2001}. The arguments above then extend, to show that a suitable change of coordinates exists to give f the form

$$j^k f + O(k+2001).$$

By this route we can remove any finite number of higher terms. This line of argument does *not* show that we can remove the whole Tayl, but it does at least render it unsurprising. We have:

Theorem 8.1 *A function $f: \mathbb{R}^n \to \mathbb{R}$ is strongly k-determinate if and only if every homogeneous polynomial of degree $k+1$ can be expressed in the form*

$$\overline{Q_1 j^{k-1}\left(\frac{\partial f}{\partial x_1}\right)}^{k+1} + \cdots + \overline{Q_n j^{k-1}\left(\frac{\partial f}{\partial x_n}\right)}^{k+1}$$

with polynomials $Q_1, \ldots, Q_n$ in $(x_1, \ldots, x_n)$ of order ≥ 2.

This is a straightforward algorithmic criterion, and we give examples of its computational use in Section 13 and in many of the chapters on applications. For a full proof of the 'if' part *see* any of the exact sources; for the 'only if' *see* Siersma [33].

4   Weaker Determinacy Conditions

Evidently strong k-determinacy implies ordinary k-determinacy, since the latter just relaxes a condition on the derivative of the coordinate change. Thus the algebraic condition of Theorem 8.1 implies k-determinacy. However, k-determinacy does not imply the condition. What does it imply?

Let us consider examples. If we allow more general coordinate changes, though still fixing the origin, we must work with the matrix (8.2) of Section 2, without putting $\alpha = 0$. The directions of possible motion increase. For a point on the r-axis,

$$\begin{bmatrix} 1+2\alpha & 0 & 0 \\ 2\beta & 1+3\alpha & 0 \\ 2\gamma & 3\beta & 1+4\alpha \end{bmatrix}\begin{bmatrix} 0 \\ 0 \\ r \end{bmatrix} = \begin{bmatrix} 0 \\ 0 \\ (1+4\alpha)r \end{bmatrix}$$

so we can move the point up and down. For a point in the (q, r)-plane

$$\begin{bmatrix} 1+2\alpha & 0 & 0 \\ 2\beta & 1+3\alpha & 0 \\ 2\gamma & 3\beta & 1+4\alpha \end{bmatrix}\begin{bmatrix} 0 \\ q \\ r \end{bmatrix} = \begin{bmatrix} 0 \\ (1+3\alpha)q \\ 3\beta q+(1+4\alpha)r \end{bmatrix}$$

so if $q \neq 0$ we can move $(0, q, r)$ in any direction in this plane by suitable choices of α, β. Similarly we can move (p, q, r) in any direction at all (Fig. 8.5). These tangent spaces have bases $\{x^4\}$, $\{x^3, x^4\}$ and $\{x^2, x^3, x^4\}$ respectively.

Integrating these possible motions decomposes I_1^4 as in the last chapter, except that the $q < 0$ and $q > 0$ parts of the (q, r)-plane are distinguished. This reflects the fact that we cannot *continuously* turn x^3 into $-x^3$ by a time-dependent change of variable; the distinction is not important for present purposes or those of the previous chapter. (Notice that we can *revolve* $x^3 + y^2$ into $-x^3 + y^2$.)

Analysis following that of Section 2, but now with a time-dependent change of variable of the appropriate form

$$(y_1^t(x), \ldots, y_n^t(x)) = (x_1 + tq_1(x), \ldots, x_n + tq_n(x))$$

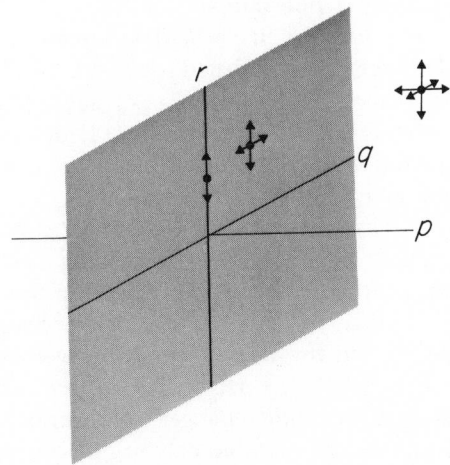

Fig. 8.5

where we require only that $q_1(0) = \cdots = q_n(0) = 0$, shows that we get a tangent space to the orbit of $j^{k+1}f$ under G_n^{k+1} consisting of vectors

$$\sum_{i=1}^{n} \overline{Q_i j^k \left(\frac{\partial f}{\partial x_i} \right)}^{k+1}$$

where this time $Q_1, \ldots, Q_n$ lie in J_n^{k+1}, that is, have no constant term.

Therefore this time what we find in the tangent space *does* depend on the terms of order $k+1$ in $j^{k+1}f$. For example, in I_1^4 we can move $(0, 0, r)$ up and down (Fig. 8.5) if and only if $r \neq 0$. (And indeed we would not expect to change the 4-jet of rx^4, $r \neq 0$ to 0 by any smooth change of coordinates.)

Putting $k + 1 = 4$, the fact that M_1^4 is contained in (and in fact *equals*) the tangent space to the orbit under G_1^4 of x^4, both spaces being just $\{\lambda x^4 \mid \lambda \in \mathbb{R}\}$, could hardly be expected to imply that x^4 is 3-determinate. However, k-determinacy of f does imply that $j^k f + M_n^{k+1}$ is contained in the orbit, hence in particular that the tangent space M_n^{k+1} at $j^{k+1}f$ to $j^{k+1}f + M_n^{k+1}$ is contained in that to the orbit. Thus we have

Theorem 8.2 *If $f: \mathbb{R}^n \to \mathbb{R}$ is k-determinate, then every homogeneous polynomial of degree $k+1$ can be expressed in the form*

$$\overline{Q_1 j^k \left(\frac{\partial f}{\partial x_1} \right)}^{k+1} + \ldots + \overline{Q_n j^k \left(\frac{\partial f}{\partial x_n} \right)}^{k+1}$$

with polynomials $Q_1, \ldots, Q_n$ in $(x_1, \ldots, x_n)$ of order ≥ 1.

This is proved in most of the exact sources. It follows trivially that if f is k-determinate, then every homogeneous polynomial of degree $k+2$ can be expressed in this way with Q_i of order ≥ 2. Hence by Theorem 8.1, f is *strongly* $(k+1)$-determinate, though it may not be strongly k-determinate.

In particular, if f is k-determinate then the algebraic condition of Theorem 8.1 is satisfied for $k' = k + 1$. Thus if the lowest k for which f satisfies that criterion is k_0, say, then f is k_0-determinate (in fact strongly so) and *may* be $(k_0 - 1)$-determinate (if it does not fail the condition of Theorem 8.2), but it cannot be $(k_0 - m)$-determinate for $m > 1$.

We have seen that Theorem 8.2 cannot be strengthened to 'if and only if', any function x^n providing a counter-example. But it can be shown that if M_n^{k+1} is contained in the tangent space to the orbit of $j^{k+1}f$ then the same is true, not necessarily for *all* points in $j^{k+1}f + M_n^{k+1}$, but for all such points near enough to $j^{k+1}f$. Consequently a piece of $j^{k+1}f + M_n^{k+1}$ lies in the orbit of $j^{k+1}f$, and arbitrary *sufficiently small* changes of order $k+1$ in f may be removed or produced by suitable changes of coordinates. (This need not mean that $j^{k+1}f$ can be reduced to $j^k f$, for the terms of order $k+1$ may not be sufficiently small.)

This leads us to define a function $f: \mathbb{R}^n \to \mathbb{R}$ to be *locally k-determinate* if there exists $\varepsilon \in \mathbb{R}$ such that every function $f + g$, where g is of order $k+1$ and satisfies

$$\left| \frac{\partial^{k+1} g}{\partial x_1^{k_1} \cdots \partial x_n^{k_n}} (0) \right| < \varepsilon \qquad \text{for all } k_i \text{ with } \quad k_1 + \cdots + k_n = k+1,$$

is locally equivalent to $j^{k+1}f$ by a smooth change of coordinates. We then have

Theorem 8.3   *A smooth function $f: \mathbb{R}^n \to \mathbb{R}$ is locally k-determinate if and only if every homogeneous polynomial of degree $k+1$ can be expressed in the form*

$$Q_1 \overline{j^k \left(\frac{\partial f}{\partial x_1} \right)}^{k+1} + \cdots + Q_n \overline{j^k \left(\frac{\partial f}{\partial x_n} \right)}^{k+1}$$

for polynomials $Q_1, \ldots, Q_n$ of order ≥ 1.

The proof of this result may be found in Martinet [31]. By patching together the small pieces of $j^{k+1}f + M_n^{k+1}$ that it says are each contained in some orbit, one can show that all of $j^{k+1}f + M_n^{k+1}$ is (under extra hypotheses) contained in the *same* orbit, and arrive at a recently published theorem of Stefan [37]:

Theorem 8.4   *A smooth function $f: \mathbb{R}^n \to \mathbb{R}$ is k-determinate if and only if for every homogeneous polynomial P of degree $k+1$, all homogeneous polynomials of degree $k+1$ can be expressed in the form*

$$Q_1 \overline{j^k \left(\frac{\partial}{\partial x_1} (f+P) \right)}^{k+1} + \cdots + Q_n \overline{j^k \left(\frac{\partial}{\partial x_n} (f+P) \right)}^{k+1}$$

with polynomials $Q_1, \ldots, Q_n$ of order ≥ 1.

This result is a convenient tool in deciding whether a function is k-determinate when it satisfies the criterion of Theorem 8.2 (if it doesn't, the theorem says it isn't) but not that of Theorem 8.1, which would guarantee k-determinacy via strong k-determinacy. Of course f is then strongly $(k+1)$-determinate, which may be more useful in some applications. We give an example of the way it is used in Section 13.

5   Transformations that Move the Origin

No function, obviously, can ever be 0-determinate. Once we are concerned with k-determinacy for $k \geq 1$, no addition γ to $j^k f$ of order greater than k will ever require us to move the origin to transform back to $j^k f$. If f is 1-determinate, that is, non-critical, the origin remains a nondegenerate point at which $f + \gamma$ takes the value $f(0)$. If it is k-determinate only for $k > 1$, it has a critical point at 0 which is necessarily isolated (Section 7), and does not move when we add higher terms. So the group G_n^k exploited in the last section comes from the most general transformations relevant to determinacy. But *stability* is another matter. When we perturb a function (or family of functions) we add terms of all orders, and may need to move the origin to get back the function or family we started with.

Returning to functions $\mathbb{R} \to \mathbb{R}$, consider a new family of changes of coordinates

$$y^t(x) = x + tq(x) = tq_0 + (1 + tq_1)x + tq_2 x^2 + \cdots.$$

This time the tangent vector at $j^{k+1}f$ that we get is

$$j^{k+1}\left(q\frac{df}{dx}\right) = q_0 j^{k+1}\left(\frac{df}{dx}\right) + q_1 x \overline{j^{k+1}\left(\frac{df}{dx}\right)}^{k+1}$$

$$+ \cdots + q_{k+1} x^{k+1} \overline{j^{k+1}\left(\frac{df}{dx}\right)}^{k+1}$$

$$= Q \overline{j^{k+1}\left(\frac{df}{dx}\right)}^{k+1}$$

where Q is an arbitrary polynomial, perhaps with a non-zero constant term q_0. The analogue for n variables is of course the vector

$$Q_1 \overline{j^{k+1}\left(\frac{\partial f}{\partial x_1}\right)}^{k+1} + \cdots + Q_n \overline{j^{k+1}\left(\frac{\partial f}{\partial x_n}\right)}^{k+1},$$

where the polynomials Q_i may have non-zero constant terms.

This is the most general sort of tangent vector we can find for $j^{k+1}f$, so we abuse language slightly and call the space of such vectors the *tangent space to* $j^{k+1}f$, suppressing reference to the orbit and the group. In mathematical language it is *generated by* the n jets

$$j^{k+1}\left(\frac{\partial f}{\partial x_1}\right), \ldots, j^{k+1}\left(\frac{\partial f}{\partial x_n}\right),$$

and the differentiation that produces them leads to the notation $\Delta_{k+1}(f)$ for it. It can equally be described as the space of $(k+1)$-jets of elements of $\Delta(f)$, where this consists of all locally defined functions of the form

$$q_1 \frac{\partial f}{\partial x_1} + \cdots + q_n \frac{\partial f}{\partial x_n}$$

where the q's are arbitrary smooth functions. (This is called the *Jacobian ideal* of f.) But we are concerned only with the finite-dimensional space $\Delta_{k+1}(f)$.

We can now express more succinctly the algebraic criteria above. Let the product $\overline{AB}^{k+1}$ of two subspaces A, B of any of E_n^{k+1}, J_n^{k+1} or I_n^{k+1} be defined as

$$\{\overline{a_1 b_1}^{k+1} + \cdots + \overline{a_s b_s}^{k+1} \mid a_i \in A, b_j \in B\},$$

the space of all linear combinations of truncated products of one polynomial from each. Then the tangent space to the orbit of $j^{k+1}f$ under G_n^{k+1} is $\overline{J_n^{k+1} \Delta_{k+1}(f)}^{k+1}$, and that to $\tilde{G}_n^{k+1}$ is $\overline{I_n^{k+1} \Delta_{k+1}(f)}^{k+1}$. Thus the algebraic conditions of Theorems 8.1 to 8.4 become

$$M_n^{k+1} \subseteq \overline{I_n^{k+1} \Delta_{k+1}(f)}^{k+1} \tag{8.1}$$

$$M_n^{k+1} \subseteq \overline{J_n^{k+1} \Delta_{k+1}(f)}^{k+1} \tag{8.2, 8.3}$$

$$M_n^{k+1} \subseteq \overline{J_n^{k+1} \Delta_{k+1}(f+P)}^{k+1} \quad \text{for all } P \in M_n^{k+1}, \tag{8.4}$$

respectively, and we use this notation henceforth.

Note that since $\Delta_{k+1}(f)$ is defined in terms of the $(k+1)$-jets of partial derivatives of f, we have by comparison of degrees of terms,

$$\Delta_k(f) = \Delta_k(j^{k+1}f),$$

but *not* necessarily

$$\Delta_k(f) = \Delta_k(j^k f).$$

It was the implicit assumption of the latter false identity that led to the erroneous instruction in Poston and Stewart [25] p. 98 to work with $j^k f$ in testing transversality. This error, which can occasionally give wrong answers, was pointed out to us by Robert Magnus.

Since multiplication by polynomials and truncation removes the effect of some terms, we have

$$\overline{I_n^{k+1} \Delta_{k+1}(f)}^{k+1} = \overline{I_n^{k+1} \Delta_{k+1}(j^k f)}^{k+1},$$

so that condition (8.1) above depends only on $j^k f$, as previously noted, whereas

$$\overline{J_n^{k+1} \Delta_{k+1}(f)}^{k+1} = \overline{J_n^{k+1} \Delta_{k+1}(j^{k+1} f)}^{k+1}$$

only, so that Theorems 8.2 and 8.3 depend, as we saw, on the terms in $j^{k+1}f$ of order $k+1$.

6 Tangency and Transversality

Rather than draw orbits for various f to which the 'tangent vectors' in $\Delta_{k+1}(f)$ are actually tangent (a project which, perhaps surprisingly, would take a lot more work and involve new concepts), we

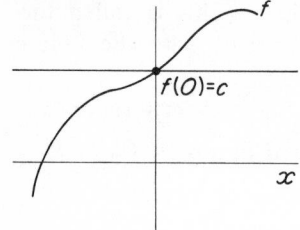

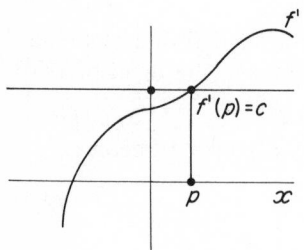

Fig. 8.6

illustrate the tangency by returning to the context of families of functions, as studied in Chapters 6 and 7.

Consider a function $f : \mathbb{R} \to \mathbb{R}$ with $\left.\dfrac{df}{dx}\right|_0 = 1$. Then

$$j^{k+1}\left(\frac{df}{dx}\right) = 1 + a_1 x + \cdots + a_{k+1} x^{k+1}.$$

Since this is not 0 in some neighbourhood of the origin, the function

$$x \mapsto \left(j^{k+1}\left(\frac{df}{dx}\right)\right)^{-1}$$

is smooth in this neighbourhood, and we may Taylor expand it to order $k + 1$, finding a polynomial

$$R(x) = 1 - a_1 x + (a_1^2 - a_2)x^2 + \cdots + q_{k+1}x^{k+1},$$

say, such that

$$\overline{R(x)j^{k+1}\left(\frac{df}{dx}\right)}^{\,k+1} = 1.$$

Since therefore 1 is in $\Delta_{k+1}(f)$, so is every polynomial P of degree $\leqslant k + 1$, because it can be given in the required form

$$\overline{Qj^{k+1}\left(\frac{df}{dx}\right)}^{\,k+1}$$

for a polynomial Q of degree $\leqslant k + 1$: we just take $Q = \overline{PR}^{\,k+1}$.

We interpret this to mean that every direction in which we can perturb f is one which we can reproduce, or reverse, by a change of coordinates. Indeed (Fig. 8.6) if $f(0) = c$, and f' is a small perturbation of f, we need only move the origin to the new locally unique point $p = f^{-1}(c)$ and smoothly reparametrize, and we obtain f again, with the same form and the same value at 0. Non-critical points are very stable, in that small perturbations may be removed using only reparametrizations of the domain.

Next take a function

$$f = c + x^2 + O(3).$$

A similar argument shows that $\Delta_{k+1}(f)$ is just J_1^{k+1}, the space of polynomials in x with zero constant term, for any $k \geqslant 2$. Correspondingly, any small perturbation can be corrected locally by a change of coordinates (perhaps moving the origin) *except that the value at the critical point may change* (Fig. 8.7). The absence of constant terms from $\Delta_{k+1}(f)$ means that a change in the critical value is not a direction in which we can move by changing x-coordinates alone. To get stability in Chapter 4 Section 5 we had to allow also changes in this 'constant'.

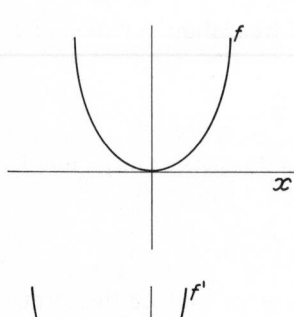

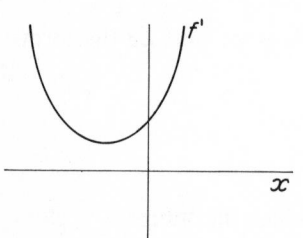

Fig. 8.7

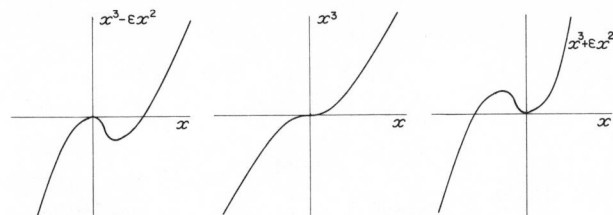

Fig. 8.8

Now consider $f(x) = x^3$ (or more generally any $f : \mathbb{R} \rightarrow \mathbb{R}$ reducible to this form: the tangency calculations give the corresponding answers, as they should, before and after change of coordinates). Evidently $\Delta_{k+1}(f)$ contains $x^2 = \dfrac{1}{3}\dfrac{\mathrm{d}f}{\mathrm{d}x}$, hence all of I_1^{k+1}. Tangency in the x^2-direction to x^3 can be most clearly seen if we embed a 'tangent family' $x^3 + sx^2$ (Fig. 8.8) in a larger family

$$f_{t,s} = x^3 + sx^2 + tx.$$

If we draw the catastrophe picture for this family (Fig. 8.9) we see that the s-axis is tangent to the curve in the (t, s)-plane above which the folds lie. Indeed, if we define

$$y = x + s/3$$
$$\tau = t - s^2/3$$

we find that

$$y^3 + \tau y - s^3/27 = x^3 + sx^2 + tx = f_{t,s}(x).$$

So $f_{(t,s)}$ reduces, by a change of (t, s)-coordinates and (t, s)-dependent change of x-coordinate, to the standard 'fold catastrophe' of Chapter 7 Section 2. The effect of s is reduced to a shear term $s^3/27$ which has no effect on the values of y for which $y^3 + \tau y - s^3/27$ is critical, or on the type of critical point.

If we embed $x^3 + sx^2$ in a still larger family

$$x^3 + sx^2 + t_1 x + t_0,$$

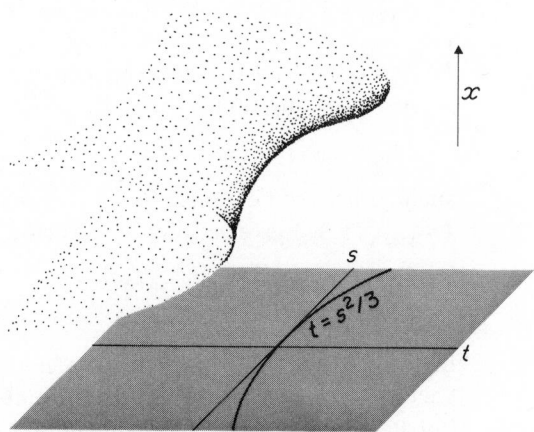

Fig. 8.9

we could set

$$y = x + s/3, \qquad \tau_1 = t_1 - s^2/3, \qquad \tau_0 = t_0 - s^3/27$$

and get

$$y^3 + \tau_1 y + \tau_0 = f_{t,s}(x)$$

exactly, with reference to s eliminated. This reflects the fact that x and 1 are the two polynomial directions 'missing' from $\Delta_{k+1}(x^3)$, for any k. But to forbid 'variable constants' like $s^3/27$ would prevent us from reducing perfectly typically encountered 1-parameter families like

$$x^3 + tx^2 + tx = (x + t/3)^3 + (t - t^2/3)x - t^3/27$$

to the standard local form. (Here this is achieved by setting $y = x + t/3$, $\tau = t - t^2/3$; notice that *globally* this includes two pure cube points $t = 0$, $t = 3$, corresponding to the way that a line $s = t$ in Fig. 8.9 would meet the fold curve twice.) So in order to make everything typically encountered by a single-parameter family equivalent locally to the three standard forms (linear, Morse, fold catastrophe) we must allow shear terms in the equivalence we use. This is reasonable in most applications, as we shall see. Indeed, in Chapter 11, on fluid flow, the physical interpretation is such that the effect of such a term is exactly nothing, so we may as well use it when it simplifies the results. Whenever the 'true' constant term really matters, we can put it back.

The proofs and the algebra work in exactly the same way whether we include such a term or not – indeed, they are slightly simpler when we do, since we can then leave the shear term out of the equivalence relation. Including it raises all codimensions except that of a non-critical point by 1, since it means working in E_n^k rather than J_n^k. Martinet [31] takes codimension in this sense – raised by one (numerically) and by 1 (as cobasis element). So does the computer routine of Appendix 1.

To allow systematic suppression of such constant terms, we write

$$J^{k+1}f = j^{k+1}f - j^0 f$$

for the $(k+1)$-jet of f with its constant term removed.

Now take any family $F_{t_1, \ldots, t_N}$ of functions around

$$F_{0, \ldots, 0}(x) = f(x) = x^3$$

such that some line $\{(st_1, \ldots, st_N) \mid s \in \mathbb{R}\}$ for a particular $t = (t_1, \ldots, t_N)$ gives a 'direction vector'

$$v = J^{k+1}\left(\lim_{s \to 0} \frac{F_{st_1, \ldots, st_N}(x) - x^3}{s}\right)$$

in the jet space J_1^{k+1} which does *not* lie in the space $\Delta_{k+1}(x^3)$ of vectors tangent to x^3. (It thus represents the 'one missing direction', but it need not be x: it might merely have non-zero linear term.)

The great theorem of the subject (Theorem 8.6 in the next section) guarantees the following. By a smooth change of coordinates from $(t_1, \ldots, t_N)$ to $(s_1, \ldots, s_N)$, and a $(t_1, \ldots, t_N)$-dependent smooth change of coordinates from x to y, the family $F_{t_1, \ldots, t_N}$ can be reduced to the form

$$F_{s_1, \ldots, s_N}(y) = y^3 + s_1 y,$$

up to a shear term. The change of coordinates has turned the tangent directions to the collection of fold points into axes, and straightened the collection to lie neatly along them. If F does not meet the condition above, as $F(s, x) = x^3 + sx^2$ does not, we can embed it as above in a family that does, straighten, and project: if $\tau(s) = -s^2/3$ and $y = x + s/3$ then $y^3 + \tau(s)y$ gives F perfectly, up to a shear term. We have reduced the description of F to that of a special curve in the variable t in the standard family $x^3 + tx$.

Even for this simplest non-trivial case, for functions of one variable, there seems to be no simple proof like those given for the determinacy criteria of Chapter 4 Section 3. Any attempt to prove the above 'barehanded', even with only one t-variable, runs into the Malgrange preparation theorem (exact sources) or, equivalently, the division theorem. Indeed, investigating such cases is a good way to gain insight into these subtle theorems. *See* Martinet [31] for such a discussion: it is quite outside our present scope. It should be remarked that the need for the preparation theorem in this context was first perceived by Thom. Malgrange initially did not believe it was true, and only considerable pressure from Thom first convinced him that it was and then persuaded him to prove it. (This illustrates why Section 7 of the previous chapter is titled 'Thom's theorem'. Thom did not publish the first proof – he has not published any proof – but he perceived the result and orchestrated the proving of it.) Thus a central result of the theory could, when first proposed, be disbelieved by a major expert in the field. Catastrophe theory, as we have insisted throughout this book, is an extension of the calculus, or a development within it, and not (as sometimes claimed) a departure like Newton's from previous descriptions of the world. But it is by no means a trivial or routine development.

7 Codimension and Unfoldings

We can now give a precise definition of the *codimension* of $f : \mathbb{R}^n \to \mathbb{R}$ at 0 as

$$\text{cod}(f) = \max_{k \geqslant 1} (\text{codimension of } \Delta_{k+1}(f) \text{ in } J_n^{k+1})$$

$$= \max_{k \geqslant 1} (\dim J_n^{k+1} - \dim \Delta_{k+1}(f)),$$

the number of 'missing polynomial directions'. The number $\dim J_n^{k+1} - \dim \Delta_{k+1}(f)$ can only increase with k: if it never stops increasing, we set $\text{cod}(f) = \infty$.

If $M_n^{k+1} \subseteq \Delta_{k+1}(f)$, which is certainly true by Theorem 8.2 if f is k-determinate, no directions of order $k+1$ are missing from $\Delta_{k+1}(f)$. It follows that no polynomial of order k is missing from $\Delta_{k+m}(f)$, appealing once more to Nakayama's lemma. Thus all the missing directions are represented by polynomials of degree $\leq k$. The codimension of f at 0 is then given exactly by that of $\Delta_k(f)$ in J_n^k, or $\Delta_{k+1}(f)$ in J_n^{k+1}.

If f is not k-determinate for any k, there are missing polynomials of all orders, and the codimension is infinite. Thus finite determinacy and finite codimension are equivalent conditions.

One importance of codimension, as defined above, is that it corresponds exactly to the kind of geometric codimension used in transversality arguments in the previous chapter. That codimension can thus be computed by this definition, rather than by finding geometrical descriptions such as the cone and the umbilic bracelet. The computations are quicker and more systematic, and are thus the standard method in the exact sources. Thom's transversality theorem guarantees that an r-parameter family typically includes only functions which have codimension $\leq r$ at each point, as long as there are no continuous families of types (such as the quartics classified by cross-ratio in Chapter 7 Section 6) which as *collections* have codimension $\leq r$. It is easy to compute that any $f(x, y)$ of order 4 has codimension at least 8; but if a 7-dimensional family includes such a point, then typically every nearby family does too – though not necessarily a point equivalent to the first one. (We said this in Chapter 7 Section 6, but it is so regularly misunderstood that it is worth saying again.) The *set* of singular points with the first three derivatives vanishing in two directions has codimension 7, that with the first two derivatives vanishing in three directions has codimension 6, but individuals in these sets have codimensions at least 8 and 7 respectively. (We do not give general methods to compute codimensions of sets like this, since the general problem of dividing jet spaces up into such sets is unsolved: however Thom has proved that such a subdivision or 'stratification' exists.)

The set of functions that are not finitely determined, though, has codimension ∞ *as a set*, as do its individual members. Thus a typical r-parameter family, for any finite r, includes only functions everywhere k-determinate for finite k. Since a k-determinate singularity is isolated (the polynomial $(x_1^2 + \cdots + x_n^2)^k$, which vanishes only at $(0, 0, 0 \ldots 0)$ must be in $\Delta_{2k}(f)$, thus expressible in terms of $\partial f/\partial x_1, \ldots, \partial f/\partial x_n$, which therefore cannot vanish simultaneously except *at* 0) non-isolated singularities share the extreme atypicality of indeterminate ones. This justifies our remarks about their scarcity in Section 1 of Chapter 4.

This is one rôle for the codimension, and an important one (though subject to the usual cautions on typicality arguments of Section 3 of Chapter 6). It may be thought of as working *inwards* to the type of function we can stably find, given the dimension r of the family that we have. (For example, one function at each point in an embryo, at each moment of its history, gives a 4-parameter family. This will typically, and stably, include only the original 'seven elementary catastrophes' of codimension ≤ 4.) Most of Zeeman's published applications argue in this direction.

However in physics, chemistry, engineering, etc., theory often gives us a *particular* function, at least up to order k. If it comes from a special situation, or analysis of an 'ideal' system, it may or may not be typical or stable, but we must be able to study it. This leads us to work *outward* from the given function, analysing (for example) perturbations of it.

We have seen that a small perturbation of a Morse critical point leads to only one critical point near the original one. An x^3 point can break up into two (a minimum and a maximum) or none, an x^4 point into three, and so on. These facts illustrate a beautiful result:

Theorem 8.5 *Suppose $f: \mathbb{R}^n \to \mathbb{R}$ has codimension c at 0. Then any sufficiently small perturbation f' of f has at most $c + 1$ critical points near 0.*

We shall not attempt a proof, or even a plausibility argument for this (though the reader accustomed to the Fundamental Theorem of Algebra will easily construct one in the one-variable case); a proof is given in Palamodov [38], together with precise replacements for our 'small' and 'near 0'.

This theorem illustrates how codimension measures the complexity of a critical point. Attempted or assumed symmetries generally increase codimension. For instance, suppose $f(x, y)$ is symmetric under a 90° turn, as is common in the theory of an 'ideal' system. If f is not Morse at $(0, 0)$ then it is of order 4 in (x, y), since the quadratic part if not Morse vanishes by symmetry, and so does the cubic term. Then f has codimension at least 8. The simplification obtained by analysing the 'ideal' system is at the expense of far more complicated behaviour associated with small 'imperfections' as the function is changed a little. This is particularly striking when a *continuous* symmetry leads to *infinite* codimension, as in the context of floating oil-rigs in Chapter 10 Section 10.

Theorem 8.5 tells us that somehow inside the degenerate singularity f are $\operatorname{cod}(f) + 1$ points waiting to be shaken loose into separate criticalities by a small perturbation.† More systematically, we can apply a smooth *family* of perturbations, such as the perturbations $x^3 + tx$ of x^3 and $x^4 + t_2 x^2 + t_1 x$ of x^4 already studied. Such a family is called an *unfolding* of f, which notion, like so much else, is due to Thom. We have seen that the unfolding $x^3 + tx$ captures the effects of all possible unfoldings, and *a fortiori* of all small perturbations. Similarly the cusp catastrophe gives all the types near anything equivalent to x^4, and the geometry of how they develop as we

† Whether all $\operatorname{cod}(f) + 1$ can be achieved by a *real* perturbation seems to be an open question. For the *complex* case they can.

continuously perturb the function. We can construct a similar family for any function which is finitely determined at 0.

To be more exact about this, we make some definitions. An *r-unfolding* of a function $f : \mathbb{R}^n \to \mathbb{R}$ is a function

$$F : \mathbb{R}^{n+r} \to \mathbb{R}$$

such that

$$F(x_1, \ldots, x_n, 0, \ldots, 0) = f(x_1, \ldots, x_n).$$

We often, as above, denote $F(x_1, \ldots, x_n, t_1, \ldots, t_r)$ by $F_{t_1, \ldots, t_r}(x_1, \ldots, x_n)$ and think of F as a family of functions $\mathbb{R}^n \to \mathbb{R}$, parametrized by t. We call $x_1, \ldots, x_n$ the *internal variables*, $t_1, \ldots, t_r$ the *unfolding variables*, r the *unfolding dimension*, and $\mathbb{R}^r$ the *unfolding space*.

A *d*-unfolding $\bar{F}$ of f is *induced* from F by three mappings:

(a) a smooth mapping

$$e : \mathbb{R}^d \to \mathbb{R}^r$$

$$(s_1, \ldots, s_d) \mapsto (e_1(s), \ldots, e_r(s));$$

(b) an *s*-dependent local change of coordinates in $\mathbb{R}^n$, that is, a mapping

$$y : \mathbb{R}^{n+d} \to \mathbb{R}^n$$

$$(x, s) \mapsto (y_1(x, s), \ldots, y_n(x, s))$$

such that for each s the map

$$y_s : x \mapsto y(x, s)$$

is a local diffeomorphism around 0; and

(c) a shear function

$$\gamma : \mathbb{R}^d \to \mathbb{R},$$

provided

$$\bar{F}(x, s) = F(y_s(x), e(s)) + \gamma(s)$$

around 0 in $\mathbb{R}^{n+d}$.

Hence each small perturbation $\bar{F}_s$ is a smooth reparametrization, plus a constant, of a perturbation we already have as $F_{e(s)}$. If e is a diffeomorphism, this is just the equivalence of families of functions that we have used for some time (Chapter 6 Section 1). But it includes more general situations; for instance, we saw above that the unfolding $x^3 + sx^2$ could be induced from $x^3 + tx$ by setting

$$e : \mathbb{R} \to \mathbb{R}, \qquad s \mapsto -s^2/3$$

$$y : \mathbb{R}^{1+1} \to \mathbb{R}, \qquad (x, s) \mapsto x + s/3$$

$$\gamma : \mathbb{R} \to \mathbb{R}, \qquad s \mapsto -s^3/27.$$

An *r-unfolding* of f is *versal* if all other unfoldings of f can be induced from it. It is *universal* if r is the smallest dimension for which a versal r-unfolding of f exists. Two universal unfoldings are thus equivalent (but not uniquely).

If F is an r-unfolding of f, set

$$v_1^k(F) = \frac{\partial}{\partial t_1}(J^k F_{t_1, 0, \ldots, 0})$$

$$v_2^k(F) = \frac{\partial}{\partial t_2}(J^k F_{0, t_2, 0, \ldots, 0})$$

$$\ldots$$

$$v_r^k(F) = \frac{\partial}{\partial t_r}(J^k F_{0, \ldots, 0, t_r}).$$

Let $V^k(F)$ be the subspace of J_n^k spanned by $v_1^k(F), \ldots, v_r^k(F)$. (Recall that $J^k F$ is the k-jet of F minus its constant term.)

Theorem 8.6 *An r-unfolding F of f, where f is k-determinate, is versal if and only if $V^k(F)$ and $\Delta_k(f)$ are transverse subspaces of J_n^k. It is universal if and only if it is versal and $r = \text{cod}(f)$.*

Again we refer to the exact sources for proof. However, in the next section we illustrate in finite dimensions the close relation between *trans*versality and *uni*versality.

As a corollary, we see that any two r-unfoldings F and G of f, for which both $V^k(F)$ and $V^k(G)$ are transverse to $\Delta_k(f)$, are equivalent as families of functions. Moreover, if F is an r-unfolding and G is an s-unfolding, $r > s$, both being versal, then F is equivalent to

$$\hat{G} : \mathbb{R}^{n+r} \to \mathbb{R}$$
$$(x_1, \ldots, x_n, t_1, \ldots, t_r) \mapsto G(x_1, \ldots, x_n, t_1, \ldots, t_s).$$

This allows us to remove the *dummy* external variables $t_{s+1}, \ldots, t_r$, as we did in Chapter 7 Section 4. In fact, we can always cut down to a universal unfolding, removing the maximum number of dummy variables, in this way.

Algebraically we may easily *construct* a universal unfolding for a k-determinate f of codimension c by choosing a cobasis (see Chapter 2 Section 2) $v_1, \ldots, v_c$ for $\Delta_k(f)$ in J_n^k and setting

$$F(x_1, \ldots, x_n, t_1, \ldots, t_c) = f(x) + t_1 v_1(x) + \cdots + t_c v_c(x).$$

For example, consider a function

$$f(x, y) = x^2 y + y^3 + O(4)$$

such as (by Chapter 7 Section 5) we may expect to find stably in a three-parameter family. We have

$$\frac{\partial f}{\partial x}(x, y) = 2xy + O(3), \qquad \frac{\partial f}{\partial y}(x, y) = x^2 + 3y^2 + O(3).$$

Thus

$$\overline{I_2^4 \Delta_4(f)}^4$$

consists of all linear combinations of

$$
\begin{aligned}
j^4(x^2(2xy + O(3))) &= 2x^3y \\
j^4(xy(2xy + O(3))) &= 2x^2y^2 \\
j^4(y^2(2xy + O(3))) &= 2xy^3 \\
j^4(x^2(x^2 + 3y^2 + O(3))) &= x^4 + 3x^2y^2 \\
j^4(xy(x^2 + 3y^2 + O(3))) &= x^3y + 3xy^3 \\
j^4(y^2(x^2 + 3y^2 + O(3))) &= x^2y^2 + 3y^4.
\end{aligned}
$$

Clearly these combinations include all homogeneous polynomials in M_2^4, hence by Theorem 8.1 f is strongly 3-determinate. Work therefore from now on in coordinates such that f takes exactly the form $x^2y + y^3$, to save reference to higher terms. *In these coordinates,*

$$\frac{\partial f}{\partial x} = 2xy \qquad \frac{\partial f}{\partial y} = x^2 + 3y^2$$

and $\Delta_3(f)$ consists of all linear combinations of

$$
\begin{aligned}
J_1 &= 2xy \\
J_2 &= x^2 + 3y^2 \\
J_3 &= \tfrac{1}{2}x(2xy) = x^2y \\
J_4 &= \tfrac{1}{2}y(2xy) = xy^2 \\
J_5 &= x(x^2 + 3y^2) - \tfrac{3}{2}y(2xy) = x^3 \\
J_6 &= \tfrac{1}{3}y(x^2 + 3y^2) - \tfrac{1}{6}x(2xy) = y^3.
\end{aligned}
$$

Hence $\Delta_3(f)$ contains all cubic polynomials, and two dimensions out of the three quadratic ones. There is no canonical choice of cobasis, but a convenient one is

$$v_1(x, y) = x \qquad v_2(x, y) = y \qquad v_3(x, y) = x^2 + y^2.$$

Then we have all linear terms as combinations of v_1 and v_2, and all quadratic terms, since

$$x^2 = \tfrac{1}{2}(3v_3 - J_2) \qquad xy = \tfrac{1}{2}J_1 \qquad y^2 = \tfrac{1}{2}(J_2 - v_3).$$

(Notice that $x^2 + 3y^2$ would *not* have worked as a choice for v_3.) Thus a universal unfolding is

$$x^2y + y^3 + t_3(x^2 + y^2) + t_2y + t_1x$$

which we met in Section 5 of Chapter 7.

When we look at special cases like this, the equivalence of *all* versal unfoldings can become a little surprising. For example, $x^3 + y^3$ is another cubic in x, y with one real, two complex, root lines, hence is equivalent to x^2y+y^3 by a linear change of variable by Chapter 2 Section 6. Hence also it is 3-determinate, so we may work with this polynomial form. This time $\Delta_3(f)$ consists of linear combinations of

$$J_1 = \frac{\partial f}{\partial x} = 3x^2 \qquad J_4 = \tfrac{1}{3}yJ_1 = x^2y$$

$$J_2 = \frac{\partial f}{\partial y} = 3y^2 \qquad J_5 = \tfrac{1}{3}xJ_2 = xy^2$$

$$J_3 = \tfrac{1}{3}xJ_1 = x^3 \qquad J_6 = \tfrac{1}{3}yJ_2 = y^3$$

and the obvious cobasis is

$$v_1 = x \qquad v_2 = y \qquad v_3 = xy.$$

Indeed Thom [1] chooses the standard form x^3+y^3 and unfolds it this way to give the hyperbolic umbilic catastrophe

$$x^3 + y^3 + t_3xy + t_2y + t_1x.$$

This time the quadratic unfolding term is *indefinite*, a *saddle* form, as against the positive definite term that we took before. When the unfolding parameters acquire physical meaning (for instance, a term $t_1(x^2+y^2)$ will often in engineering reflect variation of *load*) these differences can appear rather strange. However, if we set

$$e : \mathbb{R}^3 \to \mathbb{R}^3$$
$$(p, q, r) \mapsto (A, B, C)$$

where

$$A = -2^{2/3}p$$
$$B = 2^{-2/3}(q + r\sqrt{3})$$
$$C = 2^{-2/3}(q - r\sqrt{3}),$$

also

$$y : \mathbb{R}^5 \to \mathbb{R}^2$$
$$(x, y, p, q, r) \mapsto (u, v)$$

where

$$u = 2^{-1/3}(y + x/\sqrt{3} + 2p/3)$$
$$v = 2^{-1/3}(y - x/\sqrt{3} + 2p/3),$$

and

$$\gamma : \mathbb{R}^3 \to \mathbb{R}$$
$$(p, q, r) \mapsto -4p^3/27 + 2pq/3$$

then multiplying out we find that

$$u^3 + v^3 + Auv + Bu + Cv - \gamma(p, q, r)$$
$$= x^2y + y^3 + p(x^2 + y^2) + qy + rx.$$

It is easy to check that y defines a (p, q, r)-dependent diffeomorphism $\mathbb{R}^2 \to \mathbb{R}^2$ (globally, not just near 0) and that e and y define a diffeomorphism from the whole of $\mathbb{R}^3 \times \mathbb{R}^2$ to itself, mapping one polynomial model of the hyperbolic umbilic into the other.

This transformation is linear, and therefore not hard to find *once you know it must be there*. Without the theorem one might never suspect the equivalence, at least not till more has been learned about each form. (They seemed very different to engineers we have talked to.) Given that by universality the transformation exists, it can be found by a little experiment, and the analysis of one form can be translated directly into the other. Particularly since in low dimensions there are few possible types, this makes a knowledge of the geometry of the standard forms usefully transferable into descriptions of particular examples where they occur. (We discuss some 'catastrophe-spotting' methods by which the types can be identified using simple calculations, in Section 10.) For instance in Chapter 15 the fact that two laser problems with very different boundary conditions reduce to the same standard form (a result surprising from the physics, but natural in the light of Chapter 7) permits a lot of quantitative information to be shuttled between them.

8 Transversality and Universality

We cannot prove Theorem 8.6 at the level of this book, but we can again give a finite-dimensional analogue, showing how transversality guarantees that things very like unfoldings can be transformed into each other. To do this we could take the group G_1^4 of matrices of the form in equation (8.1) above (*see* Section 2), or the general case of G_n^k acting on J_n^k, or indeed any Lie group acting on a manifold; but the method may be illustrated equally well for the simpler situation of the group G of matrices

$$\begin{bmatrix} 1+a & 0 \\ b & 1+a \end{bmatrix} \quad (a \neq -1)$$

acting on $\mathbb{R}^2$. Readers familiar with Lie group actions will find it an easy but worthwhile exercise to generalize to the wider setting.

Since

$$\begin{bmatrix} 1+a & 0 \\ b & 1+a \end{bmatrix}\begin{bmatrix} x \\ y \end{bmatrix} = \begin{bmatrix} x+ax \\ y+bx+ay \end{bmatrix}$$

any two points with x non-zero can be transformed to each other; and the non-zero points on the y-axis, and the origin, give the other two 'types' of points under these transformations. Consider a curve

$$f : \mathbb{R} \to \mathbb{R}^2$$

with $f(0) = (0, Y)$, a point on the y-axis (Fig. 8.10). We wish to

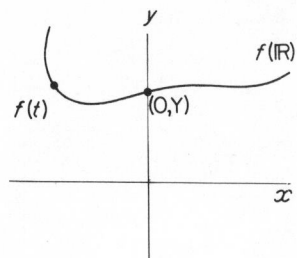

Fig. 8.10

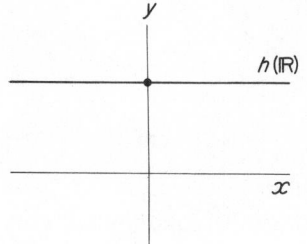

Fig. 8.11

transform f into a 'standard' curve through the y-axis (Fig. 8.11)

$$h : \mathbb{R} \to \mathbb{R}^2, \qquad t \mapsto (t, 1),$$

by a smooth invertible change of coordinates $t \mapsto e(t)$ on $\mathbb{R}$, and a t-dependent matrix $g_t \in G$ such that

$$g_t(f(e(t))) = h(t).$$

(In the analogy with an unfolding, g_t represents a t-dependent change of x-variable (what we previously called y) and the addition of a t-dependent 'shear'.) To prove this can be done (not uniquely) define

$$F : \mathbb{R}^3 \to \mathbb{R}^2$$

$$(a, e, t) \mapsto \begin{bmatrix} 1+a & 0 \\ 0 & 1+a \end{bmatrix} \begin{bmatrix} f_1(e) \\ f_2(e) \end{bmatrix} - \begin{bmatrix} h_1(t) \\ h_2(t) \end{bmatrix} = \begin{bmatrix} (1+a)f_1(e)-t \\ (1+a)f_2(e)-1 \end{bmatrix}.$$

The Jacobian of this is

$$\begin{bmatrix} f_1(e) & (1+a)\dfrac{df_1}{de}\Big|_{(a,e,t)} & -1 \\[2ex] f_2(e) & (1+a)\dfrac{df_2}{de}\Big|_{(a,e,t)} & 0 \end{bmatrix}$$

which reduces at $(a, 0, 0)$ to

$$\begin{bmatrix} 0 & (1+a)\dfrac{df_1}{de}\Big|_{(a,0,0)} & -1 \\[2ex] Y & (1+a)\dfrac{df_2}{de}\Big|_{(a,0,0)} & 0 \end{bmatrix} = \begin{bmatrix} 0 & P & -1 \\ Y & Q & 0 \end{bmatrix}, \text{ say.}$$

Now the linear equation

$$\begin{bmatrix} 0 & P & -1 \\ Y & Q & 0 \end{bmatrix} \begin{bmatrix} a \\ e \\ t \end{bmatrix} = \begin{bmatrix} 0 \\ 0 \end{bmatrix}$$

defines a line L (rather than a plane) provided Y or Q is non-zero, so that the matrix has rank 2. The line L (Fig. 8.12) can be described as the graph of a linear function

$$\mathbb{R} \to \mathbb{R}^2$$

$$t \mapsto (-Qt/PY, t/P)$$

if and only if both P and Y are non-zero. If they are, then the Implicit Function Theorem (Chapter 3 Section 9) guarantees that near the point $((1-Y)/Y, 0, 0)$ where F vanishes (that is,

$$\begin{bmatrix} 1/Y & 0 \\ 0 & 1/Y \end{bmatrix} \begin{bmatrix} 0 \\ Y \end{bmatrix} = \begin{bmatrix} 0 \\ 1 \end{bmatrix})$$

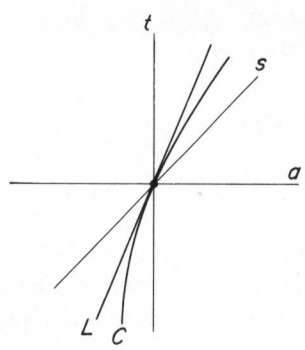

Fig. 8.12

there is a unique smooth function

$$\mathbb{R} \to \mathbb{R}^2$$

$$t \mapsto (a(t), e(t))$$

defined around 0, whose graph (the curve C in Fig. 8.12) is tangent to the line L, and such that

$$F(a(t), e(t)) = 0$$

for all t around 0, that is,

$$\begin{bmatrix} 1+a(t) & 0 \\ 0 & 1+a(t) \end{bmatrix} \begin{bmatrix} f_1(e(t)) \\ f_2(e(t)) \end{bmatrix} - \begin{bmatrix} h_1(t) \\ h_2(t) \end{bmatrix} = \begin{bmatrix} 0 \\ 0 \end{bmatrix}.$$

If we set

$$g_t = \begin{bmatrix} 1+a(t) & 0 \\ 0 & 1+a(t) \end{bmatrix}$$

we have exactly $g_t(f(e(t))) = h(t)$, as required.

Now the conditions $Y \neq 0$, $P \neq 0$ reduce to the conditions $Y \neq 0$, $\dfrac{\mathrm{d}f_1}{\mathrm{d}t}\bigg|_0 \neq 0$, which are exactly what we require for f to pass transversely through the orbit of $f(0) = (0, Y)$. (If $Y = 0$, the orbit is the origin, and no curve in $\mathbb{R}^2$ can pass through this transversely; and the condition $\dfrac{\mathrm{d}f_1}{\mathrm{d}t}\bigg|_0 \neq 0$ says precisely that f is going *through* – regardless of its speed 'up' or 'down' – at non-zero speed.) This is perfectly general: if a Lie group acts smoothly on $\mathbb{R}^n$ (indeed on any manifold M) and $h : \mathbb{R}^r \to M$ has $h(0) = x$, we can use the Implicit Function Theorem to show that h is universal in a sense generalizing that above, *if and only if* it is transverse to the orbit of x. The general proof involves little more than rewriting the above argument in more stylish notation.

The proof of Theorem 8.6, however, is not as close an analogue to this as those of Theorems 8.1 to 8.4 are to the finite-dimensional discussions preceding them. At least, not yet. In some ways the Malgrange preparation theorem has the flavour of an 'implicit function theorem', and considerable efforts are being made to formulate and prove an appropriate infinite-dimensional implicit function theorem whose use would make the proof of Theorem 8.6 look *just* like the above. Such a formulation would be a powerful tool for extending the theory, as well as possessing important pedagogical advantages.

9   Strong Equivalence of Unfoldings

Theorem 8.6 guarantees in particular the following. If F is an r-unfolding of f, where f is k-determinate and has codimension r,

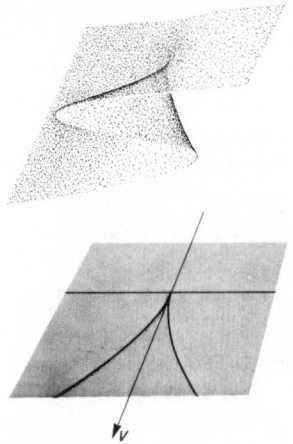

Fig. 8.13

and if $V^k(F)$ and $\Delta_k(f)$ are transverse subspaces of J_n^k, then F is equivalent to the 'truncated' unfolding, which we denote by $\bar{F}^k$, where

$$\bar{F}^k(x_1, \ldots, x_n, t_1, \ldots, t_r) = j^{k+1}f + t_1 v_1^k(x) + \cdots + t_r v_r^k(x),$$

since they are universal unfoldings of equivalent functions. (We could use $j^k f$ on the right if the result is versal, by the *separate* test of $\Delta_k(j^k f)$ being transverse to $V^k(f)$. Or we could apply $v_1^k, \ldots, v_r^k$ not to x, but to new coordinates $y(x)$ which remove the terms of order $k+1$ from f: by k-determinacy this can even be done using polynomials.) But, as in Section 1 where we introduced *strong* k-determinacy, there are important directions in the unfolding space, and we may wish to keep track of them. For instance, in most applications of the cusp catastrophe the direction v of the cusp branching (Fig. 8.13) is crucial. For exact applications, therefore, the following concept is computationally useful.

Two r-unfoldings of $f: \mathbb{R}^n \to \mathbb{R}$ are *strongly equivalent* if one is induced from the other by mappings e, y, γ as above (Section 7), such that the local diffeomorphism

$$\mathbb{R}^{n+r} \to \mathbb{R}^{n+r}$$

$$(x_1, \ldots, x_n, t_1, \ldots, t_r) \mapsto (y_1(x, t), \ldots, y_n(x, t), e_1(t), \ldots, e_r(t))$$

has derivative the identity at $(0, \ldots, 0)$.

Theorem 8.7 *Suppose $f: \mathbb{R}^n \to \mathbb{R}$ strongly k-determinate, $k \geq 3$. As a corollary of Theorem 8.1, one of the following holds:*

    (a) $M_n^{k-1} \subseteq \Delta_{k+1}(f)$

    (b) $M_n^k \subseteq \Delta_{k+1}(f)$

    (c) $M_n^{k+1} \subseteq \Delta_{k+1}(f)$.

Then a versal unfolding F of f is strongly equivalent to the unfolding

$$j^p f(x) + t_1 J^q\left(\frac{\partial}{\partial t_1} F_{t_1, 0, \ldots, 0}\right) + \cdots + t_r J^q\left(\frac{\partial}{\partial t_r} F_{0, \ldots, 0, t_r}\right)$$

if, correspondingly,

    (a) $p \geq 2k - 3, q \geq k - 2$,

    (b) $p \geq 2k - 2, q \geq k - 1$,

    (c) $p \geq 2k - 1, q \geq k$.

Note that the calculations, to *check* the conditions, involve no more derivatives than before. It is in *using* the result that we must compute with terms up to orders p and q. This result is due to Magnus [101]. We shall illustrate in the sequel (Chapter 12 Section 1, Chapter 14 Section 1) its utility for applications.

10 Numbers Associated with Singularities

As remarked in Section 7, it is much easier to find a transformation reducing a catastrophe (arising from the theory of some system) to standard form if one knows in advance which form to aim at. Otherwise one requires all the polynomial algebra needed for the classification. In fact, the same *linear* criteria that apply in Theorems 8.1 to 8.4 produce numerical invariants associated with the singularities that the catastrophes unfold, which are independent of the choice of coordinates, and which distinguish the lower cases almost completely.

We have already defined the *corank, codimension* and *determinacy* of $f : \mathbb{R}^n \to \mathbb{R}$. The last, by Theorem 8.4, may be defined algebraically as

$$\min \{k \mid M_n^{k+1} \subseteq \overline{J_n^{k+1} \Delta_{k+1}(j^k f + P)}^{k+1} \quad \text{for all} \quad P \in M_n^{k+1}\}.$$

Analogously, we use Theorem 8.1 to define the *strong determinacy* of f to be

$$\min \{k \mid M_n^{k+1} \subseteq \overline{I_n^{k+1} \Delta_{k+1}(j^k f)}^{k+1}\}.$$

These four numbers are listed in Table 8.1 below. As may be seen, the distinct cases up to codimension 5 are completely separated by these numbers, except for the pairs $D_4^{\pm}$, $D_6^{\pm}$. (This is not surprising, for in these cases the numbers are exactly the same as in the

TABLE 8.1. *Invariants of the Catastrophes of Codimension* $\leqslant 5$

Symbol	Corank	Codimension	Determinacy	Strong determinacy
A_2	1	1	3	3
A_3	1	2	4	4
A_4	1	3	5	5
A_5	1	4	6	6
A_6	1	5	7	7
D_4^-	2	3	3	3
D_4^+	2	3	3	3
D_5	2	4	4	4
D_6^-	2	5	5	5
D_6^+	2	5	5	5
E_6	2	5	4	4

There are many other numbers – right–left codimension, Coxeter number, Dolgachev numbers, and so forth – which become significant in higher codimensions, but we do not here include any that we have not specified how to calculate. An extensive survey is given in Arnol'd [40], taking the complex classification up to codimension 14. The reader is strongly recommended to look this paper up, if only in order to see (a) how complicated the classification becomes, and (b) that it is still possible.

complex version of the theory, where these pairs merge into D_4, D_6.) For completeness we indicate here a 'low level' procedure to distinguish these cases.

Often a little extra information, readily available, is enough, since the phenomenologies of the $+$ and $-$ cases are so different. For example, D_4^+ and D_6^+ admit small perturbations of f that have no zeros of Df near the origin, whereas D_4^- and D_6^- do not. Berry, as described in Chapter 12, Section 9, identified a D_4^- arising in scattering theory by the section with a 'corner', shown by no other catastrophe of codimension $\leqslant 3$. Some characteristic slices of the unfoldings of D_6^- and D_6^+ are shown in Fig. 8.14. Their conspicuous family relationship with D_4^- and D_4^+, for whose geometry see the next chapter, will be analysed in Poston, Stewart and Woodcock [39].

Analytically one may proceed as follows. Reduce f to the form

$$f(x, y) \pm z_1^2 \pm \cdots \pm z_m^2$$

by the normalization algorithm of Section 12. If f is $D_4^\pm$, then $j^3 f$ is a homogeneous cubic with at least one linear factor. Divide out by this factor to obtain a quadratic: if this is positive or negative definite, f is D_4^+, if indefinite, D_4^-.

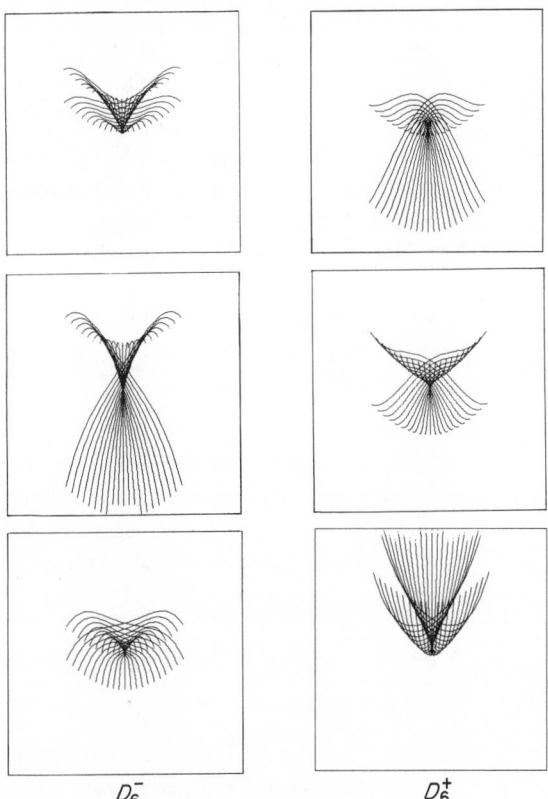

Fig. 8.14. Unfolding sections of D_6^- (left column) and D_6^+ (right column) (courtesy of A. E. R. Woodcock)

D_6^- D_6^+

If f is $D_6^{\pm}$, normalization gives

$$j^5 f = f_3(x, y) + f_4(x, y) + f_5(x, y)$$

where f_i is homogeneous of degree i. By an easily found linear coordinate change, f_3 becomes $u^2 v$. In these coordinates $j^5 f$ becomes

$$u^2 v + u^2 P(u, v) + auv^3 + uQ(u, v) + bv^5,$$

where P and Q are quadratic and quartic respectively in (u, v). Then f is D_6^+ or D_6^- according as $b - a^2 > 0$ or $b - a^2 < 0$.

11   Inequalities

Both the determinacy $\sigma(f)$ and codimension $\mathrm{cod}\,(f)$ are finite if either is, and

$$\mathrm{cod}\,(f) \geq \sigma(f) - 2.$$

It is occasionally useful to know this: the proof requires Nakayama's lemma. More important is the relation between $\mathrm{cod}\,(f)$ and the corank m of f. Take f in its 'split' form, given by the Splitting Lemma (Chapter 4), and computable by the algorithm of the next section,

$$\pm y_1^2 \pm \cdots \pm y_{n-m}^2 + \tilde{f}(y_{n-m+1}, \ldots, y_n),$$

where $\tilde{f}$ is a polynomial of order ≥ 3 and degree $\leq \mathrm{cod}\,(f)$. For $1 \leq i \leq n - m$, each $y_i = \dfrac{1}{2} \dfrac{\partial f}{\partial y_i}$, so all polynomials in J_n^k with a factor y_i of this type lie in $\Delta_k(f)$. Thus only polynomial directions involving only $y_{n-m+1}, \ldots, y_n$ can be 'missing' from $\Delta_k(f)$, hence affect the codimension. We may therefore work with $\tilde{f}$.

Now there are m linearly independent linear monomials in $y_{n-m+1}, \ldots, y_n$, and $\frac{1}{2} m(m+1)$ quadratic ones. Since $\tilde{f}$ has no quadratic terms, the tangent directions

$$j^k \left(\frac{\partial \tilde{f}}{\partial y_i} \right) \qquad (n - m \leq i \leq n)$$

have no linear terms. Thus only these tangent vectors and their linear (not polynomial) combinations give linear or quadratic directions. Thus we have *at most* m such directions (fewer if some terms are linearly dependent) and we need at least

$$(m + \tfrac{1}{2} m(m+1)) - m$$

extra to obtain all linear and quadratic directions from an unfolding. Thus

$$\mathrm{cod}\,(f) \geq \tfrac{1}{2} m(m+1).$$

Recent work with Robert Magnus has brought to light another

inequality: we have not found it in the literature. Some combinatorics with the rules in Section 7 yield the relation

$$\sigma(f) > \sqrt{m}.$$

This is fairly crude, not detecting for instance that there are no 3-determinate functions of 4 essential variables (for example, $x^3 + y^3 + z^3 + w^3$ is only 4-determinate). It would be interesting to refine it, and to know the asymptotic behaviour of the true lower bound on $\sigma(f)$ with m.

However, even in this form it shows that a topologically exact description of a system with 10^{12} essential variables by a Taylor polynomial involves *at least* all terms up to the millionth order. This suggests (though a proof would require other methods) that in a system with an infinite number of buckling modes, k-determinacy at the buckling point is impossible for finite k. This casts doubt on *all* techniques using polynomial approximation, from catastrophe theory to finite element methods. Other topological tools would seem to be required to clarify the situation.

12 Summary of Results and Calculation Methods

We repeat here the bare bones of the computational techniques discussed, for those who prefer to omit the above discussion and as a convenient reference. In some cases theorems have been collapsed into 'definitions' for brevity: the full story being found in the earlier sections. We assume notation from Chapter 3 Section 7.

Definitions   *For any smooth function $f : \mathbb{R}^n \to \mathbb{R}$,*

> *$j^k f$ is the Taylor expansion of f to order k.*
> *$J^k f$ is $j^k f$ minus its constant term.*
> *f is k-determinate at 0 if any smooth function $f + g$, where g is of order $k + 1$ at 0, can be locally expressed as $f(y(x))$ where $y : \mathbb{R}^n \to \mathbb{R}^n$ is a smooth reversible change of coordinates.*
> *f is strongly k-determinate at 0 if, further, y can always be chosen such that*

$$\left. \frac{\partial y_i}{\partial x_j} \right|_0 = \delta_{ij} \quad (where\ \delta_{ij} = 0\ if\ i \neq j,\ 1\ if\ i = j).$$

> *f is locally k-determinate at 0 if there exists $\varepsilon > 0$ such that for any smooth function g with*

$$\left| \frac{\partial_g^k}{\partial_{x_1}^{k_1} \cdots \partial_{x_n}^{k_n}} \right| < \varepsilon \quad for\ all\ k_1, \ldots, k_n\ with\ k_1 + \cdots + k_n = k$$

the function $f + g$ can be expressed locally as $f(y(x))$, where $y : \mathbb{R}^n \to \mathbb{R}^n$ is a smooth reversible local change of coordinates.

E_n^k *is the vector space of polynomials in* $x_1, \ldots, x_n$ *of degree* $\leq k$.
J_n^k *is the subspace of* E_n^k *of polynomials with zero constant term.*
I_n^k *is the subspace of* J_n^k *of polynomials with zero linear term.*
M_n^k *is the vector space of homogeneous polynomials in* $x_1, \ldots, x_n$
of degree k.

$\Delta_k(f)$ *is the subspace of* J_n^k *spanned by all* $\overline{Qj^k\left(\dfrac{\partial f}{\partial x_i}\right)}^k$, *where*
$1 \leq i \leq n$ *and* $Q \in E_n^k$. *It equals*

$$\Delta_k(j^{k+1}f)$$

by comparing terms.

$\overline{J_n^{k+1} \Delta_{k+1}(f)}^{k+1}$ *is the subspace of* J_n^{k+1} *spanned by all polynomials of the form*

$$\overline{Qj^{k+1}\left(\frac{\partial}{\partial x_i}(j^{k+1}f)\right)}^{k+1}$$

where Q is a monomial $x_1^{\alpha_1} \cdots x_n^{\alpha_n}$ *with* $1 \leq \alpha_1 + \cdots + \alpha_n \leq k+1$.

$\overline{I_n^{k+1} \Delta_{k+1}(f)}^{k+1}$ *is the subspace of* J_n^{k+1} *spanned by all polynomials of the form*

$$\overline{Qj^{k+1}\left(\frac{\partial}{\partial x_i}(j^{k}f)\right)}^{k+1}$$

where Q is a monomial $x_1^{\alpha_1} \cdots x_n^{\alpha_n}$ *with* $2 \leq \alpha_1 + \cdots + \alpha_n \leq k+1$.

The codimension cod (f) of f at 0 is the codimension of $\Delta_k(f)$ in J_n^k, for any k for which f is k-determinate.

An r-unfolding of f at 0 is a function

$$F : \mathbb{R}^{n+r} \rightarrow \mathbb{R}$$

$$(x_1, \ldots, x_n, t_1, \ldots, t_r) \mapsto F(x, t) = F_t(x)$$

such that $F_{0,\ldots,0}(x) = f(x)$, defined in a region around $(0, \ldots, 0)$.

A d-unfolding $\bar{F}$ is induced from F by three mappings, defined in a region around the origin,

$$e : \mathbb{R}^d \rightarrow \mathbb{R}^r, \qquad (s_1, \ldots, s_d) \mapsto (e_1(s), \ldots, e_r(s)),$$

$$y : \mathbb{R}^{n+d} \rightarrow \mathbb{R}^n, \qquad (x, s) \mapsto (y_1(x, s), \ldots, y_n(x, s)),$$

$$\gamma : \mathbb{R}^d \rightarrow \mathbb{R},$$

provided

$$\bar{F}(x, s) = F(y_s(x), e(s)) + \gamma(s).$$

(For a more informative account see Section 7.)

Two r-unfoldings F and G of f at 0 are strongly equivalent if they can be induced from each other as above, with

$$\left.\frac{\partial e_i}{\partial t_j}\right|_0 = \delta_{ij}.$$

An *r-unfolding of f at* 0 *is versal if all other unfoldings of f at* 0 *can be induced from it.*

It *is* universal *if it is versal and* $r = \text{cod}(f)$.

If F is an unfolding of f, set

$$v_1^k(F) = \frac{\partial}{\partial t_1}(J^k(F_{t_1, 0, \ldots, 0})),$$

$$\cdots$$

$$v_r^k(F) = \frac{\partial}{\partial t_r}(J^k(F_{0, \ldots, 0, t_r})),$$

$$V^k(F) = \text{subspace of } J_n^k \text{ spanned by } v_1^k(F), \ldots, v_r^k(F).$$

Theorem 8.1 *f is strongly k-determinate if and only if*

$$M_n^{k+1} \subseteq \overline{I_n^{k+1} \Delta_{k+1}(f)}^{k+1}.$$

Theorem 8.2 *If f is k-determinate, then* $M_n^{k+1} \subseteq \overline{J_n^{k+1} \Delta_{k+1}(f)}^{k+1}$.

Theorem 8.3 *f is locally k-determinate if and only if*

$$M_n^{k+1} \subseteq \overline{J_n^{k+1} \Delta_{k+1}(f)}^{k+1}.$$

Theorem 8.4 *f is k-determinate if and only if, for all* $P \in M_n^{k+1}$,

$$M_n^{k+1} \subseteq \overline{J_n^{k+1} \Delta_{k+1}(j^k f + P)}^{k+1}.$$

Theorem 8.5 *Suppose* $f : \mathbb{R}^n \to \mathbb{R}$ *has codimension c at* 0. *Then any small perturbation f' of f has at most* $c + 1$ *critical points near* 0.

Theorem 8.6 *An r-unfolding F of f, where f is k-determinate, is versal if and only if* $V^k(F)$ *and* $\Delta_k(f)$ *are transverse subspaces of* J_n^k.

(Note that $\Delta_k(f)$ is the same as $\Delta_k(j^{k+1}f)$ but not necessarily the same as $\Delta_k(j^k f)$, so the use of this criterion involves $j^{k+1}f$ and not $j^k f$, as stated by error in Poston and Stewart [25] p. 98, first prints.)

Corollary *If f is k-determinate, then a universal unfolding for f may be constructed by choosing a cobasis* (*Chapter 2 Section 2*) $v_1, \ldots, v_c$ *for* $\Delta_k(f)$ *in* J_n^k *and setting*

$$F(x_1, \ldots, x_n, t_1, \ldots, t_c) = f(x) + t_1 v_1(x) + \cdots + t_c v_c(x).$$

Theorem 8.7 *A versal unfolding F of f is strongly equivalent to the truncated unfolding*

$$j^p f(x) + t_1 J^q \left(\frac{\partial}{\partial t_1}(F_{t_1, 0, \ldots, 0})\right) + \cdots + t_r J^q \left(\frac{\partial}{\partial t_r}(F_{0, \ldots, 0, t_r})\right)$$

if f is strongly k-determinate, $k \geqslant 3$, and

$$p \geqslant 2k - 3, \quad q \geqslant k - 2 \quad \text{when} \quad M_n^{k-1} \subseteq \Delta_{k+1}(f)$$

$$p \geqslant 2k - 2, \quad q \geqslant k - 1 \quad \text{when} \quad M_n^k \subseteq \Delta_{k+1}(f)$$

$$p \geqslant 2k - 1, \quad q \geqslant k \quad \text{when} \quad M_n^{k+1} \subseteq \Delta_{k+1}(f).$$

At least one of these cases must hold.

Reduction Algorithm for splitting quadratic and degenerate parts. The Splitting Lemma (Chapter 4 Section 5) guarantees the existence of a convenient separation of variables for a function of corank m. An algorithm for finding one could be based on the proof we gave there; Thompson and Hunt [23] give an excellent description of another method; Magnus [101] describes explicit techniques for the infinite-dimensional case. (Once something is known to exist, all sorts of methods exist for finding it: hence the practical power of apparently 'unquantitative' existence proofs.) We give here one systematic method for use later, without claiming that it is always the best.

(a) If f is k-determinate, then all terms of degree $> k$ arising in the following operations may be immediately dropped. Further, if by applying Theorems 8.1 to 8.4, the *normalized* form of $j^k f$ is k-determinate, then such truncation in the course of normalization is justified *ex post facto*.

(b) By a linear change of variables to

$$u_i(x) = \sum_{l=1}^n \alpha_{il} x_l,$$

diagonalize the Hessian of f in the form

$$2 \begin{bmatrix} d_1 & & & & & & \\ & \cdot & & & 0 & & \\ & & \cdot & & & & \\ & & & \cdot & & & \\ & & & & d_j & & \\ & & & & & 0 & \\ & 0 & & & & & \cdot \\ & & & & & & \cdot \\ & & & & & & & 0 \end{bmatrix}$$

where $d_1 d_2 \cdots d_j \neq 0$. (The d_i can be made ± 1 if convenient, as in Chapter 2 Section 5.) This gives

$$j^k f = d_1 u_1^2 + \cdots + d_j u_j^2 + F(u_1, \ldots, u_n)$$

as functions of $(x_1, \ldots, x_n)$. Now F is still a polynomial in *all* the u_i, but with no term of degree < 3.

(c) Remove first the cubic cross-term, as follows. Collect all the cubic terms in which u_1 appears, and write their sum in the form

$$Q_1 = 2d_1 u_1 R_1(u_1, \ldots, u_n),$$

so that R_1 is a homogeneous quadratic in $u_1, \ldots, u_n$. Of the remaining cubic terms, collect those in which u_2 appears, and write their sum as

$$Q_2 = 2d_2 u_2 R_2(u_2, \ldots, u_n),$$

and so on up to

$$Q_j = 2d_j u_j R_j(u_j, \ldots, u_n).$$

Then

$$F(u_1, \ldots, u_n) = Q_1 + \cdots + Q_j + \hat{F}(u_1, \ldots, u_n)$$

where $\hat{F}$ has no cubic terms involving $u_1, \ldots, u_j$.

(d) Define

$$v_i(x) = \begin{cases} u_1(x) + R_i(u_i(x), \ldots, u_n(x)) & 1 \le i \le j \\ u_i(x) & j < i \le n. \end{cases}$$

Then for $i \le j$,

$$d_i v_i^2 = d_i u_i^2 + 2d_i u_i R_i + d_i R_i^2.$$

Thus

$$j^k f = d_1 v_1^2 + \cdots + d_j v_j^2 + \bar{F}(v_1, \ldots, v_n) \tag{8.5}$$

where

$$\bar{F}(v_1(x), \ldots, v_n(x)) = F(u_1(x), \ldots, u_n(x))$$
$$- \sum_{i=1}^{j} d_i (R_i(u_i(x), \ldots, u_n(x)))^2$$

has no cubic terms in $v_1, \ldots, v_j$ since the R_i^2 are all quartic. Unfortunately the right-hand side is still in terms of the old coordinates, and most of the work comes in:

(e) Find $u_1(v), \ldots, u_j(v)$ (to the order necessary to give R_i^2 and F to order k) in terms of $v_1(x), \ldots, v_n(x)$ around 0, and substitute in $\bar{F}$ to express (8.5) wholly in terms of the v_i.

(f) Next remove quartic terms involving $v_1, \ldots, v_j$ from (8.5) by collecting those containing v_1 into the sum

$$\bar{Q}_1 = 2d_1 v_1 \bar{R}_1$$

with $\bar{R}_1$ a homogeneous cubic, setting

$$\bar{v}_1 = v_1 + \bar{R}_1,$$

and so on.

(g) Inductively, this gives a polynomial change of coordinates whose effect is to give $j^k f$ the normal form

$$j^k f(x) = d_1 w_1^2(x) + \cdots + d_j w_j^2(x) + F(w_{j+1}(x), \ldots, w_n(x)). \tag{8.6}$$

If f is k-determinate, by the theorems above, in either x or w coordinates, there is a smooth change of coordinates $y(x)$ giving f itself the normalized form (8.6).

In the local y-coordinates this is an *exact* expression for f, not merely an approximate one. Thus all local properties (such as the corank of f, whether $f(0)$ is a local minimum, and the effect of small perturbations) may be accurately studied using the normalized expression.

Siersma's trick   This is often convenient when dealing with functions of two variables: one applies it in practice to $j^k f(x, y)$ for some k, if f is not polynomial. It should not be used as a *replacement* for the rules above but, keeping them carefully in mind, as a shorthand for them. A more elaborate and powerful version is due to Arnol'd.

The method is simplest when

$$f_1(x, y) = \frac{\partial f}{\partial x}(x, y)$$

and

$$f_2(x, y) = \frac{\partial f}{\partial y}(x, y)$$

each have only one term. For example,

$$f = x^2 + y^4; \qquad f_1 = 2x, \qquad f_2 = 4y^3.$$

Draw the 'shadows' cast by x and y^3 in a diagram of the following type (Fig. 8.15). The truncated shadows ($\sqrt{}$-shaped) contain only polynomials in $I_n^{k+1} \Delta_{k+1}(f)$ for some k. Between them they contain all the fifth layer, so f is 4-determinate by Theorem 8.1.

The whole shadows contain polynomials in $\Delta_k(f)$ for some k. Just y and y^2 are left outside, and neither can be expressed in terms of the other. Thus cod $(f) = 2$, and the obvious universal unfolding is

$$F(x, y, a, b) = f(x, y) + ay^2 + by.$$

Similarly for $x^4 + y^4$: from Fig. 8.16 one sees at once that it is 4-determinate, and of codimension 8 with universal unfolding

$$x^4 + y^4 + a_8 x^2 y^2 + a_7 x^2 y + a_6 xy^2 + a_5 x^2 + a_4 xy + a_3 y^2 + a_2 x + a_1 y.$$

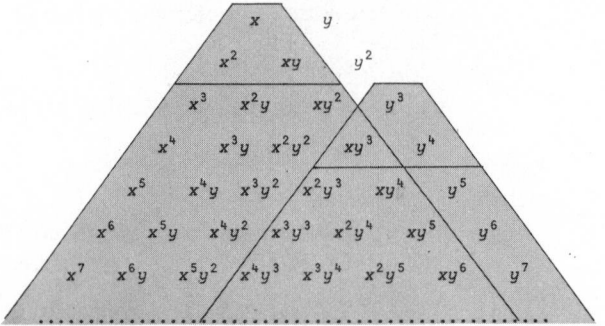

Fig. 8.15

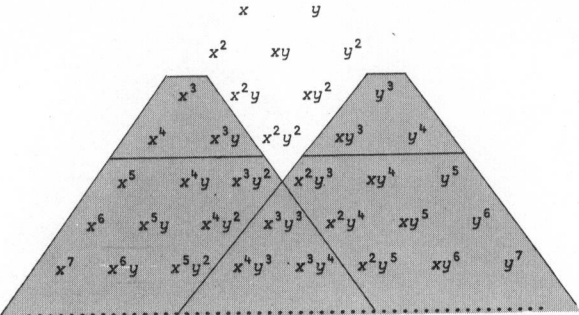

Fig. 8.16

Things are a little less simple when f_1, f_2 are compound expressions. For instance, if

$$f(x, y) = x^2 y + y^3/3, \qquad f_1(x, y) = 2xy, \qquad f_2(x, y) = x^2 + y^2.$$

We have to find combinations of f_1 and f_2 to cast shadows as well, like

$$x^3 = xf_2 - \tfrac{1}{2}yf_1, \qquad y^3 = yf_2 - \tfrac{1}{2}xf_1.$$

Since these involve already multiplying f_1 and f_2 by linear factors, their shadows contain combinations of quadratics times f_1, f_2 as soon as we are one level down. So we can 'truncate' them higher when looking for determinacy than we can the shadows cast by f_1, f_2 themselves. By Fig. 8.17 f is three-determinate. Its codimension is only 3, since $x^2 = f_2 - y^2$ and thus $\{x, y, x^2, y^2\}$ is not a cobasis, since it is not minimal. The set $\{x, y, q\}$ where q is any quadratic *not* a scalar multiple of $x^2 + y^2$ (e.g. x^2, y^2, $x^2 - y^2$, $x^2 + 2y^2$) *is* minimal, so cod $f = 3$ and a universal unfolding is

$$x^2 y + y^3/3 + a_3 q + a_2 x + a_1 y.$$

Another example:

$$f(x, y) = x^2 y, \qquad f_1 = 2xy, \qquad f_2 = x^2 \quad \text{(Fig. 8.18)}.$$

Now M_n^k is not a subset of $\overline{J_n^k \Delta_k(f)}^k$ for any k; and f is indeterminate, of codimension ∞.

Notice that for $x^2 y + y^k$ the codimension is finite, and equal to k ($k \geqslant 2$), as needed for $k \geqslant 3$ in Chapter 7, Section 5.

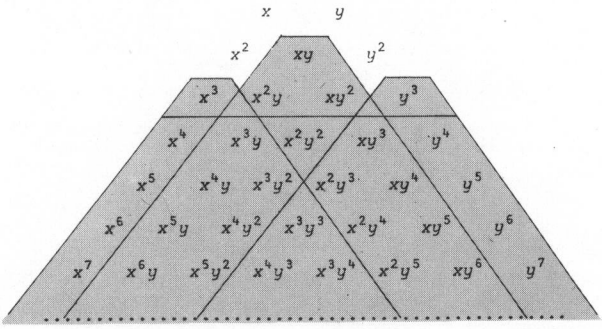

Fig. 8.17

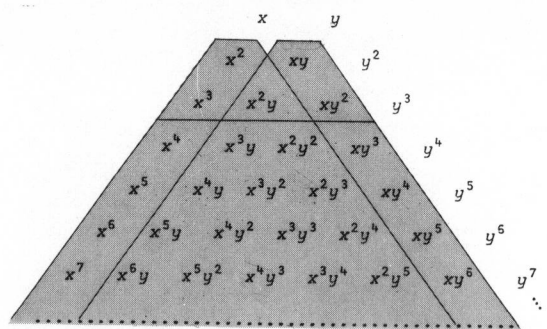

Fig. 8.18

13 Examples and Calculations

(a) If f is critical at 0, but its Hessian is nondegenerate, make a linear change of coordinates to express f as

$$x_1^2 + \cdots + x_i^2 - x_{i+1}^2 - \cdots - x_n^2 + O(3).$$

Then, with $k = 2$,

$$\overline{I_n^{k+1} \Delta_{k+1}(j^k f)}^{k+1}$$

consists of all sums of multiples of $j^2\left(\dfrac{\partial}{\partial x_i}(j^2 f)\right) = \pm\frac{1}{2}x_i$ by quadratic polynomials, hence of all cubics. Thus

$$M_n^{2+1} \subseteq \overline{I_n^{2+1} \Delta_{2+1}(j^2 f)}^{2+1}$$

and by Theorem 8.1 f is strongly 2-determinate. Theorem 8.1 is thus a generalization of the Morse Lemma.

(b) $n = 1$, $j^k f(x) = x^k$. The 1-variable version of Siersma's trick yields Fig. 8.19 since $df/dx = kx^{k-1}$. Hence f is k-determinate, as we proved more laboriously in Theorem 4.4, and has universal unfolding

$$x^k + t_{k-2}x^{k-2} + \cdots + t_1 x,$$

as we claimed in Chapter 7 (the general cuspoid catastrophe).

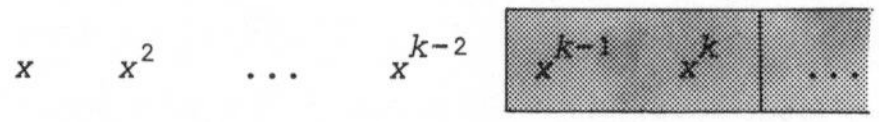

Fig. 8.19

$$x \quad x^2 \quad \cdots \quad x^{k-2} \quad \boxed{x^{k-1} \quad x^k} \quad \cdots$$

(c) $n = 2$, $f(x, y) = x^5 + y^5$. Using Siersma's trick we obtain Fig. 8.20. The monomial $x^3 y^3$ is not contained in $\Delta_6(f)$, let alone the smaller spaces

$$\overline{J_2^6 \Delta_6(f)}^6 \quad \text{or} \quad \overline{I_2^6 \Delta_6(f)}^6,$$

so f is not 5-determinate, by Theorem 8.2. It *is* strongly 6-determined, and its codimension is 15.

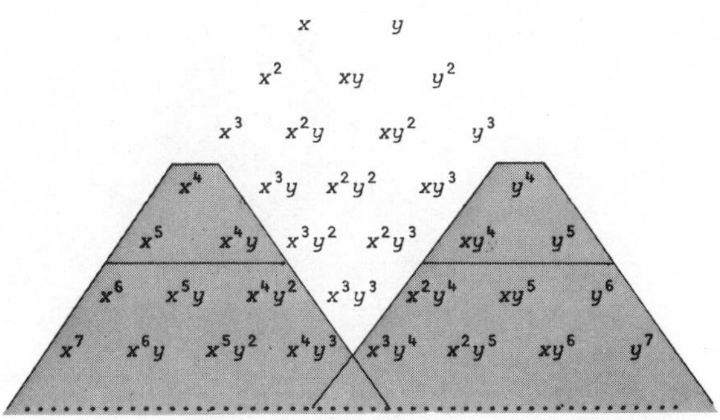

Fig. 8.20

This illustrates the falsity of the idea one sometimes meets, that nondegeneracy of the first non-zero term above order zero, a homogeneous polynomial of degree k, is enough to guarantee k-determinacy. The inadequacy of such tests as the discriminant for determinacy results is further pointed up by the next example.

(d) $f(x, y) = \frac{3}{2}x^2 + x^3 - 3xy^2$. This has degenerate quadratic part, hence is not 2-determinate. Test for 3-determinacy:

$$J_1 = \frac{\partial f}{\partial x} = 3(x + x^2 - y^2)$$

$$J_2 = \frac{\partial f}{\partial y} = -6xy.$$

Truncating above degree 4 we have

$$x^2 = \tfrac{1}{3}J_1 - x^3 + xy^2$$

$$= \tfrac{1}{3}J_1 - \frac{x^2}{3}(J_1 - x^2 + y^2) - \frac{y}{6}J_2$$

$$= \tfrac{1}{3}J_1 - \frac{x^2}{3}J_1 + \frac{x^3}{9}J_1 + \frac{xy}{18}J_2 - \frac{y}{6}J_2,$$

$$y^3 = -\frac{y}{3}J_1 + xy + x^2 y$$

$$= -\frac{y}{3}J_1 - \frac{1}{6}J_2 - \frac{x}{6}J_2$$

Thus the Siersma picture has shadows as in Fig. 8.21. The 'truncated' shadows do not include y^4, and writing out all the products truncated above degree 4;

$$\begin{array}{ll}
x^2 J_1 = 3(x^3 + x^4 - x^2 y^2) & y^2 J_2 = -6xy^3 \\
xy J_1 = 3(x^2 y + x^3 y - xy^3) & x^3 J_1 = 3x^4 \\
y^2 J_1 = 3(xy^2 + x^2 y^2 - y^4) & x^2 y J_1 = 3x^3 y \\
x^2 J_2 = -6x^3 y & xy^2 J_1 = 3x^2 y^2 \\
xy J_2 = -6x^2 y^2 & y^3 J_1 = 3xy^3;
\end{array}$$

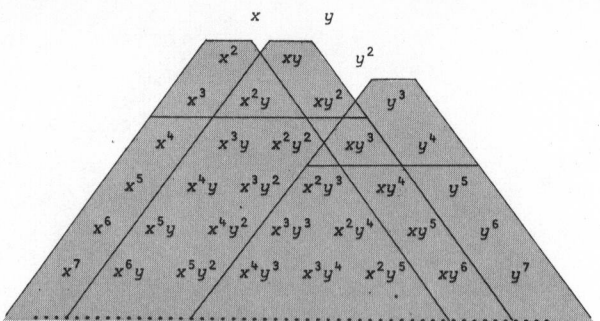

Fig. 8.21

it is clear that y^4 is not in $\overline{I_2^{3+1} \, \Delta_{3+1}(f)}^{3+1}$. Thus f is not strongly 3-determinate. But

$$M_2^{3+1} \subseteq \overline{J_2^{3+1} \, \Delta_{3+1}(f)}^{3+1}$$

as we see by truncating the shadows one step higher, so Theorem 8.2 does not tell us that f is not ordinarily 3-determinate. We could try Theorem 8.4 at this point (recommended to the reader as an exercise), but instead we test for 4-determinacy.

Truncating above degree 5,

$$x^2 = \frac{1}{3} J_1 - \frac{x^2}{3} (J_1 - x^2 + y^2) - \frac{y}{6} J_2$$

$$= \frac{1}{3} J_1 - \frac{x^2}{3} J_1 + \frac{x^3}{9} (J_1 - x^2 + y^2) + \frac{xy}{18} J_2 - \frac{y}{6} J_2$$

$$= \frac{1}{3} J_1 - \frac{x^2}{3} J_1 + \frac{x^3}{9} J_1 - \frac{x^4}{27} J_1 - \frac{x^2 y}{54} J_2 + \frac{xy}{18} J_2 - \frac{y}{6} J_2,$$

$$y^3 = -\frac{y}{3} J_1 - \frac{1}{6} J_2 - \frac{x}{6} J_2.$$

So in the Siersma picture (Fig. 8.22) we see that f *is* strongly 4-determinate. There are three monomials out of the shadow, but since x^2 is in $\Delta_5(f)$ and

$$x = (\tfrac{1}{3} J_1 - x^2) + y^2$$

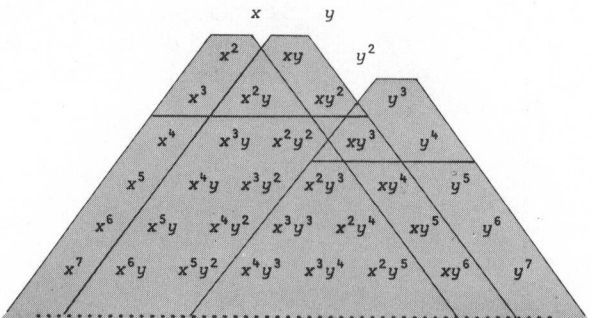

Fig. 8.22

either $\{x, y\}$ or $\{y, y^2\}$ provide a cobasis for $\Delta_5(f)$ in J_2^5. (Clearly no one polynomial does so.) Thus cod (f) is 2, and hence f is equivalent to $\pm x^2 \pm y^4$, since by Table 8.1 *only* functions equivalent to these forms have this codimension! In particular f is not 3-determinate, since $\pm x^2 \pm y^4$ is not.

By Sylvester's Law of Inertia applied to the Hessian, the type must be $+x^2 \pm y^4$. With the $+$ sign for y it would have a strict local minimum at $(x, y) = (0, 0)$; but along the real curve

$$y^2 = \tfrac{1}{2}x + \tfrac{1}{3}x^3$$

we have

$$f(x, y) = \tfrac{3}{2}x^2 + x^3 - 3x(\tfrac{1}{2}x + \tfrac{1}{3}x^3) = 0.$$

So the type must be $x^2 - y^4$, and a versal unfolding of f will be equivalent to the dual cusp catastrophe.

If it is necessary to find a specific reduction to this form, apply the normalization algorithm as follows. The Hessian

$$2\begin{bmatrix} 3/2 & 0 \\ 0 & 0 \end{bmatrix}$$

is already diagonal, so set

$$Q_1 = 2 \cdot \tfrac{3}{2} \cdot x\left(\frac{x^2}{3} - y^2\right), \qquad R_1(x, y) = \left(\frac{x^2}{3} - y^2\right)$$

$$v_1 = x + \left(\frac{x^2}{3} - y^2\right)$$

$$v_2 = y.$$

Express x, y in terms of v_1, v_2:

$y = v_2$

$x =$ the unique solution that is zero when v_1 and v_2 are of the equation $x^2 + 3x - 3(v_1 + v_2^2) = 0$,

$$= \frac{-3 + \sqrt{(9 + 12(v_1 + v_2^2))}}{2}$$

$$= \tfrac{3}{2} + \tfrac{3}{2}(1 + \tfrac{1}{2} \cdot \tfrac{4}{3}(v_1 + v_2^2) - \tfrac{1}{8} \cdot \tfrac{16}{9}(v_1 + v_2^2)^2 + \tfrac{1}{64} \cdot \tfrac{64}{27}(v_1 + v_2^2)^3 + \cdots)$$

$$= v_1 - \tfrac{1}{3}v_1^2 + v_2^2 + \tfrac{1}{18}v_1^3 - \tfrac{2}{3}v_1 v_2^3 + \cdots.$$

Therefore

$$j^4 R_1(v_1, v_2) = j^4\left(\frac{x^2}{3} - y^2\right)$$

$$= \tfrac{1}{3}v_1^2 - v_2^2 - \tfrac{2}{3}v_1^3 + 2v_1 v_2^2 + \tfrac{1}{9}v_1^4 - 2v_1^2 v_2^2,$$

$$j^4(R_1^2(v_1, v_2) = (\tfrac{1}{3}v_1^2 - v_2^2)^2.$$

(Thus a fourth order R_1^2 has the same expression in (v_1, v_2) as in (x, y): this could be deduced without computation, but it seems better to illustrate the calculations that would be essential if f were only 5-determinate.)

Since f has no terms in y only, we have

$$\hat{F}(x, y) = 0,$$

hence

$$\hat{F}(v_1, v_2) = 0.$$

Therefore

$$\bar{F}(v_1, v_2) = \hat{F}(v_1, v_2) - e_1 R_1^2 = -\tfrac{3}{2}(\tfrac{1}{3}v_1^2 - v_2^2)^2$$

and

$$j^4 f(v_1, v_2) = e_1 v_1^2 + \bar{F}(v_1, v_2)$$
$$= \tfrac{3}{2}v_1^2 - \tfrac{1}{6}v_1^4 + v_1^2 v_2^2 - \tfrac{3}{2}c_2^4.$$

(Notice that in these coordinates this implies

$$j^3 f(v_1, v_2) = \tfrac{3}{2}v_1^2$$

which makes it obvious that f is not 3-determinate: no change of coordinates can make $\tfrac{3}{2}v_1^2 + v_2^4$, with its strict minimum at 0, look like $\tfrac{3}{2}v_1^2 - v_2^4$.)

Now remove the quartic terms in u_1:

$$\bar{Q} = 2 \cdot \tfrac{3}{2} \cdot v_1(\tfrac{1}{3}v_1 v_2^2 - \tfrac{1}{18}v_1^3), \qquad \bar{R}_1(v_1 v_2^2) = \tfrac{1}{3}(v_1 v_2^2 - \tfrac{1}{6}v_1^3)$$
$$w_1 = v_1 + \bar{R}_1 = v_1 + \tfrac{1}{3}(v_1 v_2^2 - \tfrac{1}{6}v_1^3)$$
$$w_2 = v_2.$$

Rather than solve the cubic equation

$$v_1^3 - 6(3 + w_2^2)v_1 + 18w_1 = 0$$

for $v_1(w_1, w_2)$, observe that $\bar{R}_1^2$ is homogeneous of sixth degree in (v_1, v_2) and is therefore of sixth order in (w_1, w_2) and cannot influence the form of $j^4 f$. (It is clear that for higher k, the algebra can become more laborious. But when no general series solution for the previous coordinates in terms of the new ones is to hand, in the final resort there is always force. To find the coefficients up to order r (say), substitute a general expansion to order r in the equations and equate coefficients. Questions of the general term, convergence, etc. have no bearing on the results.)

Thus we have

$$j^4 f(w_1, w_2) = \tfrac{3}{2}w_1^2 - \tfrac{3}{2}w_1^4$$

and since f is 4-determinate by previous calculation (or by a very easy use of (b) above with $j^4 f$ in this form), a smooth adjustment to the coordinates gives f the local form

$$\tfrac{3}{2}y_1^2 - \tfrac{3}{2}y_1^4.$$

(e) $f(x, y) = x^3 + xy^3$. This is not 3-determinate, either by Theorem 8.2 or by Theorem 8.4 (with $P = 0$), for which the calculations are the same: with $P = 0$,

$$\overline{J_2^4 \, \Delta_4(j^3 f)}^{\,4} = \overline{J_2^4 \, \Delta_4(f)}^{\,4}.$$

Now

$$\frac{\partial f}{\partial x} = 3x^2 + y^3 \qquad \frac{\partial f}{\partial y} = 3xy^2,$$

so $\overline{J_2^4 \, \Delta_4(f)}^{\,4}$ is spanned by

$$3x^3 + xy^3$$
$$3x^2 y + y^4$$
$$3x^4$$
$$3x^3 y$$
$$3x^2 y^2$$
$$3x^2 y^2$$
$$3xy^3$$

and clearly does not contain y^4; so $M_2^4 \not\subseteq \overline{J_2^4 \, \Delta_4(j^3 f)}^{\,4}$, and f is not 3-determinate (or even locally 3-determinate).

We use Theorem 8.4 to try for 4-determinacy. Since adding $P \in M_2^5$ only alters the fifth degree and higher terms in

$$\overline{J_2^5 \, \Delta_5(j^4 f + P)}^{\,5}$$

(hence *only* the fifth degree) the calculations are not too messy. We have, for any $P \in M_2^5$,

$$\frac{\partial(f + P)}{\partial x} = 3x^2 + y^3 + \frac{\partial P}{\partial x} \qquad \frac{\partial(f + P)}{\partial y} = 3xy^2 + \frac{\partial P}{\partial y},$$

and the P-derivatives are of degree 4. Then $\overline{J_2^5 \, \Delta_5(j^4 f + P)}^{\,5} = \overline{J_2^5 \, \Delta_5(f + P)}^{\,5}$ is spanned by

$$3x^3 + xy^3 + x \frac{\partial P}{\partial x} \qquad\qquad 3x^2 y^2 + x \frac{\partial P}{\partial y}$$

$$3x^2 y + y^4 + y \frac{\partial P}{\partial x} \qquad\qquad 3xy^3 + y \frac{\partial P}{\partial y}$$

$$3x^4 + x^2 y^3 \qquad\qquad\qquad 3x^3 y^2$$

$$3x^3 y + xy^4 \qquad\qquad\qquad 3x^2 y^3$$

$$3x^2 y^2 + y^5 \qquad\qquad\qquad 3xy^4.$$

$$3x^5$$

$$3x^4 y$$

$$3x^3 y^2$$

$$3x^2 y^3$$

Now $x\dfrac{\partial P}{\partial y}$ is of degree 5, and a multiple of x, so is a linear combination of x^5, x^4y, x^3y^2 and xy^4, all of which are listed; therefore $x\dfrac{\partial P}{\partial y}$ is in the subspace under discussion, hence also x^2y^2 is in; hence y^5 is in. Thus

$$M_2^5 \subseteq \overline{J_2^5 \Delta_5(j^4f + P)}^5 \quad \text{for all} \quad P \in M_2^5$$

and f is 4-determinate.

This last result was first noted by Siersma [33], as a counter-example to a conjecture of Zeeman that 'if and only if' held in Theorem 8.1 without 'strongly'. It is instructive to test on it the proof in Duistermaat [41] Lemma 2.1.3, of this.

(f) The calculations of Section 7, and also of Poston and Stewart [25] pp. 77–109 provide further examples. These methods will be seen 'in action' in the applications chapters, on pp. 213, 233, 244, 250, 291, 296, 302, 320ff, 331, 337, 353 among others.

14 Compulsory Remarks on Terminology

There are several competing systems of terminology for catastrophe theory, and excellent arguments against complete standardization: needs vary. (The long shelf-life of 'golden Delicious' apples does not justify the EEC ban on Cox's, which taste better: coming back to terminology, the new French law announcing a fine for using 'lightyear' is a blow against spacetime geometry.) We aim to introduce the reader to the various convenient usages current, rather than constrict ourselves to a language of which practicality forces abuse.

In the notation we stick to a family of functions

$$f : \mathbb{R}^n \times \mathbb{R}^r \to \mathbb{R}.$$

Thom christened $\mathbb{R}^n$ the space of *internal* variables, thinking of some incompletely known process happening in various copies of $\mathbb{R}^n$ governed by f and affected by the point in $\mathbb{R}^r$, the space of *external* variables over which each copy sits in our drawings. Particularly when the variables in $\mathbb{R}^r$ label points in space, as in optics or biology, these names remain the most convenient.

Zeeman, motivated by systems which you alter and see what happens (as, temperature, or top elastic point in the Zeeman machine) introduced *control* variables (or parameters) as a name for positions in $\mathbb{R}^r$, and *behaviour* for $\mathbb{R}^n$. For some applications this language is too illuminating to be abandoned. Correspondingly $\mathbb{R}$ is *control* space and $\mathbb{R}^n$ *behaviour* space.

In either terminology, $\mathbb{R}^n$ is sometimes the space of *state* variables. This must be used with caution, as 'state' means something different from one laboratory to the next.

In the strictly mathematical context of this chapter, and sometimes in applications, it is natural to call $\mathbb{R}^r$ the *unfolding space* and points in it (or their coordinates) the *unfolding variables*.

The number d is correspondingly the *external, control* or *unfolding dimension*. It is occasionally called the *codimension*, but as this conflicts with the commoner meanings we have used it for so far, we suggest avoiding this usage. (Several names for one thing need *remembering:* contradictory things for one name need *deciphering*.)

When we have used the Splitting Lemma to express a function, around a point where it has corank m, in the form

$$\hat{f}(y_1, \ldots, y_m) \pm y_{m+1}^2 \pm \cdots \pm y_n^2$$

(perhaps with parameters in $\mathbb{R}^r$ for $\hat{f}$) we may call $y_1, \ldots, y_m$ the *essential* variables, $y_{m+1}, \ldots, y_n$ the *inessential* variables. (Thompson and Hunt [23, 105] use *active* and *passive* coordinates.) The splitting is very far from unique: to avoid algebraic complications, consider the degenerate case $f(x, y) = x^2$ and new coordinates $u = x$, $v = y - x$. Then f has the form u^2, and though the 'essential axis' is given by $u = 0$ and is the v-axis (or y-axis), the 'inessential axis' given by $v = 0$ (the u-axis) is the line $y = x$ which is *not* the x-axis. With more effort we could in many cases bend the 'essential axis' as well, though it would stay tangent to the y-axis at 0.

Finally, note that most singularities met by a d-dimensional family will, even when not regular or Morse, have codimension less than d. (When $d \le 5$, there will typically be no singularities of greater codimension.) For instance, allowing the fixed end of the lower elastic to move in a circle around the Zeeman machine wheel, as extra control, brings no new phenomena. By reparametrizing we can remove its effects entirely. By Theorem 8.6 we can write a d-parameter family f, around a point where it meets transversely a singularity of codimension c, in a way in which only c controls appear. (We used this fact first in Chapter 7 Section 4.) When we have done so, we may call the coordinates on $\mathbb{R}^d$ that no longer appear, *disconnected* or *dummy* controls. This splitting of the controls is likewise non-unique (in ways that can be very important) and must be fairly explicit to be useful, as we see in Chapter 17 Section 3.

9 The First Seven Catastrophe Geometries

Starting from the standard forms of the elementary catastrophes, we now investigate the various geometrical structures associated with them. Instead of the routine 'bare hands' analysis that we used for the cusp catastrophe in Chapter 5 we shall apply a more sophisticated method due to Zeeman [7] in which the form of the function near a point on the catastrophe manifold is the central object. We shall restrict attention to the traditional seven catastrophes (not distinguishing duals, *see* Chapter 7) of codimension ≤4: partly out of expediency (higher codimensions need many more pictures, and are not at all well understood as yet, except for the cuspoids); and partly because these are all we require for almost all of the applications. A notable exception is the family of double cusp catastrophes, but these are of codimension 8, which is only accessible by way of lower codimensional catastrophes, especially those of codimension 5 and 6. We hope, in Poston, Stewart and Woodcock [39], to give a fuller description of the 'higher catastrophes'. Other work on catastrophe geometry may be found in Bröcker and Lander [9], Thom [42], Woodcock and Poston [20] and Zeeman [7]; a more sophisticated approach which yields a lot of information in a compact form is expounded by Callahan [43]. A beautiful way of thinking about the geometry of the higher-dimensional cases is described in Callahan [43a].

1 The Objects of Study

We begin by recalling the various structures associated with a catastrophe, most of which we have encountered already in examples. We start with a family of functions

$$V : S \times C \to \mathbb{R}.$$

Here S is a manifold, usually $\mathbb{R}^n$, and C is another manifold, usually $\mathbb{R}^r$. Making a particular choice of names from the last section of the previous chapter, we will in this one call $\mathbb{R}^n$ the *state* space, $\mathbb{R}^r$ the *control* space. The number r is the *unfolding dimension*, which in the standard *universal* forms coincides with the *codimension* at 0 of the function unfolded. (Since the classification theorem is purely local, we are really interested only in the behaviour near the origin; but for the standard forms this is mirrored by the behaviour over the whole of $\mathbb{R}^n \times \mathbb{R}^r$, as may be seen by inspection of the calculations and pictures that follow.) We use S and C with the above meanings fixed throughout this chapter.

The *catastrophe manifold M* is the subset of $\mathbb{R}^n \times \mathbb{R}^r$ defined by

$$DV_c(x) = 0,$$

where $V_c(x) = V(x, c)$; this is the set of all critical points of all the potentials V_c in the family V. (M is a manifold provided the family V is universal as an unfolding: this actually is another consequence of transversality. The example of Chapter 6 Section 2 shows that some such hypothesis is necessary. Since the elementary catastrophes are by construction universal, M is always a manifold in this case.)

The *catastrophe map* χ is the restriction to M of the natural projection

$$\pi : \mathbb{R}^n \times \mathbb{R}^r \to \mathbb{R}^r$$

for which

$$\pi(x, c) = c.$$

The *singularity set S* is the set of *singular points* in M of χ, at which χ is *singular*, that is, the rank of the derivative $D\chi$ is less than r.

The image $\chi(S)$ in C is called the *bifurcation set B*.

It is not hard to show that S is the set of points $(x, c) \in M$ at which $V_c(x)$ has a *degenerate* critical point, by computing $D\chi$. It follows that B is the set on which the number and nature of the critical points change; for by structural stability of Morse functions, such a change can only occur by passing through a degenerate critical point. This of course can also be checked by computation in any particular case.

In most applications it is the bifurcation set that is the most important, for it lies in the control space, hence is 'observable', and all delay convention jumps occur on it. But depending on the precise application we have in mind, more (or sometimes fewer) features of the catastrophe come into play.

To simplify notation, we shall here use x and y as state variables rather than u_1, u_2 as in Chapter 7; and a, b, c, d as control variables, in place of t_1, t_2, t_3, t_4.

2 The Fold Catastrophe

The method we shall use is especially simple on this, the simplest of all catastrophes. The standard unfolding is

$$V_a(x) = \tfrac{1}{3}x^3 + ax,$$

where the numerical factor has been introduced to simplify the subsequent calculations. The catastrophe manifold M is given by

$$0 = \frac{d}{dx} V_a(x) = x^2 + a.$$

This equation suggests that we use the x-coordinate as a chart for M, with the general point of M being

$$(x, a) = (x, -x^2).$$

The function on S at this point, expanded in a Taylor series, is

$$V_a(x + X) = \tfrac{1}{3}(x + X)^3 + (-x^2)(x + X)$$
$$= \tfrac{1}{3}X^3 + xX^2 + 0X - \tfrac{2}{3}x^3.$$

Now the quadratic term

$$xX^2$$

is nondegenerate if $x \neq 0$, but degenerate when $x = 0$. Thus the singularity set is given by $x = 0$, and is the point $(0, 0)$ in M only. When $x > 0$ the quadratic term is positive, and V has a minimum; when $x < 0$, V has a maximum.

The whole geometry is therefore summed up by Fig. 9.1. The catastrophe manifold is a parabola, the bifurcation set a single point: to the left of this point there are two states, one maximum and one minimum; to the right there are none.

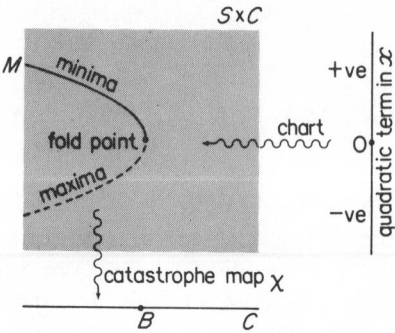

Fig. 9.1

3 The Cusp Catastrophe

We have already analysed this in Chapter 5, by the conventional routine. We can now see how the present method yields the same result. The standard unfolding is

$$V_{ab}(x) = \tfrac{1}{4}x^4 + \tfrac{1}{2}ax^2 + bx,$$

and the catastrophe manifold is given by

$$0 = \frac{d}{dx} V_{ab}(x) = x^3 + ax + b.$$

We can use (x, a) as a chart on M, the general point of M being

$$(x, a, b) = (x, a, -ax - x^3) \tag{9.1}$$

and hence the chart taking the form

$$\mathbb{R}^2 \to M, \qquad (x, a) \mapsto (x, a, -ax - x^3).$$

Taylor expanding we find that

$$V_{ab}(x + X) = \tfrac{1}{4}X^4 + xX^3 + \left(\tfrac{3}{2}x^2 + \frac{a}{2}\right)X^2 + 0X - \left(\tfrac{3}{4}x^4 + \frac{a}{2}x^2\right).$$

We let p, q, r be the coefficients of the quadratic, cubic and quartic terms, so that

$$p(a, x) = \tfrac{3}{2}x^2 + \frac{a}{2}$$

$$q(a, x) = x$$

$$r(a, x) = \tfrac{1}{4}.$$

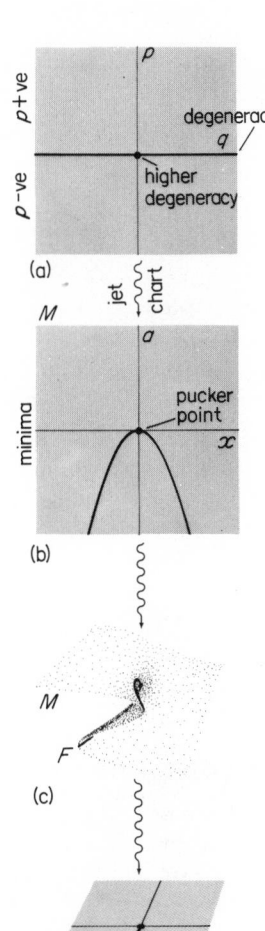

This suggests that we change the chart, and use the plane $r = \tfrac{1}{4}$, or equivalently the (p, q)-coordinates, as a chart on M, with

$$(a, x) = (2p - 3q^2, q)$$

expressing the old chart in terms of the new one. Then the quadratic term of the Taylor series is degenerate when $p = 0$, and this represents the q-axis in (p, q)-space. The image of this in M is the fold curve, expressed in terms of the (x, a) chart: when $p = 0$ we have $x = q$, $a = -3q^2$, so $a = -3x^2$. We have a local minimum when $p > 0$, a maximum when $p < 0$.

When $p = 0$ we must pass to the cubic terms, by appeal to Chapter 4, Section 3. The cubic term determines the type of critical point as long as its coefficient $q \neq 0$. If $p = 0$ and $q = 0$ (the origin in (p, q)-space) then the type is that of X^4. All this is summed up in Fig. 9.2(a) and (b).

Now the image in M of the line $p = 0$ is, as we saw, given by $a = -3x^2$; and substituting this into (9.1) we get the fold line parametrized by x as

$$(x, -3x^2, 2x^3)$$

in agreement with Chapter 5 Section 2. The bifurcation set is the image of this in C, namely the set of points

$$(-3x^2, 2x^3) = (a, b),$$

and we can either view this as a parametrization of B, or eliminate x to obtain the familiar equation

$$4a^3 + 27b^2 = 0.$$

The geometry is illustrated in Fig. 9.2(c) and (d), and may be compared with Fig. 5.6.

Fig. 9.2

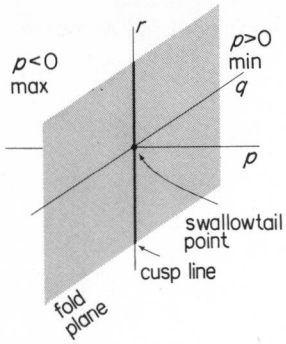

$p<0$
max

$p>0$
min

swallowtail
point

cusp line

fold
plane

Fig. 9.3

Notice how the use of the Taylor coefficients as a chart reveals the structure of M as the image of a plane, on which there is a line of *fold points* (where the function's 3-jet takes the X^3 form characteristic of the fold catastrophe) and a point on that line gives the *cusp point* (at which we get the cusp catastrophe 4-jet X^4) or *pucker point* of the catastrophe manifold. Thus the essential features of M are given by a *flag* or sequence of vector subspaces $\mathbb{R}^2 \supseteq \mathbb{R}^1 \supseteq \mathbb{R}^0$.

The bifurcation set also reflects this structure, though it has extra features due to the singular nature of the catastrophe map, such as the cusp point at 0. The precise geometry of how M sits over C also reflects the nature of the jets: the line of fold-type cubic jets actually is, geometrically, a line of folds in M. This is a consequence of the universality of the fold catastrophe as an unfolding of the X^3 function, and the phenomenon is a general one.

4   The Swallowtail Catastrophe

Here the unfolding is

$$V_{abc}(x) = \tfrac{1}{5}x^5 + \frac{a}{3}x^3 + \frac{b}{2}x^2 + cx,$$

and M is given by

$$0 = \frac{\mathrm{d}}{\mathrm{d}x}V_{abc}(x) = x^4 + ax^2 + bx + c.$$

Using (x, a, b) as a chart, mapped to M by

$$(x, a, b) \mapsto (x, a, b, c),$$

we have

$$c = -bx - ax^2 - x^4. \tag{9.2}$$

The Taylor expansion is

$$V_{abc}(x+X) = \tfrac{1}{5}X^5 + xX^4 + \left(2x^2 + \frac{a}{3}\right)X^3$$
$$+ \left(2x^3 + ax + \frac{b}{2}\right)X^2 + 0X - \left(\frac{b}{2}x^2 + \tfrac{2}{3}ax^3 + \tfrac{4}{5}x^5\right).$$

We take coordinates using the coefficients of the Taylor series as follows:

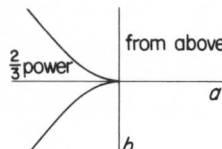

$\tfrac{2}{3}$power

from above

a

b

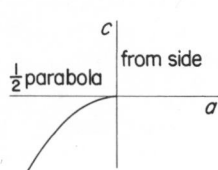

$\tfrac{1}{2}$parabola

from side

c

a

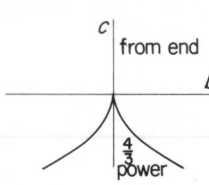

from end

c

b

$\tfrac{4}{3}$power

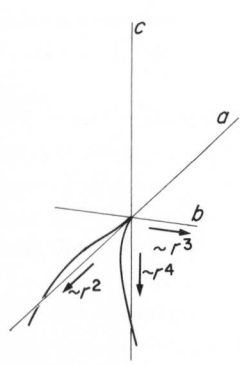

c

a

b

$\sim r^3$

$\sim r^4$

$\sim r^2$

Fig. 9.4

quadratic $p(x, a, b) = 2x^3 + ax + b/2$

cubic $q(x, a, b) = 2x^2 + a/3$

quartic $r(x, a, b) = x$

quintic $s(x, a, b) = 1/5.$

Thus the hyperplane $s = 1/5$ in (p, q, r, s)-space, hence (p, q, r)-space, can be used as a chart for M, and this relates to the (x, a, b)-chart as

$$
\left.
\begin{aligned}
x(p, q, r) &= r \\
a(p, q, r) &= 3q - 6r^2 \\
b(p, q, r) &= 2p - 6rq + 8r^3.
\end{aligned}
\right\}
\tag{9.3}
$$

The quadratic term is degenerate if and only if $p = 0$, which defines the (q, r)-coordinate plane. On this plane the 3-jet is of type X^3 unless $q = 0$ also. This defines the r-axis: on this we have a 4-jet of type X^4 or $-X^4$, unless $r = 0$. Finally, at the origin we have the original unfolded 5-jet X^5. Thus (p, q, r)-space decomposes according to the local type of function as in Fig. 9.3. Again we have a flag of subspaces

$$
\mathbb{R}^3 \supseteq \mathbb{R}^2 \supseteq \mathbb{R}^1 \supseteq \mathbb{R}^0.
$$

The swallowtail point at the origin lies in the middle of a line of cusps: dual cusps below, standard cusps above. This line runs through a plane of folds. In front of the plane we have Morse minima, behind the plane Morse maxima.

Using (9.3) we obtain the same structure on the (x, a, b)-space, and this transfers to M by way of (9.2). In fact M is given parametrically in terms of p, q, r as the set of points

$$
(r, 3q - 6r^2, 2p - 6rq + 8r^3, -2pr + 3qr^2 - 15r^4),
\tag{9.4}
$$

so the singularity set is given by setting $p = 0$, which gives

$$
(r, 3q^2 - 6r^2, -6rq + 8r^3, 3qr^2 - 15r^4),
\tag{9.5}
$$

and the higher degeneracy $q = 0$ occurs on

$$
(r, -6r^2, 8r^3, -15r^4).
\tag{9.6}
$$

(The swallowtail point of course goes to the origin.)

Hence the bifurcation set is parametrized by q, r as the set of points

$$
(a, b, c) = (3q^2 - 6r^2, -6rq + 8r^3, 3qr^2 - 15r^4).
\tag{9.7}
$$

We expect to find a line of cusps along the curve obtained by projecting (9.6) into C, namely the curve parametrized by r as

$$
(a, b, c) = (-6r^2, 8r^3, -15r^4).
\tag{9.8}
$$

It is easy to see that this curve has the general shape shown in Fig. 9.4 (for example a and c must be negative, and the growth rates with r are clear). Analysis of (9.7), taking sections for fixed values of a, gives a family of curves of the type shown in Fig. 9.5; hence the bifurcation set is as shown in Fig. 9.6. There is a line of self-intersection (where the function has two distinct inflexions corresponding to the two pieces of fold plane) which is parabolic in shape, with the vertex at the origin; the other half of the parabola, sometimes called the *whisker*, does not appear in the bifurcation set

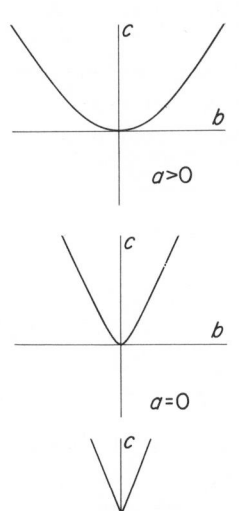

$a > 0$

$a = 0$

$a < 0$

Fig. 9.5

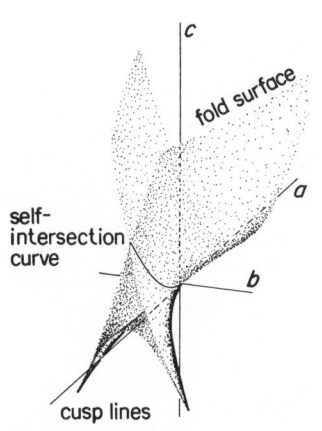

fold surface

self-intersection curve

cusp lines

Fig. 9.6

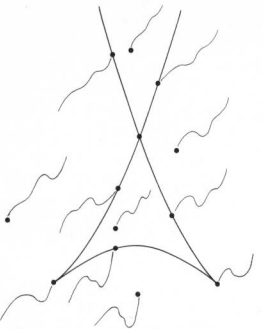

Fig. 9.7

but corresponds to certain phenomena over the complex numbers (*see* Poston and Stewart [25] p. 130).

A simple way to find the equation of the line of self-intersection is as follows. If $V_{abc}(x)$ has two points of inflexion then its derivative

$$x^4 + ax^2 + bx + c$$

is a perfect square, say

$$(x^2 + rx + s)^2.$$

Comparing coefficients we see that $r = 0$ (from the x^3 term) and hence that $b = 0$; and $c = s^2$, $a = 2s$. Eliminating s, we find the equation $a^2 = 4c$. But also, for real roots of the derivative, we must have s negative, that is a negative.

The shapes of the graphs of the functions on s for representative points in C relative to the bifurcation set are shown in Fig. 9.7; and the way the catastrophe manifold sits over the bifurcation set is shown for representative cross-sections in Fig. 9.8.

5   The Butterfly Catastrophe

The butterfly unfolding

$$V_{abcd}(x) = \tfrac{1}{6}x^6 + \frac{a}{4}x^4 + \frac{b}{3}x^3 + \frac{c}{2}x^2 + dx$$

may be treated the same way: the formulae are longer, but no different in principle; we shall not reproduce them here. The type of 6-jet defines a decomposition of the space $\mathbb{R}^4$ whose coordinates are the quadratic, cubic, quartic and quintic terms of the Taylor series;

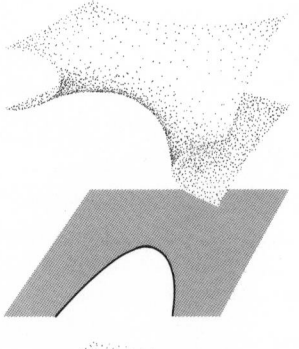

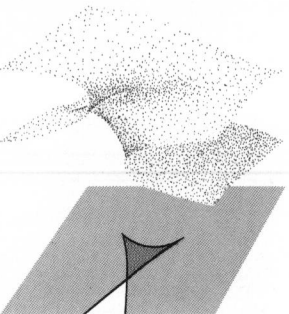

Fig. 9.8

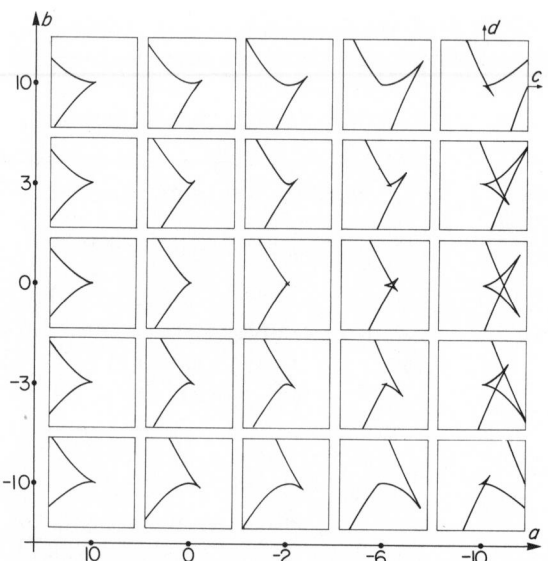

Fig. 9.9

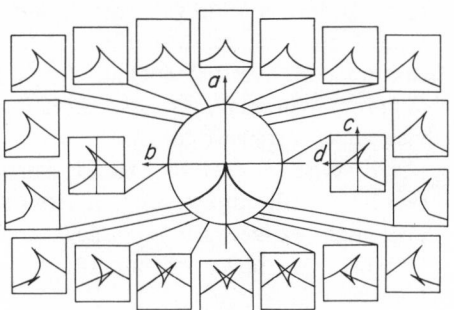

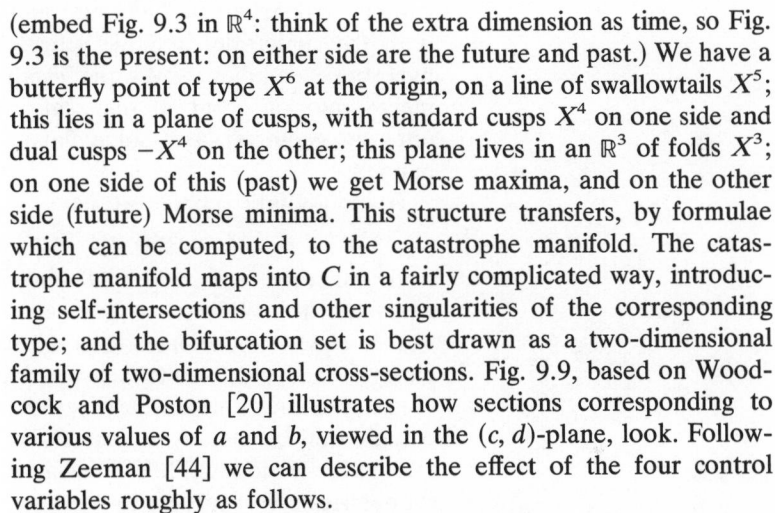

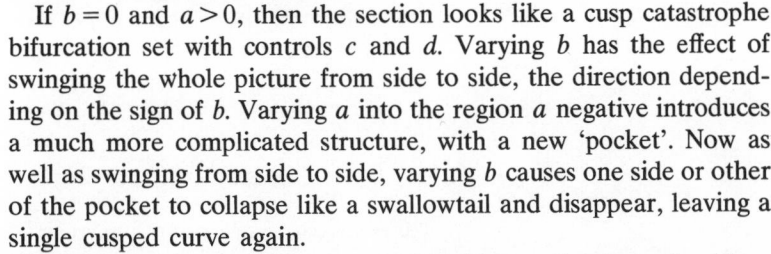

Fig. 9.10. The butterfly after K. Jänich. (From Bröcker and Lander [9].)

again we get a flag

$$\mathbb{R}^4 \supseteq \mathbb{R}^3 \supseteq \mathbb{R}^2 \supseteq \mathbb{R}^1 \supseteq \mathbb{R}^0$$

(embed Fig. 9.3 in $\mathbb{R}^4$: think of the extra dimension as time, so Fig. 9.3 is the present: on either side are the future and past.) We have a butterfly point of type X^6 at the origin, on a line of swallowtails X^5; this lies in a plane of cusps, with standard cusps X^4 on one side and dual cusps $-X^4$ on the other; this plane lives in an $\mathbb{R}^3$ of folds X^3; on one side of this (past) we get Morse maxima, and on the other side (future) Morse minima. This structure transfers, by formulae which can be computed, to the catastrophe manifold. The catastrophe manifold maps into C in a fairly complicated way, introducing self-intersections and other singularities of the corresponding type; and the bifurcation set is best drawn as a two-dimensional family of two-dimensional cross-sections. Fig. 9.9, based on Woodcock and Poston [20] illustrates how sections corresponding to various values of a and b, viewed in the (c, d)-plane, look. Following Zeeman [44] we can describe the effect of the four control variables roughly as follows.

If $b = 0$ and $a > 0$, then the section looks like a cusp catastrophe bifurcation set with controls c and d. Varying b has the effect of swinging the whole picture from side to side, the direction depending on the sign of b. Varying a into the region a negative introduces a much more complicated structure, with a new 'pocket'. Now as well as swinging from side to side, varying b causes one side or other of the pocket to collapse like a swallowtail and disappear, leaving a single cusped curve again.

Fig. 9.10 shows a different way of looking at the shape, based on Bröcker and Lander [9]. This time we vary (a, b) round the unit circle and plot sections in the (c, d)-plane; but we also mark on the (a, b)-plane the curves along which the qualitative nature of the section does not vary.

It is also useful to draw some three-dimensional sections of the bifurcation set, by stacking some of the sections of Fig. 9.9; the result is Fig. 9.11.

The types of function on X occurring for representative control points relative to B are shown in Fig. 9.12. The way in which the

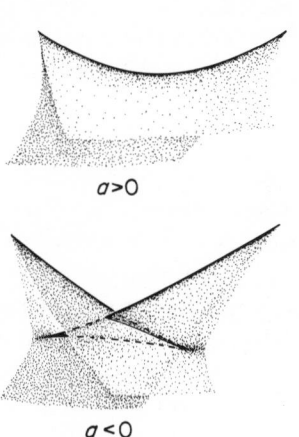

$a > 0$

$a < 0$

Fig. 9.11

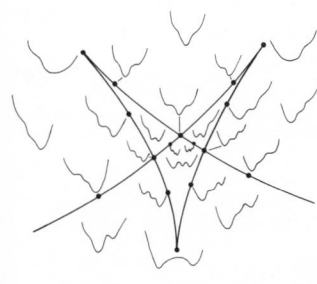

Fig. 9.12

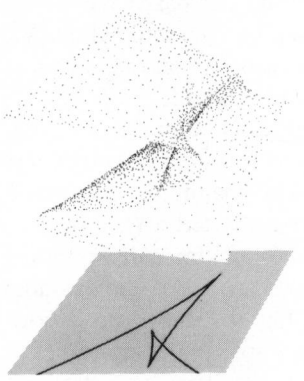

Fig. 9.13

catastrophe manifold sits over the bifurcation set is drawn, for some typical cross-sections, in Fig. 9.13. Note the five-sheeted surface that appears over the pocket, corresponding to the fact that a sixth degree polynomial may have three distinct minima, two maxima (or vice versa, for the dual case).

This type of analysis extends fairly easily to the whole cuspoid family, and the progression of cross-sections and such continues as one might expect (*see* Woodcock and Poston [20] or Zeeman [7]). However, we now turn our attention to the more interesting *umbilics*.

6   The Elliptic Umbilic

The link between the geometry of Taylor series and of catastrophes is quite striking for the umbilics; and in particular helps to explain both the similarities and the differences between the elliptic and hyperbolic umbilics. (Over the *complex* numbers these two types merge into one, but in the *real* case they exhibit dramatically different geometry. But some flavour of 'family resemblance' remains.)

For the elliptic umbilic we have

$$V_{abc}(x, y) = x^3 - 3xy^2 + a(x^2 + y^2) + bx + cy$$

as the standard form used by Thom [42]. This is different from, but equivalent to, the form we gave above in Chapter 7. It is obvious how to change coordinates to obtain $x^3 - 3xy^2$ from $x^2y - y^3$ (interchange x and y with suitable scalar factors), and the unfolding rules of the previous chapter justify the rest. (The term $x^2 + y^2$ could equally well be just x^2, or just y^2, or any number of other expressions; here we follow Thom in using $x^2 + y^2$ since this leads to useful symmetries). We analyse this particular algebraic form because it arises naturally in a major application in Chapter 11.

The catastrophe manifold M is given by the pair of equations

$$0 = \frac{\partial}{\partial x} V_{abc}(x, y) = 3x^2 - 3y^2 + 2ax + b$$

$$0 = \frac{\partial}{\partial y} V_{abc}(x, y) = -6xy + 2ay + c.$$

Therefore we can use (x, y, a) as a chart on M, where

$$(x, y, a) \mapsto (x, y, a, -2ax + 3y^2 - 3x^2, -2ay + 6xy)$$
$$= (x, y, a, b, c) \in M.$$

Now we Taylor expand about a point of M, with the result

$$V_{abc}(x + X, y + Y) = X^3 - 3XY^2$$
$$+ (3x + a)X^2 + (-6y)XY + (-3x + a)Y^2$$
$$+ 0X + 0Y$$
$$+ (-2x^3 + 6xy^2 - ax^2 - ay^2).$$

We introduce coordinates (p, q, r) for the quadratic part, so that

$$p(x, y, a) = 3x + a$$
$$q(x, y, a) = -6y$$
$$r(x, y, a) = -3x + a.$$

The quadratic term is degenerate on the discriminant cone (cf. Chapter 2 Section 5 and Chapter 7 Sections 2, 4) whose equation is

$$q^2 = 4pr$$

or

$$x^2 + y^2 = a^2/9.$$

This equation determines the singularity set in M. Since we are using charts, this set is diffeomorphic to a (double) cone. Note that this is mathematically more nasty than in the cuspoids, where the singularity set was always a manifold, indeed flat in the right coordinates: now the singularity set itself is developing singularities (the vertex of the cone). This conical structure shows up in the bifurcation set, as we shall see.

Examination of the reduction algorithm shows that when the quadratic part of the Taylor series is one-fold degenerate (two-fold degeneracy occurs only at the origin) then the cubic term in the essential variable will coincide with the restriction of $X^3 - 3XY^2$ to the straight line along which the Hessian is degenerate. This will determine the type of critical point, *unless* $X^3 - 3XY^2$ vanishes along that line, since a non-zero homogeneous cubic in one variable is automatically equivalent to x^3. If it does vanish, careful analysis is needed.

Now the cubic term has root-lines given by $X(X^2 - 3Y^2) = 0$, that is,

$$X = 0 \qquad X = \sqrt{3}\,Y \qquad X = -\sqrt{3}\,Y.$$

The quadratic term is degenerate in these directions for, respectively,

$$-q/2p = 0, \qquad \sqrt{3}, \qquad -\sqrt{3}.$$

Now the direction $p = 0$ corresponds to degeneracy in the Y-direction, which is not important, so we may assume $p \neq 0$ and solve for q, r in terms of p, getting three possible solutions for (p, q, r), namely

$$(p, 0, 0) \qquad (p, 2\sqrt{3}p, \sqrt{3}p) \qquad (p, -2\sqrt{3}p, \sqrt{3}p).$$

These are three lines in the discriminant cone (Fig. 9.14). In (x, y, a)-coordinates their equations are

$$(a/3, 0, a) \qquad (-a/6, -\sqrt{3}a/6, a) \qquad (-a/6, \sqrt{3}a/6, a).$$

The various types of critical point are marked on Fig. 9.14. On the cone but off the three lines we get fold points; on the lines we get

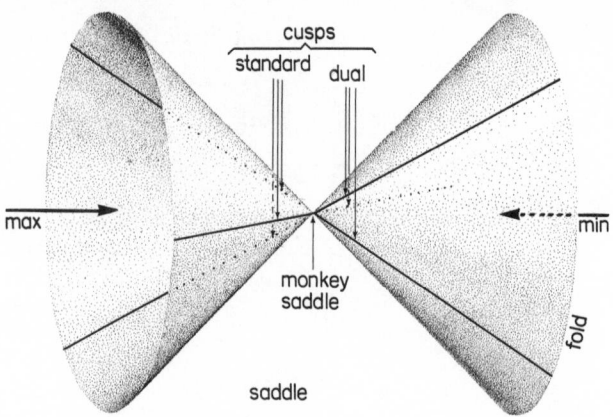

Fig. 9.14

cusps and dual cusps; at the origin the elliptic umbilic's characteristic 'monkey-saddle'. Morse maxima, minima and saddles surround the cone in the usual way.

The fact that the cusp lines involve not merely geometrical cusps in the surface, but lines of points with canonical cusp unfoldings, is quite a subtle one; where is the x^4 to unfold, among these quadratic and cubic terms? This question in fact produced example (c) of Section 13 of the previous chapter, where a jet found at such a point (with a shift of origin in (x, y)) is shown in detail to be equivalent to a cusp form, *given that its quartic part is in fact* 0 in the original coordinates. Now by Theorem 8.6 a universal unfolding of $x^3 - 3xy^2 + O(4)$ takes *exactly* the polynomial form we are studying, in suitable coordinates, so that in these coordinates we *do* know the quartic term to be 0 and can use the 4-determinacy that follows in proving the functions to be of the type claimed. (The proof that the unfolding of such a point by the two controls b, c is universal is recommended as an exercise.) A similar, more dramatic example occurs in the geometry of compound plate buckling (Chapter 13 Section 15).

Now we can find the bifurcation set by projecting into the control space, using the equations for b, c in terms of x, y and a. The three cusp lines map to the curves

$$(a, b, c) = (a, -a^2, 0), \qquad \left(a, \tfrac{1}{2}a^2, \frac{\sqrt{3}}{2} a^2\right), \qquad \left(a, \tfrac{1}{2}a^2, -\frac{\sqrt{3}}{2} a^2\right)$$

where we may think of a as a parameter. Clearly these curves are three congruent parabolae, inclined in planes at angles of 120°, as shown in Fig. 9.15. Away from these lines we have fold points only, so the rest of the bifurcation set is smooth; and along each parabola we get a line of cusps; moreover the set is the image of a double-cone. This pretty much forces the bifurcation set to look like Fig. 9.16; and we verify this as follows.

The bifurcation set is the image under projection of the cone

$$x^2 + y^2 = a^2/9$$

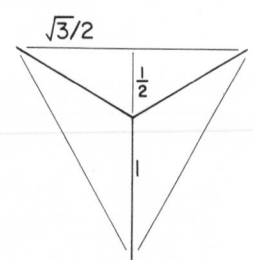

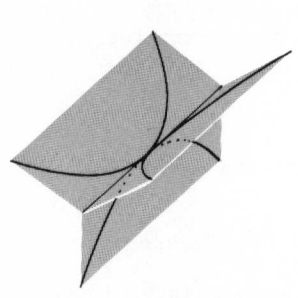

Fig. 9.15

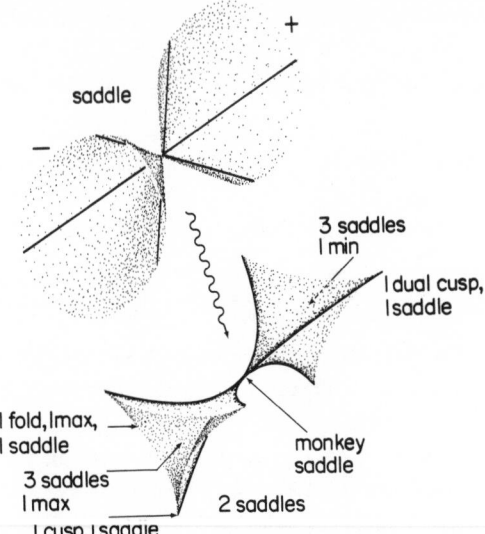

saddle

3 saddles
I min

I dual cusp,
I saddle

I fold, Imax,
I saddle

3 saddles
I max

I cusp, I saddle

monkey
saddle

2 saddles

Fig. 9.16

in M. We parametrize this cone by a, θ, as

$$(x, y, a) = \left(\frac{a}{3} \sin \theta, \frac{a}{3} \cos \theta, a\right).$$

Then the image B in control space is obtained by solving for b, c:

$$b = \tfrac{1}{3}a^2(\cos 2\theta - 2 \sin \theta)$$
$$c = \tfrac{1}{3}a^2(\sin 2\theta - 2 \cos \theta)$$
$$a = a,$$

and we continue to view a and θ as parameters. Apart from the parabolically growing factor $a^2/3$, the cross-sections for constant a parallel to the (b, c)-plane have the same θ-dependence. Now for $a = 1$ this cross-section is given by

$$b = \tfrac{1}{3}(\cos 2\theta - 2 \sin \theta)$$
$$c = \tfrac{1}{3}(\sin 2\theta - 2 \cos \theta)$$

and this is well known to be a *deltoid* or *three-cusped hypocycloid*, as shown in Fig. 9.17. Hence the constant-a cross-sections of B are similar and similarly oriented deltoids, growing parabolically.

The cusps occur in the deltoid when, writing

$$r \sin \psi = b, \qquad r \cos \psi = c,$$

we have $\partial \psi / \partial \theta = 0$. Eliminating r and differentiating, we find that

$$\sin 3\theta = -1,$$

so that $\theta = \pi/2$, $7\pi/6$, or $11\pi/6$. These lines of cusps are the same inclined parabolae that we found by studying types of critical point. We defer further study of the shape of the graphs for various

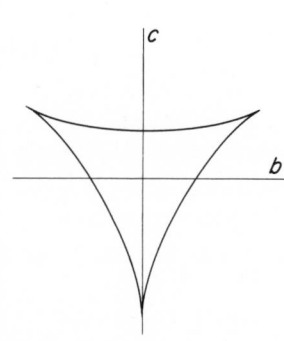

Fig. 9.17

control values until Chapter 11, where they are needed in an application. However, Fig. 9.16 has marked on it the different configurations of critical point that may occur.

7 The Hyperbolic Umbilic

For the hyperbolic umbilic we again take the standard form used by Thom [1],

$$V_{abc}(x, y) = x^3 + y^3 + axy + bx + cy$$

which (as we saw in detail in Chapter 8, Section 7) is equivalent to that in Chapter 7. The catastrophe manifold is given by

$$0 = \frac{\partial}{\partial x} V_{abc}(x, y) = 3x^2 + ay + b$$

$$0 = \frac{\partial}{\partial y} V_{abc}(x, y) = 3y^2 + ax + c,$$

so we can use (x, y, a) as a chart, by the map

$$(x, y, a) \mapsto (x, y, a, -ay - 3x^2, -ax - 3y^2) = (x, y, a, b, c) \in M.$$

Now the Taylor expansion is

$$\begin{aligned}
V_{abc}(x + X, y + Y) = X^3 + Y^3 \\
+ 3xX^2 + aXY + 3yY^2 \\
+ 0X + 0Y \\
+ (-2x^3 - 2y^3 - axy).
\end{aligned}$$

The cubic part is degenerate only in the direction $X/Y = -1$, having only one real root line. The quadratic part has natural coordinates

$$p(x, y, a) = 3x$$
$$q(x, y, a) = a$$
$$r(x, y, a) = 3y.$$

It is degenerate on the discriminant cone $q^2 = 4pr$, as before; or in (x, y, a)-coordinates,

$$36xy = a^2.$$

The cubic part only vanishes along the line of degeneracy of the quadratic when $-q/2p = -1$, that is, $q = 2p$ and $r = p$: this happens on the line parametrized by p as

$$(p, 2p, p)$$

which in (x, y, a)-coordinates is given by $(p/3, p/3, 2p)$ or

$$(x, x, 6x)$$

using x as parameter.

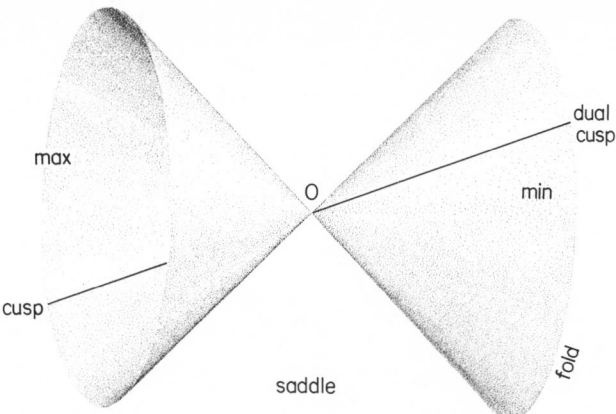

Fig. 9.18

As before, we find fold points on the cone, except for cusps and dual cusps along this line; and Morse points off the cone, as shown in Fig. 9.18.

Next we seek the images of these parts of the cone in the bifurcation set. We parametrize the cone

$$a^2 = 36xy$$

by

$$a = 6\alpha \qquad x = \alpha\xi \qquad y = \alpha/\xi.$$

This parametrization fails to take account of the lines $x = a = 0$, $y = a = 0$; but it is easy to deal with these separately. Then the singularity set S in M is parametrized as

$$(\alpha\xi, \alpha/\xi, 6\alpha, -6\alpha^2/\xi - 3\alpha^2\xi^2, -6\alpha^2\xi - 3\alpha^2/\xi^2)$$

and the image of the cusp line is found by putting $\xi = 1$, namely

$$(\alpha, \alpha, 6\alpha, -9\alpha^2, -9\alpha^2).$$

Thus B is parametrized as

$$(6\alpha, \alpha^2(-6/\xi - 3\xi^2), \alpha^2(-6\xi - 3/\xi^2)),$$

away from $\alpha = 0$. Again the constant a cross-sections are similar, and similarly oriented, with a parabolic scale factor. A typical section for $\alpha = 1$ is given in Fig. 9.19. The image of the cusp line is a parabola, inclined at $\pi/4$ to the negative b- and c-axes.

The missing lines map as follows. If $a = x = 0$, the line maps to

$$(0, y, 0, 0, -3y^2)$$

in M, hence to the negative c-axis in C; similarly the line $a = y = 0$ maps to the negative b-axis. Thus the bifurcation set has the shape shown in Fig. 9.20. Note how the surface has a self-intersection corresponding to the two exceptional lines just studied; this is where the images of the two half-cones meet.

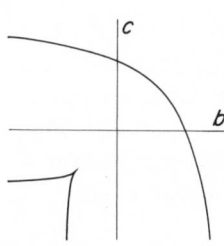

Fig. 9.19

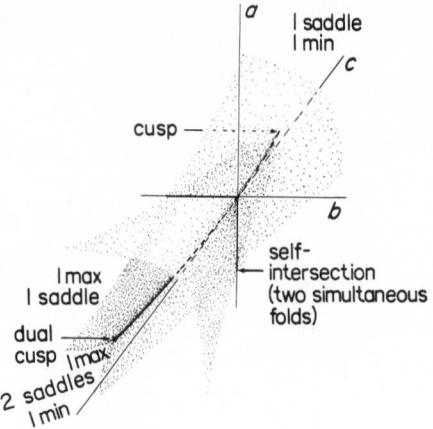

Fig. 9.20

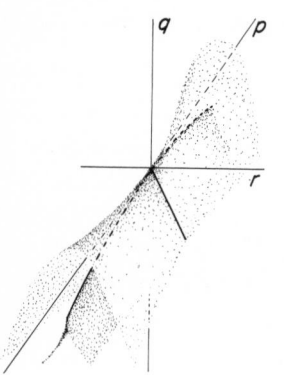

Fig. 9.21

The cone picture (Fig. 9.18) helps explain the way that the cusp curve – the image of a line under a C^∞ map – jumps from one surface to the other. Attempts to continue it within one surface lead to a curve with discontinuous derivatives.

The different types of critical point occurring are marked on Fig. 9.20. Contours for various control points are shown in Thom [42].

If in the above analysis the parameter ξ is written in the form e^θ then the analogy with the elliptic umbilic is strengthened. The formula giving the bifurcation set involves hyperbolic rather than trigonometric functions, but is otherwise analogous. One would expect this to happen because of the way that the elliptic and hyperbolic umbilics become equivalent over the complex numbers.

Fig. 9.21 shows the bifurcation set for the alternative form

$$x^2y + y^3 + p(x^2 + y^2) + qy + rx$$

for this catastrophe, deduced (by applying the transformation in Chapter 8 Section 7) from the above analysis. The similar topology is immediately visible.

8   The Parabolic Umbilic

The parabolic umbilic is decidedly trickier to handle: in fact the first edition of Thom [1] contained minor errors. The geometry was worked out in detail by Chenciner [45], Godwin [46] and computer-drawings obtained by Woodcock and Poston [20]. The following account is sketched only (since we have no particular application of the parabolic umbilic in mind) and is based on Zeeman [7].

We take the algebraic form

$$V_{abcd}(x, y) = x^2y + y^4 + ax^2 + by^2 + cx + dy.$$

The catastrophe manifold is given by

$$0 = \frac{\partial}{\partial x} V_{abcd}(x, y) = 2xy + 2ax + c$$

$$0 = \frac{\partial}{\partial y} V_{abcd}(x, y) = 4y^3 + x^2 + 2by + d.$$

Hence we use (x, y, a, b) as a chart on M, with the map

$$(x, y, a, b) \mapsto (x, y, a, b - 2ax - 2xy, -2by - x^2 - 4y^3)$$
$$= (x, y, a, b, c, d) \in M.$$

The Taylor expansion is given by

$$\begin{aligned} V_{abcd}(x + X, y + Y) = {}& Y^4 + X^2 Y + 4y Y^3 \\ &+ (y + a) X^2 + 2x XY + (6y^2 + b) Y^2 \\ &+ 0X + 0Y \\ &+ \text{constant}. \end{aligned}$$

This suggests a new chart (p, q, r, s) where

$$p(x, y, a, b) = y + a$$
$$q(x, y, a, b) = 2x$$
$$r(x, y, a, b) = 6y^2 + b$$
$$s(x, y, a, b) = 4y,$$

which relates to the old chart by

$$x(p, q, r, s) = q/2$$
$$y(p, q, r, s) = s/4$$
$$a(p, q, r, s) = p - s/4$$
$$b(p, q, r, s) = r - 3s^2/8.$$

The quadratic part of the Taylor series is degenerate when $q^2 = 4pr$, and since this equation does not depend on s, the set it defines is of the form

$$K \times \mathbb{R} \subseteq \mathbb{R}^4$$

where K is the double cone $s = 0$, $q^2 = 4pr$ and $\mathbb{R}$ is the s-axis.

Keeping in (p, q, r, s)-space, we can find the types of critical point encountered. There is double-degeneracy only on the line $p = q = r = 0$, or $0 \times \mathbb{R}$: the s-axis. If $s > 0$ we get hyperbolic umbilic points, if $s < 0$ we get elliptic umbilic points (by inspection of the 3-jet). Off $0 \times \mathbb{R}$ the degeneracy is one-fold, hence there is only one essential variable and we must get cuspoids.

The catastrophe map, in terms of the (p, q, r, s)-chart, takes the form

$$\chi : (p, q, r, s) \mapsto (p - s/4, r - 3s^2/8, -pq, s^3/8 - rs/2 - q^2/4)$$
$$= (a, b, c, d) \in C.$$

We can find the nature of the cuspoid points by computing the singularity structure of χ in this form. The result is:

folds $4pr = q^2 \neq 4p^2 s$

cusps $4pr = q^2 = 4p^2 s, \qquad p^2 \neq r$

swallowtails $4pr = q^2 = 4p^2 s, \qquad p^2 = r, \qquad p \neq 0.$

The cusps are standard for $s < 0$, dual for $s > 0$, except that in the latter case standard cusps occur between the swallowtails. To make this clearer, we shall introduce a geometric picture due to Zeeman [7].

Somehow it is necessary to collapse the set $K \times \mathbb{R}$ in $\mathbb{R}^4$ into something three-dimensional, so that we can draw it. Now, as a *set*, $K \times \mathbb{R}$ is already three-dimensional, because K is a two-dimensional surface: the trouble is that it is bent. Hence we flatten it out. Specifically, the projection

$$(p, q, r, s) \mapsto (p - r, q, s)$$

flattens each component of the complement of the origin diffeomorphically. For a single $s = \text{constant}$ cross-section, Fig. 9.22 shows how this flattening works: the end of the cone becomes a (punctured) disc. Now we can stack the discs in the s-direction, and mark on them the type of critical point. The result is Fig. 9.23.

While this figure requires careful interpretation, because of the various distortions involved, one thing that is clear is how the three lines on the cone, corresponding to elliptic umbilics, merge at the $s = 0$ level, with two disappearing and only the single line of the hyperbolic umbilic remaining. This dramatizes the parabolic umbilic's role as a transition between the elliptic and hyperbolic umbilic.

To find the bifurcation set we now substitute $q^2 = 4pr$ into the equation for the catastrophe map and see where the various types of critical point occur. This is a long job analytically, and we do not intend to do it here! The results require several different kinds of picture, each bringing out particular features.

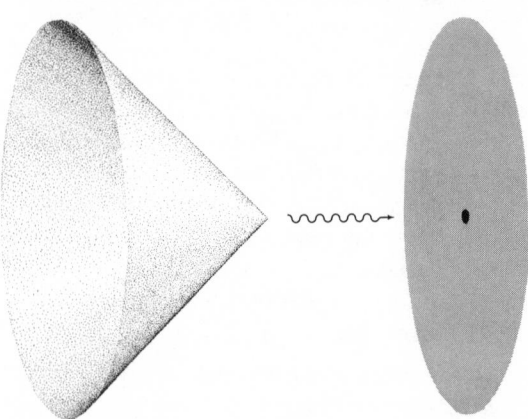

Fig. 9.22

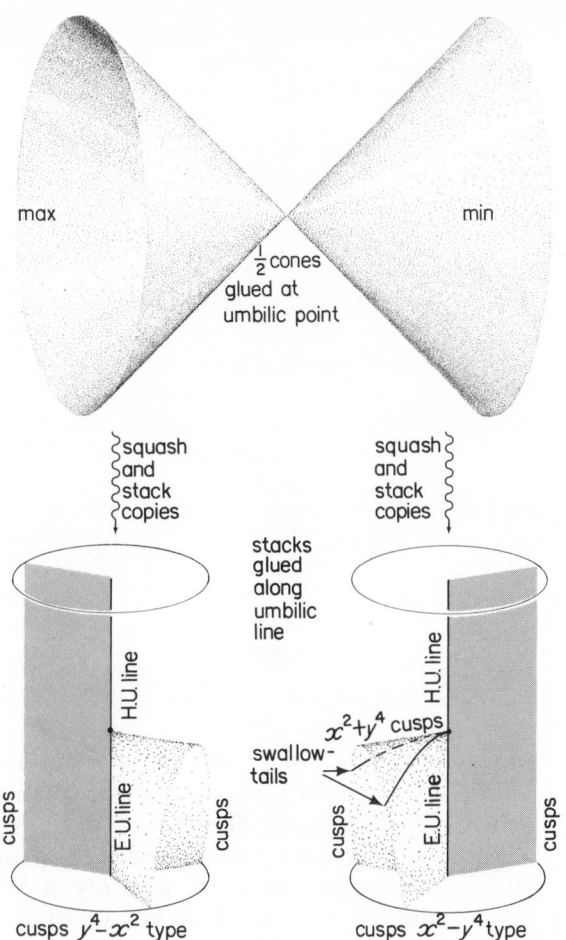

Fig. 9.23

Fig. 9.24 is based on Bröcker and Lander [9]. It shows, for the unit circle in the (a, b)-plane, the corresponding cross-sections in the (c, d)-direction. In 16 we see the usual cusp bifurcation set. At 1 a single point appears, growing in 2 to give the 'lips'. This development occurs by taking cross-sections of a line of cusps, as in Fig. 9.25, and is not a new type of catastrophe. In 3 the original cusp point penetrates the mouth; in 4 it approaches the far side of the mouth in a manner reminiscent of the hyperbolic umbilic (Fig. 9.20). At 5 we encounter a hyperbolic umbilic point: the tip of the cusp merges with the mouth at a corner, and passes through to 6 where a cusped triangle of a type reminiscent of the elliptic umbilic appears. This triangle shrinks, passing through the surrounding fold curve at 7, and shrinks to a point, 9, in an elliptic umbilic catastrophe, growing again in 10 to pierce the fold line, 11, and pass through it, 12. Now the top edge of the triangle approaches the fold curve, touches it in a 'beak-to-beak' singularity at 13, and breaks apart in 14. The beak-to-beak singularity is also just a line of cusps (Fig. 13.16). The two

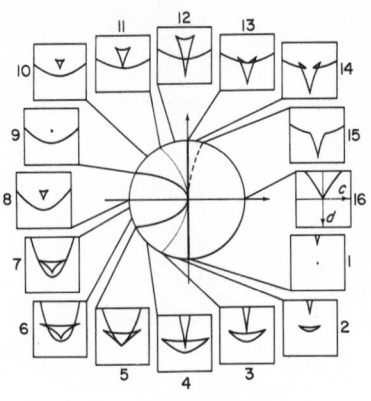

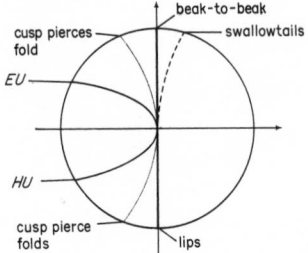

Fig. 9.24 *The parabolic umbilic after K. Jänich. (From Bröcker and Lander* [9].)

protrusions in 14 collapse in a typical swallowtail sequence at 15, and we return to the traditional cusp, 16, having come full circle.

This sequence illustrates how all the catastrophes of codimension 3 or less appear somewhere in the parabolic umbilic: of the seven in Thom's list, only the butterfly does not so appear. The seven catastrophes contain 'subcatastrophes' (that is, their unfoldings contain functions whose critical points are of the relevant type) according to the following *subordination diagram*:

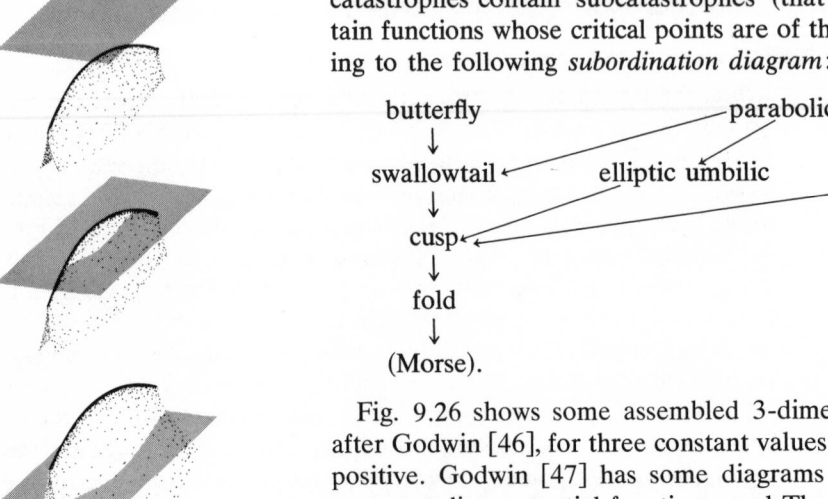

Fig. 9.26 shows some assembled 3-dimensional sections, drawn after Godwin [46], for three constant values of *d*: negative, zero and positive. Godwin [47] has some diagrams of the contours of the corresponding potential functions, and Thom [42] also shows some contours (but is slightly misleading, since they correspond to points on the non-local bifurcation set: cf. Chapter 11, Section 12). Unlike the elliptic and hyperbolic umbilics, the parabolic is not self-dual, so the

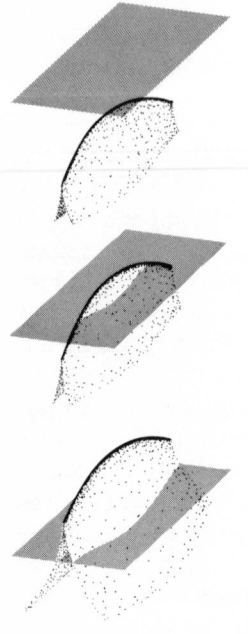

Fig. 9.25

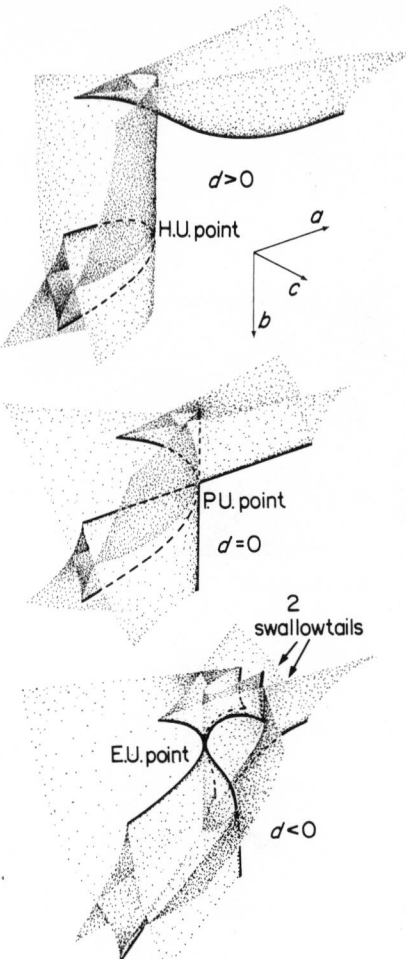

Fig. 9.26

disposition of maxima, minima, cusps, dual cusps, etc. depends on the sign encountered. In either case the region for which *any* minimum exists is quite small, as the reader may check by exploration (using the structure of catastrophes already analysed, to move from point to point).

9 Ruled Surfaces

The computer drawings of Woodcock and Poston [20] are obtained by a quite different method, which we shall indicate briefly. For example, take the cusp catastrophe, with catastrophe manifold parametrized as

$$(x, a, -ax - x^3).$$

If we fix x then, as a varies, we obtain a straight line in $\mathbb{R}^3$, whose

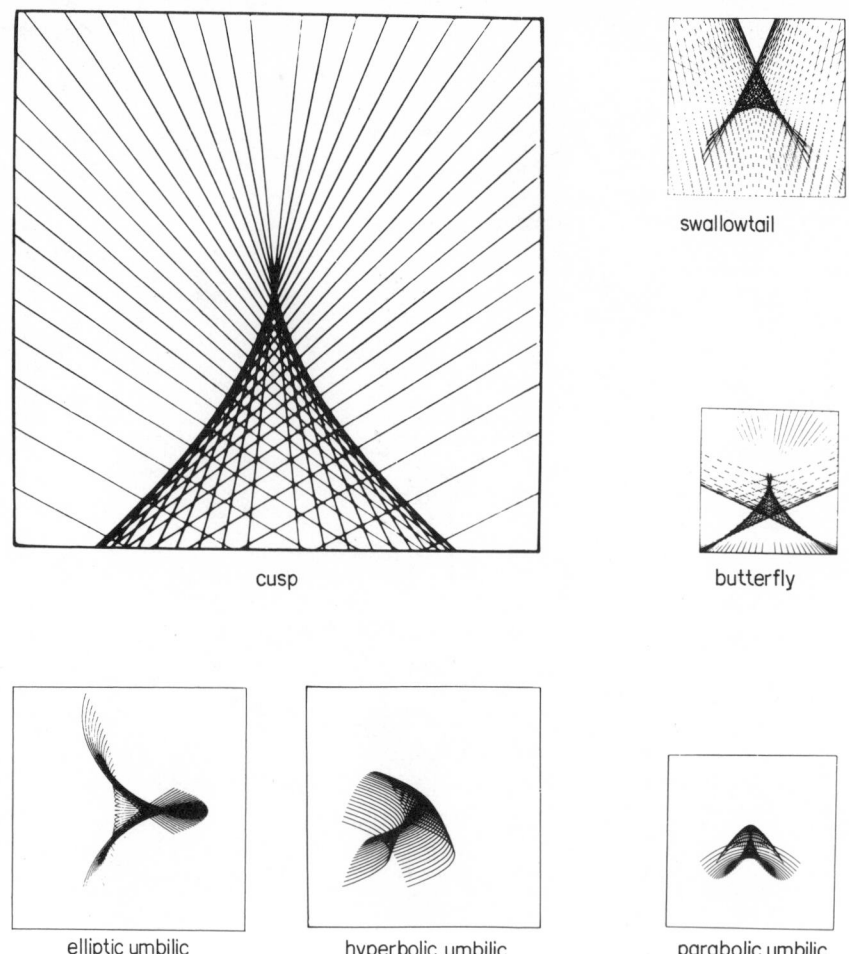

swallowtail

cusp

butterfly

Fig. 9.27   elliptic umbilic   hyperbolic umbilic   parabolic umbilic

projection onto (a, b)-space is given by

$$b = -ax - x^3.$$

Now varying x, the family of lines in $\mathbb{R}^3$ traces out a *ruled surface*, which is the catastrophe manifold, and the projection of this into (a, b)-space gives a family of lines, whose *envelope* is the bifurcation set. Fig. 5.15 illustrates this. The method is applicable generally to the cuspoids (whose catastrophe manifolds are all ruled by straight lines) and in a modified form to the umbilics (using curved lines corresponding to a family of parallel lines in a suitable chart). Fig. 9.27 shows a typical picture for each of the seven elementary catastrophes (bar the fold), reproduced from Woodcock and Poston [20]. Note how these pictures give some information on how the catastrophe manifold sits over the bifurcation set (information that must be interpreted carefully for the umbilics!)

This technique is often very useful in examining physical, non-polynomial unfoldings in the large (beyond the region where the reductions of Chapter 8 are applicable). For instance, drawing the catastrophe geometry of the Zeeman machine this way (Fig. 1.2) was not only more illuminating than drawing *only* the bifurcation set, since it made the catastrophe manifold more apparent: it was also a great deal easier, as is very clear from Poston and Woodcock [47a].

A new and unusually cheap way of getting catastrophe (and other geometric) three-dimensional pictures from a small computer is described in Rockwood and Burton [47b].

10 Stability of Ships

Sometimes a catastrophe theory approach can reformulate 'classical' results in an unexpected way; and reformulations of this type are often useful as the basis of new extensions of the older work. In order to proceed, it may be necessary to retrace one's steps a little first. An example of this is Zeeman's recent work on the stability of ships. Here we give an introduction to this, largely based on his lectures, and dealing only with *static* equilibrium. Zeeman's formulation of the problem is suited to a study of *dynamic* equilibrium, which is of clear practical significance, but it would take us too far afield to give an adequate discussion. This account should therefore be taken as a starting-point for the more comprehensive treatment to be found in Zeeman [48].

STATIC EQUILIBRIUM

1 Buoyancy

Consider a rigid two-dimensional ship S in water (Fig. 10.1(a)). (We move up to three dimensions in Section 9: much of the theory extends without essential change.) By high school physics, the upward forces on the ship exerted by the water have exactly the effect of a vertical force equal to the weight of displaced water K (Fig. 10.1(b)) acting at the centre of gravity of K. If the density of the water is constant, which we shall assume, this centre of gravity is the geometric *centroid* of the submerged region of S which replaces K. For given attitude θ, this force will be greater for lower positions of S (unless water starts pouring in somewhere). Thus if S can float at all there will be for each θ a *unique* height at which the upward force equals the weight of S. In catastrophe theoretic language, then, vertical position is an *inessential variable* and we can parametrize the position of S by θ alone, for the purpose of equilibrium studies. (Indeed, even in dynamic analysis: movement away from vertical equilibrium meets such immense restoring forces, and is so heavily damped, that only strange resonance effects could make it non-negligible.)

Thus for each θ we have a well-defined height for S and hence a well-defined geometry for the submerged shape K, with its centroid $B(\theta)$. The point $B(\theta)$ is the *centre of buoyancy* of S at position θ. The weight of S (including cargo, etc.) or equivalently that of the

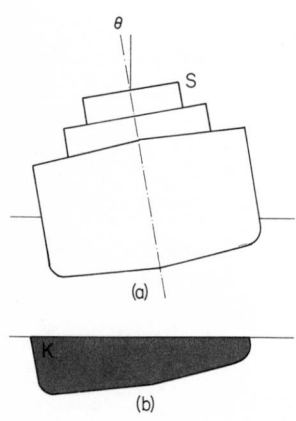

Fig. 10.1

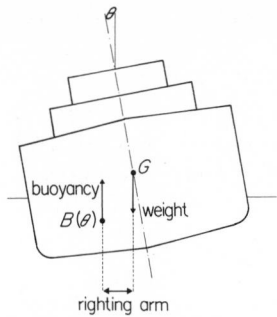

Fig. 10.2

water S displaces when in vertical equilibrium, is called the *displacement.*

2 Equilibrium

For equilibrium the weight and the buoyancy must not only be forces of equal size, as we assumed above. For rotational equilibrium they must be in the same vertical line. Forces as in Fig. 10.2 would exert a moment (turning effect) given by the product of their common value by the distance between the lines along which they act. Given a positive sign when the resulting moment tends to decrease θ (rather than using signs to distinguish clockwise and anticlockwise moments, as a mathematician might) this distance is called nautically the *righting arm.*

Stability

If $B(\theta)$ were a fixed point of support for S (as it effectively is for a submarine – why?) stability would require that the centre of gravity G of S be below $B(\theta)$. For movement from an equilibrium position like Fig. 10.3(a) would take us to Fig. 10.3(c). But in fact $B(\theta)$ moves with θ. Thus situations like Fig. 10.2 are possible, in which G is *above* $B(\theta)$, but rotation away from $\theta = 0$ produces a *restoring* moment, and the upright position is a stable equilibrium. Thus stability depends on the manner in which $B(\theta)$ varies with θ. Before making any further theoretical remarks we analyse a particular (rather common) case.

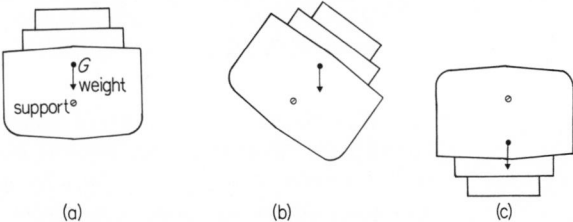

Fig. 10.3   (a)   (b)   (c)

4 The Vertical-sided Ship

Consider a symmetrical ship S of width $2w$ with straight, vertical sides AD and A′D′ (Fig. 10.4(a)). Suppose it lists to an angle θ, with

$$\tan \theta = t,$$

for which the water level stays between A and D, A′ and D′ (Fig. 10.4(b)). Choose coordinates (x, y) fixed relative to S as shown. Define l, h by requiring that the line AA′ be $y = l$, and the waterline

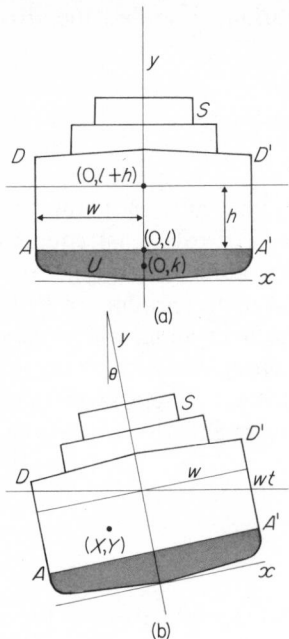

Fig. 10.4

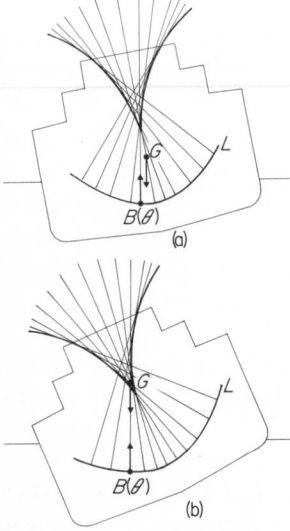

Fig. 10.5

in equilibrium at $\theta = 0$ be $y = l + h$. Let the part U of S below AA' have centroid $(0, k)$ and area V.

If S rolls to an angle θ keeping the point $(0, l + h)$ fixed at water level, equal triangles are added to and subtracted from the submerged region. Thus vertical equilibrium *requires* that $(0, l + h)$ remain at this level, and we can find the centre of buoyancy $B(\theta) = (X, Y)$ by taking moments:

(a) Parallel to the x-axis:

$$Y \text{ (submerged area)} = kV + (l + \tfrac{1}{2}(h - wt))(2w(h - wt)) + (l + \tfrac{1}{2}(h - wt) + \tfrac{2}{3}wt)(\tfrac{1}{2} \cdot 2w \cdot 2wt)$$

so

$$Y = \frac{(kV + 2lwh + wh^2)}{V + 2wh} + \frac{w^3}{3(V + 2wh)} t^2.$$

(b) Parallel to the y-axis:

$$X(V + 2wh) = -\frac{w}{3}(\tfrac{1}{2} \cdot 2w \cdot 2wt)$$

so

$$X = \frac{-2w^3}{3(V + 2wh)} t.$$

Since k, w, h and V are constants (at least for a given weight of the ship and its cargo) we can write this conveniently as

$$X = -2ct, \qquad Y = k + ct^2$$

with combined constants k and c.

Thus the set of points described by $B(\theta)$ as θ varies, which we call the *buoyancy locus L*, is the parabola

$$Y = k + \frac{1}{4c} X^2.$$

Fig. 10.5 shows how this curve lies, together with, for each θ, the line through $B(\theta)$ that is vertical when the ship is inclined at angle θ. It is clear that if the centre of gravity G of S is *below* the cusp point, then the equilibrium at $\theta = 0$ will be stable (Fig. 10.5(a)); *above*, and it will be unstable. (To see this, consider the turning moments.) In the latter case, however, there are two new stable equilibria, with the ship listing slightly to one side or the other (Fig. 10.5(b)).

This looks very like the behaviour of the parabolic roller catastrophe machine discussed in Chapters 1 and 5. To show that the geometry corresponds exactly, we must prove that the lines in Fig. 10.5 are again the normals to the parabola. Since they are defined to be vertical when the ship is at angle θ, we must show that when S is at angle θ the buoyancy locus L passes *horizontally* through $B(\theta)$. This is trivial analytically for the special case above, but for more general ships we can prove it as follows.

5   Geometry of the Buoyancy Locus

Consider an arbitrary ship S of weight W, in vertical equilibrium at some given position in the water (Fig. 10.6(a)). Its centre of buoyancy B is the position that would be occupied by the centre of gravity of the same weight W of water, in an S-shaped vessel S' (Fig. 10.6(b)). The water's surface would correspond to the water-line of Fig. 10.6(a). This is the unique minimum energy arrangement for the water in a vessel of this nature held in this way (a fact which is easy to prove, but so intuitively obvious that we do not do so). It therefore must be the arrangement for which the height of the centre of gravity of the water is least. Thus any change in attitude of S gives a new configuration (Fig. 10.6(c)) of water in S', with a new centroid B', necessarily *higher* than B in terms of the original vertical. Thus the buoyancy locus L has its unique lowest point (in terms of this vertical) at B. A minimum, if differentiable, must be horizontal, so L passes through B horizontally, which is what we set out to prove.

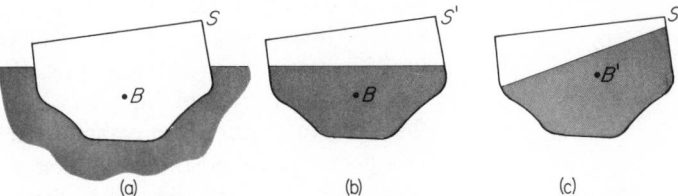

Fig. 10.6

    (a)   (b)   (c)

This argument depends on L having no corners. For an arbitrary piecewise differentiable ship geometry this will be true for almost all values of the displacement. [Technically, in two or three dimensions, the buoyancy locus is parametrized by ship attitude in a C^1 (but perhaps not better than piecewise C^2) way for all except a set of displacements of measure zero. The proof of this, however, appeals to theorems of differential topology beyond the scope of this book. (*See* Pitt and Poston [49].)

In all cases, the above argument shows L to be strictly convex, and nothing in the argument restricts it to two-dimensional ships.

6   Metacentres

We see, then, that a ship has the identical global static equilibrium properties as does a convex rolling catastrophe machine, the shape of the buoyancy locus L for a given displacement. The control is the position of the centre of gravity of the ship, and the bifurcation set is the evolute of L. (For L not C^2 we must broaden the definition of 'evolute' slightly.)

In particular the straight-sided ship at moderate inclinations has equilibrium geometry exactly modelled, through a linear change of variables, by the parabolic roller, which is governed as we have seen by the standard cusp catastrophe in its canonical polynomial form. The relevant energy function is also that of the roller. For changing the height of the centre of buoyancy does not change the energy if we take into account the energy of the water no longer displaced; hence the energy is given by the height of the centre of gravity above the instantaneous centre of buoyancy, regardless of any vertical motion of the latter.

We may compare this analysis with that of a standard text for ship designers. In *Principles of Naval Architecture (Revised), Written by a Group of Authorities,* edited John P. Comstock, and published by the Society of Naval Architects and Marine Engineers (New York 1967), we find the following (p. 70).

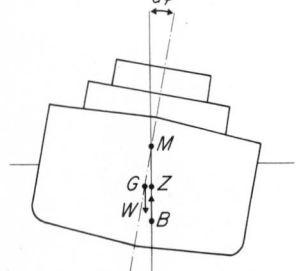

Fig. 10.7. (Courtesy of the Society of Naval Architects and Marine Engineers.)

'*The transverse metacenter and transverse metacentric height.* Consider a symmetrical ship heeled to a very small angle, dϕ, shown, with the angle exaggerated, in [Fig. 10.7]. The center of buoyancy has moved off the ship's centerline as the result of the inclination, and the lines along which the resultants of weight and buoyancy act are separated by a distance, $\overline{GZ}$, the righting arm. A vertical line through the center of buoyancy will intersect the original vertical through the center of buoyancy, which is the ship's centerline plane, at a point M, called the transverse metacenter. The location of this point will vary with the ship's displacement and draft, but, for any given draft, it will always be in the same place. Unless there is an abrupt change in the shape of the ship in the vicinity of the waterline, point M will remain practically stationary with respect to the ship as the ship is inclined to small angles, up to about 7 or, sometimes, 10 deg.'

The metacentre M for the ship when vertical is thus the centre of curvature of L at $\theta = 0$; in our example above, the cusp point. The definition quoted makes M lie always on the centre line for different θ. Thus M is not in general the same as the centre of curvature of L, which moves off along the evolute (in our example, the cusp lines). We shall nevertheless call the evolute in this context the *metacentric locus.* Symmetry of the ship implies that the curvature of L will have a local maximum or minimum at $\theta = 0$, hence the metacentric locus will always have a singularity there: typically a standard or dual cusp.

We have here an interesting illustration of a recurrent theme of this book: the nature of derivatives as approximations. For S as in Section 4, M is given by the intersection of the y-axis with the line

$$x + 2ct = t(y - (k + ct^2)),$$

vertical when $\tan \theta = t$. This gives

$$y = k + 2c + ct^2,$$

which is linearly approximated by the constant $k + 2c$, around $t = 0$. We learn that this approximation is practical here up to about

$$t = \tan 7° \sim 0.12,$$

or sometimes,

$$t = \tan 10° \sim 0.18.$$

This makes t^2 about 0.015 or 0.03. The quadratic description is exact as long as only the straight sides AD, A'D', meet the water-line. More general shapes, of course, give less simply polynomial results, and the catastrophe-theoretic tools for precise amputation of Tayls come in useful.

Great emphasis is laid, in the legal constraints on the construction and loading of ships, on the *metacentric height*: the distance by which the metacentre for $\theta = 0$ is above the ship's centre of gravity. This is essentially a measure of 'infinitesimal' stability, giving the ship's response to a small perturbation. A typical destroyer has metacentric height three-quarters of a metre, beam 10 metres and performs small oscillations with a period of about 8 seconds; a liner has metacentric height half a metre, beam 25 metres and oscillates with a period of some 24 seconds. Thus a liner (more sluggish to manoeuvre than a destroyer) reacts less violently to rolling. To compare the metacentric height with more global stability measures, we need another example.

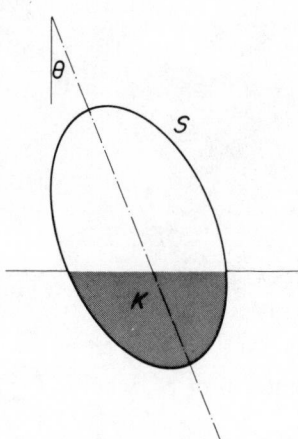

Fig. 10.8

SHIP SHAPES

7 The Elliptical Ship

This shape is not drawn from real shipbuilding practice like that of Section 4, but it is convenient for exposition in that it exemplifies without computation some important points.

Let S be an ellipse with area A. We take its displacement to be kA, for some fixed k between 0 and 1, and investigate the buoyancy locus. For an arbitrary θ as in Fig. 10.8, we want the centroid of the submerged region K defined by the condition that its top be level and its area kA.

Now, for an analogous *circular* ship S' of radius 1, it is easy to show that the submerged region is as in Fig. 10.9(a), where s is the unique solution of

$$s = \cos(k\pi + s\sqrt{(1 - s^2)}).$$

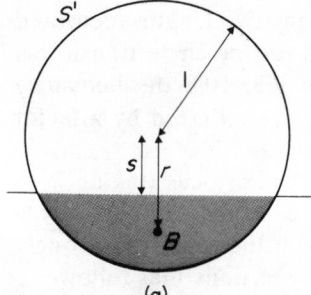

(a)

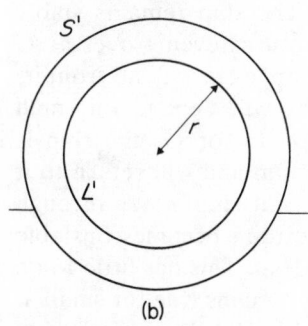

(b)

Fig. 10.9

Fig. 10.10

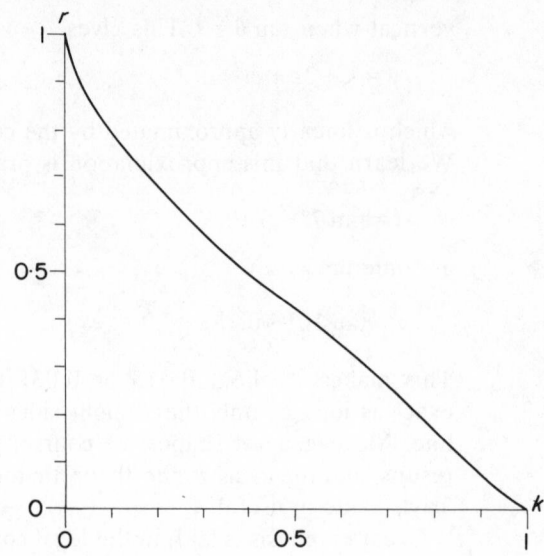

Fig. 10.11

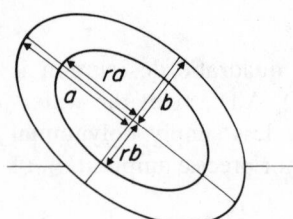

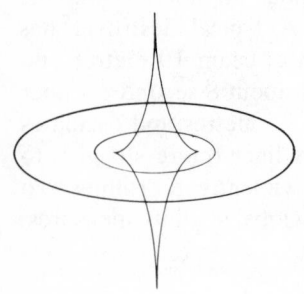

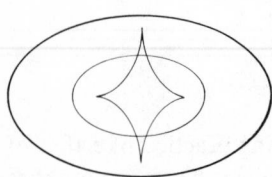

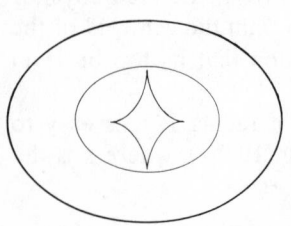

Fig. 10.12

Its centroid B is at a point r below the centre 0 of S', where

$$r = \frac{2}{3k\pi}(1-s^2)^{3/2}$$

Thus the buoyancy locus L' of S' is a circle with centre 0 and radius r (Fig. 10.9(b)), where r depends on k as shown in Fig. 10.10.

Now a circle may be mapped to any given ellipse with the same centre by a linear change of coordinates. Such a map preserves lines, ratios of areas, centroids, and the ratio of any two lengths measured along the same line. Thus the above result for the circle transforms into the fact that for the elliptical ship of Fig. 10.8 the buoyancy locus is a similar ellipse (Fig. 10.11) with size adjusted by a factor r (still given by k as in Fig. 10.10).

Fig. 10.12 shows the evolute of the buoyancy locus (taking k to make $r = \frac{1}{2}$) for elliptical ships of various eccentricities. Experiments with the elliptical roller of Chapter 1, which the reader is strongly recommended to make, will bear out the assertions that follow.

Moving the centre of gravity along the long axis of the ellipse will give the behaviour shown in Fig. 10.13. The ship remains stably upright until G reaches the metacentre P, turns evenly over as G goes from P to Q, then remains steady the opposite way up from its starting position. (Which direction it turns will depend on small factors as G passes P.) With G in position G_1 or G_2 the ship is *self-righting*: it has only one stable equilibrium and will return to it from any other position, provided it does not ship water through topside openings. Note that the upright position becomes unstable by way of a standard cusp like that of Fig. 10.6. This has little to do with the vertical character of the sides, as it remains true for small k, large r, and a floating position like Fig. 10.14. The change from stability to instability of uprightness is sudden, but the result is at first only a small inclination of the ship. We have the usual parabolic branching associated with the standard cusp (*see* Fig. 6.3(a)).

Fig. 10.13

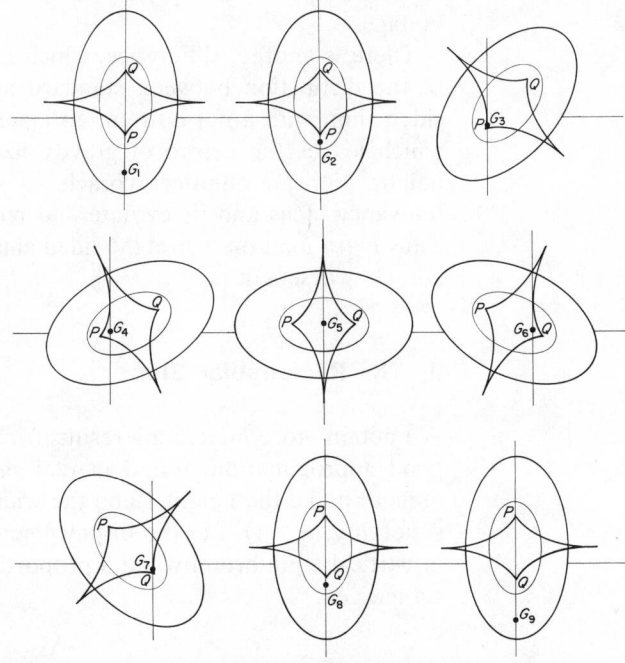

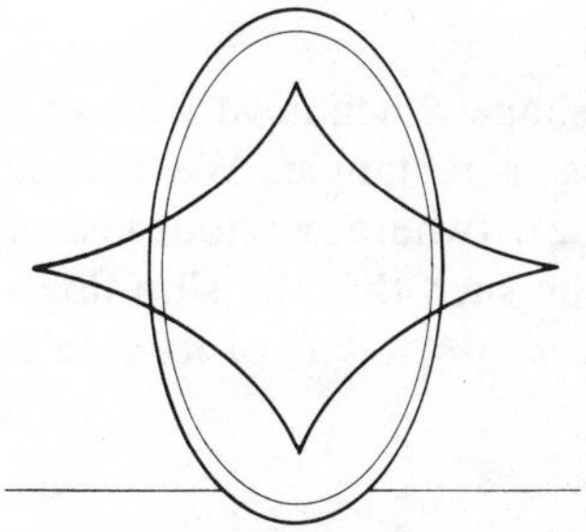

Fig. 10.14

Fig. 10.15

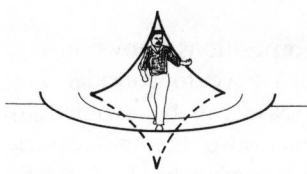

By contrast, consider someone in a coracle of elliptical cross-section (Fig. 10.15). For simplicity, assume he is a mathematician, with the whole weight of himself and the boat concentrated in his head. The boat is in stable equilibrium as shown. But suppose he leans over, just far enough to shift the centre of gravity outside the metacentric locus. There is now no stable equilibrium with the boat the right way up. Instead of tilting a little ... it turns turtle (Fig. 10.16).

More seriously, consider a ship with a cross-section of this form. A load to one side, as in Fig. 10.17, can easily shift the centre of gravity so as to overturn the ship. This kind of dual cusp instability, shown for instance by the shallow rounded boats of the Vikings and

Fig. 10.16

Fig. 10.17

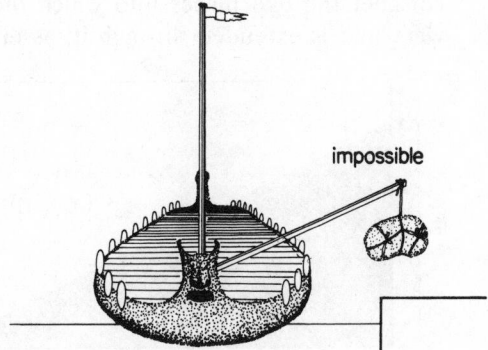

the canoe built in his teens by one of the authors (which capsized everyone who borrowed it) is in contrast to the standard cusp behaviour produced by the vertical-sided ships of more modern vintage.

There is another difference, which may prove to be characteristic of the distinction between standard and dual cusps. In a vertical-sided ship, with a flat bottom, an increase in the weight of the ship which leaves the centre of gravity fixed *increases* the metacentric height. For an elliptical coracle, a similar increase *shrinks* the buoyancy locus and its evolute and *reduces* the metacentric height. Thus extra load on a straight-sided ship improves stability and on a coracle worsens it.

8 The Rectangular Ship

We obtain more interesting results for a shape which is, at least to a good approximation, found in real ships: a rectangle. We choose units to make the height 1 and the width $2w$ (where, without loss of generality, $w \geqslant 1$). Let the displacement be such that the ship floats in vertical equilibrium when a proportion λ $(0 < \lambda < 1)$ of its area is submerged.

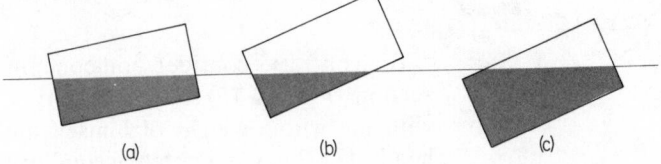

Fig. 10.18   (a)   (b)   (c)

Provided $\lambda \leqslant \frac{1}{2}$ we need analyse only two positions, shown in Fig. 10.18(a) and (b). For $\lambda > \frac{1}{2}$ we would also have to consider Fig. 10.18(c). But separate treatment of this case can be avoided by noting that the buoyancy locus (and hence also the metacentric locus) for $\lambda \geqslant \frac{1}{2}$ is found from that for λ replaced by $1 - \lambda \leqslant \frac{1}{2}$ by rotating 180° about the centre of the rectangle (which here has no effect, by rotational symmetry, but the argument holds for asymmetric shapes too), and scaling by a factor $(1 - \lambda)/\lambda$. To prove this, consider the two pieces into which the rectangle divides when the waterline is extended through it, as in Fig. 10.19. The centroids of

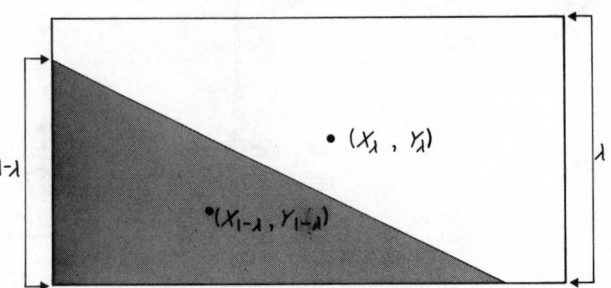

Fig. 10.19

these pieces have coordinates (Y_λ, Y_λ) and $(X_{1-\lambda}, Y_{1-\lambda})$, say, relative to the centroid of the whole rectangle as origin. Taking moments about this centroid we have

$$X_\lambda = -(1-\lambda)X_{1-\lambda}$$
$$Y_\lambda = -(1-\lambda)Y_{1-\lambda},$$

so that

$$(X_\lambda, Y_\lambda) = -\left(\frac{1-\lambda}{\lambda}\right)(X_{1-\lambda}, Y_{1-\lambda}).$$

Henceforth we shall assume $0 < \lambda \leq \frac{1}{2}$. It is easy to verify that the buoyancy locus consists of four arcs of parabolae and four arcs of hyperbolae (the latter being absent when $\lambda = \frac{1}{2}$) as shown in Fig. 10.20. The parabolae correspond to the ship being in the attitude shown in 10.18(a), the hyperbolae to 10.18(b). Using the coordinate axes shown, with origin at the centroid, the coordinates of the twelve points marked are as follows:

A $\left(\dfrac{4\lambda w}{3} - w, \dfrac{1}{6}\right)$　　　　G $\left(w - \dfrac{4\lambda w}{3}, -\dfrac{1}{6}\right)$

B $(\lambda w - w, 0)$　　　　H $(w - \lambda w, 0)$

C $\left(\dfrac{4\lambda w}{3} - w, -\dfrac{1}{6}\right)$　　　　I $\left(w - \dfrac{4\lambda w}{3}, \dfrac{1}{6}\right)$

D $\left(-\dfrac{w}{3}, \dfrac{2\lambda}{3} - \dfrac{1}{2}\right)$　　　　J $\left(\dfrac{w}{3}, \dfrac{1}{2} - \dfrac{2\lambda}{3}\right)$

E $\left(0, \dfrac{\lambda - 1}{2}\right)$　　　　K $\left(0, \dfrac{1-\lambda}{2}\right)$

F $\left(\dfrac{w}{3}, \dfrac{2\lambda}{3} - \dfrac{1}{2}\right)$　　　　L $\left(-\dfrac{w}{3}, \dfrac{1}{2} - \dfrac{2\lambda}{3}\right).$

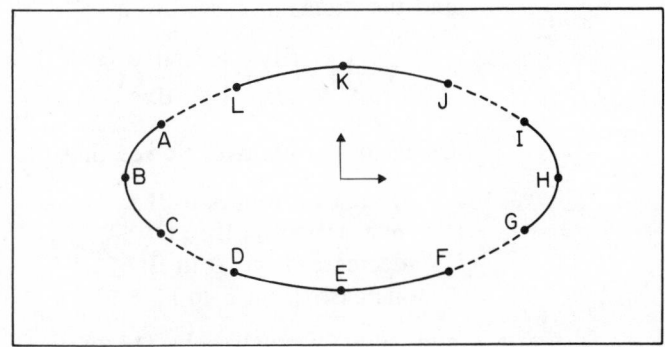

Fig. 10.20　　　———————— parabola　　　− − − − − − − −hyperbola

The equations of the arcs ABC, CD, DEF (the remainder follow by symmetry and will not be written down) are:

ABC $\quad x - \lambda w + w = 12\lambda w y^2$

CD $\quad (x + w)(y + \frac{1}{2}) = \dfrac{4\lambda w}{9}$

DEF $\quad y - \dfrac{\lambda - 1}{2} = \dfrac{3\lambda}{2w^2}\, x^2.$

It remains only to compute the metacentric locus as the evolute of these arcs, which involves a certain amount of case-by-case analysis. On the parabolic arcs the radius of curvature has minima at B, E, H, K; and the radius increases monotonically away from these points along the parabolae. On the hyperbolic arcs the behaviour depends more critically on the values of λ and w. If

$$\lambda \geqslant 1/4w$$

then no maximum or minimum occurs on the interior of a hyperbolic arc; but where this joins the parabola at C (for arc CD, and symmetrically for the rest) there is a maximum. If on the other hand $\lambda < 1/4w$ there is a new minimum at an interior point of the hyperbolic arc, where it crosses one of the lines $y = \pm x$. The latter case is more complicated, the metacentric locus having eight cusps and eight dual cusps, and we refer to Zeeman [48] for further analysis, our present purposes being illustration of the method rather than exhaustive study.

At a point (x, y) on a curve $y = y(x)$ the centre of curvature has coordinates (X, Y) where

$$X = x - \left[1 + \left(\frac{dy}{dx}\right)^2\right]\left(\frac{dy}{dx}\right)\Big/\frac{d^2y}{dx^2}$$

$$Y = y + \left[1 + \left(\frac{dy}{dx}\right)^2\right]\Big/\frac{d^2y}{dx^2}$$

and the radius of curvature is

$$R = \left[1 + \left(\frac{dy}{dx}\right)^2\right]^{3/2}\Big/\frac{d^2y}{dx^2}.$$

Using these formulae, we see that the radius of curvature

    decreases from A to B
    increases from B to D
    decreases from D to E
    increases from E to F,

and so on, symmetrically. Discontinuities in the radius of curvature would lead to straight-line segments in the metacentric locus as shown in Fig. 10.21, but an argument of Zeeman shows that these

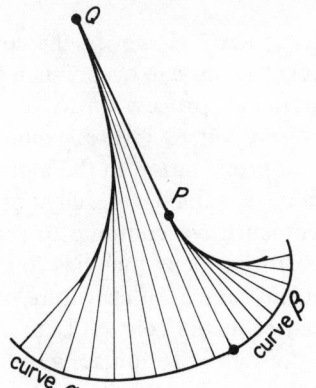

Fig. 10.21

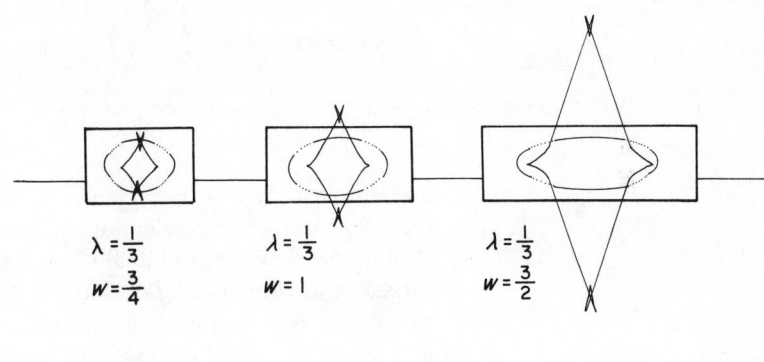

$\lambda = \frac{1}{3}$ $\lambda = \frac{1}{3}$ $\lambda = \frac{1}{3}$
$w = \frac{3}{4}$ $w = 1$ $w = \frac{3}{2}$

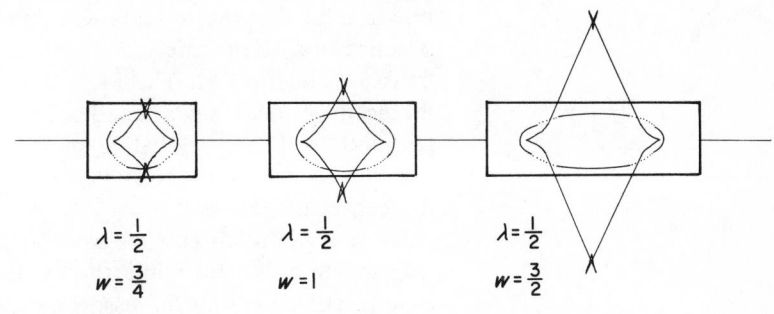

$\lambda = \frac{1}{2}$ $\lambda = \frac{1}{2}$ $\lambda = \frac{1}{2}$
$w = \frac{3}{4}$ $w = 1$ $w = \frac{3}{2}$

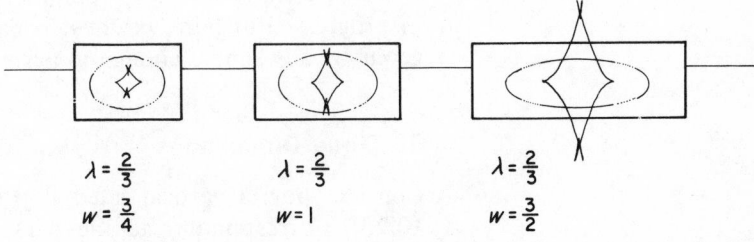

$\lambda = \frac{2}{3}$ $\lambda = \frac{2}{3}$ $\lambda = \frac{2}{3}$
$w = \frac{3}{4}$ $w = 1$ $w = \frac{3}{2}$

Fig. 10.22

do not occur here. However, the curvature need not vary smoothly, and 'atypical' cusps can occur, such as parabolic contact, with $P = Q$. (By a 'rounding' process on the corners of the rectangle the curve may be smoothed to give a genuine cusp at Q, but this gives no better an approximation to the metacentric locus. Also, the smoothing must be specified carefully, or else it introduces new cusps.) For convenience we continue to refer to a point like Q as a cusp. The metacentric locus thus has four cusps and four dual cusps: the former at the centres of curvature of the *parabolic* arcs at D, F, J, L and the latter at the centre of curvature at B, E, H, K. Computing these by the above formulae we obtain:

$$\text{at B} \quad \left(\lambda w - w + \frac{1}{24\lambda w}, 0 \right)$$

$$\text{at D} \quad \left(\frac{\lambda^2}{3w}, \lambda - \frac{1}{2} + \frac{w^2}{3\lambda} \right)$$

$$\text{at E} \quad \left(0, \frac{\lambda - 1}{2} + \frac{w^2}{3\lambda} \right),$$

and the rest follow by symmetry. Fig. 10.22 shows the resulting metacentric loci in a selection of cases.

Note the way that the dual cusp of the elliptic ship has fragmented into a cusp plus two dual cusps, in a 'butterfly' configuration. This possibility is recognized in the literature, *see* Robb [50] p. 134. It is a subtle example for stability analysis. The infinitesimal stability estimate, by metacentric height in the vertical position, would lead one to infer that the stability was satisfactory. Metacentric heights in practice are often, safely, as little as half a metre (*see* the quoted figures in Section 6). The local stability analysis, showing that the metacentric locus has a standard, not a dual, cusp at $\theta = 0$, is also favourable. But the global geometry of the metacentric locus shows the ship to be uncomfortably sensitive to sideways displacement of the centre of gravity.

Of course this discussion would be best carried out for genuine ship geometries. But while the evolute of a convex curve is fairly easy to sketch (using the association of standard cusps with curvature maxima and dual cusps with curvature minima, together with the corresponding positions of the centres of curvature) it is harder to find the buoyancy locus for a given shape and displacement. This problem is easily solved by computer techniques, and work on this is in progress. But here we must make do with examples which are geometrically simple enough to permit calculations by hand.

9 Three Dimensions

Consider now a vertical-sided ship of arbitrary cross-section (Fig. 10.23). Corresponding to the parabolic shape of the metacentric

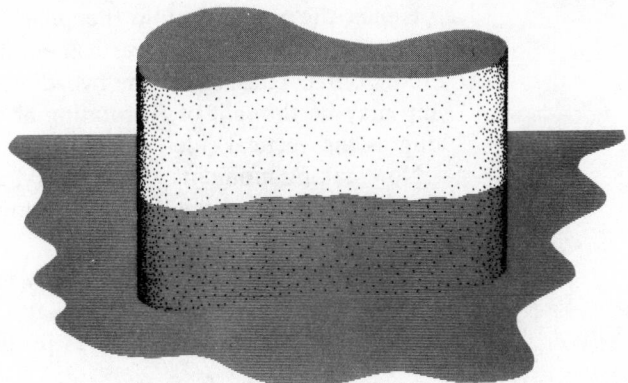

Fig. 10.23

locus in two dimensions, we have the following result: the buoyancy locus, for those attitudes where only the vertical sides meet the water, can by a *rigid* change of coordinates (not just a smooth one) be expressed in the form

$$\{(x, y, x) \mid z = ax^2 + by^2\}.$$

Thus we know the metacentric geometry around the upright position for *any* vertical-sided ship once we know the corresponding values of a and b: all the complexities of the ship's cross-section disappear.

To prove this, let $t = \tan\theta$ as before, where θ is the inclination in the y-direction, and similarly $s = \tan\psi$ where ψ is that in the x-direction. Choose coordinates fixed relative to the ship such that the centroid of the intersection S of the water level with the ship is at $(0, 0, 0)$, as in Fig. 10.24. We claim that the volume of the ship below any plane

$$P = \{(x, y, x) \mid z = sx + ty\}$$

is equal to that originally displaced. It suffices to show that the volume added equals that removed. But this may be deduced from the condition that the moment of S about the line

$$L = \{(x, y, z) \mid sx + ty = 0\}$$

vanishes (by a linearity argument), and this is just the condition for L to pass through the centroid of S, which it does by arrangement.

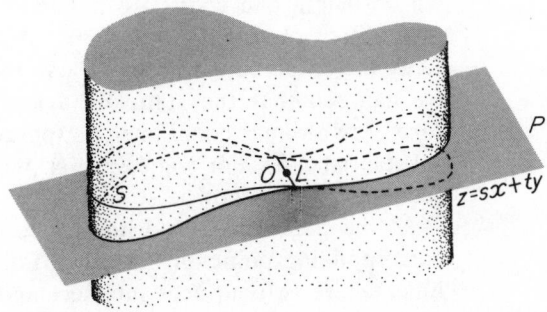

Fig. 10.24

Hence the region below the plane P is exactly the submerged region when the ship is in vertical equilibrium at an attitude making P horizontal. Thus, as for the two-dimensional ship of Fig. 10.4, this ship may be thought of as rotating about a point at water level, as long as the vertical sides meet the water line.

Now, taking moments about the (y, z)-plane, we find a change in the x-coordinate of the centre of buoyancy proportional to

$$\int_S x(sx + ty)\, \mathrm{d}x\, \mathrm{d}y = s \int_S x^2\, \mathrm{d}x\, \mathrm{d}y + t \int_S xy\, \mathrm{d}x\, \mathrm{d}y.$$

Similarly the y-coordinate moves proportionately to

$$s \int_S xy\, \mathrm{d}x\, \mathrm{d}y + t \int_S y^2\, \mathrm{d}x\, \mathrm{d}y,$$

and the z-coordinate proportionately to

$$\int_S (sx + ty) \cdot \tfrac{1}{2}(sx + ty)\, \mathrm{d}x\, \mathrm{d}y$$

$$= \tfrac{1}{2}s^2 \int_S x^2\, \mathrm{d}x\, \mathrm{d}y + st \int_S xy\, \mathrm{d}x\, \mathrm{d}y + \tfrac{1}{2}t^2 \int_S y^2\, \mathrm{d}x\, \mathrm{d}y.$$

Thus if the centre of buoyancy is (X, Y, Z) then X and Y depend exactly linearly on (s, t) in a nondegenerate way, whilst Z depends quadratically on them, for the various integrals are constant. Thus the buoyancy locus is the graph of a quadratic form, necessarily positive definite (by convexity); hence by rotation of the coordinates it may be diagonalized to have equation

$$Z = aX^2 + bY^2$$

as required.

(The axes giving it this form are actually the principal axes of the inertial ellipse E of S, and a and b are likewise given by E.)

Thus up to a scale factor (depending on the weight of the ship as well as its dimensions) the geometry of the metacentric locus of a vertical-sided ship depends only on the ratio a/b derivable from its cross-section. Fig. 10.25(a) shows its form (viewed from somewhat below) for a ratio a/b moderately greater than 1, choosing x along the long axis of the ellipse. For a large ratio a/b, that is, for a ship whose length and width are not too close, the complicated part is well up out of reach of the centre of gravity for practical purposes. Only the parabola P of cusps above the keel is then important, and we can see that the static behaviour corresponds closely to our former two-dimensional analysis (particularly since P is very flat for substantial a/b). For some special purposes though, as in the next section, a/b may not be so big.

At the points I, a cusp line intersects a surface: at H (and the corresponding point on the other side) we have a hyperbolic umbilic, where two surfaces intersect and the cusp line switches from

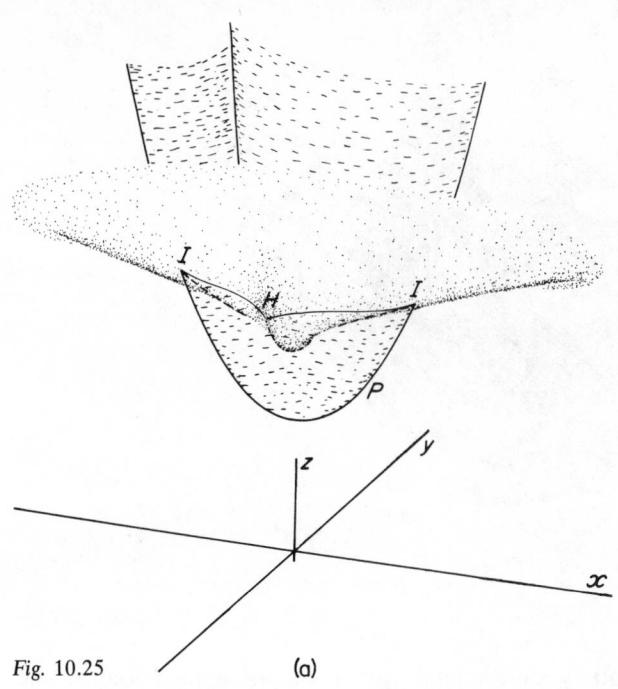

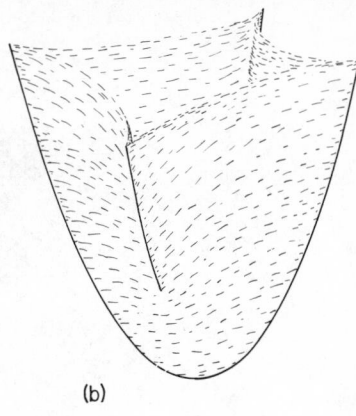

(b)

Fig. 10.25 (a)

one to the other (*see* Section 7 of the previous chapter). The reader should find it rewarding to relate the various maxima, minima and saddles at these positions for the centre of gravity, to the physical positions of the ship, in the detail forbidden us by space. Note that only in the inner region (shown separately, viewed from slightly above, in Fig. 10.25(b)) is there more than one minimum.

A similar analysis (with four hyperbolic umbilic points) applies to the ellipsoidal ship: as with the ellipse, the buoyancy locus is a similar ellipsoid, and the metacentric locus is its *centro-surface* (Fig. 10.26) analysed by Cayley [51]. A new analysis in specifically catastrophe theoretic terms appears in Banchoff and Strauss [51a].

10 Oil-rigs

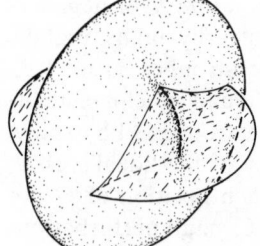

Fig. 10.26

The commonest kind of water-going vessel which is actually built with vertical sides all the way round is a floating oil-platform (Fig. 10.27). These are normally fixed to the ocean floor when on site, but they float during transport. Often they are built square. This symmetry goes through to the buoyancy locus, hence $a/b = 1$; and the buoyancy locus is a circularly symmetric paraboloid of revolution. The metacentric locus may therefore, apparently, be found by spinning the two-dimensional case (Fig. 10.28), so that the geometry of the perfectly square, vertical-sided ship is remarkably simple.

From a catastrophe theory viewpoint this simplicity is thoroughly deceptive. In suitable coordinates, the energy function takes the

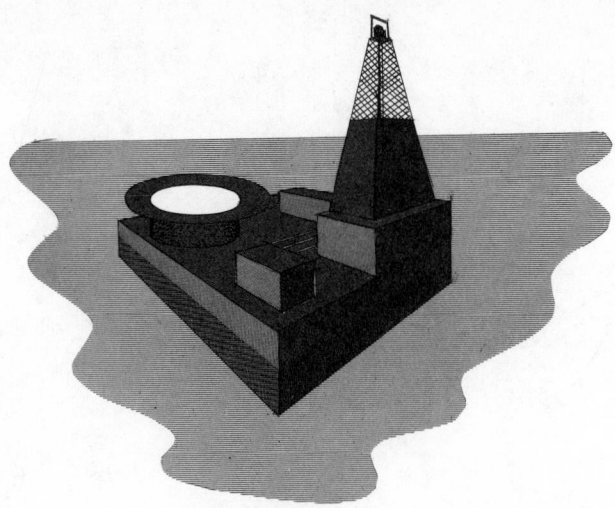

Fig. 10.27

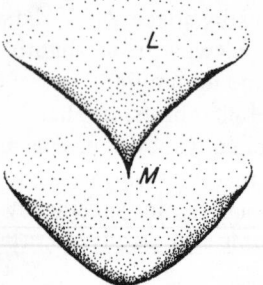

Fig. 10.28

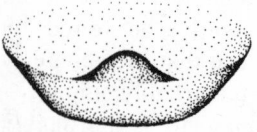

Fig. 10.29

form

$$(x^2 + y^2)^2$$

This is *not* finitely determined (by a simple application of the determinacy rules of Chapter 8) and so has infinite codimension. Hence the machinery of a universal unfolding is not available to give us a grasp of the effect of imperfections in the system.

For example, the theorem that the maximum number of critical points for a nearby function exceeds the codimension by 1 here implies that *any* number of critical points is possible. This may also be seen directly: the function $(x^2 + y^2)^2 - \varepsilon(x^2 + y^2)$ has a circular valley of degenerate critical points (of 'pigtrough' type, *see* Chapter 4 Section 1), as in Fig. 10.29; and we can break this up by ripples into arbitrarily many nondegenerate minima and saddles. Such a picture applies for all positions of *G* directly above the spun cusp point, so that the true bifurcation set is Fig. 10.30, *not* Fig. 10.28. The surface in Fig. 10.25(b) has collapsed into a line *L* of degenerate points. A small change can produce arbitrarily many umbilic points in the vicinity of *M*, and cuspoid catastrophes around *L*. (Similar degeneracy applies for the 'spun' Zeeman machine, regularly invented by people looking for higher catastrophes.) Physically, this means that the apparently simple geometry of the 'ideal' vessel, with a unique minimum energy position for each position of *G* not exactly on *L*, is violently unstable. For *G near L*, small perturbations by wind and wave can lurch and spin the leaning vessel in the most alarming fashion.

This is the first example of a *structurally unstable* metacentric geometry that we have encountered: the others have all come from the list of elementary catastrophes and hence are *ipso facto* structurally stable. On the other hand the sensitivity to imperfection, though *topologically* infinite, is in some sense *quantitatively* finite. No small

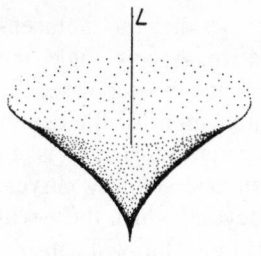

Fig. 10.30

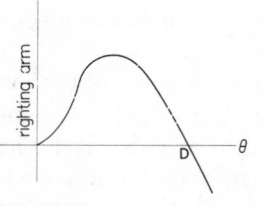

Fig. 10.31

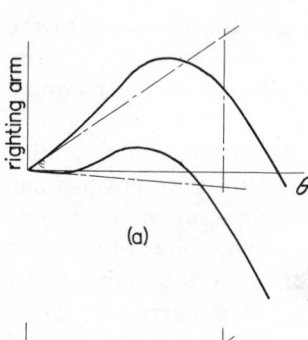

(a)

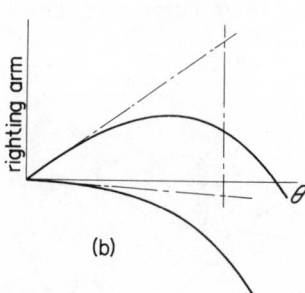

(b)

Fig. 10.32

imperfection can drop the bottom of the region *L*, in which *G* may sit to give multiple equilibria, by a sudden vast amount. (Though it may vastly increase the geometric complexity of *L*.) We have here a case where the qualitative theory is infinitely complicated, while the answers sought by a quantitative theory (like the possible fluctuations of metacentric height) consist of a few numbers. So the natural quantitative questions are actually coarser than the qualitative ones in this case. (Thus reversing the famous dictum of Rutherford: with equal (lack of) truth, 'Quantitative is just poor qualitative'.)

Finding these quantities rigorously involves a coarsening of the equivalence relation ('the same up to diffeomorphism') central to elementary catastrophe theory. It seems that a suitable relaxation of this kind may permit a description of the infinite-dimensional range of possible imperfections by nothing more pathological than a Banach space, but this is not yet certain. Even if complete, these mathematical developments would be outside the scope of this book.

11 Comparison with Current Methods

It would be absurd to expect fundamentally new information about the static equilibria of ships at this date. The relevant physical principles have been known since about the time of Archimedes, who analysed the stability of paraboloids of revolution in his work (*On Floating Bodies* Book 2) on hydrostatics, and many clever people have studied their consequences. What we have discussed above just alters the emphasis.

For example, the stability properties of a ship like that of Fig. 10.22 would not be misjudged by a competent naval architect. (S)he would compute a curve of righting arm as a function of θ (Fig. 10.31) which would show that the ship could not roll beyond angle D without capsizing. The low maximum value of the curve would indicate the ship's sensitivity to turning moments such as that produced by an off-centre load. However, at least to the inexperienced eye, Fig. 10.15 shows more clearly the direction of movement of the centre of gravity that would be most critical.

Similarly, the distinction between standard and dual cusp is well understood. Quoting from the same text as before (*Principles of Naval Architecture*, p. 96):
'The direction of curvature of the statical stability curve near the origin determines whether the ship will develop positive righting arms when the metacentric height is reduced to zero or becomes slightly negative. Two statical stability curves are shown in [Fig. 10.32], for two ships having the same metacentric height but different forms. In [Fig. 10.32(a)] which is typical of cargo and passenger types, the curve is concave upward, while the curve in [Fig. 10.32(b)] is concave downward. Assume that the center of gravity of

each ship is shifted upward the same distance, so that the metacentric height in each case becomes slightly negative. At any angle, the righting arm will be reduced by the same amount in each case, the reduction being the product of the vertical movement of the center of gravity and the sine of the angle of inclination. There is an important difference between the resulting statical stability curves; in case (a) the ship will heel to a small angle beyond which there will be some positive righting arm, while in case (b) the ship will capsize. The condition of negative metacentric height shown in case (a) may be recognized by the behavior of the ship, since there will be a list with no apparent heeling moment; the ship comes to rest with a small angle of heel either to port or to starboard but will not remain upright. It is possible for a condition of negative metacentric height to develop gradually in normal operation owing to consuming or unloading low weight, to developing a large free surface or to accumulating topside ice. In such situations, if the righting arm curve is of the type shown in [Fig. 10.32(a)], and if the existence of negative metacentric height is recognized, there will be some warning that a precarious situation is developing. With a curve of the type shown in [Fig. 10.32(b)] the only warning prior to capsizing would be a lengthening of the period of roll, which would not be apparent if the ship were in still water'.

There is little that the catastrophe-theoretic view can add to that, except perhaps some emphasis on the drastic sensitivity to asymmetry associated with the dual cusp behaviour of case (b), as in Fig. 10.15. Even when the metacentric height is still comfortably positive, as judged by the period of small rolls, a small shift of the cargo could have disastrous effect.

The discussion so far has been in terms of the geometry of the forces involved, and this is a convenient point to relate it more closely to our language in the rest of the book. If the potential energy of the ship in vertical equilibrium at inclination θ is given by $f(\theta)$, then the moment is $df/d\theta$. Thus redrawing the above righting arm curves as 'anticlockwise moment graphs' for both positive and negative θ we get Fig. 10.33. Notice that the assumption of symmetry for the ship implies that these curves have points of inflexion at the origin, so that their concavity or convexity is a higher order effect than could be found by computing their curvature. Third order terms are involved.

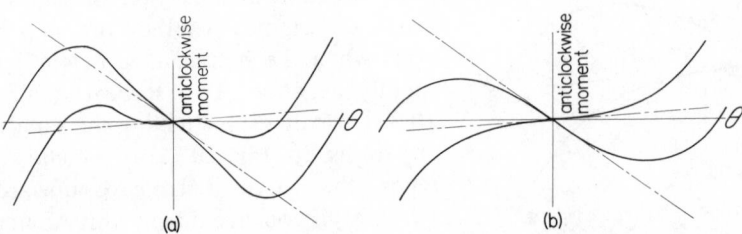

Fig. 10.33   (a)   (b)

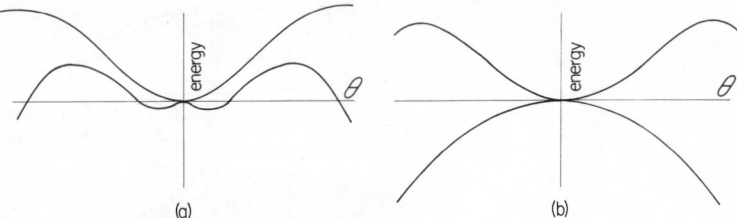

Fig. 10.34

Integrating Fig. 10.33 to get the energy (up to a constant), as in Fig. 10.34, we see around 0 the familiar changes of form for *f* associated with a control variable moving along the central axis of (a) a standard cusp and (b) a dual cusp. The positive or negative quartic terms of the energy when *G* is *at* the metacentre and the quadratic part vanishes, which unfold to give the standard or dual cusp, differentiate to give the positive or negative cubic terms which determine the direction of convexity in the righting arm diagrams.

What, then, are the advantages of using the language of catastrophe theory here? (We discount the mathematician's satisfaction in seeing geometric phenomena, which by the reasoning of Chapter 7 ought to be close to universal, turning up in yet another context.)

First, conceptual clarity and ease of exposition. Clarity is of course a subjective attribute, and no naval architect whose years of experience have given him a solid quantitative and intuitive grasp of his subject will be dazzled by this re-expression of its elementary facts. But no one is born with years of experience, and facts well known to every expert are not necessarily successfully communicated to every student – not even to every student who passes his exams. Let us quote *Principles of Naval Architecture* a third and final time (page 98):

'Stability is often, erroneously, evaluated on the basis of metacentric height alone, without the benefit of a complete righting-arm curve. This is equivalent to assuming that the righting-arm curve has the form of a sine curve, since it is assumed that the righting arm is equal to $\overline{GM} \sin \phi$. For ships with large freeboard and the type of form which produces a righting-arm curve concave upward near the origin, this practice is safe, but may result in underestimation of the ship's stability. For ships having but little freeboard and the type of form which results in a righting-arm curve concave downward near the origin, this practice may be dangerous, since it does not ensure an adequate range of positive stability or adequate residual dynamic stability'.

This danger arises exactly from identifying both standard and dual cusps as just fixed metacentres. Instructors' analysis of catastrophe geometry, with teaching aids like Fig. 10.35, might help produce safer ship designers.

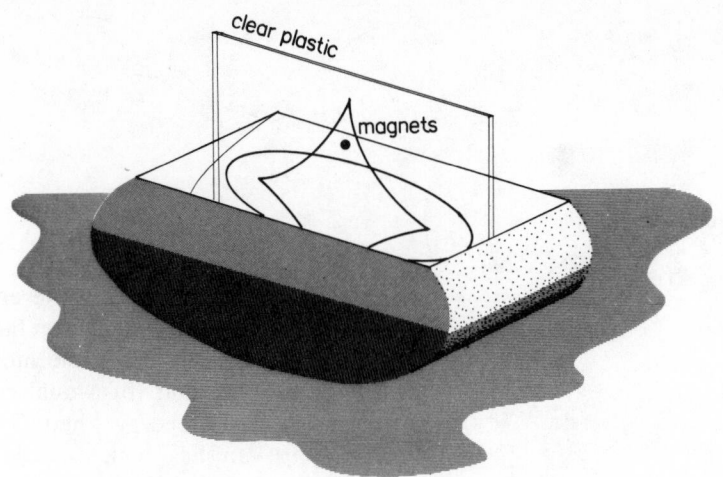

Fig. 10.35. Light plastic elliptical demonstration boat. Centre of gravity close to heavy magnets gripping central plane, on which buoyancy and metacentric loci are marked.

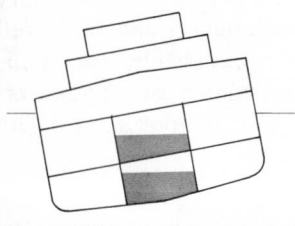

Fig. 10.36

Instruction in terms of catastrophe theory (illustrated perhaps by roller models of buoyancy loci of various ship forms at various displacements) could be an effective teaching method, even for students not previously exposed to the theory. Its concise and graphic description of the key facts of static ship equilibrium will of course be even more effective for those who have already learned something else presented in these terms. Given the number of areas in which catastrophe theory may be applied, and the explosive growth, just beginning, of its applications, this previous exposure should become increasingly likely.

Secondly, we have not touched in this elementary account on the effects of liquid cargo or fuel (Fig. 10.36), which make the entire centre of gravity mobile, or on the whole issue of *dynamic* stability. A ship at sea is in an active environment, with which it interacts in an intricate way. The problems posed, unlike that of static stability, do not admit exact solution. Not only is it complicated, for example, to allow for both the inertia of the ship and that of the water it pushes around; but the fluctuating forces of wind and waves cannot be specified completely. (We discuss some of the relevant problems at the end of Chapter 12.) Even the mechanism by which shallow water waves deliver far more force to the ship than deep water ones is far from understood. Without a theoretical understanding of the difficult dynamics involved, mechanical calculation with particular choices of assumed behaviour of a storm may be as dangerous as the statical errors discussed above. According to *Lloyd's Register*, in 1975 14 ships capsized; 12 capsized after cargo had shifted; and 130 disappeared for reasons unknown. Clearly not everything is as it should be.

Now the tool by which one studies qualitative questions (such as 'does it have a solution involving turning over?') about complicated differential equations for which complete numerical precise data are unobtainable is dynamical systems theory. This is closely related to catastrophe theory – indeed Thom, who has been a major influence here too, does not in [1] distinguish between the two. However, its study requires more topology than we have treated in this book. (A good introductory reference is Chillingworth [52]). A topological analysis along the lines we have sketched is a necessary preliminary to the use of such methods, since the dynamic problem subsumes the static one. Zeeman's work, which points to some subtle effects of the ship geometry on the dynamics, is at too early a stage to report here (though by the time this book is printed more information may be available). Much effort will be needed (not least in communication between marine engineers and mathematicians), but it is in this direction that there appear to be prospects of catastrophe theory yielding results which amount to more than an interesting reformulation of the known facts on ship stability.

11 The Geometry of Fluids

The streamlines of a moving fluid form families of curves which may, under certain physical hypotheses, be considered as the contours of a real-valued function. Stagnation points in the fluid flow then correspond to critical points of the function. The structural instability of degenerate critical points leads to instability in the topology of the streamline pattern; the unfolding rules of catastrophe theory permit the analysis of these instabilities.

This, in broad outline, is the subject of the present chapter. Some care is needed to make the physics and the mathematics mesh, so we begin by considering mathematical models of fluid flow. Our main interest is the work of Berry and Mackley [53] on the 'six-roll mill' and its technological implications, notably to the effect of dissolved polymer molecules on the flow of a fluid.

BACKGROUND ON FLUID MECHANICS

1 What We are Describing

We cannot set up here the general machinery of the theory of fluid motion: we lack space, and would need to appeal to more of the theory of differential equations than we are assuming knowledge of in the rest of this book. Our aim in the sequel is to give an account of a fascinating application of catastrophe theory, accessible to readers whose background may or may not include fluid mechanics. Those with some background in the subject may still find the discussion of interest, as the point of view is not the most usual. Detailed work developing the use of catastrophe theory will of course require a proper grounding in fluid behaviour, and not just the sketch given here.

Consider a moving fluid in two dimensions; or more exactly, one whose motion we may describe using two dimensions, since nothing changes along a third (Fig. 11.1). This situation can often be approximated with remarkable success in practice, so that the study of flows in two dimensions is important in the training of (for example) aircraft designers. Fig. 11.1(b) is very unlikely to be *exactly* true, since a whole line of static points is structurally unstable; but effectively two-dimensional flow is regularly achieved

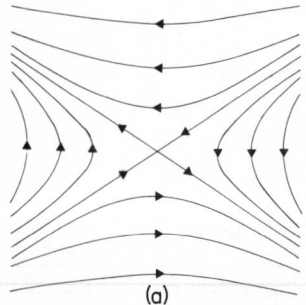

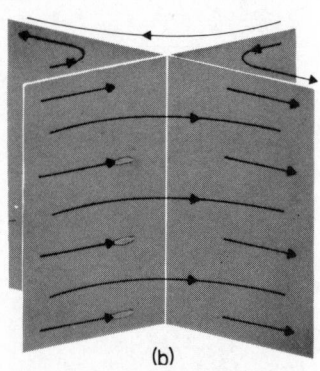

Fig. 11.1. (a) *Two-dimensional description*, (b) *three-dimensional fact (?) of fluid flow.*

experimentally to as good an approximation as is description of the flow by a velocity field.

This description supposes that the velocity of the fluid at each point x can be given by a vector $v(x)$, usually varying smoothly with x. Strictly, this is nonsense. On a small enough scale we would see molecules bouncing off each other with highly varying velocities, and empty spaces between them where no velocity could be assigned. On an even finer scale we would have a phalanx of electron clouds, quantum fields, and the like, for which description is hard and solution impossible. The idea of 'velocity at a point' is thus an approximation, fitting well the measurements we usually make, which average over regions big enough for the microscopic complications to disappear. It is a common article of faith that somehow the average behaviour of the microsystems is exactly described by the equations of fluid mechanics, but no one has proved this for a *realistic* model of the fine structure: the best proofs are for idealized models of a rarefied monatomic gas.

An equation that equates a differential at a point x to a number p says that the closer to x we restrict attention, the better p describes what we see. For the fluid equations we know this is false. For the fundamental equations of matter in spacetime, some physicists think it may be false. But even if they *knew* it to be false, for most purposes they would still use differential equations, differentiable functions and so on. The point is that physics consists in mathematically modelling the behaviour of matter, *using whatever mathematics best fits the phenomena at the scale considered.* Implied falsehoods about the structure of the next level down do not matter practically, historically or philosophically. It is false predictions about phenomena at the same scale (or sometimes the next scale up) that bring about scientific revolutions.

True predictions from a finer scale theory to larger scale behaviour are a major goal of physics, but an essentially macroscopic theory like fluid mechanics is not the less scientific. Even if some level of description is 'ultimate', and even if we find it, we cannot prove it 'ultimate' unless it obliges with tablets of stone or an equivalent. But science is concerned with *effective* description, and – so far as it remains science – leaves questions of 'ultimate truth' to those whose fundamental method of debate is to burn the unbeliever.

Hence, here as elsewhere in the book, we work with differential descriptions. Their range of applicability is in each case a partly experimental one. It cannot be decided in relation purely to theories of finer structure, even when these are well developed, as for gases: much less so when we are discussing living cells.

In particular, we discuss fluid flow in this chapter insofar as it can be modelled at any moment by a two-dimensional velocity field. This is a common and useful practice, and leads here to an application of catastrophe theory with very nice experimental results.

2 Stream Functions

Given a two-dimensional vector field v_t (illustrated in Fig. 11.2(a)) describing the flow at some time t, if $v_t(x, y)$ varies smoothly with (x, y) we can 'join up the arrows' as in Fig. 11.2(b). More precisely,

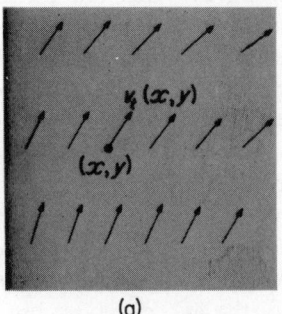
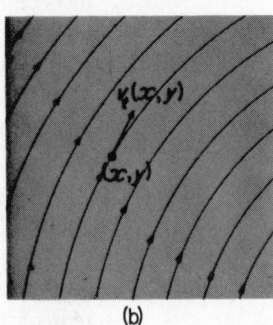

Fig. 11.2   (a)   (b)

we can fill the region U of interest with parametrized curves $\mathbb{R} \to U$, called *streamlines*, such that the tangent vector at (x, y) to the curve through (x, y) is exactly the given vector $v_t(x, y)$, for each point (x, y). (For one proof, *see* Dodson and Poston [5].) The streamlines are *not* the paths followed by particles in the fluid, unless the flow is *steady*, that is, at each point (x, y) the vector $v_t(x, y)$ is independent of t. Otherwise the curves move with time, and a particle which is moving tangent to a different streamline at each instant may actually follow none of them. Note that steadiness does not require the particles to move at constant speed: they may speed up, or slow, along turning curves. But the picture of steady flow does not change with time.

There is a very convenient way of describing a steady flow in a region U. Choose one streamline C as reference. It and any other streamline, C' say (in Fig. 11.3) may be thought of as giving a channel along which fluid moves. Supposing fluid is conserved, the same amount must be flowing through the curve A as through B, since otherwise there would be a steady build-up or loss between them. (No fluid crosses C or C', by their definition.) Hence there is a well-defined 'amount flowing between C and C'', independent of where between them we measure it. We give it a positive or negative sign according as an observer at P, looking in the direction of flow along C, sees himself on the right or the left bank of the channel.

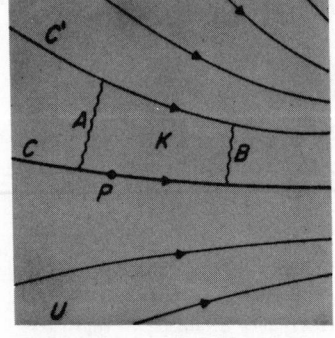

Fig. 11.3

We can use this amount as a label $f(C')$ for C'. If we now define a *stream function*

$$\phi : U \to \mathbb{R}$$

$$\phi(x, y) = f(\text{streamline through } (x, y))$$

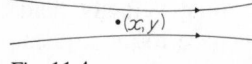

Fig. 11.4

then the streamlines become the contours of ϕ. Moreover, by taking narrower and narrower channels (Fig. 11.4) containing a point

(x, y), it is easy to see that the slope of ϕ, which is the limit of the ratio

$$\frac{\text{difference in height of } \phi\text{-contours}}{\text{distance in plane between } \phi\text{-contours}}$$

around (x, y), as the denominator goes to zero, corresponds to the limit around (x, y) of

$$\frac{\text{amount flowing through channel}}{\text{width of channel}}$$

as the width goes to zero. This should reasonably be

$$\rho(x, y)v(x, y)$$

where $\rho(x, y)$ is the density, fixed in time, of the fluid at the point (x, y). Thus we do not know all about the flow from ϕ, without further information on how v and ρ are related. But if the variation in ρ is negligible, as it often is (water is less compressible than steel), then a suitable choice of units makes $\rho = 1$, and we get u neatly from ϕ. In coordinates,

$$v(x, y) = \left(\frac{\partial \phi}{\partial y}(x, y), -\frac{\partial \phi}{\partial x}(x, y) \right).$$

Formal proof of all this is not hard, but it is not intuitive in a low level formalistic approach. We have not the space to set up the mathematical tools for the geometry rigorously (use of area, multiplied if need be by ρ, as a symplectic structure) and the reader should consult a fluid mechanics text.

The function ϕ cannot be constructed globally for the flow in an annular region shown in Fig. 11.5(a). What can be proved is that ϕ exists (uniquely up to a constant) for any smoothly varying v in a topologically simple region (e.g. a disc), even if the flow itself is rather complicated as in Fig. 11.5(b).

The argument by which we led up to the definition of ϕ breaks down if the flow is not steady, for then fluid might be temporarily accumulating in the region K between C and C', A and B. But this requires (by conservation) an increase in density somewhere in K. If changes in density are insignificant (as usually for liquids, rarely for

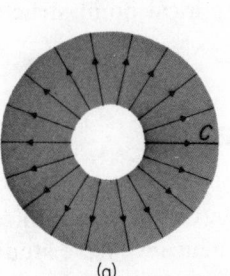
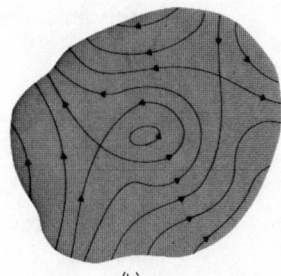

Fig. 11.5

(a)   (b)

gases) the argument goes through. Thus we may treat even *unsteady* two-dimensional *incompressible* flow as given at each instant by an appropriate function. (By *incompressible flow* we shall mean any velocity vector field giving a flow which preserves area (or volume). This may include flows that cannot be realized physically, for various reasons.) Since liquids are so much less compressible than gases, and we wish to concentrate on incompressible flow, we shall usually henceforth refer to *liquids* rather than to the more general *fluids*.

In a topologically simple region *U*, then, let us eliminate the arbitrary reference constant by choosing a point where all stream functions shall take a particular fixed value (e.g. 0). Then for each function ϕ on *U* we have an incompressible flow

$$v = \left(\frac{\partial \phi}{\partial y}, -\frac{\partial \phi}{\partial x} \right)$$

and for each incompressible flow we have a unique stream function. Natural notions of 'small perturbation' for the two correspond, and two stream functions are equivalent by a change of coordinates if and only if the corresponding flows are (by the *same* change). Thus the bifurcation theory of smooth incompressible two-dimensional flows relates to that of smooth real-valued functions of two variables, which is part of catastrophe theory. We have reached this point, incidentally, by way of perfectly classical modelling approximations to the 'real world'.

This application is quite unusual, though, in the following respect. The critical points of ϕ, where the gradient vanishes, with possibilities such as maxima, minima, saddles, monkey-saddles, etc., are the places where the flow velocity vanishes (called *stagnation points*). The 'behaviour' we are observing is not that of some system that varies over (*x, y*) maximizing or minimizing the functions involved. It is not even (as in optics) the brightness associated with critical or near-critical points, whether maximal, saddlelike or degenerate, so that we see the whole catastrophe manifold rather than just extremes. *It is the whole topological form of the stream function itself*, including such features as stagnation points, but also the way the flow winds between them. To interpret catastrophe theory purely in terms of critical point structure is myopic.

3 Examples of Flows

We have previously (Chapter 3) studied the contours of a number of functions of (*x, y*), hence we have studied a number of flow patterns. All that remains is to put arrows on them. In Fig. 11.6(a) and (b),

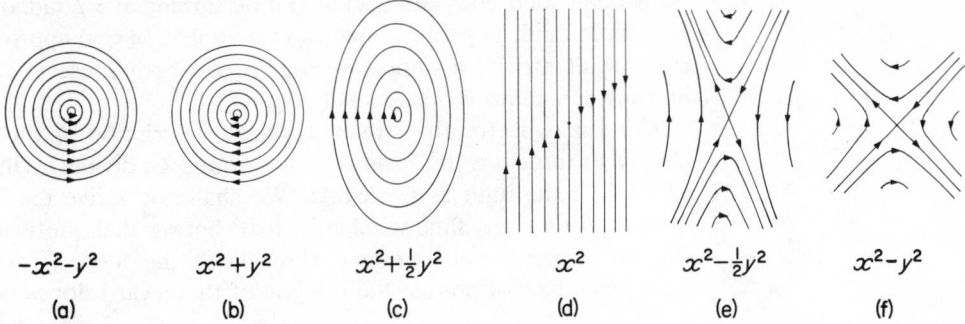

$-x^2-y^2$ x^2+y^2 $x^2+\frac{1}{2}y^2$ x^2 $x^2-\frac{1}{2}y^2$ x^2-y^2

(a)   (b)   (c)   (d)   (e)   (f)

Fig. 11.6

the liquid is rotating like a rigid body. The velocity at (x, y) in Fig. 11.6(b) is

$$\left(\frac{\partial}{\partial y}(x^2+y^2),\ -\frac{\partial}{\partial x}(x^2+y^2)\right) = (2y, -2x).$$

So the speed is proportional to the radius, as with a wheel turning on its axis. This describes, for instance, liquid in a rotating bottle B, at rest relative to B. (Any liquid with any viscosity at all, in such a bottle, will tend toward this situation, since viscosity damps out relative motions.) This is *pure rotation*. Figs. 11.6(c) to (f) all have the same vectors $(0, 2x)$ along the x-axis, and it is easy to see that the rest of the flow directions are as shown by continuity. Fig. 11.6(d) shows *simple shear*. Fig. 11.6(f) is called *pure shear* for reasons we shall come to in a moment.

4 Rotation

The local rotation, or *vorticity* $\omega(x, y)$ of a two-dimensional fluid flow v at a point (x, y) may be defined as follows. Take a small loop C around (x, y), integrate around C the anticlockwise tangential component of v, and divide by the area inside C to get an 'average'. (In Fig. 11.7 this will be negative, as the velocities towards the bottom of the figure are greater.) Then take the limit as C shrinks to (x, y), and divide by 2 (by some, not all, physicists' convention). There is work involved in proving this to be independent, in general, of the shape of C, which we omit. Taking circles around $(0, 0)$ in Fig. 11.6(a) and (b), it is trivial to show that

$$\omega(0, 0) = \pm\frac{1}{2}\lim_{r\to 0}\frac{2\pi r \cdot 2r}{\pi r^2} = \pm 2,$$

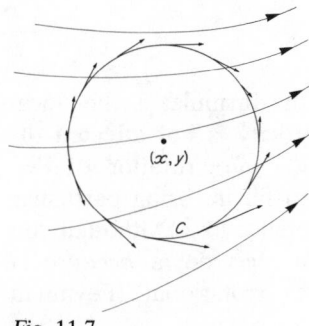

Fig. 11.7

the sign being $+$ for Fig. 11.6(a), $-$ for Fig. 11.6(b). Very little more effort shows that $\omega(x, y)$ is actually *constant*, equal to ± 2 for all (x, y), for these cases. Similarly for 11.6(d) the vorticity is everywhere -1, using square loops for convenience. Symmetry in Fig. 11.6(f) shows that $\omega(0, 0) = 0$, and so on. Clearly for 11.6(a) or

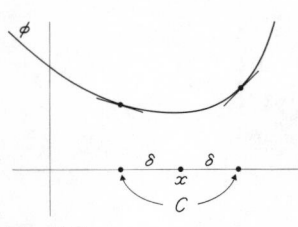

Fig. 11.8

(b) a small solid body at (x, y) would be turning at ± 2 radian/s, like the whole fluid. In general, one can prove that $\omega(x, y)$ approximates the rate of turn of a sufficiently small solid body loose in the fluid and instantaneously centred at (x, y).

The analogue for the stream function description of the flow is clearly to integrate the inward slope around C, divide by the area, and take $\frac{1}{2}$ the limit as C shrinks. We shall not derive the formula here (again see any fluid mechanics text) but we shall motivate it. In the one-dimensional analogue (Fig. 11.8) the loop C around x becomes a pair of points; the integral of the inward slopes becomes the *sum* of the two inward slopes $\dfrac{d\phi}{dx}(x - \delta)$ and $-\dfrac{d\phi}{dx}(x + \delta)$, and the area inside is replaced by the length 2δ. So the one-dimensional version of the limit is just

$$\lim_{\delta \to 0} \frac{1}{2\delta}\left(\frac{d\phi}{dx}(x - \delta) - \frac{d\phi}{dx}(x + \delta)\right) = -\frac{d^2\phi}{dx^2}.$$

The two-dimensional version turns out to be the obvious analogue of this, d^2/dx^2 being replaced by the *Laplacian* operator

$$\frac{\partial^2}{\partial x^2} + \frac{\partial^2}{\partial y^2} = \nabla^2.$$

The formula for the vorticity in terms of the stream function is thus

$$-\frac{1}{2}\nabla^2\phi = -\frac{1}{2}\left(\frac{\partial^2\phi}{\partial x^2} + \frac{\partial^2\phi}{\partial y^2}\right).$$

Using this formula it is easy to see that all the quadratic flows of Fig. 11.6 have constant vorticity, and in particular Fig. 11.6(f) is *irrotational* – its vorticity vanishes everywhere. This is the reason for calling it 'pure shear' flow: shear is happening, but rotation is not. Simple shear (Fig. 11.6(d)) is a linear combination of pure rotation and pure shear, either described by its stream function

$$x^2 = \tfrac{1}{2}(x^2 + y^2) + \tfrac{1}{2}(x^2 - y^2)$$

or by adding vectors at each point,

$$2(0, -x) = (y, -x) + (-y, -x).$$

Shear, for which we shall not discuss the formula, is the 'local relative motion' resisted by viscosity. Vorticity is not affected directly, as is seen clearly in the case of the pure rotation of Fig. 11.6(a) and (b). There are several steps to fill in, using particular properties of the system looked at, in a sentence like 'Although you put some ω in at the beginning, it soon dies down *because of viscosity* [our italics] and the flow becomes irrotational' (Feynman [54] Vol. II, 40–10).

5 Complex Variable Methods

The point of Feynman's remark, above, was to motivate a look at irrotational flow. This has historically been motivated by a theorem that in a *non-viscous* fluid (in two or three dimensions) a flow that starts irrotational remains so, however it develops in time. Thus the mathematics had a special class of exact solutions with no viscous forces and no vorticity, which were manageable and fun to study. The fact that, unlike (for instance) incompressibility, non-viscosity is a *wildly* unreasonable assumption for most fluid situations (von Neumann called it the study of 'dry water'), seems to have worried nobody before the end of the 19th century.

The spectacular advantage of a flow that is irrotational and can (through lack of viscosity) be trusted to remain so, in the two-dimensional case, is the vanishing of the Laplacian of the stream function. If ϕ is *harmonic*, that is, if $\nabla^2 \phi = 0$, then gathering x and y into a single complex variable $x + iy$ allows us to obtain ϕ as the real part of a complex analytic function $\phi + i\psi$. This allows one to rewrite textbooks on complex function theory as fluid mechanics texts, and produces all sorts of fascinating problems in complex geometry. The fact that the exact solutions of these problems gave answers that often fitted the physical flows only very approximately, and commonly not at all, was not stressed in the literature on them.

Such complex variable techniques can, sometimes, be useful. But the degree of caution appropriate is vastly greater than for the other methods discussed in this chapter. As classically discussed they often give an attractively powerful means of arriving at exact quantitative conclusions. But the conclusions may be, even qualitatively, wrong.

The irrotational approximation to a vortex is a singularity, allowing holomorphic as well as analytic complex functions. In particular

$$\text{Re}\,(\log z) = \log \sqrt{(x^2 + y^2)}$$

is a harmonic stream function describing fluid going round in circles faster and faster towards the origin (not slower, as for a wheel). Vortices where liquid moves inwards, like the one over an opened bath plug, do behave rather like that, since conservation of angular momentum forces faster rotation towards the centre. Whether the speeds 'really' go towards infinity as the logarithm would suggest, or whether velocity and vorticity are 'really' smooth functions with high finite values at the origin, is a meaningless question. Both alternatives presuppose a fine structure for the fluid, letting us discuss differentiably regions arbitrarily near the origin, which we know it does not have. For some purposes *modelling* the flow by a logarithmic singularity will be the most useful approach; for others a model using bounded, smooth (not analytic) functions will be more profitable.

STABILITY AND EXPERIMENT

6 Changes of Variable

Consider the stream function

$$x^2 + y^2 + (x^2 + y^2)^2 = r^2 + r^4$$

where r is the distance from the origin to (x, y). Plainly the flow described has circular streamlines, with velocity $2r + 4r^3$ at radius r. The angular velocity is thus $2 + 4r^2$, which is not constant. In particular, the motion is not rigid – there are local relative motions within the fluid. Thus this motion, unlike that given by $x^2 + y^2$, will be resisted by viscous forces. Unless a very special pattern of forces is acting on the fluid, the flow cannot exist as a steady one: the outer regions will slow, and/or the middle speed up. So there is a significant qualitative difference between r^2 and $r^2 + r^4$, in this physical interpretation of them.

However, these two functions are equivalent in the sense of Chapter 4. There is a global (not merely local) diffeomorphism

$$\theta(x, y) = (x\sqrt{(1 + x^2 + y^2)}, \, y\sqrt{(1 + x^2 + y^2)})$$

which transforms one to the other. This shows that in this context 'the same up to diffeomorphism' is not a strong enough relation fully to capture 'qualitatively the same'. There has been a natural tendency in some discussions to identify the two notions, because of the powerful and general results that have been proved about diffeotype. (In dynamical systems theory, where diffeotype does not yield big theorems, progressively much weaker notions of 'qualitatively the same' have been explored – and sometimes presented as True Qualitative Sameness.) The notion of 'qualitative' is in fact very variable: insofar as numbers are qualities of things – the basic hypothesis of all physical calculation – it absorbs the idea of 'quantitative'.

It turns out that the general laws (the Navier–Stokes equations) used to describe fluids, with explicit mathematical models for viscous forces, are invariant under diffeomorphisms that preserve volume (or in two dimensions, area). It follows that qualities such as describing a possible steady flow are preserved under such diffeomorphisms. There is a good case for identifying 'qualitative' in this context with 'up to volume-preserving change of variable'.

Unfortunately none of the theory of Chapters 4 to 8 goes through for this notion of equivalence. There is no such thing as k-determinacy, for instance. (This is most easily seen in one dimension, where the analogous relation is 'up to *length*-preserving change of variable'. Such a change must have the formula $x \mapsto c + x$ or $x \mapsto c - x$, where c is constant; hence it cannot even remove terms of infinite order: e.g. reduce $x^2 + e^{-1/x^2}$ to x^2.) The whole approach of Chapter 7, in which we identified standard forms to which almost everything could be reduced, fails entirely if we only allow area-preserving reductions. Indeed, it is infinitely *a*typical (in a sense we can make precise, just as for Chapter 7) for a family of functions to be reducible in this strong sense to one of the polynomial functions listed on p. 121. Likewise, the property of structural stability disappears. There is no function which we cannot, by an arbitrarily small perturbation, change to something not equivalent up to area-preserving coordinate change.

Nevertheless $x^2 + y^2$ has many features, even considered as a stream function, which a small perturbation does *not* change (like having an isolated point where the flow vanishes). There is clearly some interesting stability present. A rigorous approach to its analysis would be to work in the space of exact solutions of the full equations, but allowing general diffeomorphisms as coordinate changes. This would avoid such things as turning $x^2 + y^2$ into $x^2 + y^2 + (x^2 + y^2)^2$, since only one of them could be a solution in the presence of specified external forces, viscosity, etc. It should lead to an interesting theory.

Such a programme, however, is far from trivial. Since for any physical problem the notion of 'solution of the full equations' includes satisfying particular boundary conditions (at the walls or at infinity) we cannot confine our attention to neighbourhoods of the origin. We need to know what our functions do at the boundary, well away from 0. While catastrophe theory is relevant to this kind of problem – an example being the buckling plate analysis of Chapter 13 – there is a great deal of technical subtlety to its use. When we are not merely working with complicated boundary conditions but changing the equations themselves in completely unknown ways (as for instance below by adding polymers, which make the viscosity depend intricately on the flow, as well as vice versa) the claims of rigour and the claims of applicability this decade, part company. Even the classical Navier–Stokes equation for Newtonian flow is not well understood in the rigorous sense: no one can yet prove that it has solutions for all time with arbitrary initial data. (If it does not, presumably '*eppur si muove*': the fluid would still flow when the solution ended, but the *model* would have broken down.)

7   Heuristic Programme

However, we may get a long way in practice as follows. Consider an arbitrary stream function ϕ, or smooth family of functions $\{\Phi_c \mid c \in \mathbb{R}^s\}$ without regard to the exact solution of equations for viscous flow. Suppose that if the flow is at time t_0 given by ϕ or Φ_c, viscous drag and other forces will change it rapidly to a steady flow *near* ϕ or Φ_c. (This heuristic assumption will not always be true, and must be used with caution and due regard for experience. It does however suggest another line of rigorous approach, whereby the viscous forces, etc. within the liquid are mathematically 'turned on gradually' and solutions followed, like the Kolmogorov–Arnol'd–Moser solar system analysis which gradually turns on the mutual interactions of the planets.) Thus if ϕ or Φ is structurally stable, as a function or family of functions respectively, we expect that the true flow or family of flows will be the same up to diffeomorphism as that described by ϕ or Φ. Then we can fairly hope that such features as shear, even though not fully preserved by diffeomorphism, will not be changed *much*, so that we can use ϕ or Φ as a reasonable approximate description. This will almost certainly not be true if we start with something structurally *un*stable, for which even diffeomorphism-invariant features change with a small perturbation.

This approach is used with great success by Berry and Mackley [53]. Indeed, the discussion above is purely implicit there – as is quite appropriate, since the authors' concern is with effective description of certain complicated flows by catastrophe theoretic methods, and practical exploitation of the description. Our reason for a more elaborate treatment of the mathematical justification was primarily the general light shed on the nature and use of catastrophe theory in general – most of our readers will not be applying it in this particular context – by this significant case.

8   Experimental Realization

The pure rotation stream function $x^2 + y^2$ is easily produced experimentally, as already noted. Whatever the detailed nature of the viscous forces in the liquid, so long as their effect is always to resist relative motions within the liquid, pure rotation is the only possible steady flow in a steadily rotating container with no special external forces acting on the fluid. This is not hard to prove – the main effort lies in making it precise – and we can treat it for present purposes as intuitively obvious.

The flow of Fig. 11.6(d), simple shear, is more important. Since friction with containing walls makes a fluid's velocity relative to the walls approach zero near them, one would hope for this case to be realized by fluid between two sliding parallels (Fig. 11.9(a)). This is experimentally awkward, however, infinite planes being hard to come by. The Couette experiment (Fig. 11.9(b)) is more practical.

A strict analysis, using more machinery than we have discussed above, and the commonest mathematical model for viscous effects, gives a stream function

$$\frac{u}{1 - r_i^2} (r^2 - r_i^2 \log r)$$

up to a constant as usual. As may be seen from Fig. 11.10, this is nearly linear in r in the range $r_i < r < 1$, that is, between the two cylinders. (The flow may be thought of as a superposition of the

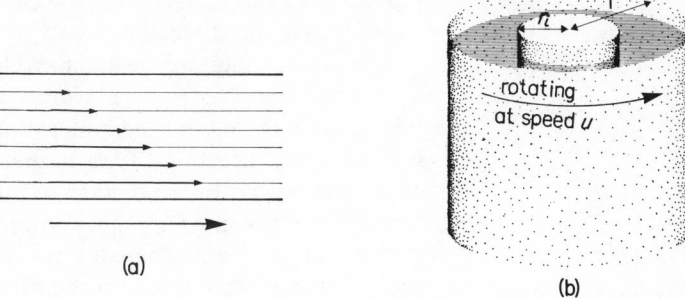

(a)

(b)

Fig. 11.9

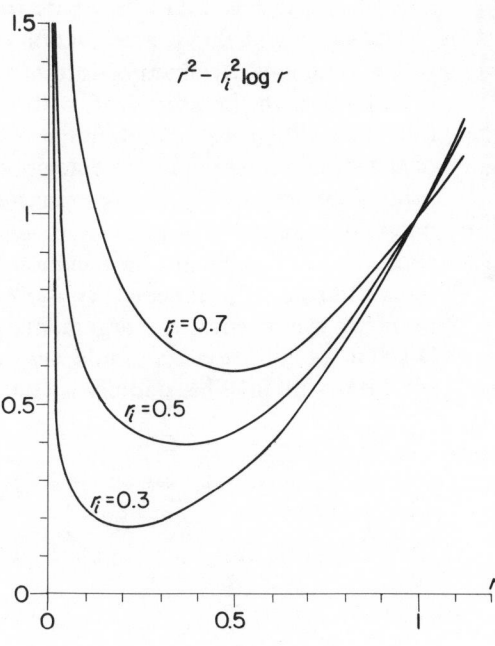

Fig. 11.10

pure rotation term r^2 and the harmonic function $\log r$ which corresponds to an irrotational flow.) Up to a smooth change of coordinates it *is* linear, and for many practical purposes it may be treated as such. Particularly if the inner radius is close to the outer one (normalized above to 1) this experimental set-up provides an excellent approximation to the 'simple shear' flow of Fig. 11.6(d).

The saddle flows of Fig. 11.6(e) and (f) present greater subtlety. There is no way that we can take a piece of the flow bounded by constant-speed streamlines and attempt to realize the flow by placing fluid between moving boundaries of the same shape. (For one thing, along *no* streamline is the speed constant.) Furthermore the pattern of viscous effects will be quite complicated, so that exact solution accounting for the energy provided by the driving force at the boundary, transmitted to the interior and dissipated by viscosity, would be a very intricate problem for a flow of this general form. In

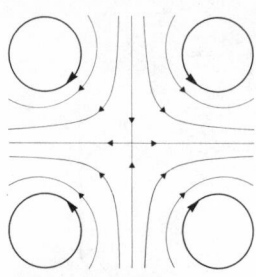

Fig. 11.11

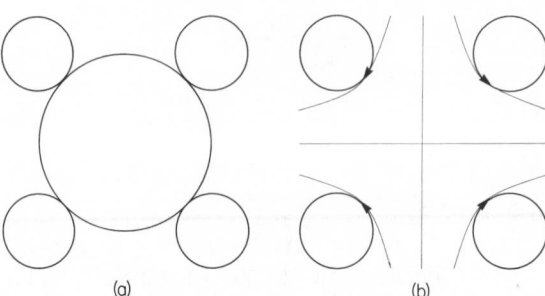

Plate 2. Streamlines in the four-roll mill observed for a 1.5% solution of polyethylene oxide in water (Crowley, Frank, Mackley and Stephenson, [55], Fig. 6).

practice, however, saddle flows can be realized to high accuracy without extreme difficulty.

The appropriate apparatus, introduced for this purpose by G. I. Taylor in 1934, has become known as the *four-roll mill*, Fig. 11.11. Four symmetrically placed rollers rotate as shown, at equal speeds in alternating directions. Clearly the surrounding flow is not exactly or globally given by a function αxy (equivalent to $u^2 - v^2$ by a 45° turn) since this has no looped contours. In practice (Plate 2) this stream function approximates the flow extremely well in a disc-shaped region fitted between the rollers Fig. 11.12(a)) if α is chosen to make the speeds of the rollers match those of the αxy streamlines shown in Fig. 11.12(b), where they touch.

The success of this approximation owes much to the character of Morse functions. Symmetry suggests a stream function whose Taylor expansion at the centre begins with αxy. For almost all values of α (all except 0) this is 2-determinate, so αxy is an exact description of the stream function in some neighbourhood V of the origin, up to some (not necessarily area-preserving) smooth change of coordinates. In practice V is quite large, and the alteration needed for the coordinates is negligible: a common, though not universal, occurrence. Further, the function xy is *stable*, so if the apparatus is not perfectly symmetrical the arguments placing a saddle *at* the origin fail, but we still expect a saddle *near* the origin, and αxy remains a good description. This stability is central to the differences between

Fig. 11.12

(a)   (b)

the four-roll mill and the six-roll mill, discussed below, which is the main concern of this chapter.

The stability of Morse saddles shows up even more clearly with the *two*-roll mill. Clearly from Fig. 11.13 there must be a stagnation point between the rollers, the simplest model for it being a stream function $Ax^2 + By^2$ with $A/B < 0$, Fig. 11.14. Thanks to structural stability, this gives a good picture of the experimentally realized flow (Plate 3). Notice the streamlines into and out of the stagnation point are curved, so that a 'curvilinear' change of coordinates would be needed to reduce the true stream function to the pure quadratic form in the middle. But the simple approximation by $Ax^2 + By^2$ in

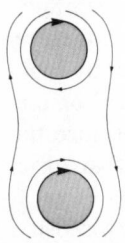

Fig. 11.13

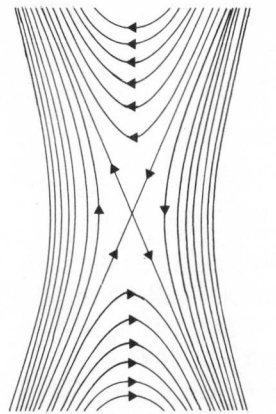

Fig. 11.14. Schematic diagram of idealized flow between two corotating rollers (Frank and Mackley [56], Fig. 1).

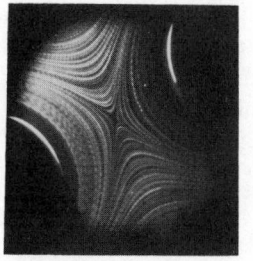

Plate 3. Streamlines of water-glycerol solution between corotating rollers (Frank and Mackley [56], Fig. 6(d)).

the original coordinates is quite sufficient for the physical calculations concerning flow around the stagnation point, as described in Frank and Mackley [56].

The strength of the property of structural stability and 2-determinacy is further dramatized when we change the physical character of the fluid. Plate 3 shows the flow of a *Newtonian* fluid (one that obeys to an excellent approximation the fluid equations that assume the simplest possible model for non-zero viscous forces). If we add polymer molecules to the fluid, the effect of viscosity becomes more complicated. Plate 4 shows the results for this experiment. The property of having a nondegenerate flow remains, but for increasing roller speeds (Plate 4(a) to (d)) the obtuse angle between the 'in' and 'out' streamlines grows from 129° to 145°, ±3° in each case. (This is in contrast to the Newtonian flow, where speeding up the rollers merely speeds up the flow without detectably changing its geometry.) The study of the relation of A and B to the other physical quantities measured sheds useful light on the behaviour of polymer molecules in flow systems, and the 2-determinacy of Ax^2+By^2 goes far to justify the neglect of higher order terms apologized for in Frank and Mackley [56]. The justification remains partly heuristic, of course, in the light of the remarks above (Section 6).

COMBING POLYMER MOLECULES

9 Non-Newtonian Behaviour

Suppose now that the liquid contains polymer (long-chain) molecules in solution, or that it *is* molten polymer. It is reasonable to suppose that different arrangements of the chains (stylized in Fig. 11.15(a) to (c)), will give different viscous effects. One would clearly expect

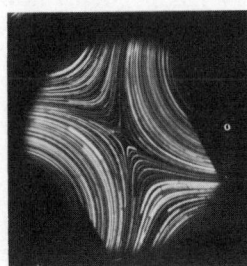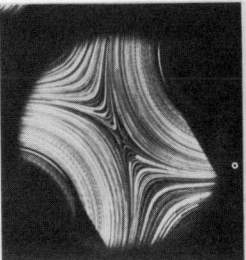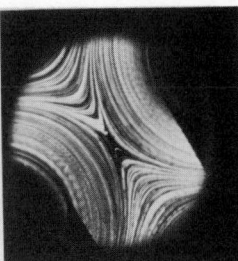

Plate 4. Streamlines of polyethylene oxide solution between corotating rollers (Frank and Mackley [56], Fig. 7).

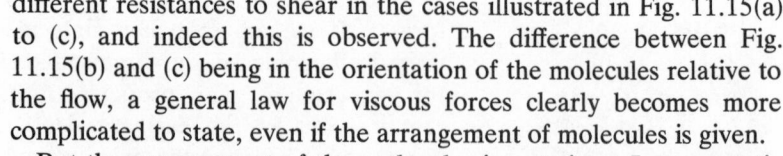

Fig. 11.15

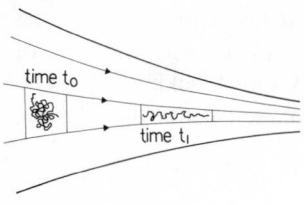

Fig. 11.16

different resistances to shear in the cases illustrated in Fig. 11.15(a) to (c), and indeed this is observed. The difference between Fig. 11.15(b) and (c) being in the orientation of the molecules relative to the flow, a general law for viscous forces clearly becomes more complicated to state, even if the arrangement of molecules is given.

But the arrangement of the molecules is *not* given. In near-static liquid many kinds will tend to tangle themselves up by elastic forces as in Fig. 11.15(a), molecules N units long taking a tangled form of order $\sqrt{N}$ across on average. But if the part of the fluid that a molecule is in stretches one way and shrinks another (as by going down a narrowing nozzle, Fig. 11.16) this will stretch the molecule, disentangling it somewhat. Since the same is happening to nearby molecules, the viscous response to shear changes over time.

Thus, *the viscous forces at a point in the liquid depend on the history of the liquid at that point*, as well as the local shearing properties of the flow. Of course, most molecules, if not continually being stretched, tend to relax back to the tangled configurations of Fig. 11.15(a), so the memory of the stretching history slowly fades. Detailed mathematical models of this kind of behaviour are discussed by Truesdell and Noll ([57], Section E). It is very hard to solve any but the most special cases for such models, and consequently deciding which model fits a particular non-Newtonian fluid is harder still. (The approximate character of any smooth model becomes even more apparent here.) Much of the 'practical' literature concerned with, for instance, the flow of extruded molten polymer through a nozzle, simply discards the history dependence of the viscosity as too hard to handle mathematically. A common device is to replace the Newtonian linear relation between shear rate and shear stress at a point by an (equally history independent) power law, choosing the exponent experimentally. This is sometimes useful, but it is often in practice *desired* that the arrangement of the chains should vary. A polymer thread from a nozzle benefits in strength from 'pre-combing' of the molecular chains in it, much as cotton or wool yarn benefits from that of its constituent fibres. A good understanding of the molecular behaviour under different flow conditions is thus useful to help build a better description of the flows themselves. Controllable flows in which the molecules have histories of different kinds are a natural research tool in studying such behaviour, and it is in providing this tool that the approach to two-dimensional flows discussed in Sections 6 to 8 comes into its own.

10 Extensional Flows

Clearly, pure rotation has no stretching effect at all. Simple shear has some (Fig. 11.17) but it also turns long molecules towards the direction 'along the flow'. Lines in that direction are *not* being stretched, so rather little net combing results.

Physically, the amount of combing may be examined by its optical effects. Light passing through a bulk of combed molecules interacts with it differently according to polarization. Electromagnetic vibration in a plane to which the molecules are parallel travels at a different speed to that in a plane orthogonal to them (Fig. 11.18), so

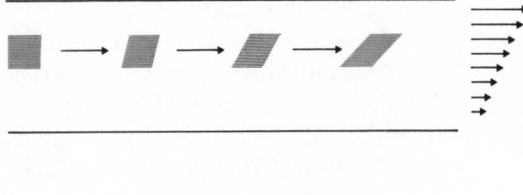

Fig. 11.17

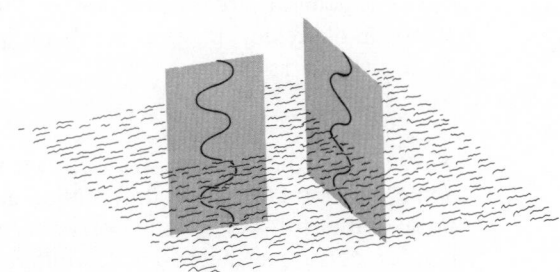

Fig. 11.18

the material has a different refractive index for the two cases, and refracts them differently: it is said to be *birefringent*. (Light polarized in other planes behaves as a superposition of these two cases.) Uncombed chains do not have this effect, so the degree of combing may be detected by the amount of birefringence. (For physical details, *see* Mackley and Keller [58].)

Simple shearing flow (as realized in Couette apparatus) shows indeed a modest birefringence, uniformly distributed in the liquid, in agreement with the moderate stretching of the polymer chains suggested by Fig. 11.17.

In pure shear flow, on the other hand, a rectangle of fluid takes, at successive times, the shapes shown in Fig. 11.19. The lines across a neighbouring streamline show the way that a partly stretched molecule turns continually toward the direction in which it will most be further stretched. This is quite unlike simple shear, where it turns so as to be stretched less. Crowley, Frank, Mackley and Stephenson [55] call this a *persistently extensional flow* for this reason.

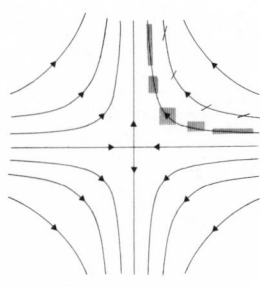

Fig. 11.19

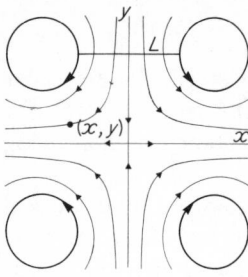

Fig. 11.20

To a good approximation, polymer molecules entering, across the line L of Fig. 11.20, the region of the four-roll mill where xy is a good description of the flow, may be supposed unstretched; for the chains' tendency to relax means that only rather fast stretching of the fluid has much effect on them. Simple calculations (Crowley, Frank, Mackley and Stephenson [55]) then show that the amount by which a fluid element at (x, y) has been stretched since passing L is inversely proportional to y, independently of which streamline (x, y) is on. Of course, the infinite value of $1/y$ at $y = 0$ is not reached by the polymer chains, which after disentangling can only stretch a little before they break. (This gives the fluid a rather special viscous effect: resistance to further stretching. Some other viscous effects, however – particularly turbulent ones – are much reduced. A very small addition of the right polymer to a flash flood will make it slip out through storm drains much faster. Likewise, polymer added to oil speeds it through pipelines – and no one quite understands why.) Infinities aside, this analysis suggests a strong maximum of birefringence localized along the outgoing axis, which is borne out by Plate 5 (from Crowley, Frank, Mackley and Stephenson [55]).

For a general stream function ϕ, these considerations lead (*see* Berry and Mackley, [53]) to the definition of the *persistent strain rate* of ϕ at (x, y) as

$$\sqrt{(-\det H\phi\,|_{(x,y)})}$$

where $H\phi$ is the Hessian. We omit the explanation here (it involves some beautiful classical differential geometry), but it should be noticed that for $x^2 + y^2$ it is everywhere $2i$, imaginary and unphysical; for x^2 (simple shear) it is 0; while for pure shear $x^2 - y^2$ it is 2.

Note that it is very much differential *geometry* (involving area if not distance) that is in action here, and not just differential *topology*. Away from a stagnation point the gradient of ϕ is non-zero, hence by Chapter 3 ϕ is, *up to a local diffeomorphism*, linear. And linear ϕ implies uniform v, with the liquid moving straight and rigidly; there

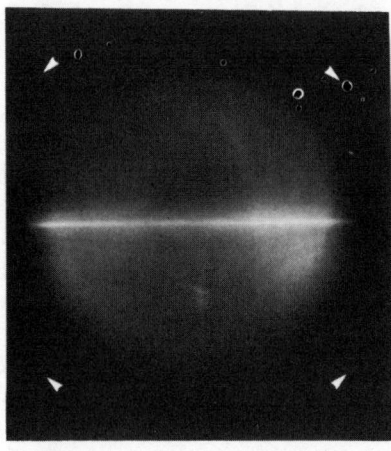

Plate 5. Localized flow birefringence in the four roll mill observed between diagonally crossed polaroids for a 1.5% polyethylene oxide/water solution. Sense of roller rotation indicated (Crowley, Frank, Mackley and Stephenson [55], Fig. 8).

is no strain, no shear, nothing happening at all; which is uninteresting in the rare cases that it is true. We are definitely appealing to the heuristic assumptions noted earlier when we apply a catastrophe model.

DEGENERATE FLOWS

11 The Six-roll Mill

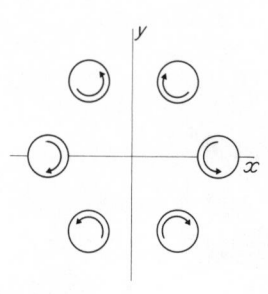

Fig. 11.21

Now consider the flows to be expected in the six-roller analogue of the four-roll mill (Fig. 11.21). If the flow is to have the symmetries that the apparatus has when speeds are equal, the expansion at $(0, 0)$ of the stream function ϕ can have no linear part. We may fix its value at $(0, 0)$ to be 0. The only quadratic function of x and y with $\frac{1}{3}$-turn symmetry is $x^2 + y^2$, and if the flow is to be symmetric under reflection in the y-axis the quadratic part must vanish also. Writing a general homogeneous cubic $q(x, y)$ as

$$\mathrm{Re}\,(\alpha z^3 + \beta z^2 \bar{z})$$

where $z = x + iy$ and α, β are complex (*see* Zeeman, [7]), the effect of a $\frac{1}{3}$-turn is to multiply z by $e^{2\pi i/3}$, and to change the cubic to

$$\mathrm{Re}\,(\alpha(e^{2\pi i/3}z)^3 + \beta(e^{2\pi i/3}z)^2(e^{-2\pi i/3}\bar{z})) = \mathrm{Re}\,(\alpha z^3 + e^{2\pi i/3}\beta z^2 \bar{z}).$$

Hence if q has $\frac{1}{3}$-turn symmetry it has $\beta = 0$ exactly, and thus is equal to

$$\alpha_{\mathrm{R}}(\mathrm{Re}\,z^3) - \alpha_{\mathrm{I}}(\mathrm{Im}\,z^3)$$

where α_{R} and α_{I} are the real and imaginary parts of α. Now reflection in the y-axis, with sign change since a reflection reverses the direction of flow, fixes the sign of $\mathrm{Re}\,z^3$ and reverses that of $\mathrm{Im}\,z^3$. Thus the general irrotational flow with the given symmetries has the form

$$\gamma(x^3 - 3xy^2) + O(4).$$

Since, for $\gamma \neq 0$, $\gamma(x^3 - 3xy^2)$ is 3-determinate, we discard the $O(4)$ Tayl and take the cubic as our local model, as we took xy for the four-roll mill. Again, the contours of the polynomial (Plate 6,1(b)) are well matched by the streamlines observed (Plate 6,1(a)).

The matching at the origin is not, however, quite so neat as that of Fig. 11.11 and Plate 2; especially since Plate 6,1(a) was obtained only after careful adjustment of the apparatus. This is because the symmetry hypotheses that led us to the function $x^3 - 3xy^2$ were not ones we could justify exactly: we cannot, in practice, expect the speeds to be exactly equal or the rollers identical. We should therefore allow, at least, for a small vorticity ω and linear flows v_x

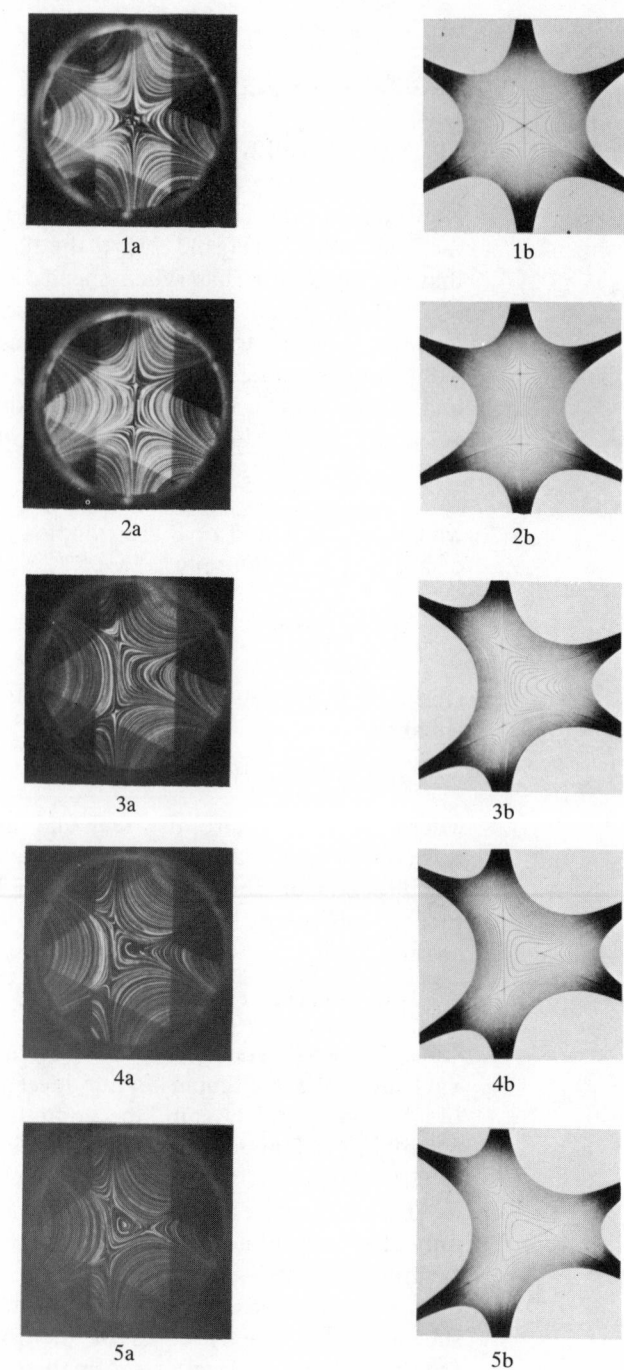

Plate 6. (a) *Flows 1–5 of glycerol in the six roll mill,* (b) *computer-drawn contours for corresponding stream functions.* (*Courtesy of Malcolm Mackley*).

and v_y in the x- or y-directions. These perturbations, which we could ignore for the four-roll mill by structural stability of xy, give us a family

$$\gamma(x^3 - 3xy^2) + \omega(x^2 + y^2) + v_x y + v_y x$$

of stream functions around the 'perfect' one $x^3 - 3xy^2$. Now this, as a three-parameter family, *is* structurally stable, and moreover we have analysed it already in Chapter 9 Section 6 (the constant γ may of course be normalized to 1). Indeed, it was by recognizing it as the elliptic umbilic that Berry perceived its stability. So, on the same heuristic grounds as before, we can neglect any other perturbations (as being equivalent to those already considered) and use this family as a model for the family of flows.

An interesting consequence of the universality of this unfolding is that although *six* roller speeds are available as controls, *three* independent controls are adequate to obtain all possible flow patterns near Plate 6,1(a). This led Berry and Mackley to an experimental arrangement in which three pairs of rollers were driven by just three independent motors. The controls were then the three roller speeds Ω_I, Ω_{II} and Ω_{III} arranged as in Fig. 11.22, all speeds being defined as positive in the sense shown.

Now we know that *almost* every 3-parameter unfolding of $(x^3 - 3xy^2)$ is (uni)versal. But suppose that any *exactly* symmetrical setting

$$(\Omega_I, \Omega_{II}, \Omega_{III}) = \Omega(1, 1, 1)$$

of the speeds produces a correspondingly symmetrical flow. (This is physically reasonable, over some interval of values of Ω at least, and holds for the analysis of Berry and Mackley [53]. We could argue topologically for it, given physical realizability of symmetry for *some* Ω, at some length.) Then for a whole line Λ in $(\Omega_I, \Omega_{II}, \Omega_{III})$-space we have umbilic points in the corresponding function (Fig. 11.23). But a transverse family with three parameters should meet only *isolated* points where the Hessian is doubly degenerate, by Chapter 7 Section 4. Indeed, we have seen this in detail for the universal unfolding studied in Chapter 9 Section 6, and hence (by the equivalence of universal unfoldings) for any 3-parameter unfolding of the symmetrical flow. Thus the unfolding by Ω_I, Ω_{II} and Ω_{III} is not universal, and we must add another control. The rather subtle one that applies in Berry and Mackley's experiments is the addition of polymer to the liquid.

How, then, are ω, v_x and v_y related to Ω_I, Ω_{II}, Ω_{III}? Obviously the strict relationship would be hard to compute explicitly, but it may reasonably be fitted in a manner like that used for the four-roll mill, though more complicated. For the linear terms v_x and v_y, Berry and Mackley assume that each pair of neighbouring rollers contributes linearly in proportion to the mean velocity indicated in Fig. 11.24, with a constant of proportionality α_+ for the inflowing streams and

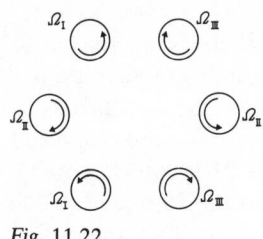

Fig. 11.22

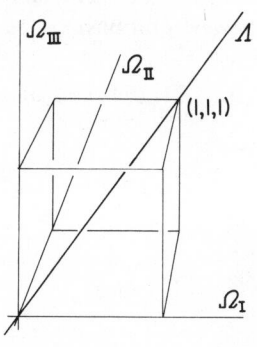

Fig. 11.23

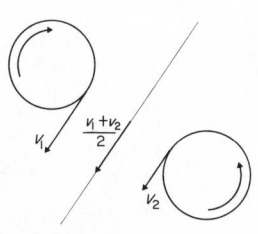

Fig. 11.24

α_- for the outflowing ones. (The reason for separating these cases will appear later.) We refer the reader to their paper for detailed arguments leading to expressions of the form

$$\frac{\omega}{\gamma} = \frac{k_1(\Omega_{III} - \Omega_I)}{\Omega_I + \Omega_{II} + \Omega_{III}} \tag{11.1}$$

$$\frac{v_x}{\gamma} = \frac{k_2\sqrt{3}(\alpha_- - \alpha_+)(\Omega_{III} - \Omega_I)}{\Omega_I + \Omega_{II} + \Omega_{III}} \tag{11.2}$$

$$\frac{v_y}{\gamma} = \frac{k_2(\alpha_- + \alpha_+)(\Omega_I - 2\Omega_{II} + \Omega_{III})}{\Omega_I + \Omega_{II} + \Omega_{III}} \tag{11.3}$$

and the computation of the constants k_1 and k_2 in terms of the radii and separations of the rollers. Notice that the 3-dimensional variation of $(\Omega_I, \Omega_{II}, \Omega_{III})$ gives only a 2-dimensional subspace of the universal unfolding parameter space (ω and v_x are proportional); this unfolding, as remarked, is not versal.

When the experimental fluid is glycerol, which is very well described by the Newtonian fluid assumptions, there is no reason to assume any difference between the inflow and outflow dynamics, nor therefore in the mechanism by which velocity at the throats is translated to v_x and v_y. This suggests setting $\alpha_+ = \alpha_- = \alpha$, say; and restricts us to the plane $v_x = 0$ (by equation (11.2)). Experimentally, adjusting the roller speeds to give, according to the formulae (11.1) to (11.3), the points 1 to 10 marked on the $(\omega/\gamma, v_y/\gamma)$-plane in Fig. 11.25, gave the results shown in Plates 6 and 7. The drawings at the right are computer plots of the contours of the polynomial stream functions of the theory. The closeness of fit may be emphasized by the following remarks.

(a) A typical plane through the origin of Figure 11.25, even one

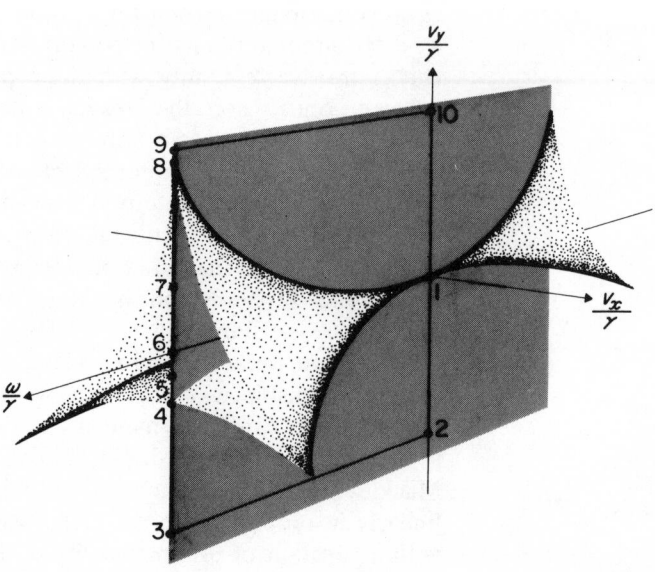

Fig. 11.25

that includes the ω/γ-axis, does not include cusp points. The closeness with which flow 8a in Plate 7 matches contour map 8b is thus confirmation of a predicted special feature of the experimental set-up. (Incidentally, a typical plane in Fig. 11.25 need not meet the origin at all. The symmetry arguments that suggested the stream function form for $\Omega_I = \Omega_{II} = \Omega_{III}$ apply only insofar as the symmetry is physically realized, in both the geometry and speeds of the system. Flow 1 is thus a triumph of experimental technique.)

(b) With $v_x = 0$ and ω/γ fixed, the cusp point occurs for a value of v_y/γ which is -3 times that for the fold point 4, irrespective of the values of k_1, k_2 and ω/γ, if formulae of the form (11.1) to (11.3) are correct. The experimental value is -2.75. Since, as is clear in Fig. 11.25, a small wobble of the $v_x = 0$ plane can lower the upper bifurcation point considerably, this is remarkable agreement – with the difference in the right direction.

(c) Symmetry under reflection in the x-axis is possessed by the function

$$\gamma(x^3 - 3xy^2) + \omega(x^2 + y^2) + v_x y + v_y x$$

if and only if $v_x = 0$. This symmetry is exact in the computer drawings; that it is so nearly exact in the photographs is evidence not only for the model in general but for the particular hypothesis $\alpha_- = \alpha_+$. The visible way in which it is not *completely* exact is related to our next remarks.

12 The Non-local Bifurcation Set of the Elliptic Umbilic

The analysis in Chapter 9 Section 6 found the points in the unfolding space of this catastrophe at which the corresponding function of (x, y) has a degenerate (hence unstable and giving rise to bifurcation) critical point. When a function has no such points, its form around any particular point is structurally stable, and hence so are its total numbers of minima, saddles and maxima (with which we are most often concerned). But a function may be *globally* unstable without any local degeneracies, by having two nondegenerate critical points at which it takes the same value.

Fig. 11.26(b) shows a graph like $(x^2 - 1)y$, with selected contours marked. It is easy to show that its only critical points are at $(\pm 1, 0)$ and that these are nondegenerate saddles, hence stable. But an arbitrarily small perturbation can make the saddle with $x > 0$ lower than the other (Fig. 11.26(a)) or higher (Fig. 11.26(c)). No diffeomorphism either on the domain $\mathbb{R}^2$ or the range $\mathbb{R}$ of these maps can make the topology of their contours match Fig. 11.26(b). Global stability of a function requires not only nondegenerate critical *points*, but also distinct critical *values*. Moreover, no diffeomorphism near the identity can transform Fig. 11.26(a) into (c)

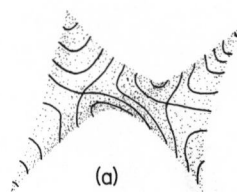

(a)

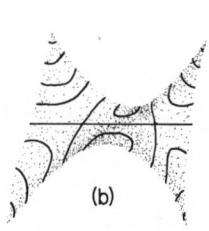

(b)

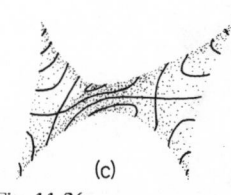

(c)

Fig. 11.26

(a reflection in the y-axis can, but no mapping that preserves handedness). A real change has taken place in crossing Fig. 11.26(b) from (a) to (c).

The contour maps 2b to 7b of Plates 6, 7 all show potentials with instabilities of this kind, apart from the *local* instability of the fold

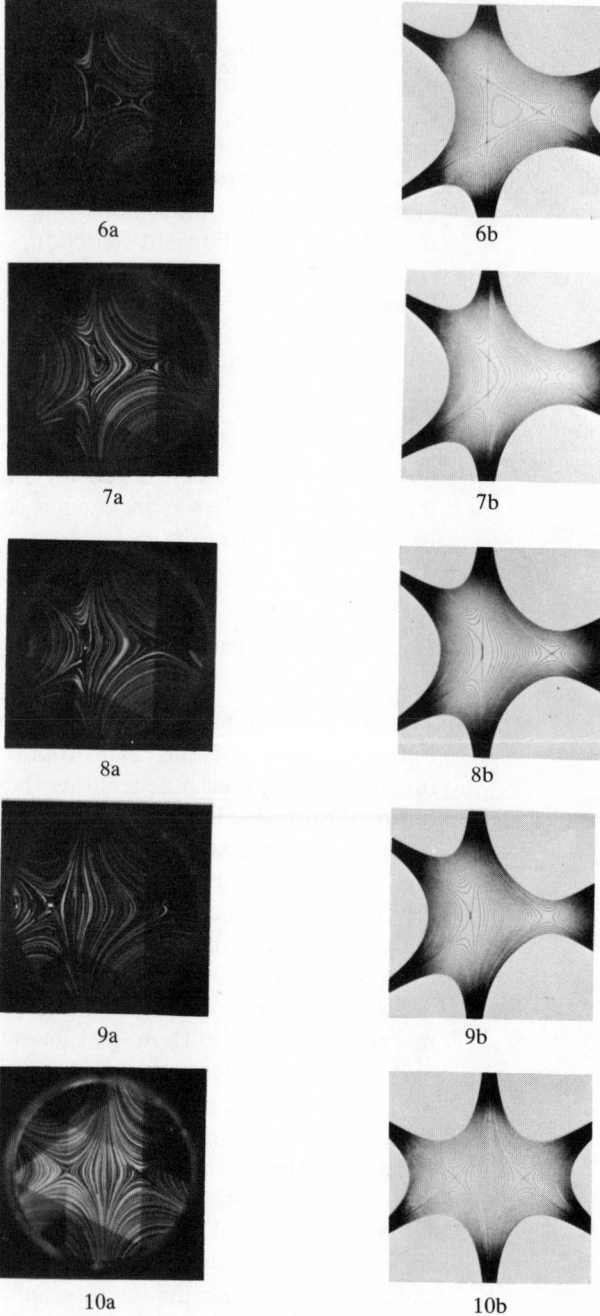

6a   6b

7a   7b

8a   8b

9a   9b

10a   10b

Plate 7. (a) *Flows 6–10 of glycerol in the six roll mill,* (b) *computer-drawn contours for corresponding stream functions.* (*Courtesy of Malcolm Mackley*).

degeneracy at the right of 4b. The corresponding photographs all show stable neighbours of the computed potentials, *globally* inequivalent to them.

To make more detailed predictions we thus must study the set of points in the unfolding space for which critical values coincide, as part of the bifurcation structure. Arnol'd [59] calls this set the *Maxwell stratum*, while Thom, who brought Maxwell's name into this, reserves the term *Maxwell set* for equality of critical values at minima (relating it to the Maxwell rule in thermodynamics, as we discuss in Chapter 14). Perhaps a useful term would be *non-local bifurcation set* for the unfolding points with coincident critical values, as against the *local bifurcation set* for those with local degeneracies. (These terms have the merit of generalizing to dynamical systems, where far more remarkable non-local bifurcations, such as the explosion into existence of 'strange attractors' are possible (Chillingworth [52]).) A point may, like point 4 of Fig. 11.25, belong to both.

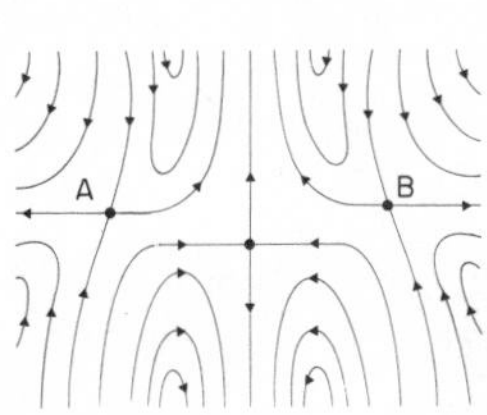

Fig. 11.27 would not become unstable *as a flow* if the corresponding stream function happened to take the same values at A and B; and we should therefore separately define the *saddle-connection set* of an unfolding as the set of points in the unfolding space where the corresponding functions have critical points joined by a contour. (If the functions are defined on $\mathbb{R}^n$ for $n \geqslant 3$, contours become (hyper) surfaces, not curves.) This is clearly a subset of the non-local bifurcation set; in the case of the elliptic umbilic the two sets coincide.

Fig. 11.28 shows the contour maps for representative values of v_x/γ and v_y/γ for fixed positive ω/γ. It is easy to show that the saddle-connection set in this plane consists of the three half-lines through the origin starting at the cusp points, as shown, and hence that the entire (local and non-local) bifurcation set looks like Fig. 11.29. Note that crossing the non-local set alters the connections between inflows and outflows; in particular the unique 'channel' joining opposite gaps (shaded in Fig. 11.30) is moved.

The experimental design of Berry and Mackley was thus an attempt to move about mostly within the non-local bifurcation set. That they came so close to remaining on this piece of surface is a measure of the closeness of fit of their theory.

Fig. 11.27

13 The Six-roll Mill with Polymer Solution

When glycerol (left column of Plates 8, 9) is replaced by a 2% solution of polyethylene oxide in water, the flows corresponding to the same control values of $(\Omega_{\mathrm{I}}, \Omega_{\mathrm{II}}, \Omega_{\mathrm{III}})$ change radically (middle column). The right-hand column, which shows birefringence intensities for these flows, helps explain why. The outflows clearly have regions of high extension, which implies different viscous effects

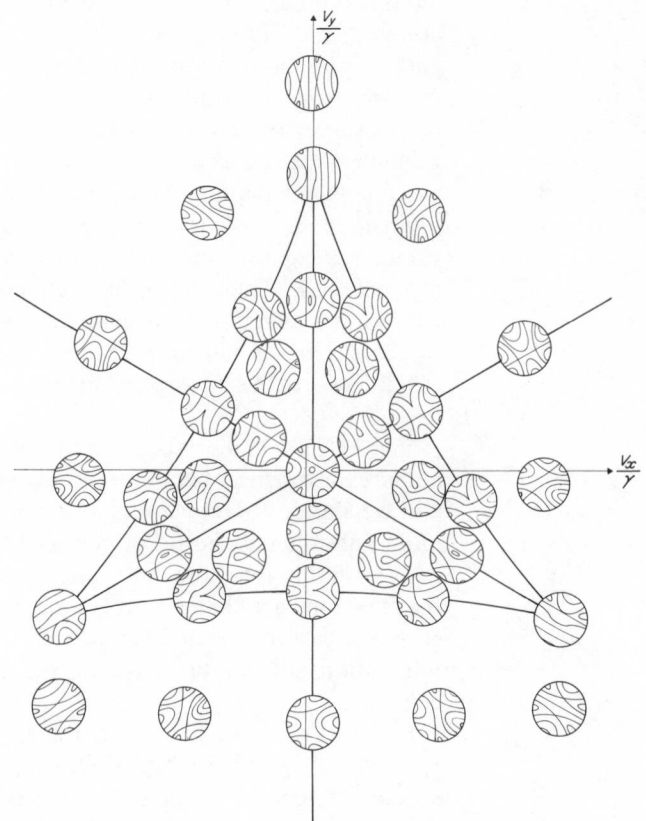

Fig. 11.28

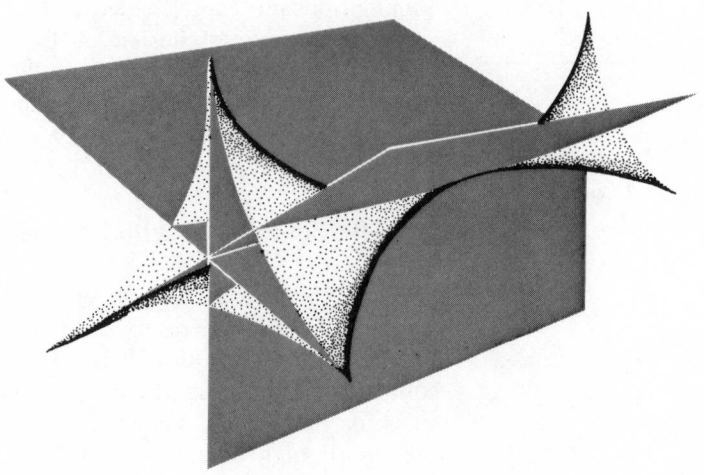

Fig. 11.29

Fig. 11.30

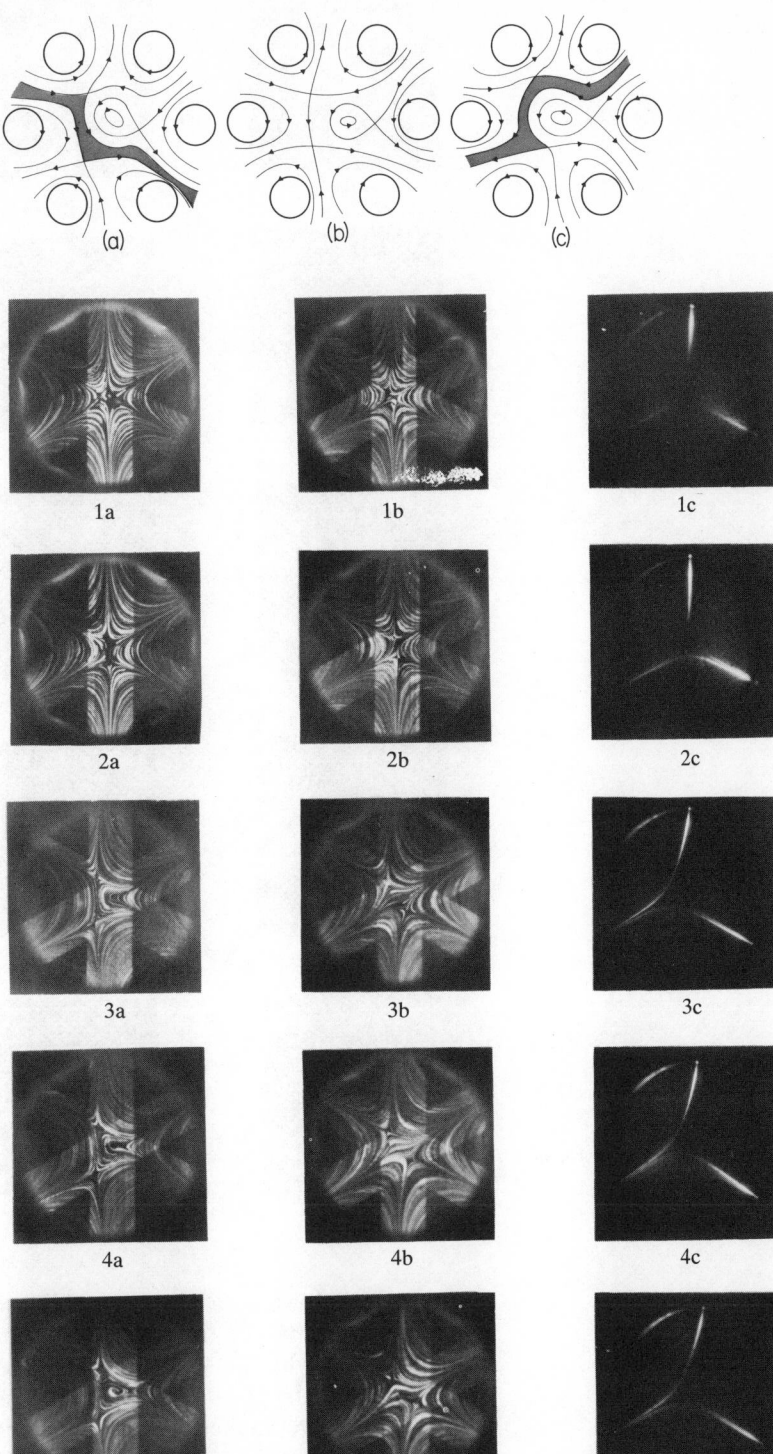

*Plate 8. The effect of polymer:
(a) glycerol, (b) 2% polyethylene
oxide solution, (c) flow birefringence for polyethylene oxide.
(Courtesy of Malcolm Mackley).*

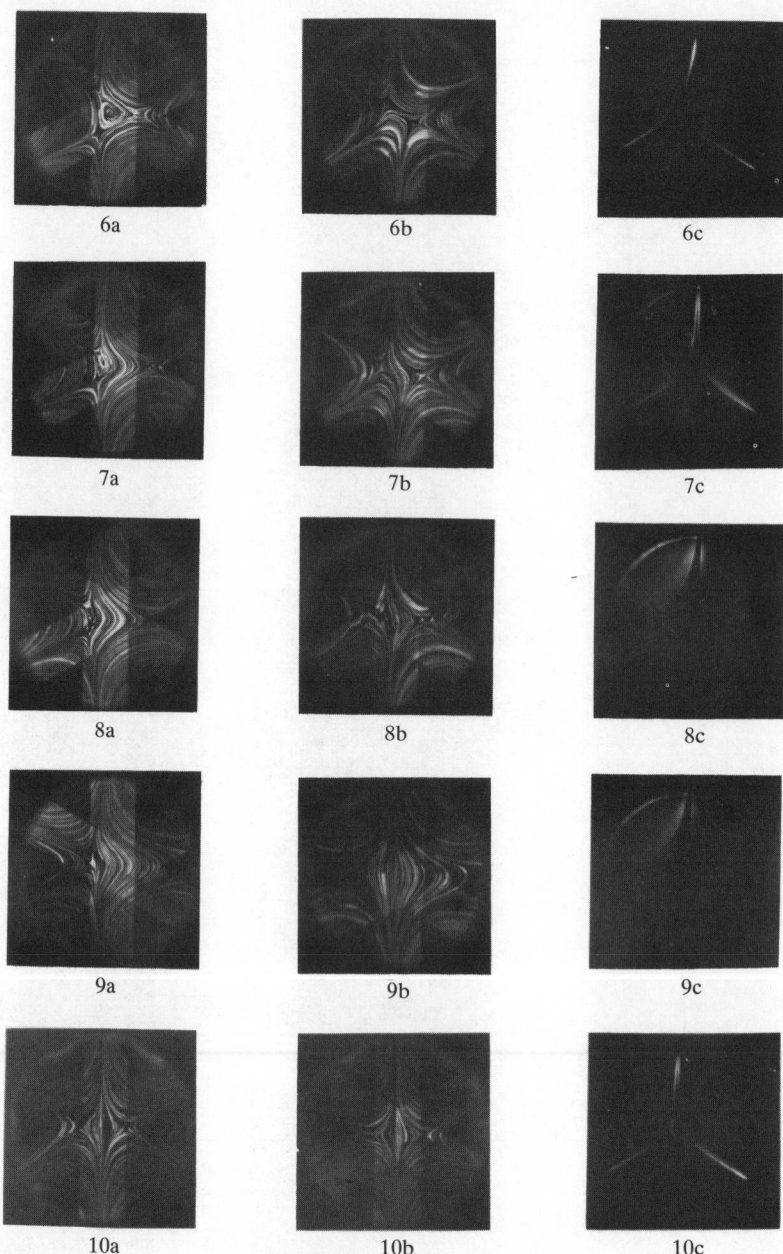

Plate 9. The effect of polymer (continued).

from the inflows, which suggests a difference between α_+ and α_-. We shall not reproduce the physical arguments of Berry and Mackley [53] as to why α_+ should exceed α_-; we merely remark that reference to the non-local bifurcation set, in addition to the local set discussed there, makes it quite clear that this conclusion is correct. It asserts that the plane in $(\omega/\gamma, v_x/\gamma, v_y/\gamma)$-space available to control is turned as shown in Fig. 11.31. Now this means that the path used

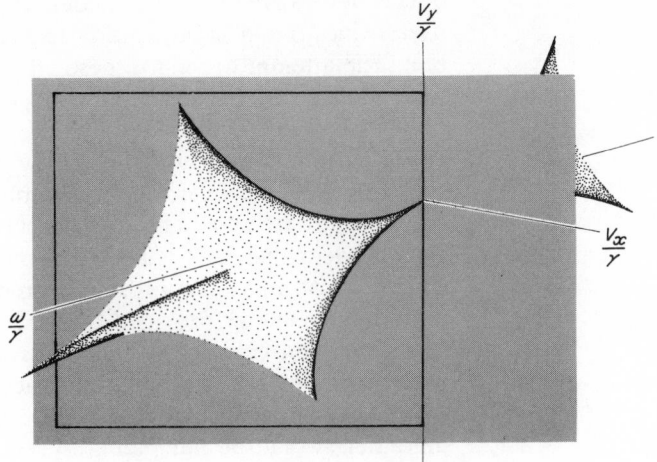

Fig. 11.31

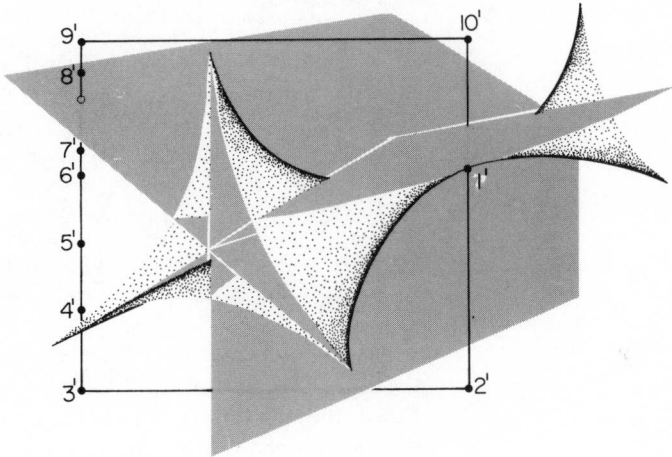

Fig. 11.32

cuts the *left* 'wing' of the saddle-connection set, as shown in Fig. 11.32, rather than the right. By reference to Fig. 11.28 these alternatives are clearly distinguishable, and comparison with Plate 8 shows Berry and Mackley are right.

Moreover, the ratio $(\alpha_+ - \alpha_-)/(\alpha_+ + \alpha_-)$ can be quantitatively estimated. If $(\Omega_{\mathrm{I}}, \Omega_{\mathrm{II}}, \Omega_{\mathrm{III}})$ is the observed control point at which the saddle connection set is met (somewhere between 7' and 8' in Fig. 11.32) then formulae (11.1) to (11.3) of Section 11 and simple geometry imply that

$$\frac{\alpha_+ - \alpha_-}{\alpha_+ + \alpha_-} = \frac{\Omega_{\mathrm{I}} - 2\Omega_{\mathrm{II}} + \Omega_{\mathrm{III}}}{3(\Omega_{\mathrm{III}} - \Omega_{\mathrm{I}})}.$$

The study of the variation of this ratio with polymer concentration would give a sensitive indication of the size of non-Newtonian effects in dissolved polymers under stretched conditions, particularly useful since the local bifurcation set was not experimentally accessible. (The attempt to reduce these effects enough to intersect this set was defeated by turbulence at concentrations of less than 0.5% of polyethylene oxide in water.)

We omit details of how this family of flows can be used to study such points of physical interest as molecular relaxation times: the reader is referred to Berry and Mackley [53]. But it should in any case be clear that here is a sensitive and practical research tool, of which catastrophe theory is an irreplaceable part. Without the unfolding geometry provided by theory, experimental investigation of the relation between roller speeds and flow topology would not easily come to grips with the intricacy evinced by Figs. 11.28 and 11.29. Further work, with one more independent speed for a full unfolding, would be interesting to see.

14   The 2*n*-roll Mill

For the natural most symmetrical generalization to $2n$ rollers, symmetry arguments like those in Section 12 lead to a model

$$\mathrm{Re}\, z^n = x^n - \binom{n}{2} x^{n-2} y^2 + \binom{n}{4} x^{n-4} y^4 - \cdots$$

as the expansion to order $2n-1$ of the stream-function to be unfolded. This is k-determinate, where

$$k = 2n - 3 \text{ for } n \geqslant 3,$$

and has codimension $n(n-2)$. These facts may be checked algebraically using the Siersma trick of Chapter 8.

Since the maximum number of critical points near 0 that a function near the given f may have is given by $\mathrm{cod}\,(f) + 1 = (n-1)^2$, we see that the complexity of the $2n$-roll mill flows, with up to $(n-1)^2$ stagnation points, increases very rapidly. (By index arguments we cannot give here, every nondegenerate flow in the family has at least $n-1$ saddles.) The ten-roll mill is probably beyond convenient experiment, but the eight-roll mill presents some interesting questions.

A full unfolding would require eight dimensions of control (this is one of the double cusp catastrophes on whose geometry work is still in progress). The eight variables provided by roller speeds would not transversely provide these, for the same reason that three speeds do not fully unfold the six-roll mill. The extra variable needed would probably be geometric, corresponding to variation of cross-ratio (Chapter 7 Section 5) among the quartics unfolded. An arrangement like that of Fig. 11.33, without a $\frac{1}{8}$-turn symmetry of the roller positions, may be expected on general grounds to give a double cusp differentially distinct from the unfolding of $\mathrm{Re}\, z^4 = x^4 - 6x^2y^2 + y^4$,

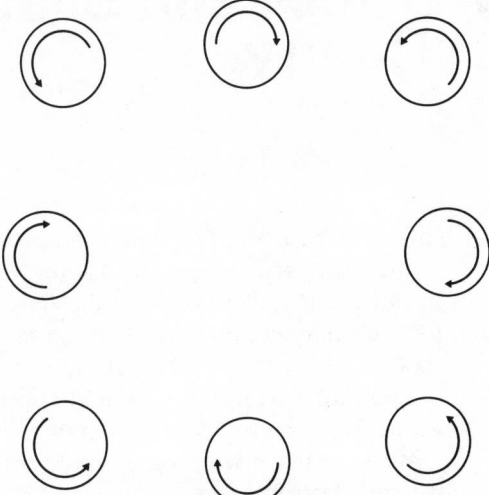

Fig. 11.33

though topologically the same. Experimental work on this would be interesting not only for whatever light it shed on non-Newtonian flows, but for its information on the physical significance of different values of the cross-ratio invariant, in a context where these values might be varied at will. (Too often, as for the buckling plate, Chapter 13, the cross-ratio term in a double cusp unfolding is difficult of experimental access.)

12 Optics and Scattering Theory

In Chapter 10 the functions whose bifurcations concerned us described potential energy; in Chapter 11 they were stream functions; in this chapter they will describe time or phase. The mathematical tools we have set up have their limitations (more complication than they can describe arises naturally, for instance, in the 'chaotic' example of Chapter 17 Section 7) but the limits are wide – often wider than experts have suspected.

We shall begin with some simple results from classical ray optics, where catastrophe theory enters in a very geometric form, serving first to unify phenomena which 'in principle' are well understood. For example, we give the fold catastrophe description of the rainbow. However, the classical ray theory is unsatisfactory, and in particular predicts infinite intensity on a rainbow curve where we know that observationally this is incorrect. Adaptation of the method to a wave theory of light (classical or quantum) leads to some very deep ideas of Maslov, Arnol'd, and others on 'oscillatory integrals', where catastrophe theory is applied in an extensive way as part of a broad mathematical development. We sketch this theory, omitting some of the deeper mathematics. The theory makes in particular predictions of intensities and interference patterns, and we illustrate these with reference to work of Berry. Finally we indicate how similar methods apply to other scattering problems, including acoustical scattering and molecular collisions, and give new applications to mirages and sonic booms.

RAY OPTICS

1 Caustics

In classical ray optics light is assumed to travel along curved paths, which in media of constant refractive index are actually straight lines. Most simple optical systems have 'piecewise constant' refractive index, and the rays are therefore composed of straight line segments. At a reflecting surface (assumed smooth) the angle of incidence θ_1 and the angle of reflection θ_2 are equal (measured from the normal to the surface at the point of incidence) as shown in Fig. 12.1(a); at a refracting interface the angles obey Snell's law

$$\mu_1 \sin \theta_1 = \mu_2 \sin \theta_2$$

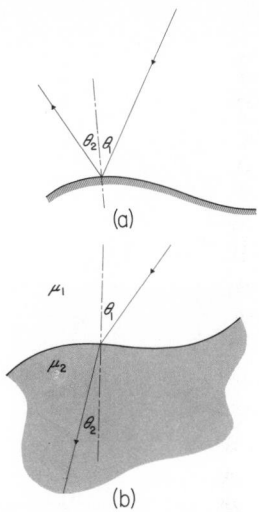

Fig. 12.1

where μ_1 and μ_2 are the refractive indices of the two media (Fig. 12.1(b)).

These two laws are special cases of *Fermat's Principle of Least Time*, formulated in deeper terms by Hamilton, which asserts that the path traversed by the ray of light is extremal (a minimum, maximum or some other type of 'critical point') with respect to the time required to traverse it, among paths with the same end points.

One of the most striking phenomena in ray optics is the formation of *caustics*: sharp, bright curves to which the light rays are tangential. ('Caustic' means 'burning', which caustic intensities can produce, if formed by sunlight with its accompanying radiant heat.) One of the easiest caustics to observe is that which forms in a cup of coffee, or an empty saucepan, when light shining into it is reflected off the cylindrical sides, yielding a caustic of the form shown in Fig. 12.2. (Note that the cusp point is brighter even than the curve.)

The explanation of this phenomenon is not new: it goes back at least to the Rev. Hamnet Holditch in 1857, and an extensive memoir on the subject was written by Cayley [60] in the same year. Consider a bundle of parallel rays incident upon a circle in the plane, as shown in Fig. 12.3, and reflected off the rear portion of the circle according to the law of equal angles. Then the reflected rays have as *envelope* the cusped curve forming the caustic. One can see intuitively why the caustic appears bright: looking along it tangentially the rays nearly coincide and so more rays occur in a smaller space than elsewhere, leading to greater brightness. Unfortunately, as we have said, a proper analysis of this idea predicts *infinite* intensity, and the theory cannot be carried too far without running into difficulties. However, as regards the *shape* and *position* of the caustic, it is quite adequate.

In detail we compute the envelope as follows. Take the unit circle in the plane as shown (Fig. 12.4) and parametrize incident rays by θ. By the law of equal angles, the equation of the reflected ray XR is

$$(y-\sin\theta)\cos 2\theta = (x-\cos\theta)\sin 2\theta.$$

Fig. 12.2

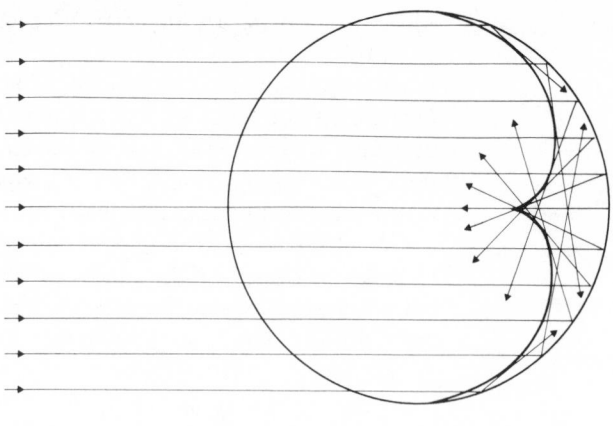

Fig. 12.3

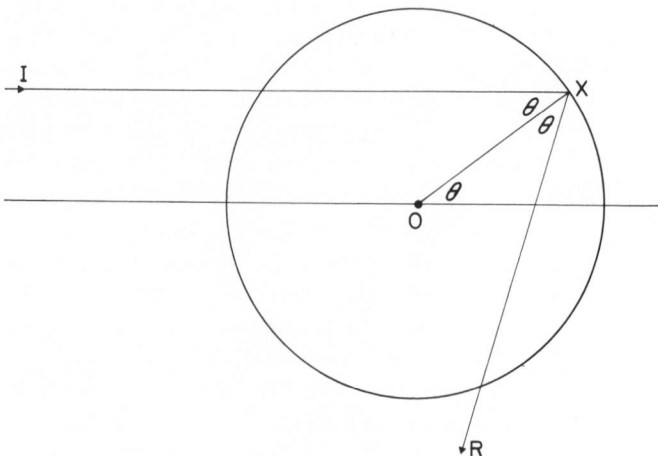

Fig. 12.4

To find the envelope, differentiate this with respect to θ, and solve the resulting equations for x and y, which leads to

$$x = \cos\theta - \tfrac{1}{2}\cos\theta\cos 2\theta$$
$$y = \sin\theta - \tfrac{1}{2}\cos\theta\sin 2\theta.$$

This is the parametric equation of a curve known as the *nephroid* (Fig. 12.5), the curve N followed by a point P on a wheel of diameter $\tfrac{1}{2}$ rolling around a fixed circle of diameter 1. (Geometrically, $\angle$DPC is a right angle, being subtended by a diameter. Thus the ray through P is tangent to the curve followed by P in rolling, with instantaneous centre D. All reflected rays are thus tangent to N, and N is conversely their envelope.) In practice only half N appears, since half is the envelope of *virtual* rays: projections backward of reflected rays like Y in Fig. 12.5. (The appearance of epicycloids in this context is classical: Cayley [60] notes that envelopes of rays which reflect several times in the circle take the form of higher epicycloids.)

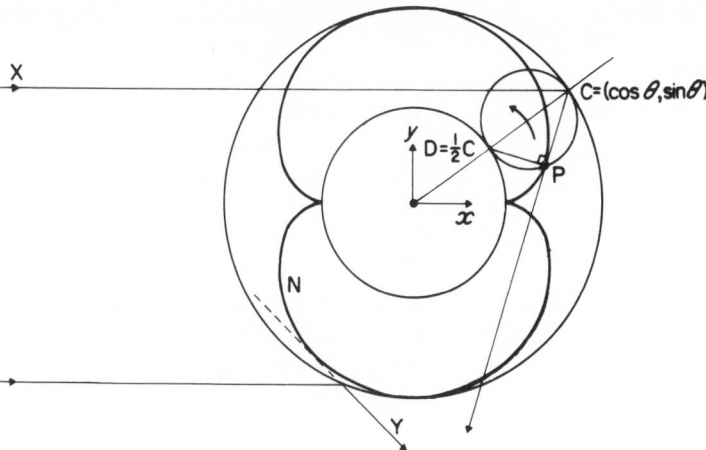

Fig. 12.5

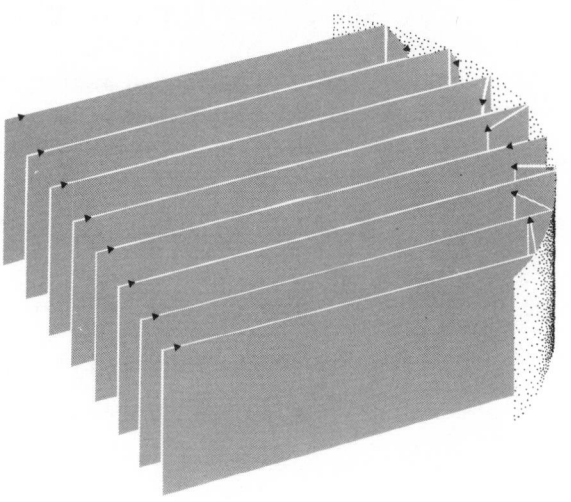

Fig. 12.6

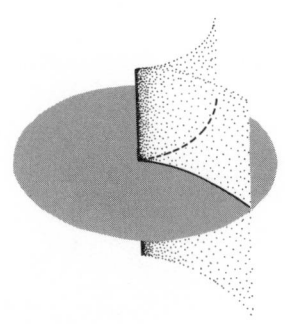

Fig. 12.7

The tip of the cusp occurs for $\theta = 0$, whence

$$(x, y) = (\tfrac{1}{2}, 0).$$

In order to apply this idealized analysis to a real cup, we must take account of the third dimension, and the fact that the light rays are inclined at an angle. However, the axial symmetry of the cylindrical reflector allows us to consider incident planes of light as shown in Fig. 12.6: these, by the above calculation and symmetry, envelop a caustic *surface* as shown in Fig. 12.7. We observe a horizontal section of this (corresponding to the surface of the coffee or the base of the saucepan). This also explains why the remainder of the cup does not interrupt the paths of the light rays, as it should if we take Fig. 12.3 too literally: since the light source is above the rim of the cup, rays can pass over it. Often this leads to extra 'edge effects', where bands of light are superimposed on the caustic. It should be noted that these are due entirely to the fact that the cup has a rim, and are not extra caustics. All of this can be checked experimentally using household equipment.

It is instructive to use the machinery of Chapter 8 and Fermat's principle on this problem. Consider a restricted family of paths $P(X, Y, y)$ from a very distant point D on the negative x-axis in Fig. 12.8 to the point at height y on the right-hand side of the circle, and reflecting on to a point (X, Y). For each (X, Y) inside the circle we have a 1-parameter family of paths, parametrized by y. Which paths are possible rays? (Clearly not the one shown.) A little thought shows that any 'once-reflected' path from D to (X, Y) satisfying the criticality condition must belong to this family. In fact a little apparatus shows that it is critical relative to all nearby paths if and only if it is a path $P(X, Y, y_0)$ in this family with

$$\frac{\partial}{\partial y}(\text{length of } P(X, Y, y))(X, Y, y_0) = 0.$$

If $D = (-d, 0)$ is far enough away to make lines from it essentially parallel in the region we are observing (the situation to which the previous analysis applies) then the length of the path $P(X, Y, y)$ may be taken as

$$f_{XY}(Y) = (d - \tfrac{1}{2} + \surd(1 - y^2)) + ((X + \tfrac{1}{2} - \surd(1 - y^2))^2 + (Y - y)^2)^{1/2}.$$

Expanding this to order 4 in y, order 1 in (X, Y) – a useful exercise for the reader – gives

$$F_{XY}(y) = -\tfrac{1}{4}(1 + 5X)y^4 + \tfrac{1}{2}Yy^3 + Xy^2 - 2Yy + (d + 1 - X).$$

Setting $X = Y = 0$, this becomes just $-\tfrac{1}{4}y^4$ plus a constant. Symmetry

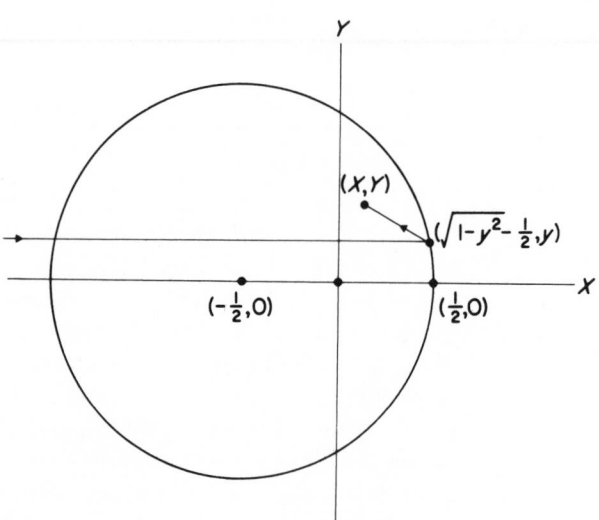

Fig. 12.8

about the Y-axis shows in fact that

$$J^5 f_{(0,0)} = -\tfrac{1}{4} y^4,$$

so we can apply Theorem 8.7 to deduce that the family f_{XY} is strongly equivalent, in some neighbourhood of $(0, 0)$, to the universal unfolding F of $-\tfrac{1}{4} y^4$. This implies a local cusp geometry (exercise: reduce F and/or f (harder) locally to the canonical form $-(x^4 + ax^2 + bx)$ up to a shear function) pointing in the direction we have already found by classical geometric methods. Note that it is a dual, not a standard cusp; of the three rays that reach a point inside the cusp region (Fig. 12.9) only the middle one takes *least* time even among near neighbours. The others are longer than neighbouring paths in the family we have considered, shorter than neighbours which are not straight.

A similar analysis holds for the cusped surface caustic of Fig. 12.7, but now we have three control parameters (X, Y, Z) of which Z is *disconnected*. The obvious families of paths are now two-dimensional, parametrized by a cylinder instead of an arc, but suitable (X, Y, Z, y)-dependent vertical translations turn the height at which the path meets the cylinder into an *inessential* state variable.

By treating ray optical caustics in a grand generalization of this approach it is possible to show that the structurally stable caustics (suitably defined) in $\mathbb{R}^n$ are the bifurcation sets of elementary catastrophes with n control variables, with corresponding typicality theorems. Thus in $\mathbb{R}^3$ we will typically observe caustics corresponding to the fold, cusp, swallowtail, elliptic umbilic and hyperbolic umbilic; and *only* these (locally). (However, as Berry [61, 62] has remarked, symmetry conditions can lead to the observation of atypical caustics as well.) Even at this classical level, only catastrophe theory allows one to come to grips with the typical caustics of nature, instead of the caustics of artificial optical systems (such as optical instruments). Recent work of Berry [61–64] and Berry and

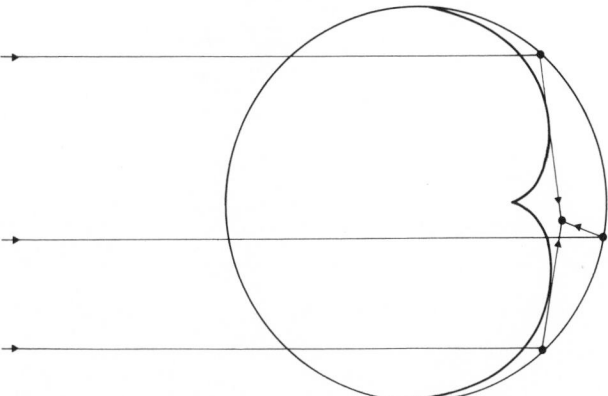

Fig. 12.9

Nye [65], some of which we shall discuss here, confirms this assertion.

2 The Rainbow

The most familiar caustic of all is undoubtedly the rainbow, which is an instance of the fold catastrophe. For light of constant wavelength the rays emerging from a spherical raindrop (after one internal reflection) envelop a caustic which is smooth, and (in cross-section) almost straight more than a few diameters away from the drop (Fig. 12.10). This caustic corresponds to a fold catastrophe (Fig. 12.11). Allowing for the circular symmetry, which revolves the caustic around the line from the raindrop to the sun, the effect is essentially as if, for the given wavelength, the raindrop emits a bright cone of light, whose axis points towards the sun. Varying the wavelength alters the vertex angle of the cone, because of the variation of refractive index with wavelength; so we can imagine the drop as emitting a set of concentric, coloured cones as in Fig. 12.12.

An observer on the ground will see light of a given colour only in a direction which corresponds to the vertex angle for that colour; and this yields the familiar multihued circular arc of the rainbow, as illustrated in Fig. 12.13. Seen from a suitable height, rainbows can exhibit complete circular arcs (one appeared in the film *The Hindenburg*). Rainbows with more than one arc are, of course, the result of rays which experience several internal reflections.

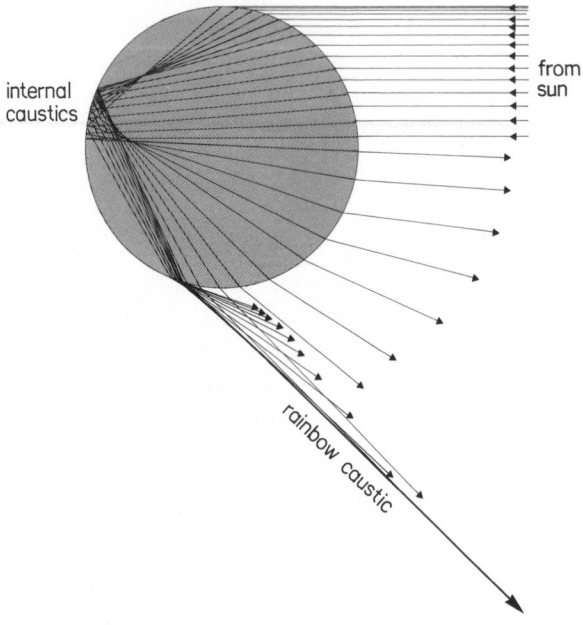

Fig. 12.10

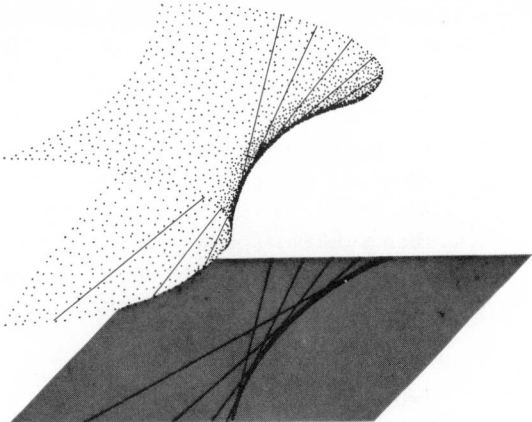

Fig. 12.11

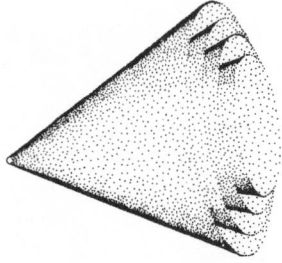

Fig. 12.12

This description is of course completely standard; although it is quite different from the spurious explanation often given in elementary texts. This treats the emergent rays of a given colour as all parallel, within the central plane. In fact, rays of all colours emerge in any given direction inside the cone, and it is only the extra intensity given to different colours in different directions by the fold caustics that stops them averaging out back to white light. This is why the inside of a rainbow is brighter than the outside, against a dark background of mountain or cloud.

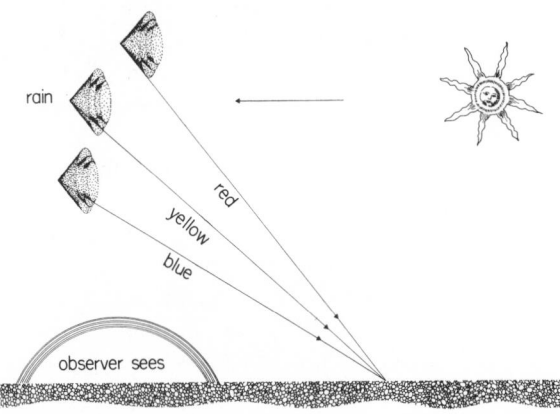

Fig. 12.13

3 Variational Principles

Figs 12.3 to 12.9 would apply equally if the rays were the paths of billiard balls on a frictionless table, bouncing elastically off its circular rim. Of course, we would have to take one path at a time to avoid collisions, or consider particles that, like photons or Newtonian 'light corpuscles', can pass through each other. In this context, the Fermat/Hamilton principle of Section 1 develops into various analogous laws, like the one (mis)called the 'Law of Least Action'. ('Action' is a quantity defined at each point of a possible path for a particle – or system – in the space of positions and momenta, and integrated along the path.) Exact analogies with the dual cusp of Section 1 occur, so that the law strictly refers to critical, or *stationary*, action. (Even 'extremal' suggests maxima and minima, rather than saddles.) This worries some people, even those who can accept a *least* whatever principle as the Divine Economy of Nature. The reason it works at all has much to do with the reason it fails to be the complete answer.

Quantum mechanically, the (complex) probability of a particle's presence follows all possible paths. But if nearby paths with the same end point P give probabilities at P with different phases, the results cancel (interfere) and the physical probability of the particle arriving by that route (the modulus of the complex probability) is low. But if a path is stationary for arrival phase (for light rays this is given by time, up to multiples of 2π times the period, or equivalently just time), then nearby paths have the same arrival phase to first order. The results add (*constructively* interfere), so the physical probability of arriving by this path *or one very near it* becomes high. In applications the arrival phases are given by time, action, or some similar variable. To give the physical laws as being of 'least time, action, etc.' is to replace the physically significant condition of *stationary* time, action, etc. by a physically unimportant statement which implies but is not equivalent to it. (Indeed, the habit causes confusion. Most people know that a geodesic is the 'shortest distance between two points' but not that a timelike geodesic in spacetime is locally *longest* instead, as explained geometrically in Dodson and Poston [5].) For a fuller discussion of the meaning of variational principles from this viewpoint, see Feynman [54].

Thus the ray optics approach consists exactly of treating the *most probable* path (given by the condition of being stationary) as the path of the particle. As we go to the classical limit, by taking wavelength to zero, the bunch of nearby almost-equally-probable paths shrinks, the probability peak rises, and in the limit the description is exact. (We can approach this limit either by looking at higher frequencies, or formally by letting Planck's constant go to zero, or by looking at larger 'classical sized' particles. The quantum wavelength of a billiard ball is about 10^{-40} cm. Thus the exact applicability of variational methods to classical mechanics is explained by quantum mechanics, in which the variations have a physical meaning.)

Even where the wavelengths are finite, the ray optical approximation of the true path by the stationary path is a good one, as long as the action, time, or whatever has a *nondegenerate* (that is, Morse) critical point at the stationary path. It is good enough for all physical calculations, if the stationary path is comfortably isolated. But at a

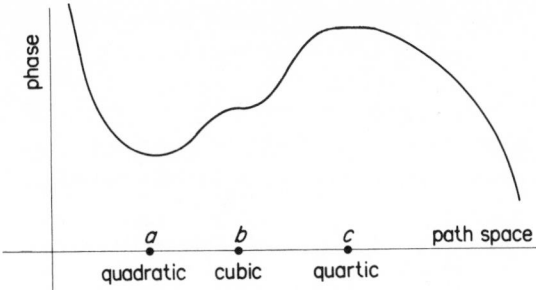

Fig. 12.14

degenerate critical point (b and c in Fig. 12.14) the band of paths with nearly the same phase is much wider. We cannot approximate the behaviour by a single One True Path without predicting unphysical infinities. Thus it is exactly around catastrophe points like the cusp in a teacup and the fold in the rainbow that ray optics breaks down. We can find these points, and the stable forms possible for their geometry, by catastrophe theory. We may then study what physically happens at such catastrophes by working with a wave description of the light, rather than the particle picture implicit in ray optics. This approach is developed in Section 5, after we have indicated in the next section some of the more physical systems to which the analysis applies.

4 Scattering

Instead of optical scattering we may consider the general problem of scattering theory, where a beam of particles (waves, wavicles) passes through some scattering centre, and the deflection is measured from a large distance away (Fig. 12.15). In practice this distance is so great that the spatial position of a trajectory on a fixed screen is almost exactly the same as the *direction* in which the trajectory emerges, and bright patches on the screen correspond to bundles of parallel trajectories. Again caustics can form; a famous one is the *quantum-mechanical rainbow* (Berry [61]) in which a scattering

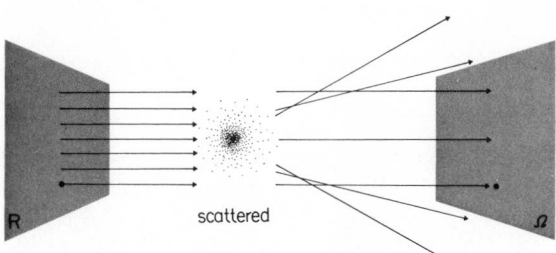

Fig. 12.15

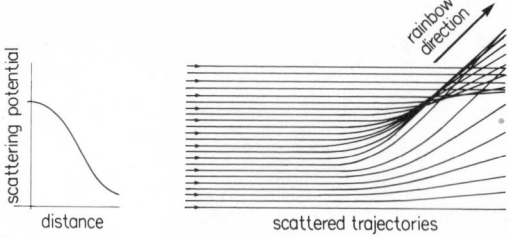

Fig. 12.16

potential of the form shown in Fig. 12.16(a) leads to scattering as shown in Fig. 12.16(b).

The detailed application of the variational approach of Section 3 yields the following. Let the initial position of a particle (in *phase space* – i.e. its physical position and momentum) be R, and its final position or direction after scattering be Ω. Then the Hamiltonian formalism gives a smooth *action* function $\phi(R, \Omega)$ such that the rays with final direction Ω are those with initial positions R such that

$$\frac{\partial}{\partial R} \phi(R, \Omega) = 0$$

for that Ω. This looks very much like the catastrophe theory set-up, with control variables Ω and state variables R. However, the action depends on the path from R to Ω, so that the state space is strictly the infinite-dimensional manifold of paths. In practice this can often be reduced to finite dimensions, by showing (as we assumed for the analysis of Section 1) that for a given Ω only one path from each R need be considered – or, sometimes, only a finite-dimensional family of paths from each R. In the following we shall ignore this difficulty, because an alternative formulation in terms of 'Lagrange manifolds' avoids it entirely, as far as applications to caustics go.

If the critical point of ϕ is degenerate, then physically we have *angular focussing* in which a whole bunch of rays in R-space end up at nearby points in Ω-space (Fig. 12.17). Hence bright caustics are observed on the bifurcation set of the catastrophe (in Ω-space). Now by Thom's theorem, if Ω is $\mathbb{R}^2$ (physically, a screen) the caustics are typically folds or cusps (locally). Fig. 12.18 shows a typical set of caustics for refraction through an irregular lens.

Many systems yield neatly to this approach. A 'perfect' lens is one that refracts all incident rays parallel to its axis (at least if close enough) through a single point, the *focus* (Fig. 12.19(a)). This is

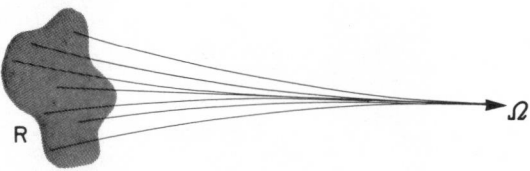

Fig. 12.17

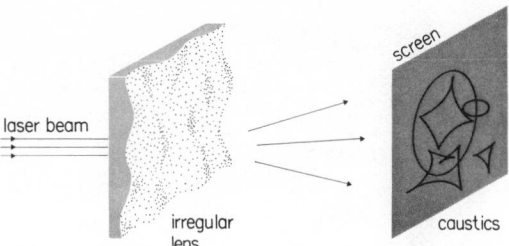

Fig. 12.18

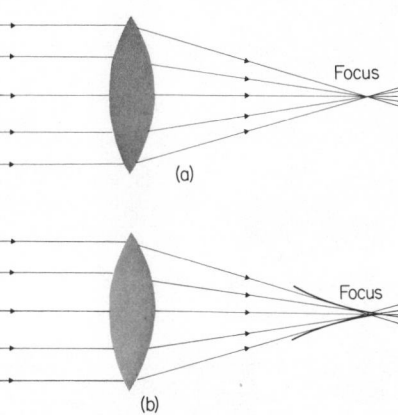

Fig. 12.19

clearly an atypical situation, and in practice what we get is Fig. 12.19(b), with a cusp caustic whose tip is at the focus. This effect in lenses is called *spherical aberration*. In actual fact these pictures have to be revolved about the axis, yielding a 'catastrophe with symmetry' of infinite codimension (in the sense that infinitely many unfolding parameters are required for symmetry-breaking perturbations). It is recognized in optics that lenses suffer many other kinds of aberration (the most important, after spherical aberration, being known as *coma, astigmatism, Petzval curvature* and *distortion*) which presumably correspond in some way to the more important terms in the unfolding. However, little work on this sort of question has yet been done.

Light refracted through an asymmetrical droplet of water on glass typically yields caustics of the form of Fig. 12.20; the caustic corresponding to parts of the droplet at which the Gaussian curvature is zero. Plate 10, due to Michael Berry of the University of Bristol, shows a photograph of just such a caustic formed by shining a laser beam through a droplet.

Though the shape of the caustic is gratifying, we have not yet explained the distribution of light in Plate 10. The various diffraction fringes present make it quite clear that a ray theory is inappropriate, and a wave theory is required. The theory that now exists is distinctly sophisticated, and we devote several sections to it.

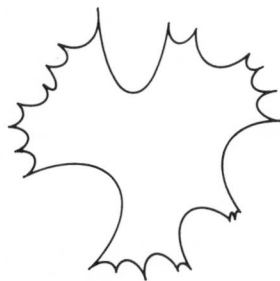

Fig. 12.20

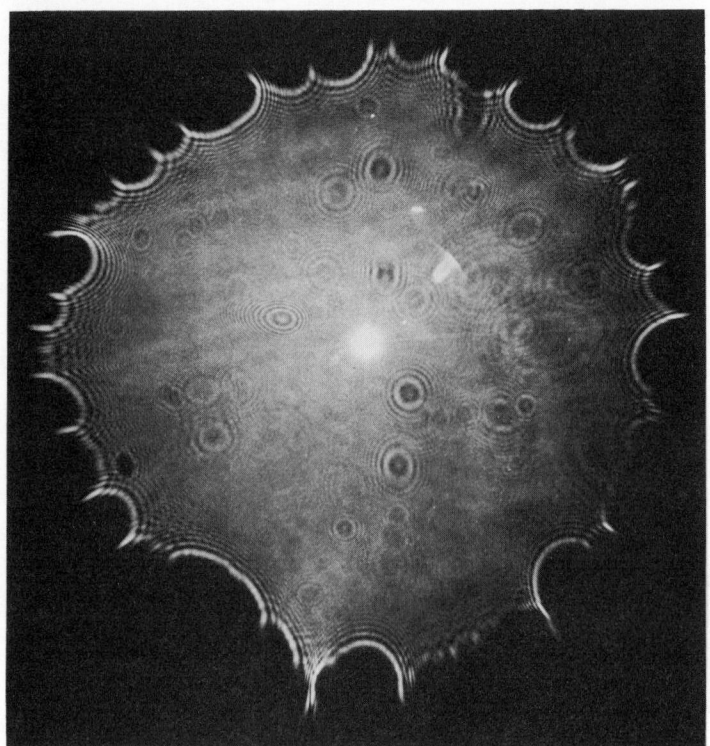

Plate 10. *Cusped caustics formed by shining a laser beam through an irregular droplet. (Courtesy of Michael Berry).*

WAVE OPTICS

5 Asymptotic Solutions of Wave Equations

The intensity problem is approached by way of Maslov's theory of 'oscillatory integrals' (Maslov [66]) which led Arnol'd [26] to his version of the classification of elementary catastrophes. Maslov's ideas are inspired by previous methods used on similar problems, and we shall not attempt to provide too much motivation for the precise formulation given. Some of the more highbrow parts of the theory are omitted (despite being essential to a proper treatment) because they require a deep and specialized background. The exposition is based on Chazarain [67] to which the interested reader should refer for further discussion. (*See also* Duistermaat [41].)

In attempting to solve the wave equation

$$\frac{\partial^2 v}{\partial t^2} = \nabla^2 v \left(\text{where } \nabla^2 = \frac{\partial}{\partial x_1^2} + \cdots + \frac{\partial}{\partial x_n^2} \right)$$

one often wishes to find solutions that do not change, but only oscillate, of the form

$$v(x, t) = e^{i\tau t} u(x) \qquad (t \in \mathbb{R}, x \in \mathbb{R}^n)$$

where τ is a real parameter. These are sometimes called *stationary* solutions, but we wish to avoid confusion with other uses of the word (as in variational principles). They correspond to harmonics of a vibrating string, where every point vibrates in the same cycle, $\sin(\tau t) = \mathrm{Re}\,(e^{i\tau t})$, with only an amplitude factor $u(x)$ depending on the point observed. (Where u changes sign, the vibration could be said to change phase by π.) Substituting this general form in the equation we are led to

$$(\tau^2 + \nabla^2)u = 0. \tag{12.1}$$

Similarly in the Schrödinger equation

$$i\hbar\,\frac{\partial v}{\partial t} + \frac{\hbar^2}{2m}\,\nabla^2 v - V(x)\cdot v = 0$$

we seek unchanging solutions of the form

$$v(x, t) = e^{itE/\hbar}u(x)$$

of energy E (here $2\pi\hbar$ is Planck's constant); and then

$$\frac{\hbar^2}{2m}\,\nabla^2 u + (E - V(x))\cdot u = 0 \tag{12.2}$$

which has certain essential properties in common with (12.1) provided the role of τ is played by $1/\hbar$. The general theory applies to a class of differential operators of the form

$$P\!\left(\tau, x, \frac{1}{i}\,\frac{\partial}{\partial x}\right) = \sum_{k=0}^{m} \tau^{m-k}\, P_k\!\left(x, \frac{1}{i}\,\frac{\partial}{\partial x}\right)$$

where P_k is an operator of degree k or less, with smooth coefficients; and the corresponding equation is

$$Pu = 0.$$

Instead of seeking *exact* solutions of this, it turns out to be useful to seek *asymptotic solutions* $u(x, \tau)$ satisfying

$$P\!\left(\tau, x, \frac{1}{i}\,\frac{\partial}{\partial x}\right)u(x, \tau) = O(\tau^{-\infty}), \tag{12.3}$$

where $O(\tau^{-\infty})$ indicates a function which vanishes rapidly for large τ (that is, faster than any τ^{-n} for positive integer n). Two asymptotic solutions u, v are *equivalent*, written $u \sim v$, if $u-v = O(\tau^{-\infty})$. The interpretation of this is that for large τ an asymptotic solution is very nearly an exact solution, and equivalent solutions are very nearly equal. For some problems (like diffraction around a sharp corner) this approach is not entirely successful, but it often succeeds triumphantly. Physics as an activity consists in considerable measure of obtaining difficult equations and then finding some substitute for solving them.

By analogy with (12.1) and (12.2) we try to find asymptotic solutions of the form

$$u(x, \tau) = e^{i\tau\phi(x)}a(x, \tau)$$

where $\phi: \mathbb{R}^n \to \mathbb{R}$ is a smooth function called the *phase* and $a: \mathbb{R}^n \times \mathbb{R} \to \mathbb{R}$ is a smooth function called the *amplitude*. An expression of this kind does not describe every point as oscillating in step, like the 'harmonics' above: one point is ahead of another by the difference in their phases. The points where u has a particular phase θ, that is

$$\theta = \arg(u) = \arg(e^{i\tau t} e^{i\tau\phi(x)}a(x, t))$$
$$= \arg(e^{i\tau(t+\phi(x))}) = \tau(t + \phi(x))$$

(up to integer multiples of 2π), move with speed $(\nabla\phi)^{-1}$ when ϕ is non-singular. This gives (except at singularities) a clear notion of the 'direction in which the oscillation is moving', which corresponds to the direction of a ray through x. If in the ray picture several rays are going through x, with no caustics nearby, the corresponding solution to be sought around x is a superposition

$$u(x, t) = e^{i\tau t}(u_1(t, x) + \cdots + u_r(t, x))$$
$$= e^{i\tau t}(e^{i\tau\phi_1(x)}a_1(x, \tau) + \cdots + e^{i\tau\phi_r(x)}a_r(x, \tau)).$$

Assuming that the amplitude a has an *asymptotic expansion* of the form

$$a(x, \tau) \sim \tau^\mu \left[a_0(x) + \frac{1}{\tau} a_1(x) + \frac{1}{\tau^2} a_2(x) + \cdots \right],$$

said to be of *degree* μ, we substitute this expression into (12.3). The result is a differential equation for ϕ, and a system of equations expressing each a_j in terms of $a_0, \ldots, a_{j-1}$. In principle these differential equations may be solved by known methods; for ϕ the equation is of 'Hamilton–Jacobi' type and is soluble *typically*, provided a certain transversality condition holds; for the a_j the equations are *linear* and in practice only small values of j need be considered.

6 Oscillatory Integrals

A problem with the above method is that it leads to solutions only in a small region of $\mathbb{R}^n$. And *global* solutions of this type may not be possible, because the equation for ϕ cannot be solved around a caustic. (Geometrically, the equation gives a 'graph' like Fig. 12.21, which does not define a function – it is 'multivalued'.) To remedy this, Maslov [66] suggested the use of *oscillatory integrals* of the form

$$u_{\phi a}(x, \tau) = \int_{\mathbb{R}^N} e^{i\tau\phi(x,\alpha)}a(x, \alpha, \tau)\,\mathrm{d}\alpha \qquad (12.4)$$

where $\alpha \in \mathbb{R}^N$ is a parameter. Intuitively this represents an *infinite* superposition analogous to the finite one above, and corresponds to allowing in all the extra, almost-equally-probable rays that Section 3 tells us to expect. Unlike the finite superpositions, in consequence, it can extend past caustics to a global solution of similar type. The proper global set-up is the theory of *Lagrange manifolds* (Duister-maat [41], Arnol'd [26, 59]) but we do not have the necessary concepts to express this formally: and in any case the global problem rapidly reduces to a local one again (once it is set up correctly!).

We continue to refer to ϕ as the (*N*-parameter) phase, and a as the (*N*-parameter) amplitude. We assume that a has an asymptotic expression

$$a(x, \alpha, \tau) \sim \tau^\mu \left[a_0(x, \alpha) + \frac{1}{\tau} a_1(x, \alpha) + \cdots \right]$$

and generalize the classical method to this parametrized form.

The most important question is to describe the asymptotic behaviour, for large τ, of the function (12.4). It can be shown that this depends only on the Taylor series of ϕ and a on the set

$$C_\phi = \left\{ (x, \alpha) \;\middle|\; \frac{\mathrm{d}}{\mathrm{d}\alpha} \phi(x, \alpha) = 0 \right\}.$$

The reasons for this are approximately as follows. The integrand of (12.4) oscillates more and more rapidly for increasing τ. Away from C_ϕ these oscillations tend to cancel out. The approximation of the integral (12.4) by the sum

$$\sum_{j=1}^r e^{i\tau\phi(x,\alpha_j(x))} a(x, \alpha_j(x), \tau),$$

where $\alpha_1(x), \ldots, \alpha_r(x)$ are the solutions for the given x of the equation

$$\frac{\partial}{\partial\alpha} \phi(x, \alpha) = 0$$

is called the *method of stationary phase*, or *Wentzel–Kramers–Brillouin (WKB) method*. Around a point x where $\alpha_1(x), \ldots, \alpha_r(x)$ are all Morse singularities of ϕ_x, the results of Chapter 4 show that they are locally (as the notation suggests) functions of x, and this method – essentially equivalent to ray optics – works well. But around degenerate singularities, we need the whole integral.

Now the definition of C_ϕ suggests a catastrophe formalism: think of α as state variable, x as control variable and ϕ as the potential: then C_ϕ is the catastrophe manifold (when it *is* a manifold!) defined by the critical points of ϕ. This formalism is quite a fruitful one, as it allows us to put the integral into a standard form and to compute many required properties by considering a small number of cases.

The behaviour of (12.4) near a point $x_0 \in \mathbb{R}^n$ can be considered for three distinct cases.

(a) *The shadow zone.* No real value of α exists with $(x_0, \alpha) \in C_\phi$, that is, x_0 does not lie below (in the image under projection of) C_ϕ. Solution involves only the complex roots of $(\partial/\partial\alpha)\phi(x, \alpha) = 0$, and such features as the 'whisker' on a swallowtail catastrophe (Poston and Stewart [25] p. 130) where *complex* roots coincide require investigation. It transpires – not easily – that they are not physically significant, at least in the swallowtail case.

(b) *The illuminated zone.* Now there do exist points α with $(x_0, \alpha) \in C_\phi$, but the relevant critical points are all Morse: x_0 lies below the projection of the catastrophe manifold but not in the bifurcation set.

(c) *The caustic.* One or more critical point is degenerate, and so x_0 lies in the bifurcation set.

Fig. 12.21 shows these three regions for the case where ϕ defines a fold catastrophe: historically this was the first case to be well understood.

The 'shadow' intensities vanish rapidly as τ or distance from the caustic increases, and we shall not consider them further. To Morse singularities of the illuminated zone, which may be treated locally as a function $(\alpha_j(x))$ of x, we apply the method of stationary phase, with attention to the approach to the limit as $\tau \to \infty$. Inserting the asymptotic expression assumed above for a, we get a contribution

$$u_j(x, \tau) \sim \frac{\tau^\mu (2\pi/\tau)^{N/2}}{\det H_{\alpha_x}\phi} \left(a_0(x, \alpha_j(x)) + O\left(\frac{1}{\tau}\right) \right) e^{[i(\tau\phi(x,\alpha_j(x)) + \pi\sigma/4)]}$$

where $H_{\alpha_x}\phi$ is the Hessian of $\phi(x,)$ at $\alpha(x)$, and σ is its signature when considered as a quadratic form. (The phase factor $\pi\sigma/4$ is constant over any sheet of Morse points, and becomes significant

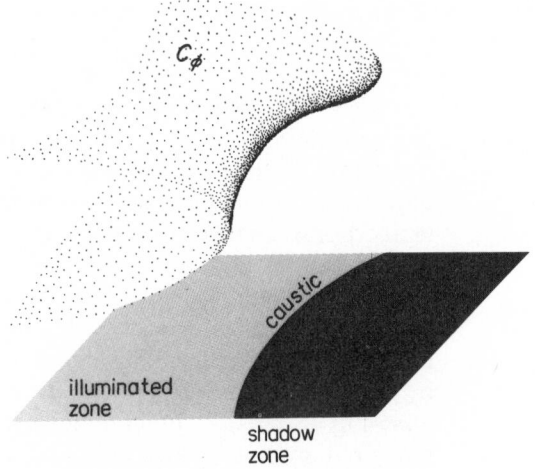

Fig. 12.21

only when we wish to combine and compare the contributions of different sheets attached to the same singularity.) Evidently for fixed τ this blows up if the Hessian approaches a singularity as we move x, so we cannot use this estimate close to the singularity. *How* close we can use it is a problem to be solved separately for each catastrophe.

There remains the caustic case. The estimates here of course depend on the type of caustic. We define the *order* of the caustic to be the least upper bound of the set of numbers $\nu \in \mathbb{R}$ such that

$$u(x, \tau) = O(\tau^{\nu})$$

near x_0 for all u of the given type. The problem of computing the order is local, and the unfolding techniques of catastrophe theory may be applied.

7   Universal Unfoldings

We can concentrate on 'typical' cases. We consider those ϕ which arise in the typical families prescribed by the classification theorems of catastrophe theory, that is, ϕ equivalent to a standard catastrophe form. Then ϕ will be an unfolding of

$$g(\alpha) = \phi(x_0, \alpha)$$

around the point $x = x_0$. Suppose that the universal unfolding in the sense of Chapter 8 is

$$\psi(y, \alpha) = g(\alpha) + y_1 g_1(\alpha) + \cdots + y_p g_p(\alpha)$$

where $y_1, \ldots, y_p$ is a new set of control variables. We may also replace g by any equivalent function (in the catastrophe theory sense) and hence assume that ψ is one of the elementary catastrophes in standard form, with polynomial functions $g, g_1, \ldots, g_p$.

Now the asymptotic behaviour of the integral (12.4) can be proved to be completely determined by that of a similar integral with ϕ replaced by ψ. (In fact, if we include in the unfolding a constant term $y_{p+1} g_{p+1}$, where $g_{p+1}(\alpha) = 1$, then the behaviour is identical; but then we can take this constant term outside the integral and get rid of it if we so wish. Unfoldings with constant terms are what we need with the strict form of right equivalence discussed in Chapter 8 Section 5, whereas for the elementary catastrophes we use the weaker version, which does not introduce constant terms into the unfolding. Either version can be used here, with mathematically minor but physically important changes in the detailed formulae.)

Thus we reduce the problem to considering standard integrals of the form

$$G(y, \tau) = \int_{R^N} \exp \left[i\tau(g(\alpha) + y_1 g_1(\alpha) + \cdots + y_p g_p(\alpha)) \right] d\alpha,$$

where $g(\alpha) + y_1 g_1(\alpha) + \cdots + y_p g_p(\alpha)$ is an elementary catastrophe in standard form. Because the g's are then polynomials, it follows that this integral, for fixed y, has an asymptotic expansion of the form

$$\sum_{\gamma} \sum_{k=0}^{N-1} C_{\gamma k} \tau^{\gamma} (\log \tau)^k.$$

Now it remains to make precise the values of the exponents and the coefficients, and how everything depends on y. We can do this for the list of elementary catastrophes, as far as it is established, by taking each member of the list in turn. Here we shall consider only the seven catastrophes of codimension $\leqslant 4$.

8 Orders of Caustics

For a caustic of fold type we may take

$$\psi(y, \alpha) = -\frac{\alpha^3}{3} + \alpha y$$

where $\alpha, y \in \mathbb{R}$. Then

$$G(y, \tau) = \int_{\mathbb{R}} \exp\left[i\tau(\alpha y - \alpha^3/3)\right] d\alpha.$$

This may be evaluated by standard methods of mathematical physics. We change the variable to

$$\beta = \tau^{1/3} \alpha,$$

and obtain

$$G(y, \tau) \sim \tau^{-1/3} \mathrm{Ai}\,(\tau^{2/3} y) \tag{12.5}$$

where Ai is the *Airy function* (Airy [68], Abramowitz and Stegun [69]) defined by

$$\mathrm{Ai}\,(z) = \int_{\mathbb{R}} \exp\left[i(z\beta - \beta^3/3)\right] d\beta.$$

This function can be expressed as a power series, and may also be defined by a second-order linear differential equation, and its behaviour is well understood. Its graph is shown in Fig. 12.22. The dotted line shows its 'asymptotic' form (obtained by using the method of stationary phase on the Morse critical points) which blows up on the ray-optical caustic. The region of good agreement between the ray-optical stationary phase approximation and the Airy function is remarkably well defined. Calculations of this kind of fit, necessary for physical applications, have been made also for the cusp caustic (Holford [70], Berry [61]) and work is in progress on

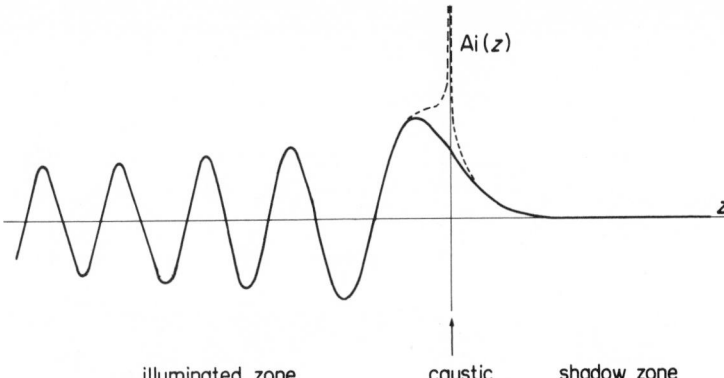

illuminated zone   caustic   shadow zone

Fig. 12.22

the higher catastrophes. Note in Fig. 12.22 the sharp decay in the shadow zone, the peak near (but not at) the ray-optical caustic point – where the intensity, so far from being infinite, is about two-thirds of its peak value – and the oscillations in the illuminated zone. As τ increases, the formula (12.5) shows that the peak of $G(y, \tau)$ rises and approaches the ray-optical caustic. This graph should be compared with Plate 11 (part of a photograph due to Berry [61]) which shows experimental observations of intensities across a fold caustic.

The Airy function is *bounded* for all z, and this fact allows one to compute the order of the fold caustic: it is 1/6. This result, along with much else, has been obtained by 'bare hands' methods by Ludwig [71].

At a cusp caustic we can take

$$\psi(y, \alpha) = \frac{\alpha^4}{4} + \tfrac{1}{2} y_1 \alpha^2 + y_2 \alpha,$$

or any other convenient form. The Airy function is replaced by the *Pearcey function* (Pearcey [72])

$$\mathrm{Pe}\,(z_1, z_2) = \int_{\mathbb{R}} \exp\left[i(\beta^4/8 - z_1(\beta^2/2) + z_2\beta)\right] \mathrm{d}\beta.$$

(*See also* Berry [61].) The non-standard coefficients here are those used by Pearcey. Fig. 12.23 shows the contours of $|\mathrm{Pe}\,(z_1, z_2)|^2$, which gives the intensity distribution in applications to optics.

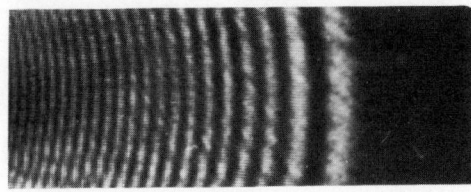

Plate 11. Airy diffraction fringes at a fold caustic, enlarged from Plate 13. (Courtesy of Michael Berry).

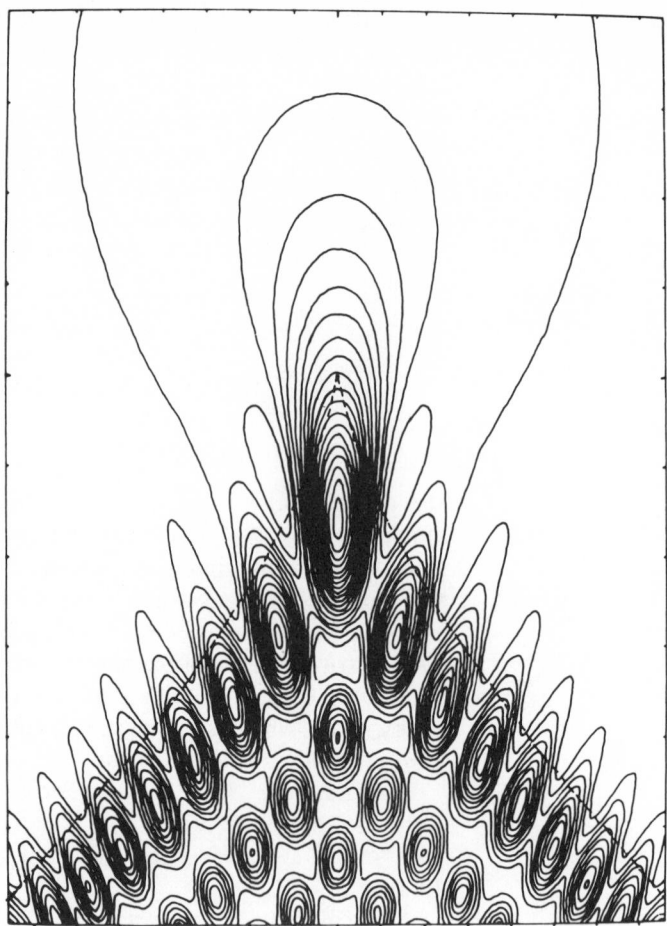

Fig. 12.23

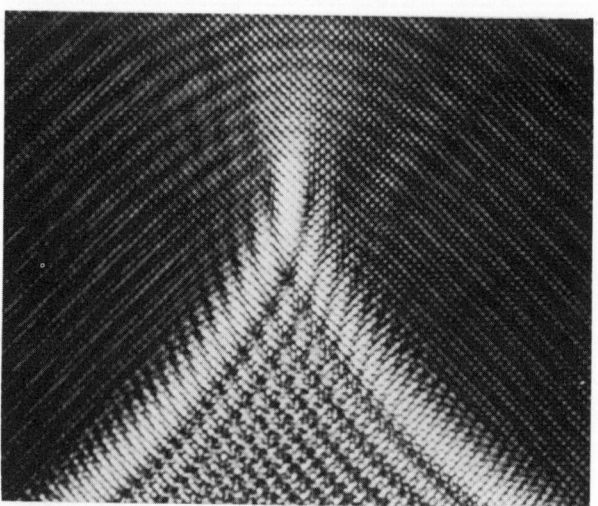

Plate 12. *Fine structure of a cusp caustic, showing quantization into diffraction spots, Airy patterns at fold caustics and Pearcey pattern at the cusp point* (Berry [61] Fig. 7(b)).

Graphs giving the actual value (a complex number) of the function in the form $re^{i\theta}$ may be found in Pearcey [72] and are reproduced in Connor [73] and Holford [70]. Fig. 12.23 should be compared with Plate 12, due again to Berry [61], showing experimentally observed diffraction fringes near a cusp caustic.

The Pearcey function is again bounded, and the order of the caustic may be computed as 1/4. The orders of the seven caustics corresponding to catastrophes of codimension ≤4 are as follows:

fold	1/6
cusp	1/4
swallowtail	3/10
butterfly	1/3
elliptic umbilic	1/3
hyperbolic umbilic	1/3
parabolic umbilic	3/8.

Arnol'd [74] tabulates the orders of caustics to higher codimension, under the name *degree of singularity* or *singularity index.*

Note that the order of the cusp is half again that of the fold: this is connected with its greater brightness in Fig. 12.2.

Much computation remains to be performed for the higher caustics. An attractive feature of the aspect of Thom's theorem that asserts that 'almost everything is stable and locally of these forms' is that these particular calculations, once completed, will be useful over and over again. 'Atypical' forms, apart from those associated with common symmetries, will be very rare. That there would be such solid grounds for concentrating on a finite list would not have been expected even 15 years ago.

APPLICATIONS

9 Scattering from a Crystal Lattice

Now we show how the theory sketched above may be used predictively, following a paper of Berry [61] on the scattering of atoms by a crystal. The structure of a crystal is a periodic lattice. Since atoms do not exhibit quantum effects to too great an extent, we can use so-called 'semiclassical' mechanics, and treat the atoms as elastic particles, and the crystal surface as a perfectly reflective undulating surface; or equivalently replace atomic trajectories by light rays and the crystal by a perfectly reflective undulating mirror.

Take coordinates (x, y) in the horizontal plane, and let $f(x, y)$ denote the height of the crystal surface above this plane (Fig. 12.24). Then f is a doubly periodic function: for simplicity we take the case of a rectangular lattice, so that f is periodic in x and y

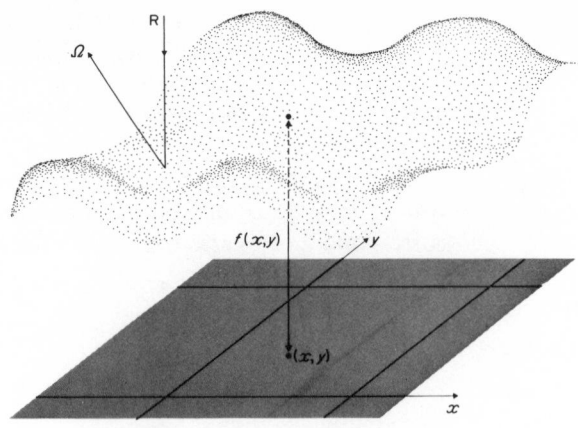

Fig. 12.24

separately, say

$$f(x + \eta, y) = f(x, y) = f(x, y + \zeta)$$

for fixed periods η, ζ.

Classically, the intensity observed in a direction Ω is given by

$$I = |J(\Omega)|^{-1}$$

where J is the Jacobian (with respect to x and y) and

$$\Omega = \frac{-2\nabla f(x, y)}{1 + (\nabla f(x, y))^2}$$

where $\nabla f = (\partial f/\partial x, \partial f/\partial y)$. This leads to

$$I(\Omega) = \sum \frac{1}{H(R_i)}$$

where H is the Hessian of f and the sum is over all rays, starting at R_i, emerging along Ω.

The reader may easily understand the rôle here of the Hessian by considering 2-dimensional scattering of vertical rays from the graph of $\sin x$. Peaks and troughs reflect vertically: scattering angle 'turns back' towards the vertical where the second derivative vanishes, even though the slope is non-zero. Scattering of light *through* a refracting surface is closely analogous.

Exactly as in the usual catastrophe set-up with finitely many state variables, we expect a caustic to be formed at the image under the map $\mathbb{R} \to \Omega$ of the parts of the crystal surface at which f has zero Hessian.

In the simplest case, the topology of the crystal surface will resemble Fig. 12.25. The solid dots indicate maxima (corresponding to the sites of atoms in the crystal lattice, which exert strong repulsive forces), the open dots minima (as far as possible away from atoms); and the crosses are saddles. The broken lines indicate the cell structure of the crystal lattice. The heavy curves are where

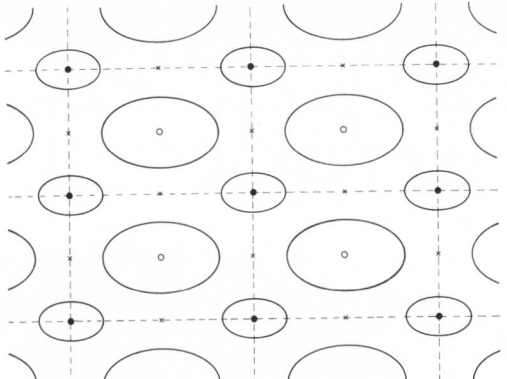

Fig. 12.25

the Hessian is zero: typically we get a closed curve around each maximum and each minimum.

If we observe the scattering from a large distance, we can restrict attention to a single cell of the lattice: we then find exactly two closed curves of zero Hessian, and hence expect the caustic to consist of two closed curves (possibly with singularities).

The simplest case for calculations is when

$$f(x, y) = A \cos\left(\frac{2\pi x}{\eta}\right) + B \cos\left(\frac{2\pi y}{\zeta}\right).$$

However, this is *not* typical. The zero-Hessian loops touch at their corners and form a rectangular lattice of lines, leading to a caustic in the form of a closed rectangle (Fig. 12.26), because the images of the two loops coincide.

Hence we must perturb f slightly, to obtain a typical case, say to

$$g(x, y) = f(x, y) + \varepsilon \tilde{f}(x, y)$$

for small ε. However, instead of picking $\tilde{f}$ and doing the calculations, we apply a little catastrophe theory. We have found an atypical caustic, with corners; what is the simplest catastrophe

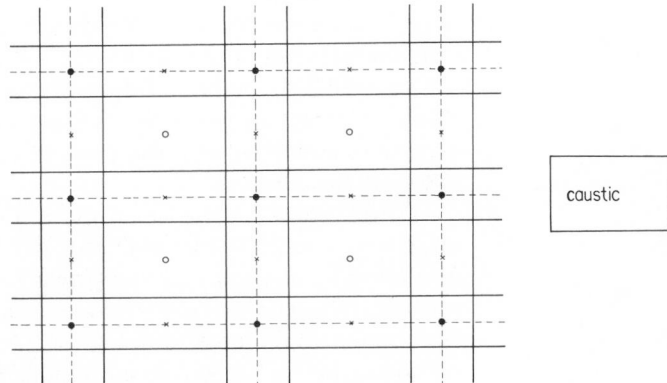

caustic

Fig. 12.26

Fig. 12.27

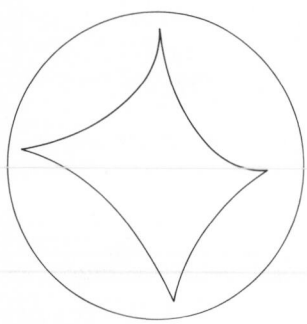

Fig. 12.28

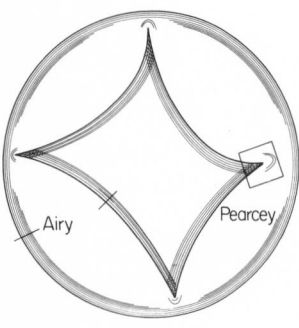

Fig. 12.29

containing it? (We cannot rigorously argue that the square, coming as it does from an atypically symmetric model, must be part of a low-dimensional typical structure, at least without careful argument as to what is typical *within the class of symmetric things*. But it is sensible to look at lowest cases first.) Inspection of the geometry (Chapter 9) shows that only the hyperbolic umbilic has a corner; and under perturbation this unfolds as shown in Fig. 12.27. Hence, taking the rectangular symmetry into account, we expect a caustic of the form of Fig. 12.28. Now we have the expected two curves, which separate off from the original single rectangle as the perturbation takes effect.

Calculation of the order of the caustic as $\frac{1}{3}$ (Berry, private communication) shows that the only candidates are the butterfly and the elliptic and hyperbolic umbilics (all higher catastrophes are known to have higher orders, even where not computed explicitly yet), making the use of corner geometry a rigorous argument, and this prediction solid. Detailed analysis of the particular case

$$f(x, y) = \cos\left(\frac{2\pi x}{\eta}\right)\cos\left(\frac{2\pi y}{\zeta}\right),$$

according to Berry [61], confirms it. Thus one may reasonably predict the form of Fig. 12.28 for scattering from a very wide range of exact doubly periodic forms of surfaces, given the structural stability of the hyperbolic umbilic catastrophe.

But there is more. The oscillatory integral method allows us, in a phrase of Berry's, to 'sew the quantum flesh on the classical bones'. In Fig. 12.28 the caustic consists of fold lines and cusp points (not just in the geometric sense, but in terms of the type of critical point that the potential function of the catastrophe possesses, as in Chapter 9). Across folds we expect Airy diffraction fringes, and at cusps a Pearcey pattern. Taking into account the direction in which folds occur (to locate shadow zones correctly) we can predict that the diffraction pattern observed in this type of crystal scattering should resemble Fig. 12.29.

The problem now arises of testing this prediction experimentally. It is not easy to obtain sufficient detail in observations for atoms scattered by crystals, especially near the cusp points. Berry uses an optical analogue, suggested by a remark (made to him by John Barker) that patterns like Fig. 12.28 may be seen in bathroom windows (an effect which the reader may well have noticed, and

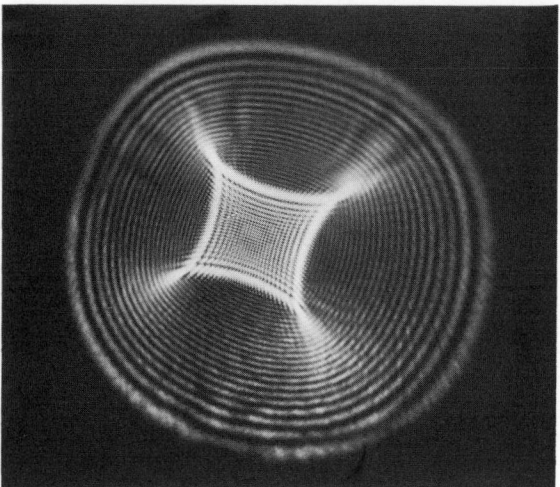

Plate 13. *Four linked hyperbolic umbilics, partially unfolded, in a caustic formed by refraction in periodically frosted glass using laser light* (*Berry* [61] *Fig.* 7(a)).

should watch out for if not). The frosted glass commonly used in bathrooms has a periodic lattice structure, and refractive scattering *through* glass is mathematically analogous to reflective scattering *off* a mirror. It is relatively simple to manufacture closely ruled glass, and shine a laser through it. Plate 13 (Berry [61]) is the result. It should be compared with Fig. 12.29: the caustic and diffraction effects are very close to prediction. In addition, there is a finer pattern (resembling knitting) due to Bragg diffraction (the effect of a large number of lattice cells instead of the approximation used above of a single cell).

Berry's paper [61] on this work contains a considerable amount of extra information, including the random motion of atoms within the crystal lattice.

10   Other Caustics

The scattering of light by ultrasound (Berry [75]) leads to a (theoretically) infinite sequence of cusp caustics. The number of light rays through x, and the number of caustics near x, increase without limit as x moves to the right (Fig. 12.30). The chaotic situation approached, with the discontinuities classically attached to caustics everywhere dense, illustrates Berry's idea that 'ergodicity is the ubiquity of catastrophe'. Cusp caustics arise acoustically when pressure variations cause sound waves to bend parabolically upwards and bounce off the surface of the sea (*see* Fig. 12.31 and Holford [70]), an effect which is important for sonar detection. Pearcey-type diffraction formulae are important here, since they give the distribution of the intensity of sound, and permit calculation of transmission losses.

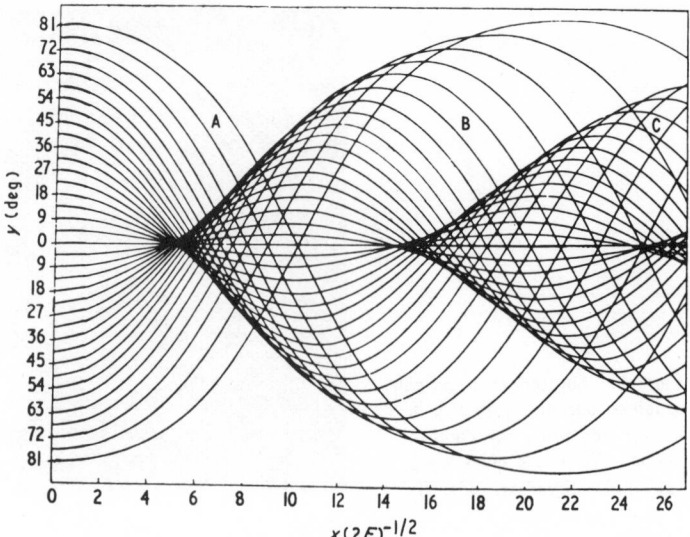

Fig. 12.30

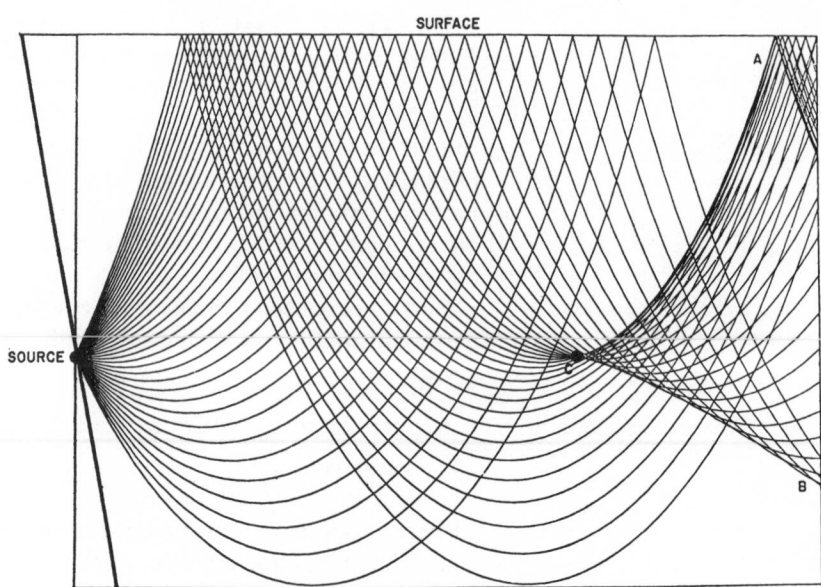

Fig. 12.31

A number of optical effects involve caustics which are not structurally stable (in the usual sense), but which arise naturally because of symmetries of the system. An example is the *glory*, caused by back-scattering from spherical droplets. This may sometimes be observed from a slight rise, with the sun directly behind the observer and mist or cloud in front; and takes the form of a series of concentric halos around the shadow of the observer's head. (For photographs *see* Bryant and Jarmie [76]; for further analysis *see*

Berry and Mount [77], Berry [62], and Khare and Nussenzveig [77a].) Other examples are *spiral scattering* (an infinite sequence of glories, Berry and Mount [77]) and the *forward diffraction peak* (Berry and Mount [77]). One way to see that these caustics are not any of the standard catastrophes is to compare their orders, which turn out to be too large. We cite these examples in case the reader is tempted to place too much reliance on Thom's classification theorem: we repeat that it lists *almost* all catastrophes, and all the 'typical' ones, but that special considerations may lead to atypical effects (indeed one of the functions of technology is to seek atypical effects, and find ways of making them happen). A well developed theory of 'equivariant' catastrophes (having symmetries of specified kinds) would be very useful – but is also very hard to obtain. Some progress has been made (Field [78], Poènaru [79–81]; Wasserman [81a] is the most directly applicable) but much remains to be done.

The above methods apply equally well to other scattering problems, and the papers of Connor contain material on molecular collisions; in particular Connor [82] gives a good survey and applies the hyperbolic umbilic to atom–molecule and atom–surface scattering.

11 Mirages

Optical catastrophes also play a role in the formation of certain types of *mirage* (which seems not to have been stated explicitly in the literature, but is easily deduced from theories which are). But here the catastrophes do not appear as caustics: their role is more unusual.

Mirages are produced when temperature gradients in the atmosphere refract light rays. For a detailed description of the effects involved, *see* Fraser and Mach [83]: we give both their diagrammatic explanation and a catastrophe version. If the temperature gradient is as in Fig. 12.32(a), then light rays entering the observer's eye travel along the curved paths shown (exaggerated) in Fig. 12.32(b). There is a fold-type envelope, but more important is the associated multivalued nature of the catastrophe manifold, which

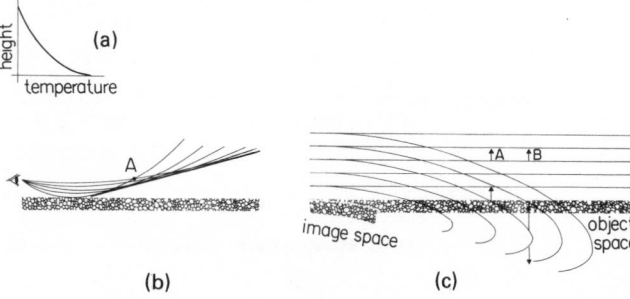

Fig. 12.32

Plate 14. *Two-image inferior mirage (Fraser and Mach* [83]). © 1978, *Alistair B. Fraser.*

takes visible form: for an object in position A, *two* of the emergent rays enter the observer's eye. A good way to visualize the effect is to deform 'real' space in such a way that the light rays travel in straight lines, since this is how the eye interprets the image it sees; thus obtaining a fictitious 'image space'. This is illustrated in Fig. 12.32(c). For the object at B there will be two copies visible in image space, of which the lower is upside down. Plate 14 shows a photograph of such a mirage, taken from Fraser and Mach [83], and termed a *two-image inferior mirage* because the ground appears to slope down. If the rays bend down instead of up, we obtain a *superior mirage.*

If instead of a fold catastrophe we have a cusp, we can obtain a *three-image mirage.* For a superior mirage, the temperature must have a point of inflexion as shown in Fig. 12.33(a). The rays entering the observer's eye are as in Fig. 12.33(b), and the result in image-space terms is Fig. 12.33(c). An object at position A is

Fig. 12.33

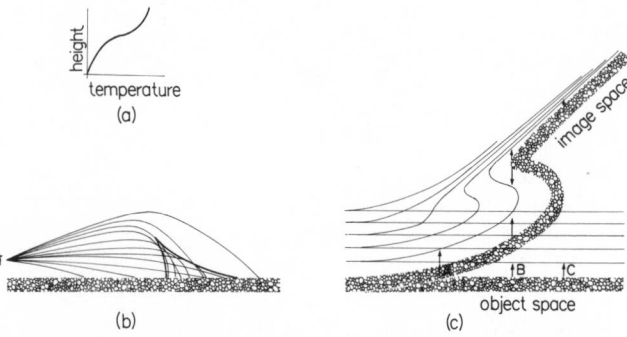

displaced up slightly. One at B is displaced further, and appears three times, the middle upside down, and the top severely shrunk. One at C is seen still higher, but as a single, very foreshortened image.

Notice that although the figures display envelopes, these appear only when we select the final position for the observer: they are not physical caustics.

12 Sonic Booms

Optical caustics make mirages: Mirages make sonic booms: booms make shock caustics, or *superbooms*.

We have had to omit the applications of catastrophe theory to the *formation* of shock waves for reasons of space, which is a pity since their history explains the name *Riemann–Hugoniot catastrophe* for the cusp (Thom [84]); the shorter name 'fronce' (pleat), like 'queue d'aronde' (swallowtail), being due to Bernard Morin. Moreover, the rigorous theory (*see* Guckenheimer [85] and Golubitsky [86]) is so far developed only for simpler equations than any in real physics, though the appearance of catastrophe geometry in 'real' problems is nicely illustrated by Fig. 12.34. However we can display here the relevance of catastrophe theory to the *evolution* of shock waves far from their source. We start by giving a highly simplified description of the cause of sonic booms, before studying their geometry.

An aircraft at each moment makes disturbances in the air (engine noise, pushing the air aside, etc.) which spread out radially from that point. At low speeds they spread out in front as well as behind (Fig. 12.35(a)), with the aircraft only partly catching them up. At supersonic speeds it outruns them (Fig. 12.35(b)) and their envelope forms a *shock wave*. Near the aircraft the structure is far more complicated than we show, with separate main shocks from nose and tail, etc., but at a greater distance such effects are less significant.

We are not concerned with treating the shock as an envelope: rather we treat it as a wavefront, and apply geometrical acoustics to *it*. In most optical problems the idea of a 'wavefront' is not very useful, being an approximation equivalent to ray theory but less informative about steady caustics. No flash is short enough to approximate a wavefront – but a sonic boom is. It has a complicated fine structure, but its propagation through the atmosphere can be – and routinely is – computed by the analogue of geometric optics.

Now Fig. 12.35(b) shows the shape the shock wave would have in a homogeneous still atmosphere: a perfect cone. But outside the laboratory the air is *not* homogeneous; it is warmer nearer the ground, so that the speed of sound is higher. Consequently the lower part of the wavefront travels faster than the upper, which

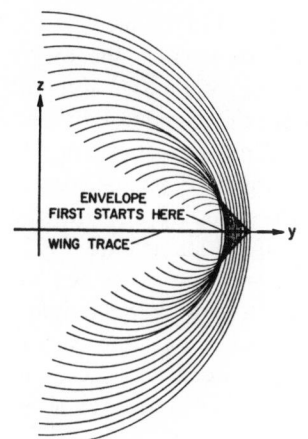

Fig. 12.34. Mach cone envelopes in the plane $x = 0.5$ for a wing planform given by $y^* = 0.50x^* + (x^*)^2$, $y^* > 0$ (Davis [87]).

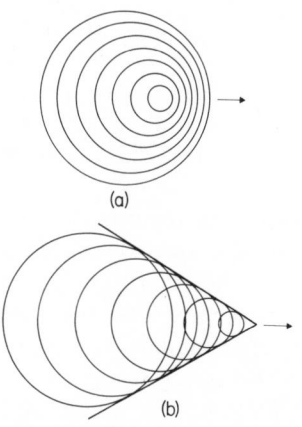

Fig. 12.35

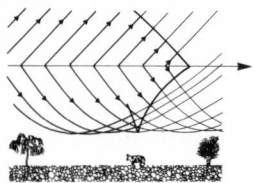

Fig. 12.36

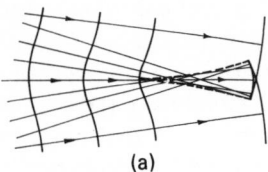

(a)

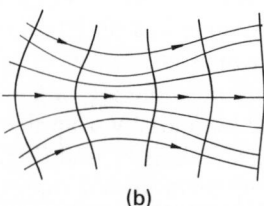

(b)

Fig. 12.37. *Focusing of a weak shock:* (a) *according to geometrical acoustics (linear);* (b) *according to shock dynamics (nonlinear).* (*Sturtevant and Kulkarny* [88].)

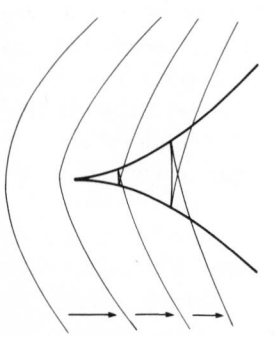

Fig. 12.38

bends it forward and eventually *up* (like the light speeded up by hot air near the ground in Fig. 12.30(a)). Thus if the aircraft is high enough, and cruising not faster than about Mach 1.3, then the boom may never reach the ground at all. Fig. 12.36 shows the 'ray acoustics' involved; the 'rays of shock emitted by the aircraft' at each moment (at right angles to the initially conical wave) are bent upwards as just described, forming a horizontal fold caustic. We have drawn heavily the parts of the rays already traversed, and heaviest the consequent form of the shock wave, at some particular moment.

Unfortunately, this caustic cannot in general be kept above ground level (consistently with the current aims of operators of civil and military aircraft) and the feature that gives caustics their name becomes important: high intensity.

As in optics, the ray theory intensity prediction is, incorrectly, infinite. But correct predictions are much harder, since the linear superposition of solutions presupposed in the oscillatory integral methods described above (good enough approximations to be taken as axioms in wave optics and quantum mechanics) are *not* true for shock waves. The Airy function for correct optical intensities around a fold caustic dates from 1838, but the corresponding analyses for sonic boom caustics were still plagued with infinities around 1972! Corrections beyond the techniques of Sections 5 and 6 are needed to obtain pictures like Fig. 12.37(b) for the 'true local behaviour'. (Indeed, the geometry of the shock some distance past the cusp is approximately that predicted by ray theory for 'weak' shocks, quite different for 'strong' ones: sonic booms at a distance from an aircraft, however, are usually 'weak'.)

Now, oscillatory integral theory is hardly trivial, and the developments needed here are clearly post-quadrivium. We can only refer the more expert reader to theoretical and experimental work such as Obermeier [88] and Sturtevant and Kulkarny [89]. Plate 15 is reproduced from the latter paper to show the geometric richness of the cusp (there called the *arête*); weak shock waves evolve just as ray theory suggests (Fig. 12.38), but strong ones far more intricately.

Fold caustics can reach the ground in various ways: Fig 12.39 shows a rare (but experimentally convenient) manoeuvre. Steady turn produces similar effects (Fig. 12.40), even in homogeneous air. The *beginning* of a turn produces a cusped caustic (Fig. 12.41(a)), as does that of a dive (Fig. 12.41(b)).

The codimension 1 nature of fold caustics makes them fairly easy for experiment; they must occur in *curves* on the ground, so that a dense line of microphones can easily meet one. Cusps however occur in curves in three dimensions, *points* at ground level (like the points C in Fig. 12.41(c)), which gives the three-dimensional result of the manoeuvre in Fig. 12.41(b)). Since with varying winds and so on it is hard to predict ground positions to within a few kilometres, experiment becomes difficult. (It is expensive to fill several square

M_s

1.005

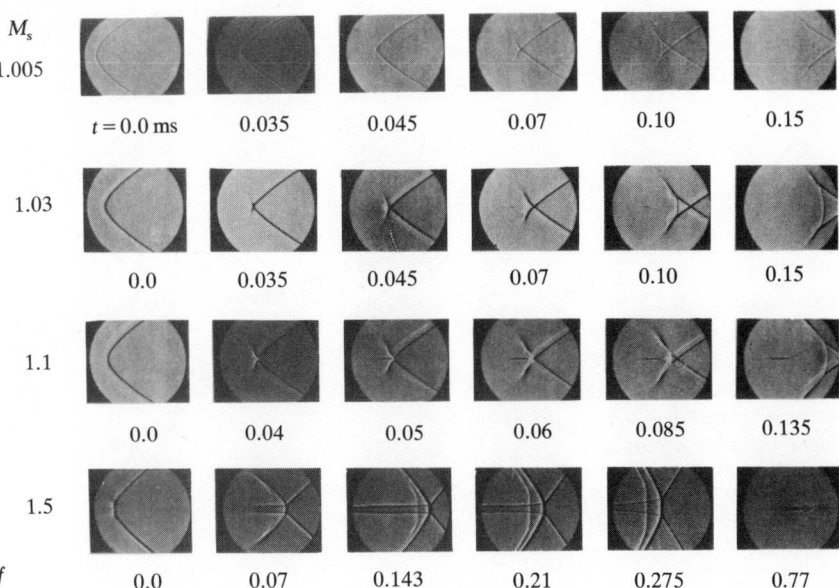

| $t = 0.0$ ms | 0.035 | 0.045 | 0.07 | 0.10 | 0.15 |

1.03

| 0.0 | 0.035 | 0.045 | 0.07 | 0.10 | 0.15 |

1.1

| 0.0 | 0.04 | 0.05 | 0.06 | 0.085 | 0.135 |

1.5

| 0.0 | 0.07 | 0.143 | 0.21 | 0.275 | 0.77 |

Plate 15. *Successive stages of focus for four different shock strengths: arête (cusp). Mach number in left-hand column; evolution at times t shown. (Sturtevant and Kulkarny [88] Fig. 17.)*

kilometres densely enough with microphones! However, since large areas densely supplied with *people* are common, the high intensities at 'hyperboom' cusp caustic points are important.) The French Working Group, experimenting with a Mirage IV, succeeded in getting a cusp point astonishingly near a line of microphones in an exercise aptly named 'Jericho-Carton' (roughly, Jericho-Bullseye). Fig. 12.42 shows their recordings along the line in one experiment; the swallowtail shape of the shock is plain, as is the fine structure, with higher intensities at the fold caustic (cusped wavefront) points. It is interesting that where two sheets *cross* they add more or less linearly: where they *interfere* at a caustic, more complication and higher intensities result.

Now, everything described above was done without catastrophe theory: all we have done is to interpret it in these terms. But the rôle that catastrophe theory can play is apparent from the following quotation (typical of the field) from Sturtevant and Kulkarny [89]:

'Three kinds of foci may be identified: caustics, arêtes and perfect foci'.

(By *caustics* and *arêtes* they mean the folds and cusps we have been discussing.) Now, as we saw in Section 4, perfect focussing (even perfect *line* focussing, as with a spherical lens) is a phenomenon of infinite codimension, which can be 'stabilized' only by the condition of a continuous symmetry. The microscopic rotational symmetry of water droplets that produces the glory is ineffective here; sonic booms are bigger than photons, and do not notice the droplets' shape. Thus a perfect focus can reasonably be dismissed as a pure

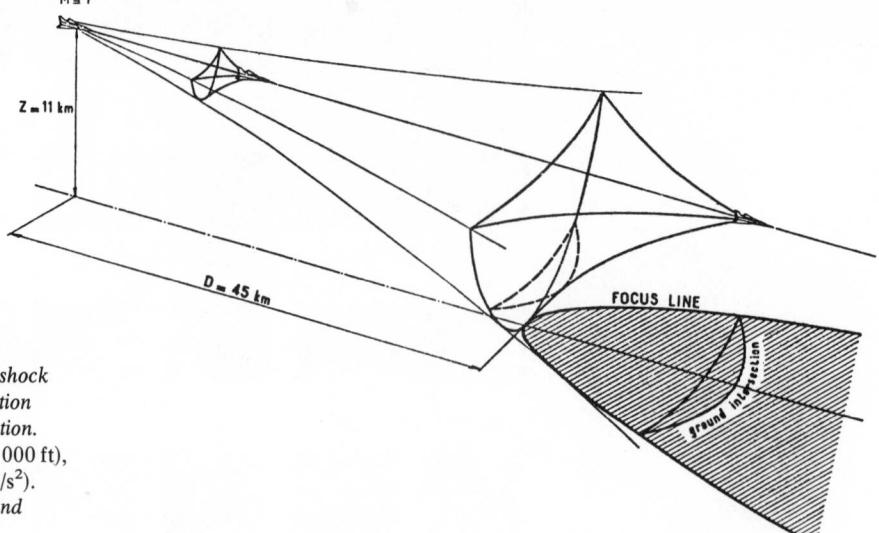

Fig. 12.39. Shape of the shock wave and ground intersection during rectilinear acceleration. Flight level 11 000 m (36 000 ft), *acceleration* 1 m/s² (3.3 ft/s²). (*Wanner, Vallée, Vivier and Théry* [90]).

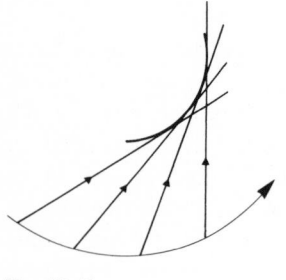

Fig. 12.40

laboratory phenomenon, where lathe-turned precomputed shock reflectors can approximate them to the degree that ray acoustics approximates shocks.

Folds and cusps are the only caustic singularities to be expected exactly on the two-dimensional ground: but (a) a city is somewhat three-dimensional, (b) a singularity is physically spread in its effects over a certain *volume*. Thus the three catastrophes that can stably occur at points in three dimensions (the swallowtail and first two umbilics) have a finite chance of being 'airbursts' close enough to ground or buildings for the associated intensities – higher even than the cusp – to matter. (Indeed one should properly take account of a hierarchy of singularities that 'almost happen' in *three* dimensions, but we cannot develop this here. Compare Berry [90a].)

Strictly, we are not concerned with the caustic geometry itself but with the way the wavefront evolves through it; a slightly more refined problem since there exist unstable (and atypical) ways through a stable caustic. Arnol'd [91] classifies the typical, stable wavefront evolution singularities in dimensions less than 6, and gives an application (in a somewhat simple physical model due to Zeldovič) to the shapes of galaxies. Here our concern is with dimension 3; and Fig. 12.43 shows typical wavefronts evolving through the swallowtail and elliptic and hyperbolic umbilics, and the two stable ways through a cusp line. (We have already met the first in Fig. 12.41.) In optics the intensities at the latter two points would be as for the ordinary cusp, but only more elaborate treatment will show whether this is equally true for shocks. All five cases require distinctly difficult analysis. Note the role of catastrophe theory here: not providing the Complete Answer to anything, but (since this is a

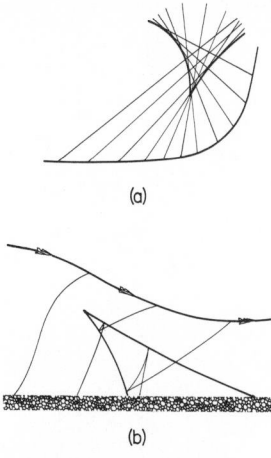

(a)

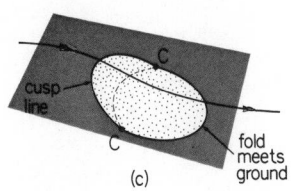

(b)

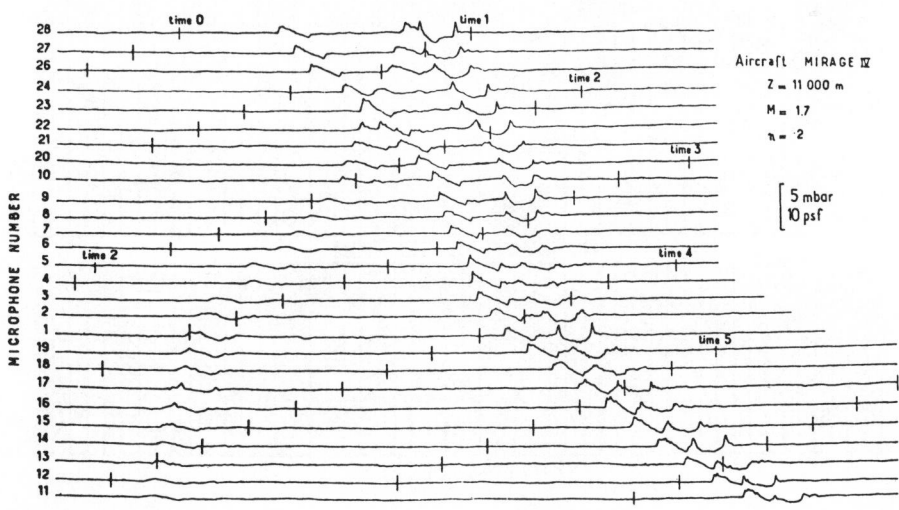

(c)

Fig. 12.41

place where typicality arguments are very reasonable) giving new information as to *what are the important cases* for which more detail must be sought, with the encouragement that they are finite in number.

Just as *global* inhomogeneities in the air (Fig. 12.36) refract shock waves, incidentally, so do *local* ones – notably turbulent weather, as in Fig. 12.44. Geometrically this seems closely analogous to the problem of optical refraction through a randomly rippling water surface, for which such remarkable results hold (Berry and Hannay [92a]) as that umbilics occur in the ratio 73.2% hyperbolic, 26.8% elliptic.

13 Giant Ocean Waves

Acoustic shock waves, above, are harder to analyse than linearly additive optical, acoustic or quantum oscillations. Harder still are water surface waves. Even a regular wavetrain shows nonlinearity; the wave profile is not a sine curve (despite the etymology of 'sine'!) and waves twice as high with the same wavelength are not the same shape with the height doubled. (In the small height limit, though, sinusoidal shape is approached.) However, in the large they have much in common with other waves. In particular, they form caustics.

Caustics can occur for a variety of reasons, such as reflection from a curved coast or focussing by a reduced speed (= 'high refractive index') region of shallow water. But the most dramatic cases appear

Fig. 12.42. Record of the three sheets of the shock wave in a turn entry manoeuvre (Wanner, Vallée, Vivier and Théry [90]).

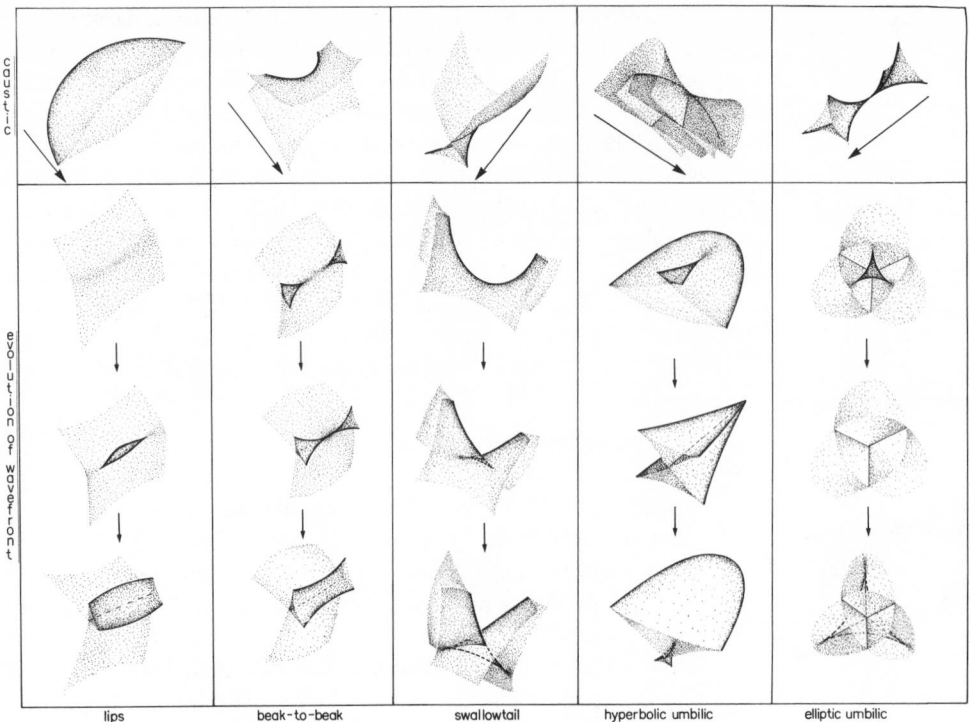

Fig. 12.43

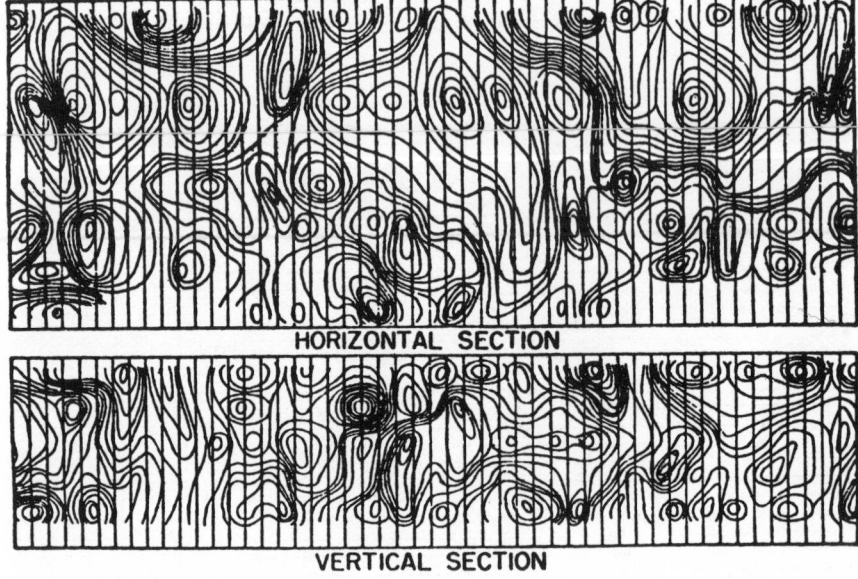

Fig. 12.44. *Gust structure near the ground as indicated by isotachs (Lumley and Panofsky [92]).*

to be due to *currents*. Particularly, waves travelling 'upstream' along
a narrow current are slowed most in the middle, where it is fastest,
and hence are bent inwards. (For nonlinear reasons, waves in this
situation *gain energy* from the current, rather than suffering transmission losses.) This leads at the simplest to two lines of fold caustic
(Fig. 12.45(a)), though the unevenness with which the waves will
generally enter the current (as when coming from the south-west in
Fig. 12.46) would more naturally lead to the kind of cusped caustic
we have seen for other channels (Fig. 12.45(b)). Indeed, upstream of
enough such cusps Berry's motto (Section 10) suggests that the
motion of the sea would have to be treated ergodically. The fold
caustic, however, is bad enough to be getting on with, and is likely
to be met more often than cusp points, by codimension. A single
ship's course can stably cross a fold, while the higher intensities
around a cusp are localized in a small region. (Of course, all this is
smeared out and hence somewhat reduced by 'chromatic aberration', even at the large scale where ray theory is useful. Real waves
come in a great mixture of wavelengths, which individually would
form caustics at different places.)

Fig. 12.46 shows an area where this may have been happening.
Since the closure of the Suez Canal in 1967, many ships (particularly
oil tankers) have been going that way. The supertankers built since
then are too big for the Canal, so must still go that way despite its
reopening. Time is expensive for a supertanker, so when travelling
westwards (full of oil) they like to ride down the Agulhas current.
(Politics, reduced demand and oversupply of tankers has led to some
strange economics recently, including slow steaming; but they still
follow this course.) When weather conditions are driving big waves

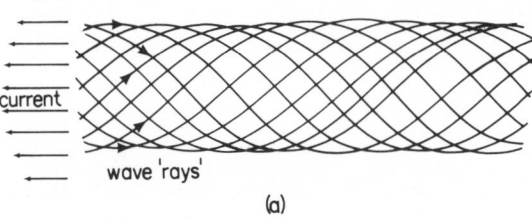

(a)

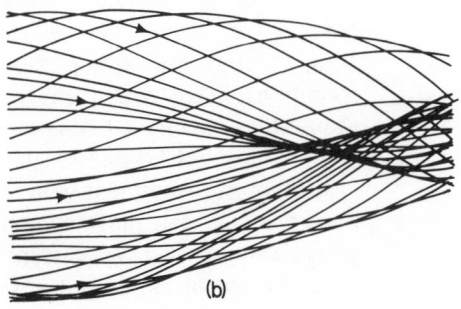

(b)

Fig. 12.45

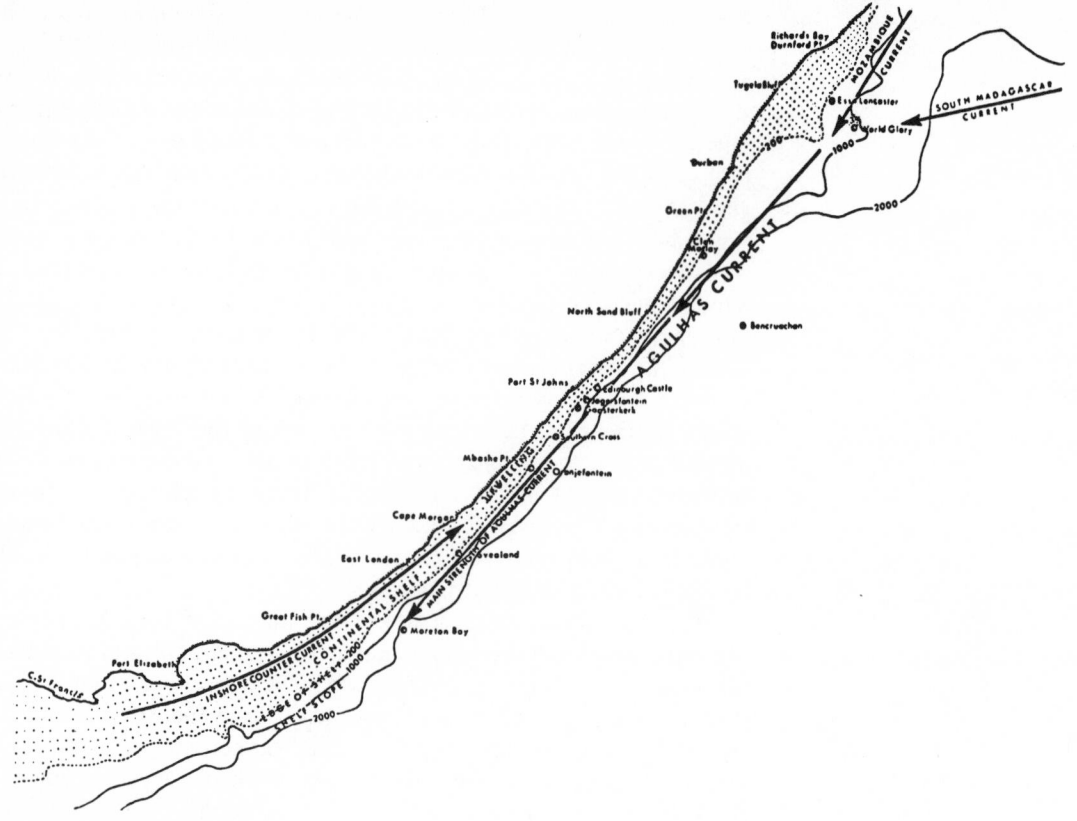

Fig. 12.46. *Chart of south-east coast of South Africa showing continental shelf, shelf edge and continental slope, Agulhas current, inshore counter-current and coastal upwelling, positions of ships which have encountered an abnormal wave. Depth in metres. (After Mallory [93].)*

more or less straight up the current (notice that the focussing effect gives a confined 'beam' without their having to start exactly straight – within 30° or so is probably enough for the velocities and wavelengths concerned) this practice has led to disaster. Captain Mallory [93] lists eleven encounters with giant ocean waves in the area, all but one riding the current and all in these weather conditions; most with damage, and with one ship broken in half. Many ships also have been lost without trace along this coast, which may or may not be due to giant waves.

Asymptotic treatment with oscillatory integrals like those of Sections 5 and 6, but more complicated with the nonlinearity and the extra features such as current (note that waves *down*stream are refracted *out* of the current – we have nonlinear 'birefringence') has been done for the fold caustic in deep water by Smith [94]. This indicates that the wave-height amplification, highly localized to the caustic region, could exceed a factor of 4. Moreover, a ship coming from the shadow zone (thus with no warning of the bigger waves ahead) is liable to encounter *asymmetrical* waves with the short steep side approaching it. It can thus find itself steaming at full speed down a long slope, towards something like a brick wall which is rushing towards it (Fig. 12.47). Tankers are built to rise above normal, round waves – even big ones, met slowly – but not this, at full speed.

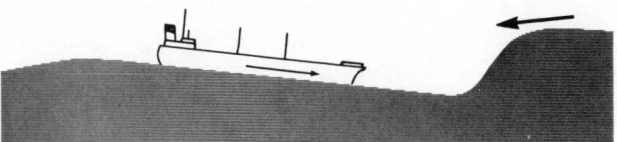

Fig. 12.47

13 Elastic Structures

In this chapter we consider the longest established counter-example to the claim that 'catastrophe theory is the first mathematical method capable of treating divergent phenomena'. The buckling of a beam one way or the other as stress increases is just such a phenomenon, and was first analysed by Euler in 1744 [95]. Many similar problems of elasticity may be formulated and solved in catastrophe language, but most examples in this area of the *lower* catastrophes were so well understood in their own terms by engineers before 'catastrophe theory' existed that for these we cannot expect the new mathematics to bring new practical information; only a reformulation–found unhelpful by many engineers.

Catastrophe theory will, however, be useful in this field; but it will shed new light only on complicated cases, whose study from this viewpoint has only begun. This posed problems for the planning of this chapter. A rigorous, direct attack on these cases, set in the appropriate Sobolev spaces, would be accessible only to a restricted audience (comprising specialists in the field, who will encounter the methods in any case as rigorous application becomes widespread). But our main purpose here is to show catastrophe theory in action to a *general* scientific audience.

We therefore begin with simple examples, where the engineering casts light on the catastrophes (rather than vice versa), and move on to more complicated ones where catastrophe theory is starting to illuminate the buckling problems. Even omitting details where we feel it possible, this gives a long chapter; and the multitude of topics we have been forced to omit, in this area which has for longest been anticipating catastrophe theory, make it clear that a monograph is needed. For the same reasons of space, we use the terminology and viewpoint of the rest of the book, rather than that of engineering practice, because the latter would require more introduction. We hope that we may thereby achieve our twin goals of demonstrating some new possibilities to the mathematical engineer, while displaying some important examples of catastrophe geometry to the general reader.

Finally we remark that time and space forbade discussion of applications to the variational theory of *plastic* buckling, which we had originally hoped to include. It seems clear that the bifurcations in the variational problems discussed by Sewell [96, 97] are susceptible to topological techniques–certainly a topological view of the singularities in the 'yield surface' would be of interest–but this could not be developed here. The importance of plastic buckling in practice does not diminish that of elastic analysis; to quote Butterworth [98]:

'... plasticity effects may become apparent as deformations increase, either before or after theoretical elastic buckling occurs. This possibility only serves to underline the necessity for accurate nonlinear *elastic* analysis, for how else can the material stresses be found so that it may be known when yield stress has been reached?'

We trust that catastrophe theory will contribute in due course to elastic–plastic problems, and wish in any case to draw the attention of differential topologists of all kinds to the area, where once they have mastered the notation they will find many delightful questions after their own hearts.

GENERAL THEORY

1 Objects under Stress

Consider a general engineered structure S–a bridge, say, or a crane–bearing a load. Perfectly rigid structures are impossible, but the shape of S is unaltered for most practical purposes as long as the only effect of the load (including the weight of S itself) is slightly to stretch or compress the components of which S is made. Significant change occurs when these components change shape radically–when they *break* or *buckle*. We shall not be concerned with breaking here, as its initiation and dynamics involve essentially the microphysics of the materials used. (That is not to say that the mathematical methods of this book are inapplicable in that field–they are probably useful–but such applications have not yet been developed.) Broadly, there are two classes of buckling: *elastic* or reversible change of shape under stress, and *plastic* or irreversible change.

Most of the common household 'plastics' are essentially elastic up to near breaking point. Experiment with a 'plastic' ruler: it is hard to bend it permanently out of shape; if you try you are liable to break it. Most household metals are elastic only for rather small deformations. Bend a television aerial a little way, as a wind does, and it springs back; bend it further and it stays bent.

As these examples illustrate, the distinction is between elastic and plastic *behaviour* and not between *materials*. Moreover it is far from absolute: there are materials that can be deformed and left lying apparently changed in shape, but a day later (like the mind of a bureaucrat you thought you had convinced of something) appear as if they had never been disturbed, having reverted slowly but inexorably to the original position. On a large timescale this behaviour might be considered perfectly elastic, but on the timescale of bridge collapse one would think of it as plastic.

Rather than enter into the various possible constitutive equations for the behaviour of bulk solids, and their classification, we shall take it here that an *elastic* body B has a unique *elastic energy* $\Phi(C)$ associated with any configuration C in which it may be found. Such a configuration C is most naturally described as a function $B \to \mathbb{R}^3$ specifying the position $C(b)$ in three-dimensional space of each point b of the body, and the collection $\mathscr{C}$ of all possible configurations is clearly infinite-dimensional. The range of positions over which the 'elastic' hypothesis is a useful approximation will obviously never be the whole space of mappings $B \to \mathbb{R}^3$; so $\mathscr{C}$ will be a *region* in this function space rather than the function space itself.

If we treated B as a vast-but-finite set of point atoms with forces between them, $\mathscr{C}$ would become vast-but-finite-dimensional... and the sums impossible. Treat the atoms more realistically with quantum mechanics, and $\mathscr{C}$ is infinite-dimensional again. We here treat B as a continuous domain with a fine structure that lets us differentiate C with respect to $b \in B$. Just as in the discussion for fluids at the beginning of Chapter 11 we know that strictly this is untrue. The key entities of the theory of elasticity, stress, strain and the rest, are formally *defined* as limits, by considering arbitrarily small parts of the body, but are only *meaningful* as representing average behaviour over regions large in comparison to atoms. These remarks are made not as an attack on the standard mathematical models of continuum mechanics used here, but as an illustration of the general separation of the thing described and even the most well-established and practical model of it. This separation has as consequence the practical futility of distinctions between 'exact' and 'approximate' solutions within a model, drawn finer than the fit of the model to the thing–which can never be perfect. It bears on the point that much published criticism of catastrophe theory refers rather to mathematical models in general, *including those that the critic is accustomed to consider as simply 'true'*. At the time of writing no critique which is really of catastrophe theory as such has been published: regrettably there have been only criticisms which apply to all mathematical models, or which apply just to straw men drawn from short popular articles. There *have* appeared useful attacks on particular models suggested by catastrophe theorists; but the lack of methodological assaults based on an understanding of the principles (like that of Bishop Berkeley [99] on the differential calculus) has allowed certain fungi to grow on the subject. A body of *informed* criticism should prove very valuable, and is greatly to be encouraged.

We consider, then, a space $\mathscr{C}$ of 'configuration' functions with an elastic energy $\Phi(C)$ associated with each $C \in \mathscr{C}$. In the language we have been using, Φ is a function from $\mathscr{C}$ to $\mathbb{R}$. In the early days of the calculus of variations, a function whose domain was a function space was christened a *functional* because it seemed a new kind of thing (and 'function of a function' was already in use, misleadingly, for *composite* functions taking x, say, to $f(g(x))$). This gave us the subject named 'functional analysis', by apposition of nouns rather than attaching an adjective to a noun: a grammarian would guess that it means 'analysis which works'. In mathematics, 'functional' has come to mean usually 'real-valued function defined on a Banach space', including the cases where the Banach space is finite-dimensional or not a function space. In mechanics it has become a signal that an 'exact' infinite-dimensional formulation is in use, rather than a finite-dimensional 'approximate' one. It is used even where the configurations lie not in a vector space but in a manifold (as the Zeeman machine wheel takes positions in a circle, not a line). In neither usage is it indispensable, and since this chapter is in the debatable lands between applications and pure analysis, we drop it. 'Function' in this chapter shall *include* all meanings sometimes reserved for 'functional'. Correspondingly we shall impute to the domain $\mathscr{C}$ enough structure to allow most of the finite-dimensional differential calculus to apply, becoming more technical about this as the chapter advances.

2 Elastic Equilibria

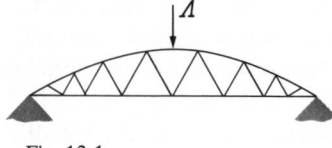

Fig. 13.1

Consider a complicated elastic structure subject to a load Λ (Fig. 13.1), and suppose that it is in stable equilibrium. We identify this supposition with the mathematical hypothesis that its configuration C lies within the range of $\mathscr{C}$ over which the elastic energy function is defined and descriptively useful, and that the total potential energy function

$$\phi = \Phi + \Gamma : \mathscr{C} \to \mathbb{R}$$

has a locally strict minimum at C. (Here Γ represents most often the *gravitational* potential energy associated with the position of the structure and the load. Other conservative forces may of course contribute.)

Increase the load Λ slightly, and the structure will move a little: much of the time its configuration will vary smoothly with the load. In the mathematics, $\mathscr{C}$ can usually be given enough structure for us to decide when the minimum of ϕ at C is a *Morse* minimum, and to apply an infinite-dimensional Implicit Function Theorem to deduce this smooth behaviour, as in Section 5 of Chapter 6, with the single external parameter Λ. (This theorem does *not* hold in infinite dimensions with hypotheses much weaker than $\mathscr{C}$ being a Banach manifold, so a certain caution is in order.) Schematically, this is usually represented as in Fig. 13.2, where the heavy curve is the *equilibrium path* given by the equilibrium $C(\tilde{\Lambda})$ corresponding to each load value $\tilde{\Lambda}$. The load Λ is taken as height in the diagram; c_1 and c_2 are representative coordinates describing the configuration of the structure, standing in for the infinite set of numbers needed to specify a point in $\mathscr{C}$ completely. The study of equilibrium paths and

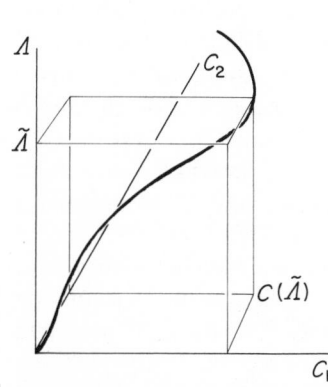

Fig. 13.2

their bifurcations is central to the literature on elasticity theory. In the engineering literature appeal to the Implicit Function Theorem for their existence is usually itself implicit, as indeed is much of the computational practice of catastrophe theory. The systematization and clarification of such computations by topological techniques is only now approaching sufficient integration with the great amount of difficult analysis already complete to permit genuinely new developments.

Equally, the engineering literature raises explicitly topological questions: the general theorems about rising equilibrium paths discussed in Thompson and Hunt [110] p. 62 have only recently been refined and proved for arbitrary finite dimensions by Kuiper (private communication to Chillingworth), by an essential use of algebraic topology.

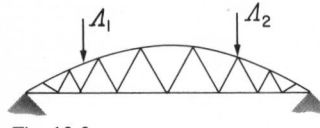

Fig. 13.3

The situation of Fig. 13.1 is of course somewhat unusual; it is commoner for a load to be *distributed*. If there are two points at which independent loads may be applied (Fig. 13.3) then 'load' becomes a two-dimensional quantity $\Lambda = (\Lambda_1, \Lambda_2)$ and the equilibrium path is replaced by an equilibrium *surface* parametrized by Λ around a Morse minimum of ϕ. If the points at which the loads Λ_1, Λ_2 are applied may also vary, then the specification of the load Λ becomes higher-dimensional still, and we have an equilibrium *manifold*. A load varying continuously from point to point must be specified by a function, and the space L of possible loads becomes infinite-dimensional. If (as is usual) L can be described as a Banach manifold then the arguments giving an equilibrium path through a Morse minimum generalize even to this case, and we get the local existence of. an infinite-dimensional equilibrium Banach manifold.

Clearly we are in the setting of catastrophe theory, with Λ as 'external parameter(s)' or 'control', and the equilibrium path/surface/manifold as catastrophe manifold. Things become more complicated as we approach a singularity, however. It usually happens in mechanics that the way in which the minimum in question becomes non-Morse (as Λ varies) is like the finite-dimensional case: the Hessian acquires a finite number of directions in which it is degenerate, and the other directions can be separated neatly from these in an infinite-dimensional analogue of Chapter 2 Section 5. Around a point where this happens, moreover, the Splitting Lemma in the parametrized form of Chapter 6 Section 4 goes through – indeed, although that proof of the finite-dimensional version uses quite elementary methods, the first published proof of *any* version was for the infinite-dimensional case and not until 1969 (Gromoll and Meyer [100]). The Implicit Function Theorem part of the splitting (which gives the local existence of the submanifold of essential variables without uniformizing the quadratic part on the rest) is about 70 years older, as *Liapunov–Schmidt reduction*, and can do the same practical work for many purposes. Their truth has been responsible for the practical success of elasticity calculations based on linearization and substituting back, as in Section 5 below, since at least 1744 (Euler [95]); a technique which works because it approximates a possible reduction to finite dimensions. This is a closer link between topology (also pioneered by Euler) and mechanics than even that greatest mind ever produced by Switzerland may have anticipated.

3 Infinite-dimensional Peculiarities

When the limit of stability is reached in the above nice way, the reduction to a finite number of essential variables in $\mathscr{C}$ permits the whole of the rest of the machinery of finite-dimensional catastrophe theory to go through. (Alternatively, under the same conditions on the degeneracy of the Hessian results like Theorem 8.6 can be proved directly for this case, *see* Magnus [101–103].) But one result of the proof that under these conditions everything is nice is a clearer recognition of the way things can go wrong. These are associated with the new ways to be singular that arise for the Hessian of a function defined on an *infinite*-dimensional space. They are best described in terms of the Hessian considered as a linear operator: readers with no exposure to functional analysis will please forgive us any incomprehensibilities in the next two paragraphs, as their background would take us another chapter.

One complication is simply that the Hessian H may be degenerate in infinitely many directions. If the problem can be set in a nice space, and 0 is 'isolated in the spectrum of H', this is tame in certain ways (H must have 'closed range') but means that the space of *essential* variables becomes infinite-dimensional and the codimension clearly infinite. It cannot be ruled out on typicality grounds (particularly for 'ideal' systems on drawing-boards) but the questions treated in this book for finite-dimensional systems cannot be handled for it in the same way.

Another, less intuitively obvious possibility, is a Hessian which is singular not by possessing *any* directions in which it is degenerate – as an operator it remains injective – but by the failure of surjectivity while it still has topologically dense range. We have not identified any examples of this in specific engineering problems (probably through ignorance of where to look). However, a mathematical analysis of an inhomogeneous 'elastic string' in which the local relation of extension to energy is not at all points in the string bijective, shows *typically* and *stably* just this phenomenon (Magnus and Poston [104]). There is no reduction possible to a one-dimensional essential variable (or *buckling mode* in the terminology below) around the bifurcation point: the Hessian does not single out a direction in configuration space along which energy does not increase. But there is a well-defined manifold of equilibria, stable and unstable, whose geometry is stable and corresponds to that of the fold catastrophe. It is not yet clear how far this example generalizes to a classifiable family of 'unsplittable' catastrophes related to those of Chapter 7, in which typicality and the Splitting Lemma let us lose variables like money.

Thus we cannot appeal to the mathematical 'typicality' results outlined in Chapter 7, except where we have evidence for nice behaviour of the Hessian. However, we are in any case in this chapter considering *designed* systems, for which (as discussed in Chapter 7 Section 3) typicality arguments are apt to prove a buckled reed if not a broken one. The interesting aspects of catastrophe theory for engineering purposes lie more in the computational tools of Chapter 8, by which specific questions can be answered about those specific problems of the mathematics of static elasticity which do satisfy the hypotheses of the Splitting Lemma. Since such problems constitute the vast bulk of the literature, this condition is not an unduly restrictive one.

EULER STRUTS

We start our explicit examples with the strut, not because it is the last word on buckling phenomena but because it is the first – just as we start the next chapter with van der Waals equation. All new developments in elasticity theory, analytical or numerical, are first tested on this old favourite, precisely because the example is so well understood as to shed light on the methods. We begin, in fact, with the simplest possible version.

4 Finite Element Version

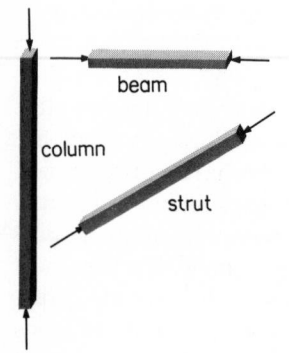

Fig. 13.4

We are modelling the behaviour of a long thin object under compressive forces at the ends. Nomenclature varies according to position relative to the vertical (Fig. 13.4); we shall include all cases under *strut*, which seems to be the most general term.

We suppose that the lengthwise compressibility is negligible in comparison to the flexibility of the strut, and that the ends are *simply supported;* that is, the ends are constrained to move on a particular line, but are not clamped in any special direction. Furthermore, we assume that the strut is confined to a plane; for many systems this is reasonable, as movement out of the plane can easily be proved an inessential variable, in the language of Chapter 8 Section 14. (For some of course it is not.)

First, let us concentrate all the flexibility at one point. Namely (Fig. 13.5) we replace the strut by two *rigid* rods, flexibly jointed at A, B, C and with a spring at B which tries to hold them straight. If this is a linear spring, it will exert a force $\gamma\beta$ proportional to the

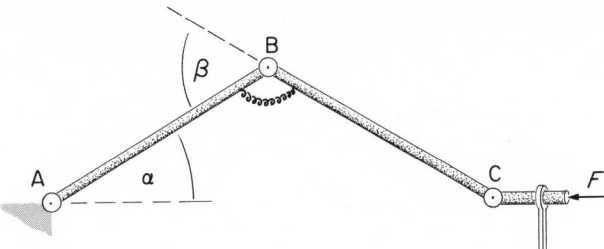

Fig. 13.5

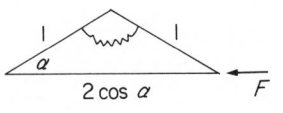

Fig. 13.6

angle β of Fig. 13.5, and contain elastic energy $\frac{1}{2}\gamma\beta^2$, where γ is the spring's 'elastic constant'. Suppose that the two parts are of length 1 as in Fig. 13.6. Combining elastic energy from the spring, and potential energy from the position of the force F, we have total energy (up to a constant)

$$U = \tfrac{1}{2}\gamma(2\alpha)^2 + 2F\cos\alpha.$$

Hence

$$j^4 U = 2\gamma\alpha^2 + 2F\left(1 - \frac{\alpha^2}{2} + \frac{\alpha^4}{24}\right)$$

and

$$J^4 U = (2\gamma - F)\alpha^2 + \frac{F}{12}\alpha^4.$$

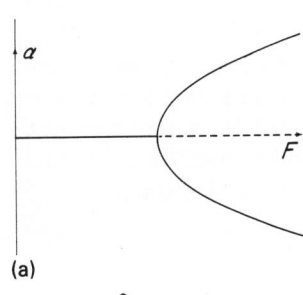

(a)

When F is less than 2γ, U has a Morse minimum at $\alpha = 0$ (and it is easy to show that this is the only minimum). When $F = 2\gamma$, U has a degenerate minimum there which is immediately recognized as a *standard cusp* point, since the coefficient $F/12 = \gamma/6$ of α^4 is positive. Varying F through 2γ, say setting $F = 2\gamma + b$, gives us an unfolding

$$\frac{\gamma}{6}\alpha^4 + \frac{b}{12}\alpha^4 - b\alpha^2.$$

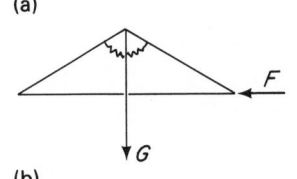

(b)

This is universal among even functions (though we have not given the algebra for establishing this) and in fact the bifurcation of the original U is *strongly* equivalent around the point of interest to

$$\frac{\gamma}{6}\alpha^4 - b\alpha^2,$$

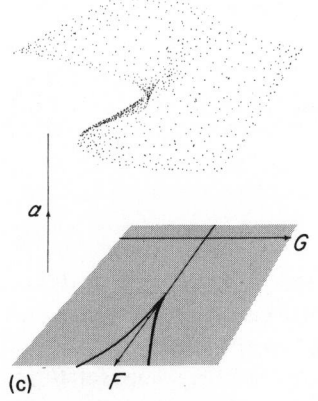

(c)

Fig. 13.7

with the catastrophe diagram of Fig. 13.7(a). Of course symmetry is unreasonable as an absolute restriction here, but the addition of almost any asymmetrical force (Fig. 13.7(b)) gives us the usual cusp picture (Fig. 13.7(c)). Indeed, taking G as shown we get

$$U = \tfrac{1}{2}\gamma(2\alpha)^2 + 2F\cos\alpha + G\sin\alpha$$

so

$$j^5 U = 2\gamma\alpha^2 + 2F\left(1 - \frac{\alpha^2}{2} + \frac{\alpha^4}{24}\right) + G\left(\alpha - \frac{\alpha^3}{6} + \frac{\alpha^5}{120}\right)$$

and

$$J^5 U = \frac{\gamma}{6}\alpha^4 - b\left(\alpha^2 - \frac{\alpha^4}{12}\right) + G\left(\alpha - \frac{\alpha^3}{6} + \frac{\alpha^5}{120}\right).$$

By Theorem 8.7 it follows that U is strongly equivalent to

$$\frac{\gamma}{6}\alpha^4 - b\alpha^2 + G\alpha$$

which, up to linear change of variable, is the standard cusp.

This, unlike the symmetrical analysis considered first, is a *structurally stable* description. Perturb U by changing it to $U + tV$, where V is any smooth function of α, F, G and any other 'control' parameters included in the description (length difference of the two parts, sideways movement of load G, angle between G and the perpendicular to F...). Then for small enough values of t and the extra parameters, not merely the same picture but the same formula applies, up to smooth change of variables. (Stronger, more 'uniform' statements hold, but are more technically complicated.) The cusp point, cusp direction, and so on move smoothly as a function of t and the extra variables. All this is a consequence of the universality of transverse unfoldings, and the stability of transversality.

5 Classical (1744) Variational Version

Now spread the flexibility back over the length of the strut, returning temporarily to the symmetrical case (Fig. 13.8). That is, to $G = 0$ in the above: notice that this assumes that the horizontal strut of Figs 13.5 and 13.8 is weightless. 'Qualitatively', that is, topologically, *no* $G \neq 0$ is negligible, though quantitatively a small G may be. ('Qualitative' distinctions, as remarked in Chapter 10 Section 10, may be finer than quantitative ones.) We still suppose that the strut is incompressible in length, which we fix as 1, and assume an elastic energy distributed over the strut proportional to the square of the curvature at each point by the same constant. We assume that the position may be described by a function

$$s \mapsto (f^x(s), f(s))$$

from the unit interval to the plane and ignore (as is standard) the rôle of f^x, which is small for small deflections, except for the end position $f^x(1)$. Since the ends are confined to the x-axis, f vanishes there; since there is no bending moment at the ends, so does d^2f/ds^2. (An entirely separate analysis is needed for the *clamped* strut, with

Fig. 13.8

the angles at the ends controlled.) Write

$$f'(s) = \frac{df}{ds}(s), \qquad f''(s) = \frac{d^2f}{ds^2}(s), \qquad \text{etc.}$$

and define

$$\theta(s) = \sin^{-1}(f'(s))$$

(Fig. 13.9). Then

$$f'(s) = \sin \theta(s)$$
$$f''(s) = \theta'(s) \cos \theta(s)$$

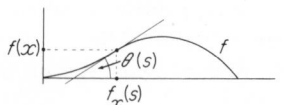

Fig. 13.9

by the chain rule. The curvature is $\theta'(s)$ by definition, hence

$$\text{curvature at } s = f''(s)/\cos \theta(s)$$
$$= f''(s)/\sqrt{(1-(f'(s))^2)}$$
$$= f''(s)(1 + \tfrac{1}{2}(f'(s))^2 + \tfrac{3}{8}(f'(s))^4 + \cdots).$$

Hence we take the elastic energy as

$$\int_0^1 \tfrac{1}{2}\mu(\theta'(s))^2 \, ds = \frac{\mu}{2} \int_0^1 (f''(s))^2 \, ds + \frac{\mu}{2} \int_0^1 (f''(s))^2(f'(s))^2 \, ds + \cdots$$

where μ is the elastic modulus per unit length, corresponding to γ in the finite element version. Similarly, the distance between the ends, previously proportional to $\cos \alpha$, is now

$$f^x(1) = \int_0^1 \cos(\theta(s)) \, ds = \int_0^1 \sqrt{(1-(f'(s))^2)} \, ds$$
$$= \int_0^1 (1 - \tfrac{1}{2}(f'(s))^2 - \tfrac{1}{8}(f'(s))^4 - \cdots) \, ds.$$

Thus, up to a constant, the potential energy associated with movement of the ends is

$$-\frac{F}{2} \int_0^1 (f'(s))^2 \, ds - \frac{F}{8} \int_0^1 (f'(s))^4 \, ds - \cdots.$$

Now, we expect as usual that bifurcation will occur when the quadratic part of the total energy,

$$H(f) = \frac{\mu}{2} \int_0^1 (f''(s))^2 \, ds - \frac{F}{2} \int_0^1 (f'(s))^2 \, ds$$
$$= \tfrac{1}{2} \int_0^1 (\mu(f''(s))^2 - F(f'(s))^2) \, ds$$

becomes degenerate. In infinite dimensions we cannot test this by taking the determinant of H, but we can 'differentiate with respect to f' and see whether $DH(f) = 0$ has non-zero solutions. In finite dimensions their existence is *equivalent* to degeneracy of H (if this does not seem obvious, treat it as an exercise added to Chapter 3).

In infinite dimensions degeneracy can be more subtle, as we remarked in Section 3 above, but their existence still *implies* degeneracy of H and in this analysis we shall assume (correctly, as we see later) that this test is here complete.

So how do we differentiate with respect to f? This is not the place for a systematic account of the calculus of variations, and the wild variety of notations used in it. Without such an account the calculations involved would be opaque to some readers, routine to others, illuminating to none. Suffice it to say that the equation $DH(f) = 0$ is the Euler–Lagrange equation for the above integral with the given boundary conditions, which comes out as

$$\mu f'''' + Ff'' = 0.$$

This, being a linear ordinary differential equation in the single variable s, is easily solved, the general solution being

$$f(s) = x \sin\left(\sqrt{\left(\frac{F}{\mu}\right)}s\right) + \gamma \cos\left(\sqrt{\left(\frac{F}{\mu}\right)}s\right).$$

This satisfies the boundary conditions if and only if $y = 0$ and $\sqrt{(F/\mu)}$ is a multiple of π, as is easily checked. Since for $F = 0$ it is obvious that H is positive definite, we have a minimum of energy for $f = 0$ as F increases, until the first bifurcation point is reached with the first existence of non-trivial solutions to $DH(f) = 0$, that is the first degeneracy of H, when

$$F = \mu\pi^2.$$

To find out what the buckling looks like, we put (some of) the higher order terms back. Now, the space X (say) of essential variables is one-dimensional (because the solution space of the 'linearized' equation $DH(f) = 0$ is) and is a curve in the space of possible positions which is *tangent* to the linearized solution space, the line L of functions

$$s \mapsto x \sin(\pi s)$$

for various x. (The function $\sin(\pi s)$, since it serves as a basis for L which approximates the curved 'essential variable axis' X, is called a *buckling mode*.) So we approximate the restriction of the energy to X by its restriction to L.

There is a point calling for some caution here, which we can illustrate well in finite dimensions. Suppose

$$g(x, y) = x^2 + 4xy^2 + 2y^4.$$

The quadratic part of this is just x^2, so the 'linearization' of

$$Dg(x, y) = 0$$

is just

$$x = 0.$$

Thus L, in the notation above, is here just the y-axis. Restricted to this, g is just $2y^4$, and the positive coefficient 2 could lead us to expect the universal unfolding of g to be a standard cusp catastrophe.

Not so.

We must split off the inessential variables properly (*not* just forget them) by some means like the Reduction Algorithm of Chapter 8, before jumping to conclusions. In fact

$$g(x, y) = (x + 2y^2)^2 - 2y^4,$$

so in the coordinates $(x + 2y^2, y)$ in which g is split, the 4-jet of g in the essential variable is seen to be *negative* definite, and its unfolding will then be a *dual* cusp (which implies quite different behaviour). So substituting solutions of the linearized equation back into the original energy function does not invariably produce qualitatively correct answers.

However, in the example above it was a cubic term that did the damage. If

$$h(x, y) = x^2 + 2y^4 + (ax^3y + bx^2y^2 + cxy^3)$$

for any a, b, c, then analysis of the reduction procedure (or theoretical argument) shows that a change of variable *can* alter the 4-jet of h simply to $x^2 + 2y^4$. In this case, then, substitution of y back into h does give answers strongly equivalent to the exact ones, whatever a, b, c and (by determinacy) higher terms may be. Similar results hold in infinite dimensions (Magnus [103]) and the absence of cubic terms applies for the present example.

So, parametrizing $L = \{s \mapsto x \sin(\pi s) \mid x \in \mathbb{R}\}$ by x, we find the 4-jet of the energy, as restricted to X, correctly by computing

$$\frac{\mu}{2} \int_0^1 (-\pi^2 x \sin(\pi s))^2 \, ds + \frac{\mu}{2} \int_0^1 (-\pi^2 x \sin(\pi s))^2 (\pi x \cos(\pi s))^2 \, ds$$

$$-\frac{F}{2} \int_0^1 (\pi x \cos(\pi s))^2 \, ds - \frac{F}{8} \int_0^1 (\pi x \cos(\pi s))^4 \, ds$$

$$= \frac{\mu \pi^4}{4} \left(x^2 + \frac{\pi^2}{4} x^4 \right) - \frac{F\pi^2}{4} \left(x^2 + \frac{3\pi^2}{16} x^4 \right)$$

$$= \left(\frac{\mu \pi^6}{16} - \frac{3F\pi^4}{64} \right) x^4 + \frac{\pi^2}{4} (\mu \pi^2 - F) x^2.$$

This is degenerate, as expected, when $F = \mu \pi^2$. Setting $F = \mu \pi^2 (1 + a)$ it becomes

$$\frac{\mu \pi^6}{64} (1 - a) x^4 - \frac{\mu \pi^4 a}{4} x^2,$$

an unfolding of $\mu \pi^6 x^4 / 64$. As in Section 4, this is not universal or stable (except among even functions), and we should add an asymmetrical load. The weight of the strut would serve, for the horizontal

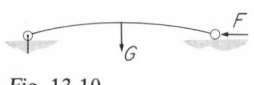

Fig. 13.10

position we have been drawing the strut in, but for something more easily experimentally varied we take a point load G in the middle (Fig. 13.10). This gives us a potential energy term $G \times$ (height of centre), which agrees to first order in x (thanks to tangency of X and L) with

$$G(x \sin (\pi/2)) = Gx$$

and by symmetry is an odd function of x. So we have an unfolding whose 4-jet is

$$\frac{\mu\pi^6}{64}(1-a)x^4 - \frac{\mu\pi^4 a}{4}x^2 + G(x + O(3)).$$

By symmetry, the 5th order term in x of the function unfolded vanishes, so the exact strut energy (restricted to the space X of essential variables) is strongly equivalent (using Theorem 8.7) to

$$\frac{\mu\pi^6}{64}x^4 - \frac{\mu\pi^4 a}{4}x^2 + Gx.$$

Up to positive constants, this is exactly what we found in Section 4, so the buckling behaviour is the same. Although the proofs are distinctly harder (discussed somewhat more in Section 7 below) the remarks on stability at the ends of Section 4 apply with equal force here. Their practical force is particularly strong in the context of imperfection sensitivity, discussed for various systems below. Here they make possible an *ex post facto* but rigorous argument for our neglect of f^x in setting up the analysis, when the formal setting discussed in Section 7 below is available.

6   Perturbation Analysis

In the asymptotic analysis of the previous chapter, we saw that around a Morse point, only the quadratic part of the phase function really mattered; the small wavelength limit therefore gave a good approximation for many practical purposes. Here the 'linearized' theory, which may be described very exactly as the small *perturbation* limit, plays a similar role. Around a Morse minimum (or saddle) it gives a good description, but as we approach a *degenerate* critical point, higher terms must be put back. And the degeneracy need only lack the special symmetries that killed the cubic terms above, or be a little more singular than x^4, for simply 'plugging in linearized solutions' to be untrustworthy (as we saw). Sometimes, a higher-order approximation than the line L of linearized solutions is needed to the curve X traced by the essential variable (or the analogues for finitely many essential variables) before reinsertion in the original energy function is a reliable technique.

Splitting procedures involving a Hilbert or Banach manifold structure on the space of positions of the system are the most rigorous and general (we discuss this further in the next section), but the topology of Banach spaces will not be standard in the engineering curricula this year or next. (Indeed, we have been impressed by engineers who calculate faster than we can, in finite or infinite dimensions, getting sensible answers because of the Implicit Function Theorem – which they tell us they have not heard of.) When the conditions for the Splitting Lemma hold, many methods can exploit its truth at an apparently lower level of analysis. One of the most important, an asymptotic expansion technique analogous in some ways to the methods of the previous chapter, was brought into engineering computational practice for elasticity problems by Sewell in 1964, and has been widely and effectively exploited since. This is the *static perturbation* method. Parametrizing X (in the above notation) by a parameter x, this consists essentially in solving successively higher linear equations to find the terms in the Taylor expansion of the function which defines the inclusion of the curve (or surface, or manifold) X into the space of positions. Without a topology on this space, technically we cannot talk about such an expansion; but because of the deeper structure possible for the problem than is actually exhibited, the method usually works very well – and translates perfectly into rigorous language.

We do not go into the computational details here, partly because they are like those of the last section (only more so), partly because they are excellently described in Thompson and Hunt [105], which anyone thinking about engineering and catastrophe theory together should in any case read, mark, learn and inwardly digest. It just predates its authors' contact with catastrophe theory, but shows how ripe their field was for the contact, by frequent, partly heuristic parallels with it. Much of their material is essential to effective exploitation of catastrophe theory in elasticity.

In the case of the Euler strut, parametrizing the curve X of essential variables by central deflection b of the strut and writing

$$w(s) = bw_1(s) + b^2 w_2(s) + \cdots$$

where the w_i are particular configuration functions, Thompson and Hunt compute ([105] pp. 29–34) that at the buckling point

$$w_1(s) = \sin(\pi s)$$

as we have seen (translating their results into our notation – which unfortunately reverses the rôle of s and x),

$$w_2(s) = 0,$$

as can be argued by symmetry reasoning,

$$w_3(s) = -(\pi_2/64)(\sin(\pi s) + \sin(3\pi s)),$$
$$w_4(s) = 0,$$

the procedure being clear from there on. (In general, b may be multidimensional.) By determinacy reasoning, convergence questions may be avoided; but the proof that something is there to be approximated requires topological methods, such as those discussed in the next section. When the problem splits properly and the equations defining the w_i can be successfully solved, substitution of the perturbation expansion to order k in b into the exact energy will give correct results if the resulting polynomial in b (or $(b_1, \ldots, b_n)$) is k-determinate. Sometimes (as indeed for the Euler strut) correct results come with less expansion than this; and the fact that the first term is often sufficient helped the whole theory get started, two centuries ago, and has helped it thrive since. A compendium of tricks for showing in advance just how much expansion is needed belongs in the monograph on computational methods referred to in the introduction to Chapter 8.

7 Modern Functional Analysis

To do all of the above 'properly', we need to be far more specific about the space $\mathscr{C}$ of functions which describe possible configurations. To *prove* in particular cases that everything happens just as for a finite element version requires information about the structure of $\mathscr{C}$. Finite-dimensional real vector spaces have identical topological and differential structures if they have the same dimension. Infinite-dimensional spaces are more varied. Roughly, the tamest are Hilbert spaces (having a 'dot product' with nice topological properties), then Banach spaces (having a nice *norm* or 'size' measure on vectors), followed by Fréchet spaces (with infinitely many not-quite-norms), followed by a wilderness of weaker structures. On the whole, most of the calculus in several variables goes through to Banach spaces, though even in Hilbert space things lurk like those discussed in Section 3. The best treatment of this subject is probably still Dieudonné [106]. Beyond, life gets steadily harder. Most of the weaker structures, we should stress, have been discovered in applications rather than invented by perverse theorists; indeed the space of *germs* ('locally defined functions') in which the theorems of Chapter 8 must be set for actual proof is so weak in structure from an analyst's point of view that none of the essential calculus results like the Implicit Function Theorem have been proved in it. (There is not even a good definition of a tangent vector to a path in it!) It is suspected of having a stronger structure than has yet been fully recognized, exactly because results like those of Chapter 8 hold – proved up to now by what to some analysts feel intrinsically round-about methods – but it is certain to remain harder to work in than a Banach space.

It is thus important, where possible, to get variational problems into a Hilbert or Banach space (or manifold) setting. Otherwise computationally helpful statements that seem intuitively obvious may be actually false, not just hard to prove. For most elasticity problems this can be done, but the methods are recent (post World War II, rather than 1744) and somewhat surprising.

The most obvious candidate for $\mathscr{C}$ in the example above would be the space of smooth functions defined on the interval from 0 to 1. Feeling more liberal, one might require less differentiability; say take the space of k times continuously differentiable functions for some k. The occurrence of second derivatives in the energy would suggest that k be at least 2, and the fourth derivatives in the linearized equation suggest 4.

These spaces can be given perfectly nice structures for some purposes, but not for this one! The structures do not tie in to the problem, and the energy function is not made tame enough for effective mathematical treatment. In order for the right kind of convergence properties and so forth to be available ('right' meaning 'we can prove things for this problem') far stranger functions must be allowed in for the convergences to be toward. The best setting for the strut problem appears to be (Chillingworth [107], Ball [108]) a Hilbert space denoted by

$$H^2 \cap H^1_0.$$

This is not easy to describe: H^2 is the space of functions on the interval whose squares, and the squares of whose *generalized* first and second derivatives, can be integrated. 'Generalized' here means that the derivatives need not exist in the classical sense at all – they are 'measurable' functions defined only up to changes on arbitrary sets of measure zero. (The powerful Sobolev embedding theorems at least permit functions in H^2 themselves to be thought of as continuous in the classical sense.) The space H^1_0 comes in to provide an equivalent of the boundary conditions $f(0) = f(1) = 0$; so much topology comes into its definition that we shall not attempt even a description here. We simply have not set up the language.

In this setting the computations of Section 5 can be given a precise justification, and the *solutions* of the problem proved to be nice, classical, smooth functions. (All such are 'included' in the space $H^2 \cap H^1_0$, but the space of smooth functions alone lacks topological elbow room for the problem – though it contains the answers. It is like a 3 cm slot with a pencil in it, which you can reach in and touch but not pull out.) For more elaborate buckling problems such a reformulation adds computational power (for examples see the papers collected in Keller and Antman [109]) as well as being necessary for theoretical understanding in complications like those mentioned in Section 3. Where these mathematically rigorous versions develop such subtleties, classical variational techniques give wrong answers – theoretically, qualitatively, and numerically wrong. The modern techniques of functional analysis are a necessary and well established development of the classical theory, with a large literature of practical applications.

They are also a dramatic illustration of the nature of modelling. We have gone beyond the casual assumption of 'configurations pretty much like smooth functions if we don't look too closely' introduced in Section 1. We are making very delicate hypotheses (like square integrable second distributional derivatives) about exactly what we would see if we *could* look arbitrarily close: hypotheses we know to be false. In what sense these 'infinitesimal' hypotheses can be said to model the local average nature of elastic solids is a dark and gleaming mystery. The results of the model, in macroscopic predictions, are very successful indeed; but it is hard to say in what sense the theory is more exact than a computer treatment of a finite element model like that of Section 4, but with many more rods and springs (Fig. 13.11). (The frequent close relation between the predictions of finite element and continuum analyses is much explored in Thompson and Hunt [105].) One model has infinitely more elements than the collection of springily attached molecules that make up a strut, the other has finitely less. A choice must be made on the basis of practicalities like cost, numerical fit and conceptual power, not of confusion between exact *calculations* and exact *theory*. (We include conceptual power firmly among the practicalities, because long calculations that you do not understand, whether by number-crunching or Banach space methods, will land you with bridges that fall down. A theory is only as good as your understanding of it.)

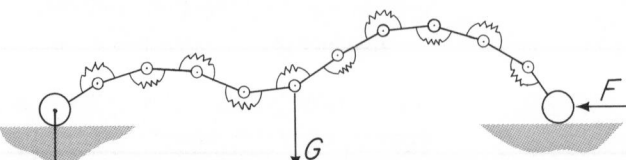

Fig. 13.11. *Finite element model (not in equilibrium).*

The Hilbert space approach to the strut calculations of Section 5 is treated in Chillingworth [107], using a model taken from Ball [108] which is close to taking the 4-jet (in f' and f'') of the energy function we described. (There is more apparent difference than this suggests, but the differences may be removed by diffeomorphism.) By 4-determinacy at the buckling point, the model in terms of curvature θ', untruncated in f' and f'', is equivalent to the reduced version, and this is the framework in which stability and universality of suitable transversal unfoldings can be proved. Chillingworth, incidentally, expounds the analysis first for the simply supported symmetrical strut, but rather than unfold the problem (as above) by a new load G he repeats the analysis for the harder problem shown in Fig. 13.12, and unfolds *that* universally by varying the angle at which the end is clamped. This problem likewise reduces rigorously

to a standard cusp (indeed in both cases the reduction is seen to be by *strong* equivalence of unfoldings, if we add Theorem 8.7 to the analysis). We will not go into the details of the Hilbert space calculations, since we spent the early part of this book setting up only the finite-dimensional calculus; but some generalities about computation are in order here.

The Reduction Algorithm of Chapter 8, of course, will not work here, being strictly a one-dimension-at-a-time approach to straightening X; nor will a computational method based on the Chapter 6 proof of the Splitting Lemma. Strictly infinite-dimensional methods are needed, and come in all shapes and flavours. Methods based directly on the infinite-dimensional Splitting Lemma proof of Gromoll and Meyer [100] are convenient once the problem has been set up explicitly in Hilbert or Banach space terms. (An explicit procedure in the context of universal unfoldings is given by Magnus [101].) Much of the functional analysis literature, in fact, is taken up by methods of rigorous reduction from infinite to finite dimensions. It is fair to remark, however (like Lenny Bruce of the preacher seeking the Kingdom of Heaven) 'how does he know what to do with it, when he gets it?'. Immense theoretical strength and computational expertise is given to infinite-dimensional analysis and reduction procedures like Ritz–Galerkin approximations, but the eventual finite-dimensional problem is treated as more trivial than it usually is. In particular, most questions are treated with only a one-dimensional control, and the singularities encountered (much more degenerate – by construction – than occur in a stable or typical one-dimensional family) are not otherwise unfolded. (Recently two- and three-dimensional unfoldings, of singularities with codimension *eight*, have begun to appear.) Questions of structural stability of the analysis, under general perturbations or under those appropriate to special features (like symmetries) of the problem, thus escape analysis.

8 The Buckling of a Spring

In Sections 4 and 5, we treated the strut as lengthwise incompressible. This is a good approximation, and in fact treating it as a long thin piece of isotropic solid, with the corresponding compressibility, leads to the same cusp catastrophe bifurcation. (Here the structural stability of the cusp is useful in letting us see why a small lengthwise compressibility 'should not' alter the problem enough to change the type of buckling. The negligibility is obvious to an experienced engineer, but as we remarked for ship design, no one is born with years of experience. Moreover, a method can be extended to new and more subtle cases faster, and with greater chance of success, than an intuition.) However, if we take something that is substantially compressible relative to its stiffness, like a spring, quite different behaviour emerges. To save space and time, we will consider a finite element version, which gives also a very elementary illustration of a design principle more general than the instability of symmetries: design choices aimed to simplify the calculations may complicate the problem. (Preliminary examination of a continuum version suggests similar catastrophe geometry, but our analysis is not complete.)

Consider then the system of Fig. 13.13: like that of Figs. 13.5, 13.6, 13.7, except that the two straight segments have variable length L. Assume that their compression is proportional to longitudinal thrust, and choose units to make the resulting energy in each segment $\frac{1}{2}(1-L)^2$. (We are using without proof the fact that difference between the strut lengths is an inessential variable near the unbuckled state; the sceptical reader should repeat the analysis with three state variables.) Other variables are as in Section 4, and the total energy is now

$$U(L, \alpha, F, G, \gamma) = (1-L)^2 + \tfrac{1}{2}\gamma(2\alpha)^2 + 2FL \cos \alpha + GL \sin \alpha.$$

Make an F-dependent change of state variable

$$l = L + F - 1,$$

inverse to

$$L = l + (1-F).$$

In the new state coordinates l, α;

$$U = (F - l)^2 + 2\gamma\alpha^2 + 2F(l+1-F) \cos \alpha + G(l+1-F) \sin \alpha.$$

Expanding in l, α we get

$$J^4 U = l^2 + \frac{F}{12}(1-F)\alpha^4 + (2\gamma - F(l+1-F))\alpha^2$$
$$+ G(l+1-F)\left(\alpha - \frac{\alpha^3}{3}\right).$$

Clearly when α and G vanish, for equilibrium l must too (which is the reason we took it as a coordinate). Around such an equilibrium the quadratic part of U is

$$J^2 U = l^2 + (2\gamma - F + F^2)\alpha^2$$

which is degenerate exactly when the coefficient of α^2 vanishes, that is, when we have longitudinal force

$$F = \tfrac{1}{2} \pm \tfrac{1}{2}\sqrt{(1-8\gamma)}.$$

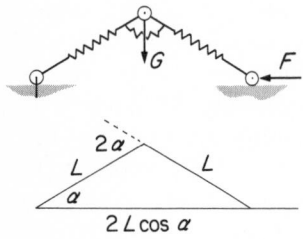

Fig. 13.13

A suitable choice of γ will tidy the formulae by avoiding square roots; the simplest perfect square less than 1 is

$$\tfrac{1}{4} = 1 - \tfrac{3}{4} = 1 - 8(\tfrac{3}{32}),$$

so choose γ as 3/32 for a convenient example. This gives as bifurcation loads

$$F = \tfrac{1}{2} \pm \tfrac{1}{4}.$$

The greatest interest is in the lowest load for which bifurcation occurs, so put $\gamma = 3/32$, $F = 1/4$, to get (still with $G = 0$)

$$J^4 U = l^2 + \tfrac{1}{48}\tfrac{3}{4}\alpha^4 - \tfrac{1}{4}l\alpha^2 = l^2 - \tfrac{1}{4}l\alpha^2 + \tfrac{1}{64}\alpha^4.$$

To examine this we must dispose of the cubic cross-term, which is all too easy:

$$J^4 U = (l - \tfrac{1}{8}\alpha^2)^2,$$

which is obviously not 4-determinate. So much for simplifying the arithmetic; by symmetry the 5th order term will be no help and we must look at the 6-jet.

$$J^6 U = (\tfrac{1}{4} - l)^2 + \tfrac{3}{16}\alpha^2 + \tfrac{1}{2}(l + \tfrac{3}{4})\left(1 - \frac{\alpha^2}{2} + \frac{\alpha^4}{24} - \frac{\alpha^6}{720}\right)$$

$$= l^2 - \tfrac{1}{4}l\alpha^2 + \frac{\alpha^4}{64} + \frac{l\alpha^4}{48} - \frac{\alpha^6}{1120}$$

$$= \left(l - \tfrac{1}{8}\alpha^2 + \frac{\alpha^4}{96}\right)^2 + \frac{\alpha^6}{240}.$$

This is 6-determinate by Chapters 4 and 6, or the rules in Chapter 8, and we observe that we have caught a butterfly. Since the sign of 1/240 is positive, it is a standard butterfly rather than a dual one. For problems like this, where the initial analysis assumes symmetry, the butterfly and its dual should be as common when two parameters may be varied (here γ and F) as the cusp and its dual are with one. (The same symmetry conditions in thermodynamics produce an abundance of 'tricritical points' – next chapter Section 17 – with two controls.) They are rare in elasticity theory (Thompson and Hunt [110]) because that theory usually varies more than one parameter only in somewhat specialized ways (structural optimization and imperfection analysis: see below), but they may not be rare in elasticity. This one fluttered naturally enough on to our calculations. Let us see how it unfolds.

Without new computations we know that the three parameters γ, F, and G will not give a universal unfolding, as the butterfly has codimension 4. Algebraically it is trivial to complete an unfolding for a k-determinate function (just complete a vector cobasis for Δ_k of it in M^k) but there is a certain art in spotting convenient 'physical'

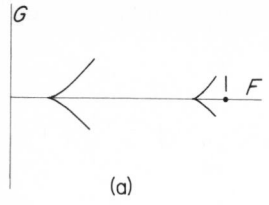

(a)

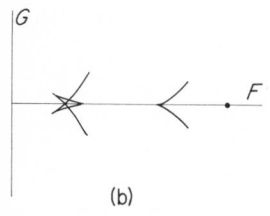

(b)

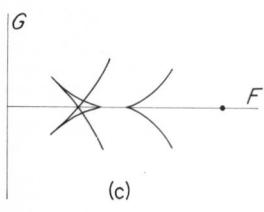

(c)

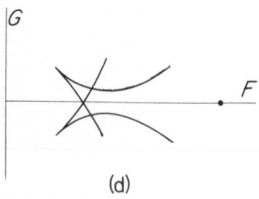

(d)

Fig. 13.14

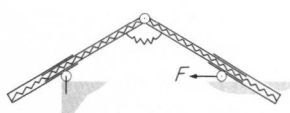

Fig. 13.15

representatives of such terms. Here, a good choice is an imperfection in the straightness produced by the central spring, so that its elastic energy becomes

$$\tfrac{1}{2} \cdot \tfrac{3}{32}(2\alpha + \varepsilon)^2 = \tfrac{3}{16}\alpha^2 + \tfrac{3}{16}\varepsilon\alpha + \text{shear term};$$

the spring was incorrectly adjusted and is trying to hold the joint at $(-\varepsilon)$ from perfect straightness. Then, setting

$$b = F - \tfrac{1}{4}, \qquad c = \gamma - \tfrac{3}{32}$$

we find the 6-jet of the total energy as

$$\left(l - \tfrac{1}{8}\alpha^2 + \frac{\alpha^4}{96}\right)^2 + \frac{\alpha^6}{240} - b\left(\frac{\alpha^2}{2} + l\alpha^2 - \frac{\alpha^4}{24} - \frac{l\alpha^4}{12} + \frac{\alpha^6}{360}\right)$$

$$+ \frac{3G}{4}\left(\alpha - \frac{\alpha^3}{6} + \frac{\alpha^5}{120}\right) + 2c\alpha^2 + \frac{3\varepsilon}{16}\alpha$$

up to shear, to first order in (b, G, c, α). It is easy to check by the rules in Chapter 8 that this unfolding is transversal, and hence equivalent to the standard butterfly studied in Chapter 9. The reader should find it a useful exercise to change the variable α so as to eliminate the 8th order term from the function unfolded, and use Theorem 8.7 to help reduce this unfolding to standard form. (It is essential to work in the *split* variables $x = \alpha$, $y = l - \tfrac{1}{8}\alpha^2 + \tfrac{1}{96}\alpha^4$.)

This, with some simpler calculations to see that this is the *only* butterfly at $\alpha = l = 0$ for any γ, and that the other bifurcation value $(F = \tfrac{1}{2} + \sqrt{(1 - 8\gamma)})$ is always a dual cusp, gives a complete picture of the buckling behaviour in the region of the unbuckled state. We describe it first for $\varepsilon = 0$, drawing the bifurcation set in the (F, G) plane for various γ, as in Fig. 13.14.

For γ small, the system is much more flexible than it is lengthwise compressible, and we get at a low value of F the standard cusp buckling that we studied in Section 4. Indeed, to first order in γ the buckling load is the same. The only effect of introducing some compressibility is to produce a dual cusp restabilization for large load, with a narrow region of stability (at right in Fig. 13.14(a)). Since $F = 1$ corresponds in the model to length 0, this will not be physically meaningful unless special experimental steps are taken (Fig. 13.15) to keep the behaviour within the elastic and linear range.

As γ increases through 3/32, we pass through the butterfly point examined above; Fig. 13.14(b) is for a value of γ just past this. The buckling point for F increasing with $G = 0$ is now a *dual* cusp, and if F reaches it the result will be 'snap buckling' as the system jumps. For γ only slightly above 3/32, however, note that it will not jump *far*: the two other sheets of minima are still close. If G is a little different from zero, the jump will be earlier; experiments with a

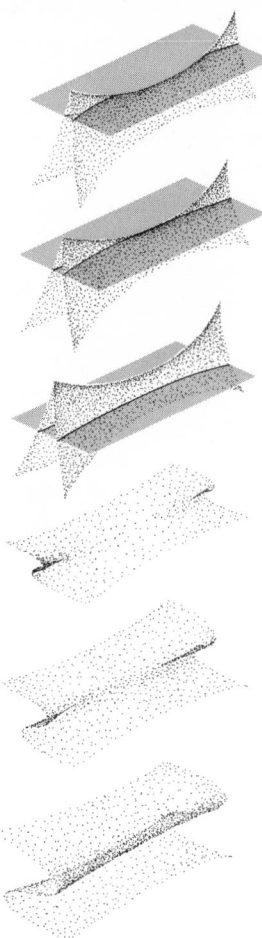

Fig. 13.16

low-friction Zeeman machine illustrate how hard it is to go right down to a dual cusp point.

The next development is only a special 'catastrophe' if one is maintaining a distinction between 'structural parameters' like (γ, ε) and 'load parameters' like (F, G), and disallowing coordinate changes that mix them. If so, then we have a 'beak-to-beak' catastrophe (Wasserman [111, 112]); if not, just a particular way of seeing, plane by plane, a curved line of cusps (Fig. 13.16). We return to this kind of distinction below (Section 11). In the present case, for $\gamma > \frac{1}{8}$ it becomes possible to compress the system down to the limit of the validity of the model, without buckling. (The smaller $\gamma - \frac{1}{8}$ is, the more delicate this operation is around the half-way point: G and C must be kept close to 0, and jerks avoided.) At this value the stiffness 'wins'.

Finally, consider the effect of non-zero ε. Around cusp points clear of the butterfly, as for the system in Section 4, a compensating push from G can effectively cancel it out, by the universality of F and G as unfolding parameters. It simply moves the cusp point off the F-axis by the amount of G needed to hold the system straight against the imperfect spring. But at the butterfly point, adjusting G to cancel the linear effect of ε leaves a cubic term uncompensated. At a cusp this term would be in Δ_k, and without effect to first order (it is *tangential* like the quadratic term sx^2 in Chapter 8 Section 6) but for a butterfly it is transverse, and changes Fig. 13.14(b) to pictures like Fig. 13.17(a) or (b), depending on its value and those of the other parameters. A detailed quantitative analysis, applying the methods of Chapter 9 to the particular unfolding arising here, is straightforward but spacefilling. We leave it to the reader. (Note that for the global problem in this instance, without truncation of the Taylor series, the variables $(\gamma, \varepsilon, l, \alpha)$ provide nondegenerate coordinates on the catastrophe manifold, in terms of which the catastrophe map can be given in closed form.)

This description is structurally stable, but note that perturbations must be 'small' up to 6th order for it to remain valid; in particular the force law of the springs must be close to linear to 5th order. (For instance, if the energy of the central spring were $\frac{3}{16}\alpha^2 - \frac{1}{240}\alpha^6$, we would see a higher order bifurcation, of codimension 6.) But the property of having *at least* a butterfly catastrophe is (given the symmetries of the 'perfect' system with $\varepsilon = G = 0$) stable against much larger perturbations, as can be seen by considering quantitatively the geometrical reasoning of Chapter 7.

9 The Pinned Strut

Consider once more an essentially incompressible strut, but now with its ends in fixed positions A, B closer together than its unbent length L (Fig. 13.18). It is thus forced to buckle, even when the load F shown is zero. What happens as F is increased?

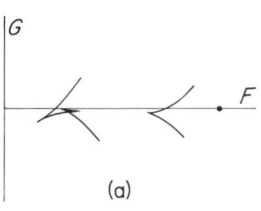

(a)

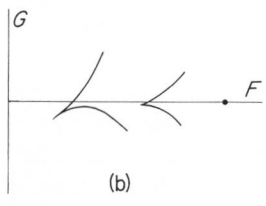

(b)

Fig. 13.17

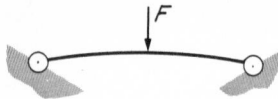

Fig. 13.18

In the standard cusp buckling of Sections 4 and 5, with $G = 0$ and F increasing, the strut ceases suddenly to be straight, but it does not *move* discontinuously. With $G \neq 0$ it passes to one side or the other of the cusp, moving differentiably as a function of F. But this new system behaves differently: it 'snaps' to a downward position, with a sudden large movement.

This problem can be treated in many ways. The most complete, whether in terms of the original 'elastic curve' model of Euler (perhaps with some compressibility added) or of the more subtle description of a rod by the 'rational elasticity theory' of Truesdell and his followers, is to work in the infinite-dimensional manifold of configurations. First, find the $F = 0$ position as exactly as is practical; we saw in Section 6 that finite deflections are not *exactly* sine curves, though small deflections (with small x) are very nearly so. Note that here we are concerned with buckling *from* an already non-zero deflection, so that non-zero contributions from $\sin(3\pi s/L)$, $\sin(5\pi s/L)$, etc. are already present. Exact solution in every model may not be possible, but a good estimate allows rigorous topological quantitative argument to go through. (We are thinking here particularly of finding estimates on the first few Taylor coefficients fine enough to permit determinacy and transversality arguments to be carried out independently of where within these estimates the 'true' values lie. But the same holds, for instance, for proving the existence of a buckled solution by Leray–Schauder degree theory and hence allowing its computation by 'direct' variational methods.) Follow this solution as F increases – analytically easier for this example if F is a continuously distributed load $F = f \sin(\pi s/L)$ than for the point load shown; which explains why this rather unphysical formulation is fairly popular. Examine the evolution of the Hessian as the equilibrium moves, and get estimates for where and how it becomes degenerate, and what the terms to order k are at that point; where k is defined by 'enough to give a nondegenerate result' and in most formulations of this 'arch' problem equals 4. Study the bifurcation around this point.

A more approximate approach to the mathematics is to restrict attention to a finite-dimensional subspace of the possible configurations; here the convenient choice, much used, is the plane P of shapes of the form

$$x \sin(\pi s/L) + y \sin(3\pi s/L).$$

The constraint that the strut must have a fixed length, of course, restricts us to a curve in P; using the quadratic approximation in f' to the length integral of Section 5, this curve is an ellipse. (Thus the configuration space with the given constraints is *not* a vector space but a manifold, topologically non-trivial except in infinite dimensions where spheres are contractible.) Within this approximation, and with the energy reduced to its quadratic part in (x, y), Zeeman [113] shows that for the right choice of physical variables the global

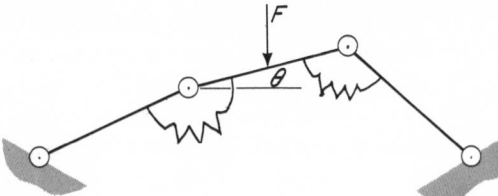

Fig. 13.19

catastrophe geometry of this system corresponds by linear change of coordinates to that of an elliptical roller (Chapter 1 Section 3) or an elliptical ship (Chapter 10 Section 7) of suitable eccentricity. This global topological structure appears to be preserved when fourth order terms are restored to the problem, though a proof of this has not yet appeared.

Finally, we may consider a finite element analogue. Fig. 13.19 is recommended to the reader as an exercise, with the suggestion that he use the angle θ made by the central segment with the horizontal for a local analysis along the lines of Section 4. (The set of possible states is topologically a circle, on which θ is not everywhere a good coordinate; for the global problem, labelling positions requires some care.)

In all the versions we have examined, the case with F sharing the designed symmetries of the arch (by being central, if a point load; symmetrically distributed otherwise) gives a 4-jet around the bifurcation point with one essential variable (say x) and resembling those found in Sections 4 and 5 except that the sign of x^4 is negative. Just as adding G there gave a universal unfolding, so does almost any measure of asymmetry here. A convenient one is the distance ε from exact centrality of the point of application of F (Fig. 13.20); another physically significant choice would be a small springiness in joint A or B, instead of 'simple support'. A look at Fig. 13.19 will suggest many more possibilities, which the reader is recommended to test. However, any *one* that together with F gives a universal unfolding is 'enough'; locally, the effects of any other perturbation of the problem can be described as a smooth reparametrization of the picture given by F and ε. These remarks acquire particular practical meaning in our next section, with certain qualifications to be found in the one after.

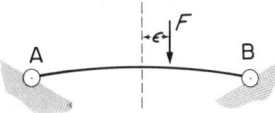

Fig. 13.20

THE GEOMETRY OF COLLAPSE

10 Imperfection Sensitivity

If one strut in a compound structure starts to buckle – more particularly, if several do simultaneously – the result may be more drastic than the behaviour we studied in Sections 4 and 5. (We shall discuss such 'compound' behaviour below.) But the arch behaviour of the

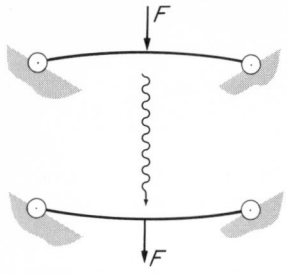

Fig. 13.21

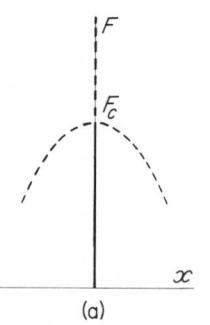

(a)

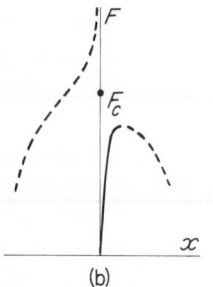

(b)

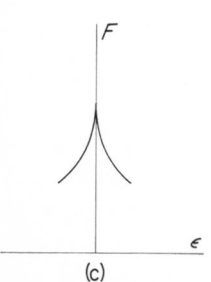

(c)

Fig. 13.22

previous section is *already* sudden and drastic; for most practical purposes the jump in Fig. 13.21 represents 'collapse'.

To see the practical effect of the dual cusp geometry of the previous section, we return to 'equilibrium path' language as in Section 2. A system for which the symmetry of the problem (though not necessarily of the solutions – cf. Chapter 14 Section 17) is perfect, will show a bifurcation of the energy reducible to

$$(-c^2)x^4 + (F_c - F)x^2$$

where F_c is the *critical load* and c is constant. Fig. 13.22(a) shows the corresponding equilibrium paths, solid for minima (stable) and dotted for unstable equilibria. This picture is of course the usual central slice of the dual cusp, dual to Fig. 6.3. It is unstable in the same way; if for a small fixed ε we consider

$$(-c^2)x^4 + (F_c - F)x^2 + \varepsilon x,$$

then Fig. 13.22(a) is replaced by Fig. 13.22(b), in which the largest F for which stable equilibria exist is substantially lower. More exactly, buckling load as a function of ε is given (in correct local coordinates) by the bifurcation set of the dual cusp catastrophe Fig. 13.22(c). Only for (F, ε) inside the cusp does a stable equilibrium locally exist. Notice the rapid fall away from F_c with $|\varepsilon|$.

This is where exact analysis of imperfection sensitivity began in 1945, with Koiter's thesis [114] obtaining such diagrams by exact solution of particular models. Attention became considerable on imperfection sensitivity *exponents*. Since Fig. 13.22(c) makes it clear that buckling load is not a smooth function of the 'imperfection parameter' ε, its singularities must be described. Here the singularity obeys the law

$$(\text{buckling load}) \sim F_c - \varepsilon^{2/3}$$

in the original coordinates (where $\sim$ has its usual physical 'limiting exponent' meaning). For catastrophe theory to prove, a quarter of a century later, that in suitable local coordinates '$\sim$' may be replaced by '$=$', is not of course a triumph in terms of practical engineering! Particularly since the partly heuristic exploitation of finite Taylor expansions had been very powerfully developed in the meantime, notably by Thompson and coworkers. *For this case* (and for dual cusp buckling in general), everything of practical importance had been said before catastrophe theory entered the scene. In particular the catastrophe manifold viewpoint (as distinct from exclusive focus on paths and their branching) was introduced for two dimensions as 'equilibrium surface' by Sewell [115], where in an example he treated the point – crucial to the topologist's 'stratification' analysis – that the fold curve is smooth in the catastrophe manifold, with the cusp singularity lying in its projection to the space of unfolding parameters. However, catastrophe theory applies to higher singularities than the cusp (though critics who have read only popularizations of the theory may be forgiven for being unaware of this). *Its promise lies in the grip that it gives on these higher singularities*, and its tests for 'enough' unfolding parameters. (Note that Fig. 13.22(a) does not show a manifold: a nondegenerate chart (cf. Chapter 4 Section 6) for the set of equilibria only becomes available when we go to the universal unfolding.)

Note too that it is the computational aspects of the theory, such as Chapter 8, that are most relevant. In engineering, where quantitative information is often freely available, catastrophe-spotting can take advantage of this. We caught our butterfly not by observing the flapping of its wings, but by calculating its 6-jet. It has never been proposed that one solves engineering problems by data-fitting to the seven catastrophes on Thom's list: firstly, because these would not be enough (see Section 13 below), and secondly, because we have available far superior computational techniques. The square tube which Croll [116] advances as an example where data-fitting might erroneously suggest a cusp is a chimæra: with a little more detail than he gives it would be possible to perform a quantitative analysis along the lines laid out above. Even with the information given, it is immediately obvious on symmetry grounds that his example is unlikely to be a cusp (see Section 13).

The range over which the Taylor approximation methods practised by engineers and systematized by catastrophe theory give a good fit in the original coordinates is often considerable. Recall that the theorems of Chapter 8 promise *exactness* only after smooth changes of coordinates – with derivative the identity at the buckling point, if required – and only in some neighbourhood, *a priori* perhaps small. Fig. 13.23 is taken from the analysis of another dual cusp (the propped cantilever) in Thompson and Hunt [105], where the exact bifurcation set is unusually easy to find in closed form. (Incidentally Sewell [96] regrets the use here of 'bifurcation set', because a fold involves no bifurcation of an equilibrium *path*. The energy *function* however bifurcates from a point of inflexion to two

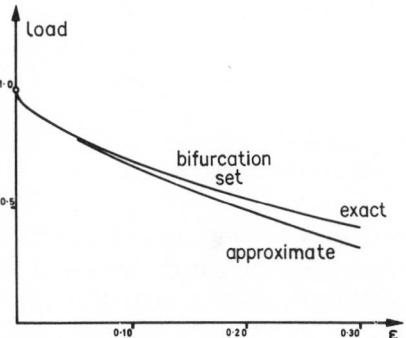

Fig. 13.23 *After Thompson and Hunt* [105].

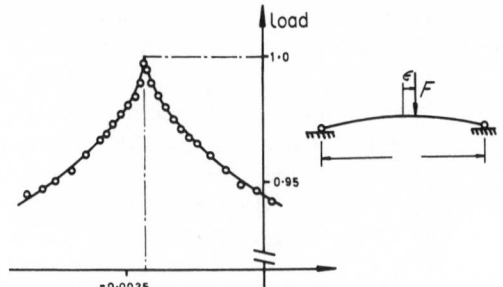

Fig. 13.24. Imperfection sensitivity curve for the Euler strut: solid line theoretical, dots experimental. (After Roorda [117].)

nearby types, with and without critical points, which recommends this usage to a topologist. We use it here for consistency with the rest of the book.)

Agreement with experiment, for cusp and dual cusp bifurcations, is often equally good. For the shallow arch of Fig. 13.18, Fig. 13.24 shows the results of experiments by Roorda [117] in 1965. Note that the 2/3 power law geometrical prediction is satisfied more exactly than is symmetry; unavoidable experimental imperfections move the cusp a little in the (ε, F)-plane, but can be compensated by a small change in coordinates. Note also the numerical consequence of those imperfections: an asymmetry equivalent to moving the load about $L/500$ sideways reduces the load at which collapse occurs by nearly 6%.

However, sometimes 'local' will mean just that. For instance, Fig. 13.14(b) shows that a dual cusp may have only local influence; the 'secondary bifurcations' represented by the two standard cusps nearby mask it, on a larger scale. As can be seen for the branching diagram for the 'perfect' $(G = \varepsilon = 0)$ system of Fig. 13.25, the jumps indicated for increasing F will be quite small near the butterfly point.

Notice how we have found the 'secondary' folds in Fig. 13.25 by following them from the 'primary' butterfly point as γ varies, rather than by a separate 'secondary bifurcation analysis' which is usually harder. When secondary bifurcations have an 'organizing centre' in this way it is worth varying a few extra parameters to exploit it, even when they are normally fixed for a particular system.

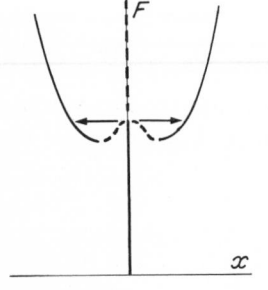

Fig. 13.25

11   (r,s)-Stability

Although the change from Fig. 13.22(a) to (b) is important, it is by no means the only perturbation of Fig. 13.22(a) possible. Notably, an arbitrarily small imperfection can tilt the tip of the cusp relative to the (ε, F)-plane as in Fig. 13.26(a). (For instance, tilting the

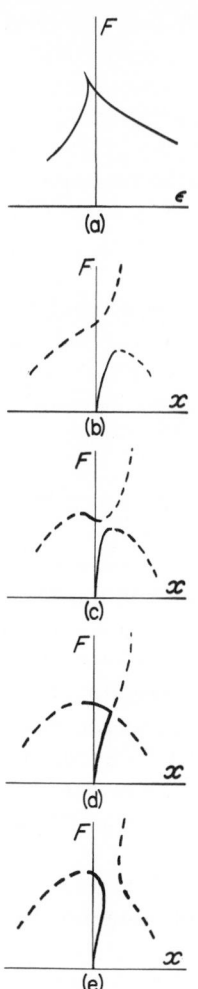

Fig. 13.26

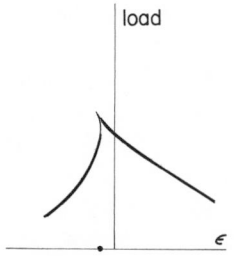

Fig. 13.27

experiment slightly has this effect.) The cusp geometry remains, by structural stability, but the pattern of branching of equilibrium paths for various fixed ε changes radically. (Figs 13.26(b) to (e) correspond to ε-values in the same left-to-right sequence in Fig. 13.26(a).) If one draws a rigid distinction between *imperfection* parameters, considered as fixed for a given physical structure, and load parameters that one varies, then transformations that mix these become forbidden. But only such transformations can make the geometries of Figs. 13.22 and 13.26 equivalent. So 'tilt' has become a new kind of imperfection parameter, whose effects are *not* reducible by the new class of allowed transformations. This is made particularly clear by Fig. 13.27, which shows that a small tilt has made 'load for which collapse first occurs' (heavy curve) a *discontinuous* function of ε, not merely a singular one. The data points of Fig. 13.24 are worth looking at with Fig. 13.27 in mind.

The sensitivity is numerically less drastic for small tilts than for small ε (an easy calculation for the canonical dual cusp shows that the drop in maximum load is quadratic in 'tilt', like metacentric height in Chapter 10 Section 6, rather than showing some fractional exponent), but this mildness is not universal for imperfections thus considered. For example, consider the fold catastrophe. If this is met transversely by load as a control parameter, the result is stable and responds only smoothly to any imperfections whatever. But if the energy reduces for the 'perfect' system to

$$x^3 + (F_c - F)x^2,$$

the results are quite different. We saw in Chapter 8 Section 6 how this family, enlarged by one term to

$$x^3 + (F_c - F)x^2 + \varepsilon x,$$

can be reduced to a standard fold (Fig. 13.28(d)). But the transformations involved are quite unreasonable here; we cannot vary the imperfection as we vary F, to make load a dummy control. Maintaining the distinction leads to the observation that while negative ε has very gentle effects (Fig. 13.28(a)), positive ε leads to collapse (Fig. 13.28(c)), with a worse lowering of maximum load than for the arch problem above; Fig. 13.28(d) shows clearly that the falloff is with $\varepsilon^{1/2}$ rather than $\varepsilon^{2/3}$. These features are invariant (under diffeomorphisms that preserve the load/imperfection distinction) along with others analysed, and substantiated with experimental data, by Thompson and Hunt [105]. The whole 2-parameter family is stable; and it is *not* typical, with one load and one other parameter, to miss it. This is why two common buckling phenomena are listed in Thompson and Hunt [110] against Thom's single 'fold': they are using a finer classification, in which the usual fold (their 'limit point') and this phenomenon (their 'asymmetric point of bifurcation') are not equivalent.

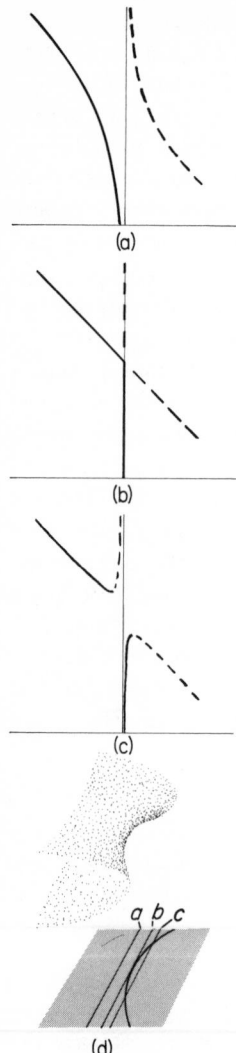

Fig. 13.28

However, the Pandora's box of catastrophe theory contains many creatures beside Thom's seven, and some of them are appropriate to this difficulty. The correct mathematical notion of equivalence for this context (as Thompson and Hunt [23] have recently pointed out) is that of (r, s)-*equivalence*, analysed by Wasserman [111, 112, 118]. This consists precisely, in the present language, of allowing coordinate changes in s imperfection parameters $\varepsilon_1, \ldots, \varepsilon_s$; and only then an $(\varepsilon_1, \ldots, \varepsilon_s)$-dependent change in r load parameters $\Lambda_1, \ldots, \Lambda_r$. On top of all this, of course, comes the usual $(\Lambda_1, \ldots, \Lambda_r, \varepsilon_1, \ldots, \varepsilon_s)$-dependent change of behaviour variable (distinguishing two kinds of 'external variable' does not reduce the need for this), so that statements in the theory become quite notationally elaborate. It seems clear that the (r, s) classification of umbilics is closely related to the geometric (not merely topological) and thus finer classification of the umbilics in Porteous [118a]. The powerful techniques of 'pure' differential geometry have been little exploited in the engineering literature.

Wasserman's work contains all the necessary theorems for applications in this field, analogous to those in Chapters 7 and 8 (except conditions for strong equivalences, which are probably a routine extension). But the computational results are stated with the full succinctness of 'ring of germs' language, as in the rigorous proofs (and even so the complexity noted above makes the analogue of the versality criterion of Chapter 8 six terms long!), which does not tend to immediate comprehensibility among those whose expertise lies elsewhere. A translation, like that for the general version in Chapter 8 (first done, we recently learned, by C. T. C. Wall years ago – but distributed only to a few high-energy physicists who failed to find a use for it) is needed to make it a tool for the general scientist. Wasserman's presentation is in fact applications-oriented, but in the topologist's sense of application to classification rather than computation (though much computation comes into his proofs) with a view to use in the 'Thom programme' as sometimes understood – 'only stable phenomena repeatably happen, so classify them and you have classified the observed world'. Hence he lists as conclusions the (1,3)-stable and (3,1)-stable phenomena, with the 3- and 1-dimensional variables being space and time. However, restrictions on the notion of equivalence make fewer things stable (as we saw vividly in Chapter 11 Section 6) and with either r or s greater than 1, we meet infinities like those of Chapter 7 Section 6 for $r > 5$. Thus (r, s)-stability is not typical for $r + s > 2$, and one leg of the rhetoric for such a version of the Thom programme collapses.

For engineering applications this collapse is irrelevant, however, as no kind of stability has ever been typical on drawing boards – hence, exactly, the need for imperfection sensitivity analysis. But space forbids inclusion of the necessary rules here; they would take another chapter, at the notational level of this book, and to develop their engineering use on examples complicated enough not to be effectively understood already would take a monograph. We must similarly omit the analysis of one-sided constraints (like stopping blocks) that can resist but not hold, push but not pull. These, not uncommon in engineered systems, lead to the *constraint catastrophes* catalogued in Chapter 16 Section 7. (The computing rules for these have their own special features, and when crossbred with (r, s)-stability will produce something almost as complicated as standard engineering notation.) We remark only that the 'snap into the downward position' at the end of Section 3 of Zeeman [113] exemplifies Figure 16.9(ii) rather than anything from Chapter 9.

It follows trivially from Wasserman's analysis that the number, i say, of imperfection parameters needed for a full description is greater than or equal to the codimension of the degenerate singularity unfolded, minus the number of load parameters. (Often, as in the above fold and cusp examples, it is more – it can easily be *much* more.) By Chapter 8 Section 11, the codimension is greater than or equal to $m(m + 1)/2$, where m is the corank of the singularity or, in the terminology above, the number of buckling modes. (Only for the fold and the first two umbilics does strict equality hold.) In most engineering analyses, load is taken as one dimension, so

$$i \geqslant \text{codimension} - 1 \geqslant \tfrac{1}{2}m(m + 1) - 1.$$

This contrasts interestingly with the rule of thumb $(i = m)$ of the engineering literature. For the ordinary fold $m = 1$, but no imperfection parameters are needed, in that none can have other than a smooth effect on collapse load. For the 'asymmetric point of bifurcation' discussed above, $i = m = 1$. For cusps whose axes are not equilibrium paths (rare in engineering analyses) the same is true. For symmetrical cusps like those of Sections 4 to 10, and for all other singularities whatsoever, i is greater than m. In most cases, it is much greater. [Added in proof: Golubitsky and Schaeffer have now systematized and are writing up this kind of analysis.]

12 Optimization

For structures intended to bear a given maximum load F, it is common to make a design with a number of adjustable parameters $d_1, \ldots, d_r$, say, and to *optimize* (for example, minimize weight or cost) over $(d_1, \ldots, d_r)$-space, subject to the condition that the collapse load be F. (Sometimes one maximizes F for given weight or cost.) An *ad hoc* safety factor is added afterward. This procedure often leads to singularities of high codimension and corank. Indeed, it so often leads to designs with more than one buckling mode that it is sometimes replaced by a caricature: the 'wonderful one-hoss shay principle', otherwise known as *simultaneous mode design*. (Roughly, if a structure buckles in mode A before mode B you have 'wasted' strength and material on mode B; thus optimization tends to suggest

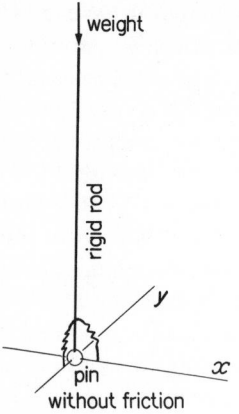

weight

rigid rod

y

pin

x

without friction

Fig. 13.29

making them buckle together, unless stability against sudden reduction of buckling load is part of the optimized quantity.) Mathematically, this principle says 'try to build umbilic catastrophe machines', or machines of higher corank still. The results are often highly sensitive to imperfections. Rather surprising effects occur: from the combination of two buckling modes each of which *separately* would be governed by a standard cusp, and buckle suddenly but continuously, can come something like a dual cusp only worse – *snap* buckling, with large movements, and severe imperfection sensitivity. For an analysis of the stiffened plate, in which equating the loads at which whole-plate buckling and between-stiffener buckling occur leads to a hyperbolic umbilic catastrophe, see Thompson and Hunt [23]. (Such plates are a standard component in box-girder bridges, whose wreckage was decorating the motorways of Britain and the rivers of Australia a few years ago: but so much else was found wrong by the enquiry that efficient cause, proximate cause, ultimate cause etc. are less than clear.) Similarly the analysis of the Augusti model (Fig. 13.29) in Thompson and Hunt [105] pp. 243–250 shows that for equal springs in the (x, z)- and (y, z)-planes – as suggested by a natural optimization procedure – the system is violently sensitive. If restricted to the (x, z)- or (y, z)-planes it would exhibit standard cusp behaviour, with onset of leaning at a certain load. But in the planes $x = \pm y$ it looks like a *dual* cusp, with sudden complete collapse and two-thirds power law imperfection sensitivity. In fact the whole bifurcation set is one of the *double cusp* catastrophes of codimension 8, whose geometries are not yet fully understood. (There is no problem in principle, and they are yielding to standard enough techniques, but eight dimensions of complication takes time to explore!) Calculations with the ordinary full unfolding suggest that the linear imperfection term investigated by Thompson and Hunt probably has the worst exponent possible here, but only an analysis by the methods discussed in the previous section can establish this.

Other systems of similar description are analysed in Thompson and Gaspar [119], though their main purpose is to exhibit the occurrence of the parabolic and symbolic umbilics in buckling problems (which, like the butterfly, are rare in the literature). The analysis is carried out with reference to the umbilic bracelet, and immediately suggests several extensions.

13 Symmetry: Rods and Shells

As we have seen repeatedly in the last three chapters (with for instance the oil rig, the six-roll mill and the glory) symmetries tend to lead to degenerate functions, with complicated unfoldings, typicality arguments notwithstanding. We discussed in Chapter 7 Section 3 the tendency of design to lead to symmetry: in *optimized* design this tendency is particularly pronounced. For example, Fig. 13.22(c) can be interpreted as giving a sharp optimum at $\varepsilon = 0$ (perfect bilateral symmetry); the 'optimum' Augusti model, with equal springs, has all the symmetries of a square. If all the symmetries to be present are specified beforehand, one can cautiously use typicality arguments *within the class of functions with those symmetries*. The reasoning by which we expected the six-roll mill to be an unfolding of $x^3 - 3xy^2$ (Chapter 11 Section 11) is an experimentally very successful argument of this kind. Similarly, the 4-jet of the energy function for the Augusti model, if perfectly symmetric, must for any load be of the form

$$a(x^2 + y^2) + b(x^4 + y^4) + cx^2y^2.$$

Typically, at a bifurcation point ($a = 0$), the quartic term is non-degenerate ($c \neq 2b$). Indeed, at bifurcation the perfect 'optimum' Augusti model has, by the computations in Thompson and Hunt [105], the 4-jet

$$\tfrac{1}{24}x^4 - \tfrac{1}{4}x^2y^2 + \tfrac{1}{24}y^4.$$

The square tube described by Croll [116] has the same symmetries – indeed, that is about all we are told about it – so we would expect bifurcation to be by one of three types of double cusp catastrophe (unfolding $x^4 + cx^2y^2 + y^4$ for, respectively, $c > 2$, $-2 < c < 2$, and $c < -2$. One other class exists, represented by $x^4 - y^4$, but cannot have square symmetry.) Thus the result of a small imperfection very probably could be, as he asserts, the appearance of a hyperbolic umbilic catastrophe; all three types contain 5-dimensional sheets of these in their universal unfoldings. (They are similarly present in those of the higher degeneracies possible with the given symmetries, if by inspired design he has found one: we cannot prove that the 4-jet is nondegenerate without more information than he gives.) The 'dangers of catastrophe theory' that Croll discusses amount only to the dangers of believing that 'everything that happens is on Thom's list', without considering symmetries, on the basis of a popular article. These dangers exist, but they are the general dangers of the misuse of uncomprehended mathematics.

Symmetries sometimes emerge where we might not, *a priori*, expect them; for example in Chapter 10 Section 9 we saw that strictly vertical sides to a ship (a continuous symmetry under the vertical translation pseudogroup!) gave the metacentric locus a particularly simple geometry, the graph of a quadratic form. Combined with more than two vertical planes of symmetry, as in an oil rig, this forces *circular* symmetry for the 'perfect' system, and thence infinitely complicated imperfection sensitivity. There is more symmetry in the problem than meets the eye, and typicality reasoning within the square symmetry class, which works for a pyramidal ship, fails for the vertical-sided square one. This kind of problem exists, and causes difficulties when it really happens.

Fig. 13.30

However, it sometimes *apparently* happens by the way we found the butterfly in Section 9; through too great a wish to 'simplify' the calculations. For example, it is a commonplace of the engineering texts that the bending properties of a rod depend only on the 'inertial ellipse' of its cross-section, and that hence if it has more than two planes of symmetry (like the square and triangular rods of Fig. 13.30) its bending properties are rotationally symmetric about the centre line. This makes the 'perfect' analysis easy, because the direction of bending can be left out of the calculations (which is why it is so readily believed), and quantitatively it is not so bad an approximation as absolutely to force reexamination. It makes an exact 'qualitative' analysis almost impossible, however, as the analogous result does for the oil-rig, by forcing infinite codimension. A topologist therefore is not so ready to believe it, since his calculations have been not simplified but complicated: symmetry-breaking imperfections could have an infinite variety of effects. On examination thus prompted, the 'inertial ellipse' theorem turns out to consist of the lowest order Taylor series truncation of the local elasticity theory of rod bending that could reasonably be tried. When the final result is determinate and of finite codimension (as it is for the buckling of rods with rectangular or elliptical cross-section) this is fine; but when truncation introduces degeneracies it is only prudent to try putting the higher terms back! We hope to publish a rigorous study of the results in due course; we merely report here that preliminary investigations are promisingly compatible with typicality arguments.

A rod manufactured to *circular* cross-section specifications, of course, *has* circular symmetry to the best of its manufacturer's ability. A merely quantitative analysis of the effects of possible imperfections is there the best that one can hope for, and even then new techniques are required for a rigorous treatment.

Similar remarks apply to the theory of spherical and cylindrical shells. These combine the problems of large corank (or 'many simultaneous modes' or 'highly multiple eigenvalues' according to dialect) and symmetry. For example, a typical cylinder theory will have several simultaneous *pairs* of buckling modes, like

$$r = 1 + x \sin n\theta$$

$$r = 1 + y \cos n\theta$$

for various n as shown in Fig. 13.31 for small x, y with $n = 1$. The rotational symmetry of the problem yields rotational symmetry in each of the (usually finitely many) (x_n, y_n)-planes. Use of the Splitting Lemma thus gives a finite-dimensional problem, but one of such degeneracy as to have infinite codimension in the topologist's sense.

Such shell problems are in fact violently imperfection-sensitive: Thompson and Hunt [120] show in an example that a particular specified imperfection, within normal engineering tolerances, can

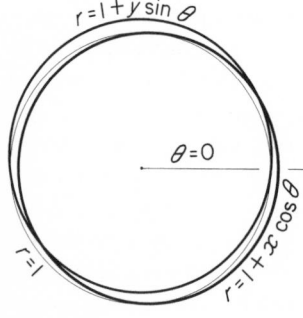

Fig. 13.31

reduce critical load by a factor of ten. (Thus explaining why the theoretical strength of the 'perfect' system has never been realized by experimentalists, and why some of the latter have so little respect for theory. Without stability, 'exact' need very definitely not equal 'correct'.) An *a priori* method of singling out a finite-dimensional family of imperfection terms among which the worst sensitivity must lie, even for problems that cannot finitely be versally unfolded, should be possible by topological techniques when the Splitting Lemma gives finitely many essential variables.

Meanwhile we can recognize that the difficulties encountered in studying cylindrical and spherical shells arise not so much from perversities of any particular mathematical formulation of the problem, be it catastrophe theory or finite element computation, but from the perversity of the problem itself. While this does not rule out the possibility of *conceptually* simple methods, it probably rules out *computational* simplicity.

BUCKLING PLATES

14 The von Kármán Equations

The curvature energy model for strut bending that we assumed in Section 5 was an approximation to the 'real' compression and stretching energies given by the elastic solid theory of a strut of finite thickness. The corresponding treatment of a thin plate requires considerable geometric subtlety, and even more skill in judging just what approximations *en route* to manageable equations are sensible. In the long term, a marriage of functional analysis with the rigorous truncation and reduction techniques of catastrophe theory may, not replace such judgement, but support it with system and proofs. We have not space here even to give a derivation, and simply give the most popular model, the von Kármán equations, in dimensionless form. (Other plate models are under study, but in many cases by functional analysts whose concerns are topological – such as the existence of solutions of various types – rather than numerical. This does not make them 'purely qualitative' any more than it does catastrophe theory, but to get numbers from any theory you have to *want* them.)

The position of the plate is modelled by a surface in $\mathbb{R}^3$, whose edges are required to remain a rectangle in the (x, y)-plane: in the absence of stress, the rectangle with corners $(0, 0)$, $(0, 1)$, $(l, 0)$, $(l, 1)$. The equations involve a function $w(x, y)$ specifying the z-coordinate of the point originally at $(x, y, 0)$, and a function $f(x, y)$ describing the stresses. A force of dimensionless magnitude ρ is applied normal to the edges $x = 0$, $x = l$, as shown in Fig. 13.32. We take the

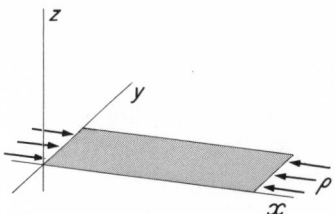

Fig. 13.32

equations in the form given by Bauer and Reiss [121]:

$$\nabla^2 f + \tfrac{1}{2}[w, w] = 0$$

$$\nabla^2 w - [w, f] + \rho w_{xx} = 0$$

(where for any functions u, v defined on the rectangle,

$$[u, v] = u_{xx} v_{yy} + u_{yy} v_{xx} - 2 u_{xy} v_{xy},$$

and subscripts indicate partial derivatives) with boundary conditions

$$w = Dw = f = Df = 0,$$

intended to represent the plate as *simply supported*, like the strut in Section 5.

We should remark here that 'simple support' is not a simple idea, either in theory – other mathematizations may be defended – or in practice. The usual experimental setup is sketched in cross-section in Fig. 13.33; it is clearer that it is exerting little leverage than, for instance, that it is *constraining* the edges to be straight. (Particularly the long sides, which the model supposes not pressed inwards!) The grooves are carefully machined and oiled: lingering traces of the paint applied by a workman who misunderstood his instructions can spoil results in this kind of experiment. (Imperfection sensitivity need not be towards collapse. It is not very easy to make a strut under lengthwise load buckle at as *low* a value as the theory gives. Partly, any friction gives something like a stiff Zeeman machine, which with the control taken straight up the axis jumps 'late': partly, rheumatic support takes the problem towards the *clamped* version, which has a higher buckling load anyway.)

Fig. 13.33

The linearized equation is simply

$$\nabla^2 w + \rho w_{xx} = 0,$$

giving the buckling modes for the specified boundary conditions

$$\sin \frac{n\pi x}{l} \sin (m\pi y).$$

It turns out that with the force on the long sides zero or nearly so, only $m = 1$ is relevant: higher m cannot give a *first* buckling mode, or an energetically *stable* equilibrium. Thus the possible buckling modes are those of the form $\sin (n\pi x/l) \sin (\pi y)$, as in Fig. 13.34. As

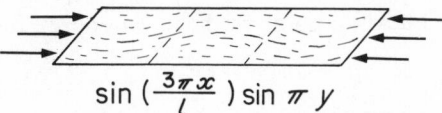

$$\sin\left(\frac{3\pi x}{l}\right)\sin \pi y$$

in the strut problem, these are only the first order descriptions of the deflection shape, but experimentally (Supple [122]) they describe the buckling forms excellently for a steel plate, up to $2\frac{1}{2}$ times the critical load.

Evidently an $l \times 1$ plate can buckle at the same load as a $2l \times 1$, or a $3l \times 1$, and so on (because the $l \times 1$ modes, like the plates, can be fitted together end to end). The lowest critical load is for a square or an integer $\times 1$ composite of squares. The general formula for the value of ρ at which $\sin(n\pi x/l)\sin(\pi y)$ becomes a solution of the linearized equation above, for a plate of unstressed length l, is

$$\rho = \frac{\pi^2(n^2 + l^2)^2}{n^2 l^2}$$

which we draw for various n in Fig. 13.35. This makes clear that for most l, increasing ρ will meet one of these curves first, and the plate will buckle in a single mode. Around such a point the standard analysis gives the usual 'stable symmetric branching diagram' of Fig. 13.7(a) above; the addition of any normal loading term (Chow, Hale and Mallet-Paret [123] take a general 'abstract' addition to the

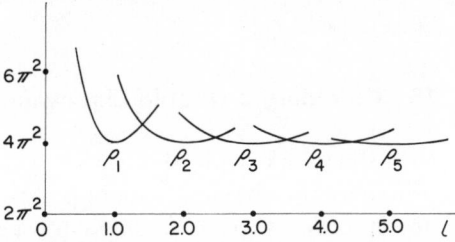

Fig. 13.35

equations: they are more specific in [124]) gives a standard cusp catastrophe. The structural stability of this implies a reasonable insensitivity to (for instance) the fact that exactly zero horizontal force on the long sides is experimentally problematic.

However, when $l = \sqrt{n(n+1)}$ for some n, increasing ρ in Fig. 13.35 meets two of these curves at once; like the Augusti model (Fig. 13.29) the bifurcation has two essential behaviour variables. In particular for $n = 1$, $l = \sqrt{2}$ and we have the two simultaneous buckling modes shown in Fig. 13.36. Transforming the von Kármán equations into an equivalent energy formulation and applying the appropriate Hilbert space reduction to the essential variables, at the

Fig. 13.36

buckling point we have an energy function

$$a\mu^4 + b\mu^2\nu^2 + c\nu^4$$

where the position labelled by (μ, ν) is (to first order in (μ, ν) and, experimentally, accurate for μ, ν fairly large)

$$\mu \sin\frac{\pi x}{\sqrt{2}} \sin(\pi y) + \nu \sin(\pi x\sqrt{2}) \sin(\pi y),$$

and by computation (giving the results to three decimal places)

$$a = 0.592 \qquad b = 6.013 \qquad c = 3.899.$$

(All results of functional analytic calculations are taken from Magnus and Poston [125].)

The symmetries of the problem (the relevant ones here being reflection in the (x, y)-plane and in the plane $x = l/2$) force a function which is even in each of μ, ν separately: subject to this and our deliberate choice of a point with corank 2, the result is as 'typical' as it can be – a nondegenerate, 4-determinate quartic. As we increase n and $l = \sqrt{n(n+1)}$, however, the problem approaches rather fast the degeneracy of the infinitely long plate problem, with its continuous (translational) symmetry. As can be seen in Fig. 13.35, the buckling modes crowd more and more together; moreover, at the double eigenvalue points the quartic coefficients converge to degeneracy (at $n = 10$ the 4-jet is of the form kx^2y^2 to about eight significant figures). New techniques are therefore needed for qualitative exactness with long plates.

15 Unfolding a Double Eigenvalue

We will work with the $\sqrt{2} \times 1$ plate, as above; and by 4-determinacy of the bifurcation point we need not consider terms above order 4. (For *strong* equivalence of unfoldings, Theorem 8.7 requires that for the unfolded jet – though not for the unfolding terms – we go to order 6. The results are absorbed into the treatment that follows.)

Defining $\rho_0 = 9\pi^2/2$, the 'perfect' plate problem becomes the 1-parameter unfolding

$$a\mu^2 + b\mu^2\nu^2 + c\nu^4 + (\rho_0 - \rho)((0.872)\mu^2 + (3.489)\nu^2)$$

up to order 4. For $\rho > \rho_0$ this function has 9 critical points; precisely the maximum given by Theorem 8.5, since by a straightforward computation as in Chapter 8 Section 13, a function with 4-jet $a\mu^4 + b\mu^2\nu^2 + c\nu^4$ has codimension 8 for these values of a, b, c. By the same token, we need 7 more terms for a full unfolding; but before we find them, let us consider the 'perfect' system.

The commonest representation of the branching of the solutions is Fig. 13.37(a), but this compresses two dimensions of behaviour into one. Fig. 13.37(b) shows the relative positions of the solutions in (μ, ν)-space more clearly, if the information is added that the branches in the planes $\mu = 0$ and $\nu = 0$ are the stable equilibria; the $\mu = \nu = 0$ branch consists of Morse 2-saddles (restricted to the essential variables, maxima) and the others 1-saddles. We sketch the contours of the energy as a function of (μ, ν) in Fig. 13.37(c); notice the small 'basins of attraction' of the minima on the

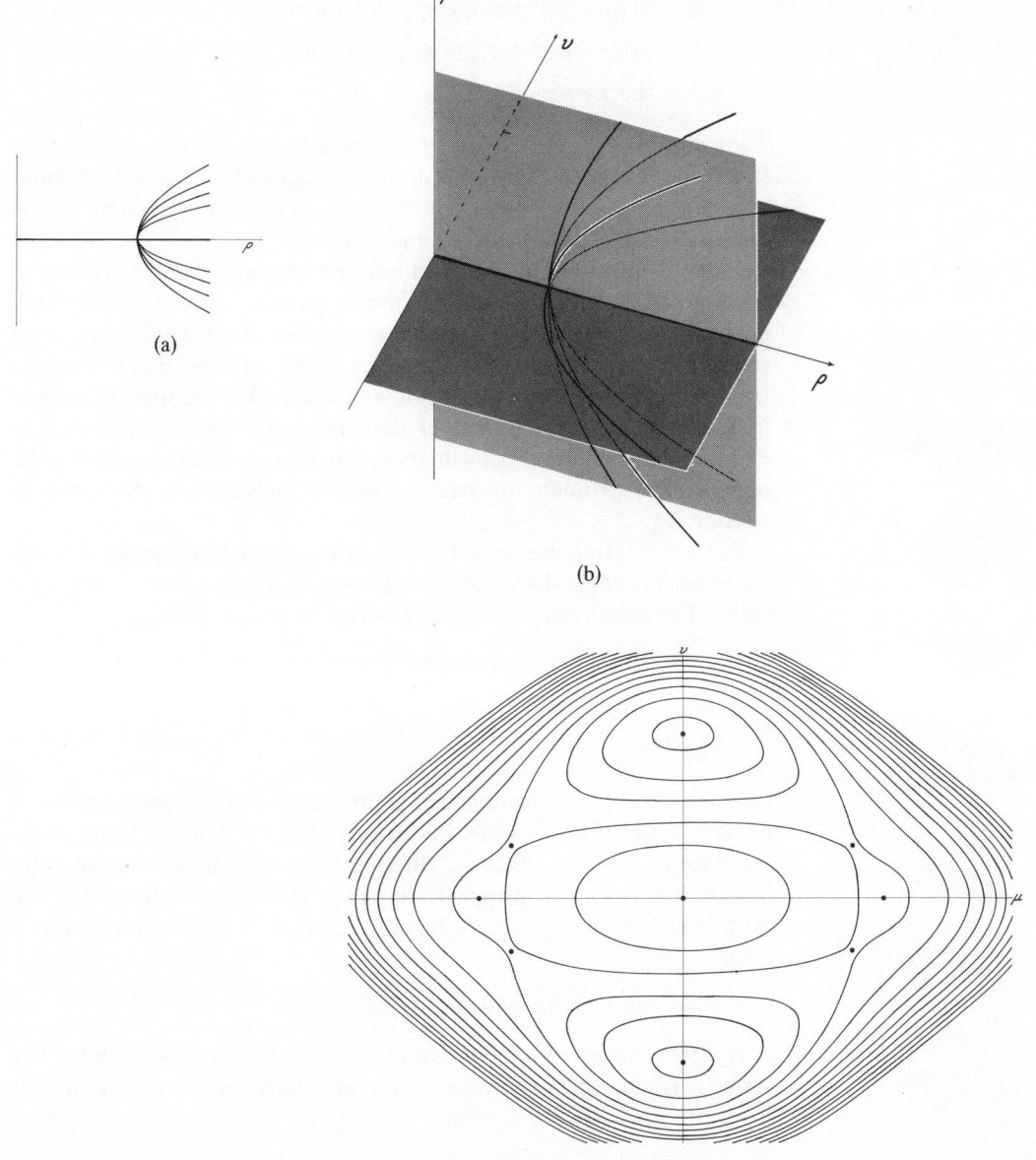

Fig. 13.37

μ-axis compared to those on the ν-axis. This feature could be changed by diffeomorphism (though near the bifurcation point, the changes of variable used in reduction are near the identity) but the relative heights of the critical points are invariant under diffeomorphisms of the state variable. Once we consider this unfolding as induced from a universal one, therefore, we can deduce that to second order in $(\rho - \rho_0)$ the energies of the critical points are, up to a common constant,

$-(0.322)(\rho - \rho_0)^2$ for the minima on the μ-axis,

$-(0.781)(\rho - \rho_0)^2$ for the minima on the ν-axis,

$-(0.301)(\rho - \rho_0)^2$ for the single-saddles,

0 for the two-saddle at $\mu = \nu = 0$.

Thus the energy needed to get from a single-hump solution to a saddle (and thence fall into a double-hump state) is about 23 times less than that needed the other way, apart from needing to be applied by a less carefully planned push.

Indeed, the proximity in position and energy of the μ-axis solutions to the saddles makes it obvious that with a fairly small perturbation we can remove these minima entirely, leaving only saddles between the basins of the two ν-axis minima. But a systematic study of this requires a full unfolding. We choose to unfold mainly with load terms (rather than imperfections) with a view to eventual experimental convenience. Furthermore, this avoids for the present the problem of two classes of parameters, discussed in Section 11.

For even terms, we take $(\rho_0 - \rho)$ as before, σ measuring an even force on the long sides, and τ a linearly varying force on the long sides. The quadratic terms contributed by these are

$(\rho_0 - \rho)((0.872)\mu^2 + (3.489)\nu^2)$

$\sigma(1.774)(\mu^2 + \nu^2)$

$\tau(0.889)\mu\nu.$

Linear and cubic terms come together, as suitable combinations of normal loads. The analysis involves a slightly more delicate argument for point rather than for distributed loads, but we choose point loading for clarity of physical meaning. The natural choice for a μ term is shown in Fig. 13.38(d); more exactly, a central load gives a term

$\beta_1(\mu - (0.0125)\mu^3 - (0.0215)\mu\nu^2).$

Any arrangement with the symmetries of Fig. 13.38(e) gives a ν term; it is convenient to take a load at points on the centre line $l/6$ from the ends, giving a term

$\beta_2((1.732)\nu - (0.0134)\mu^2\nu - (0.00329)\nu^3).$

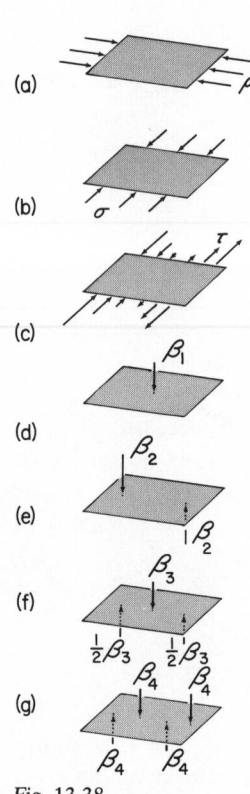

(a)

(b)

(c)

(d)

(e)

(f)

(g)

Fig. 13.38

In principle, almost any two extra normal load variables will combine with these to give the linear and cubic terms required in a full unfolding, but as the above cases illustrate the cubic contributions are rather small. A search for force combinations giving the least exiguous pure cubic terms led to the arrangements of Fig. 13.38(f) and (g), and terms

$$\beta_3(-(0.0188)\mu^3+(0.0618)\mu\nu^2)$$
$$\beta_4(-(0.0122)\mu^2\nu-(0.00785)\nu^3).$$

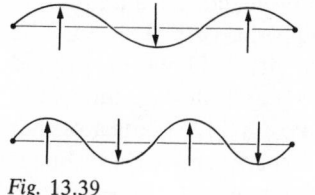

Fig. 13.39

Notice (Fig. 13.39) that these load patterns are trying to push the plate into its third and fourth buckling modes. We have not yet analysed the effect of these shapes as imperfections, but since a small deflection in these modes requires large forces compared to the first two (since the plate is still *stiff* – see Section 17 below – in the higher modes), the cubic coefficients in a corresponding imperfection analysis are probably more comparable with the others.

It remains to find a quartic unfolding term. Curiously, none of the in-plane load patterns we have examined give an unfolding term with quartic part not lying in $\Delta_4(a\mu^4+b\mu^2\nu^2+c\nu^4)$; and a 'cross-ratio' term appears only when we change the aspect ratio l of the sides from $\sqrt{2}$ to $\sqrt{2}(1+L)$. To first order in L this gives an unfolding term

$$L(6.457\mu^2-25.830\nu^2+0.386\mu^4+2.956\mu^2\nu^2+2.397\nu^4).$$

As the reader may check, the terms in ρ, σ, τ, L, β_1, β_2, β_3, β_4 now give a transversal unfolding of the quartic found by the bifurcation analysis.

This ten-dimensional geometry (counting μ, ν) is not yet fully understood. It is another of the double cusp catastrophes, but the quartic unfolded is one with all root lines complex (rather than all real as in the Augusti model) and the unfolding geometry is topologically as for x^4+y^4, partly analysed by Zeeman [7]. A full understanding of this example – and simultaneously others – will thus develop hand in hand with that of the abstract singularity geometry. Indeed, the branching analysis of Supple [122] is one of the more important studies of this catastrophe published to date.

The subsidiary bifurcations 'organized' by this one, as fold lines are by a cusp, are all of those listed at the end of Chapter 7 except for D_6^+ and D_6^-, plus the codimension 6 cuspoid A_7 which unfolds x^8. This is a little unexpected (without catastrophe theory) in an analysis in which only fourth order terms have been retained; but because the Tayl can be *removed* by suitable transformations, not just forgotten, the 8-determinacy of the quartic jets found at some points guarantees local exact reducibility to $\pm x^8 \pm y^2$ as required. (Compare the discussion of the 'cubic' cusp points in the elliptic umbilic, Chapter 9 Section 6.) All those swallowtails, butterflies, wigwams and A_7's are really and exactly in there, despite apparently requiring higher order terms than we have considered.

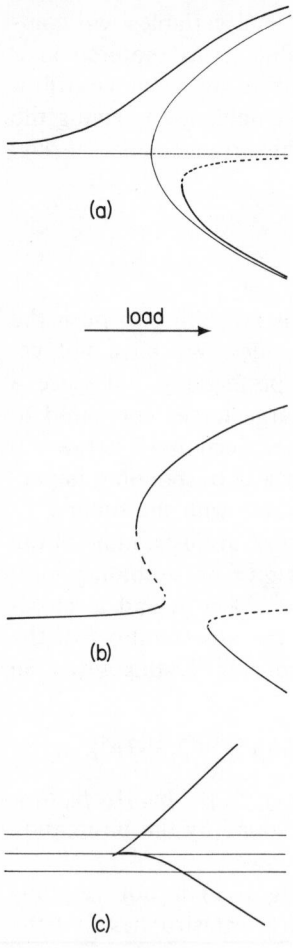

(a)

load

(b)

(c)

Fig. 13.40

(To say it again: the behaviour of Taylor series in two or more variables is much more surprising than one might guess, and nothing simpler than the rules of Chapter 8 will handle the problem.)

Physically, this complexity seems to have the following implication. For standard cusp buckling, the usual analysis gives a 'perfect' (light) and 'imperfect' (heavy) behaviour as in Fig. 13.40(a). A 'tilt' imperfection, like those examined for the dual cusp in Section 11, *can* force a small 'snap' buckling as load increases (Fig. 13.40(b)), the maximum jump growing linearly with the imperfection. But even with tilt, most loading histories will not include such a jump; in Fig. 13.40(c), only the middle line through the cusp does so. The jump probability is not zero, but not large either. However, around the *compound* buckling point we have been considering, things are more subtle. The analysis in Magnus and Poston [125] indicates that while jumps may remain rare for *rough* experiments (where linear imperfections probably dominate), refinement of experimental technique will make 'mode jumping' more and more frequent. The system behaves less smoothly, not more, as more care is exercised! Moreover, the 'cross-ratio' term turns out to have a major rôle geometrically, despite its topological insignificance: *particularly* around the 'simple' case $x^4 + y^4$. In other cases (as for the plate) it is less important as an unfolding term, but the *value* of the cross-ratio at the point unfolded greatly affects the imperfection sensitivity.

DYNAMICS

16 Soft Modes

When passing from mathematical statements to physical remarks about buckling we have – as is most usual in elastic buckling studies – assumed *perfect delay* behaviour. For many purposes this is an excellent approximation, though for no explicit dynamics, gradient or otherwise, is it exact. (It is best approximated for very slow changes of load: a *quasi-static* loading history.) But if we took the energy function that we have been examining in various cases as part of a Hamiltonian (adding a momentum term) so as to consider the conservative† mechanics of these systems, complications would arise. Linearizing the resulting equations generally reproduces the buckling modes as *vibration modes* possible before buckling. Without linearization, elaborate ergodic behaviour would clearly be apparent in most cases – particularly around the bifurcation point

---

† In the engineering literature 'conservative' often means the existence of a well-defined configuration energy – that is, elastic rather than plastic behaviour – but with large dissipation leading to minimization of energy. (This is distinct from the usage in physics, which excludes dissipative terms.) Indeed, too nearly 'physicist's conservative' behaviour makes quasi-static buckling experiments extremely difficult.

discussed in the previous section – and become more and more prominent for large vibrations. Breaking the vibration up into modes is
thus a first approximation, but a useful one in most cases, particularly when there is a reasonable amount of damping which nonetheless permits observable vibrations.

Within a mode, the restoring force towards a stable equilibrium is
proportional to the gradient of the energy function; this rule usually
defines one or the other. In the quadratic approximation to the
energy (good at a Morse energy minimum away from bifurcation
points but not *exact* since the dynamical laws are not diffeomorphism invariant) this is proportional to the amplitude x of the deflection from equilibrium. Thus if along the x-axis in deflection space
we have energy kx^2 to second order, then to first order we have
dynamics

$$\ddot{x} = -2kx + (\text{damping terms})$$

within this mode, a *damped simple harmonic oscillator*. With most
models of moderate damping, this leads to diminishing vibrations
after an initial disturbance, of period $\pi\sqrt{(2/k)}$. As k goes to
zero – that is, as a quasi-static buckling point is approached – this
period goes to infinity, as does the amplitude of vibrations with a
given initial energy. Of course, neither of these infinities is more real
than the infinite intensities of the previous chapter. As the quadratic
term shrinks, the quadratic *approximation* becomes less good, and
higher terms prevent infinities in both cases. But the smaller the
vibrations, the closer the fit. In particular, if the strut of Section 5 is
lightly struck it will vibrate in the mode $\sin(\pi s)$ with frequency
approximately $\sqrt{(\mu\pi^2 - F)/8}$, up to near the buckling value. Thus
unloaded it will go 'ting', moderately loaded 'bong', and near
buckling point 'boiinnggggg'. (This is eminently familiar to the practical engineer, but should warn the general reader to prefer soprano
structures to bass.) Similar phenomena are important in the study of
crystallographic phase transition (next chapter, Section 10), and
exactly the same analysis underlies the remarks about the period of
small rolls as warning of imminent capsize, in Chapter 10 Section
11.

17 Stiffness

The amplitude of the vibrations above is proportional to the square
root of the disturbing energy (for given k) but the name *soft mode*
used to describe those for which linearized treatment breaks down is
more clearly motivated if we return to quasi-statics with a disturbing
force. For the strut, Fig. 13.41 shows the first-order response of the
strut to applied normal force G, as the tangents to the catastrophe
surface in the G direction above various points on the F-axis. To
first order, the distance moved from the central position with given

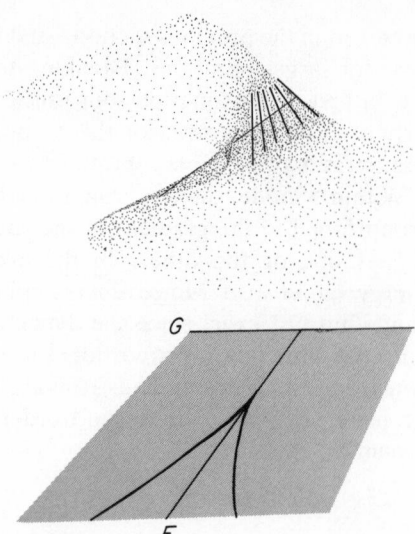

Fig. 13.41

applied force G is proportional to $1/(\mu\pi^2 - F)$; the strut feels softer, less rigid and stiff to the grip, as buckling is approached. Notice the difference here between cusp and dual cusp: reduced to standard form the first order stiffness and vibration measures of distance from the cusp point give the same answers, but for a standard cusp this point is the *nearest* buckling point. For a dual cusp, other points on the bifurcation set are much closer; for given first order stiffness, an act like Samson's can much more easily bring down the house.

14 Thermodynamics and Phase Transitions

The involvement of catastrophe theory with thermodynamics, like most thermodynamic processes, has produced more heat than light.† In this chapter we cannot aim to do more than clarify some relationships. Catastrophe theory is not the Whole Answer to phase transition physics, any more than its direct equivalence to ray optics makes it the Whole Answer to the physics of caustics (Chapter 12). Just as for caustics, however, it is our view that catastrophe theory is a necessary component in any physical treatment of phase transitions short of exact solution of the Schrödinger equation for bulk matter. Implicitly, its more elementary cases always have been, so that again no new results can be expected in low codimensions.

A complete case in favour of this view would require a systematic treatment of thermodynamics, and detailed analysis of several higher order cases, by the joint application of catastrophe theory and renormalization group theory (the appropriate 'asymptotic' method in this context, analogous to the oscillatory integral techniques of Chapter 12). This is clearly impracticable for the present book, and would be a substantial project in itself. Moreover, the technical depth required would counter the aim of demonstrating to the general scientist the varied ways that one may use catastrophe theory. We hope that experts in phase transitions will bear with our less than fully elaborated approach.

For coherence with the rest of the book we are using the phrase 'critical point' in the sense 'point where the derivative of a function vanishes', which is the usual mathematical usage. In thermodynamics it normally has a more special sense, which for the Landau theory translates into our terms as 'degenerate critical point which is also an absolute minimum (or, for entropy, maximum)'.

EQUATIONS OF STATE

1 van der Waals' Equation

As we shall see, description of phase transitions by equations of state is not always adequate, but it is a convenient starting point. We begin with van der Waals' equation (1873), the first step beyond

† Compare the Onsager relations, which since 1931 have generated a vast mass of literature treating them as the cornerstone of irreversible thermodynamics. Equally eminent physicists claim (Truesdell [126]) not only that they are ill-defined and unproved, but that there is no case on record of their being of practical use.

the Boyle's law description $PV = RT$ of the 'ideal gas'. On the basis of physical reasoning which we shall not go into (since no one now believes in the picture on which it was based) van der Waals proposed the modified equation

$$(P + \alpha/V^2)(v - \beta) = RT.$$

Here P, V and T are the pressure, volume and temperature of the sample of fluid concerned, and R is a constant (usually now written as Nk, where N is the number of molecules in the sample and k is Boltzmann's constant, 1.380×10^{-16} erg/deg). The constants α and β are chosen, nowadays, by experimental best fit, and are no longer given a physical meaning. We assume the reader knows what volume, pressure and temperature mean, though in fact they are quite subtle to define, especially the last. More extensive discussions will be found in thermodynamics texts, such as Callen [127].

In the form above, van der Waals' equation is usually represented graphically as in Fig. 14.1, by drawing the graph of P against V for various T. (A more vivid picture, perhaps, is the surface of Fig. 14.2, showing the set of points in (P, V, T)-space that satisfy the equation.) It may be seen that as the substance described experiences decreasing pressure at constant temperature, there are several possibilities.

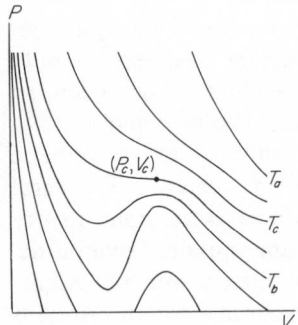

Fig. 14.1

(a) If the temperature is $T_a > T_c$, the volume can smoothly increase.

(b) If the temperature *is* T_c, then V is a continuous but not a differentiable function of P.

(c) If the temperature is $T_b < T_c$, there are pressures for which several volumes are possible, and the equation does not give volume as a function of pressure. At some point, it would appear, matter attempting to follow the curve must jump. Above the jump, the matter resists a small reduction in volume with a much greater increase in pressure than below; the hard-to-compress *liquid* suddenly becomes an easy-to-compress *gas*.

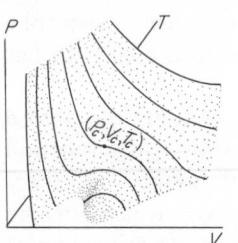

Fig. 14.2

How does the matter decide at which point to jump? Maxwell gave in 1875 an argument from the equivalence of heat and work, whose conclusion may be geometrically expressed as the *equal area rule:* part of the original curve must be replaced by a horizontal straight line as shown in Fig. 14.3, such that the two shaded areas are equal. For a clear explanation of why the physics can be given this graphical expression, *see* Callen [127] pp. 148–153. (Note that his Fig. 9.5 corresponds to a set of slices of our Fig. 7.14(b), and not to a swallowtail catastrophe.)

Now we approach the relationship with catastrophe theory by way of a standard reduction, and a change of variables from Fowler [128]. The point (P_c, V_c, T_c) which lies at the heart of what is

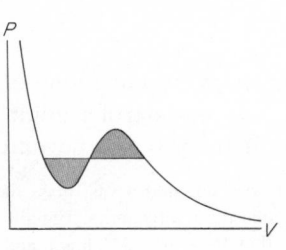

Fig. 14.3

happening is easily found as

$$\left(\frac{\alpha}{27\beta^2}, 3\beta, \frac{8\alpha}{27\beta R}\right).$$

Normalizing this point to $(1, 1, 1)$ by setting

$$P' = P/P_c, \qquad V' = V/V_c, \qquad T' = T/T_c,$$

we get the *reduced* van der Waals' equation

$$(P' + 3/V'^2)(V' - \tfrac{1}{3}) = \tfrac{8}{3}T'.$$

Replacing volume as variable by density X, that is, setting $V' = 1/X$, we get in some ways a more intuitive description of the 'state' of the matter. Finally, we transfer the origin to the point $(1, 1, 1)$ around which everything is happening, by setting

$$p = P' - 1, \qquad x = X - 1, \qquad t = T' - 1.$$

We get

$$x^3 + \frac{8t + p}{3}x + \frac{8t - 2p}{3} = 0,$$

or

$$x^3 + ax + b = 0,$$

where

$$a = \frac{8t + p}{3}, \qquad b = \frac{8t - 2p}{3}.$$

This is exactly the cusp catastrophe surface (Chapter 5 Section 2, Chapter 9 Section 3), and we draw it again in Fig. 14.4, cut off along

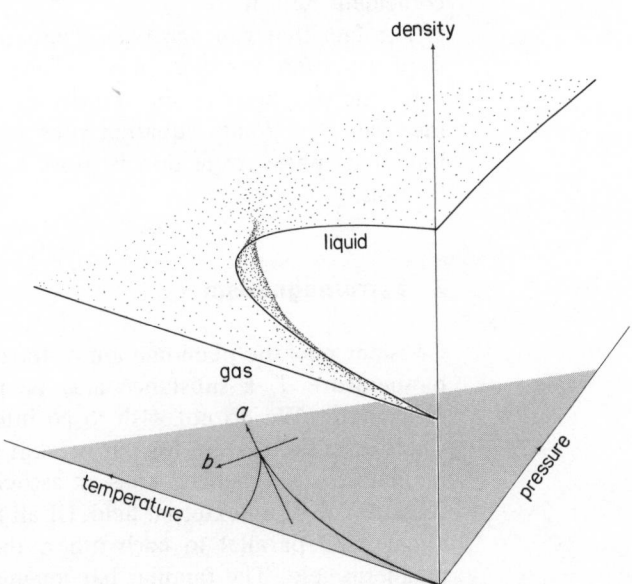

Fig. 14.4

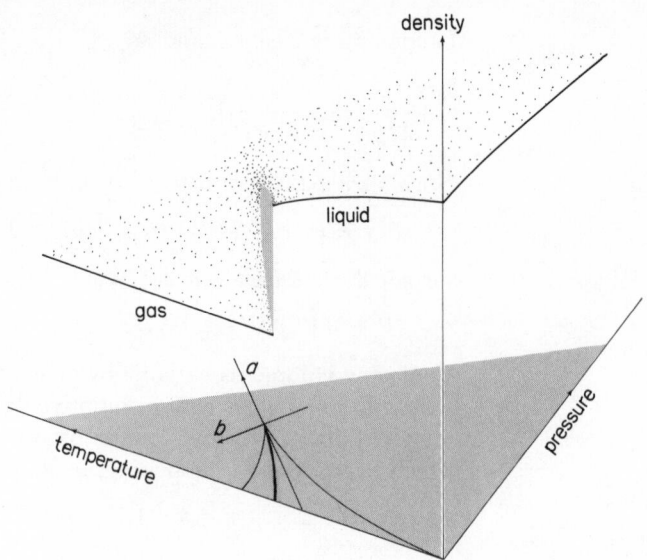

Fig. 14.5

the heavy lines by the condition that pressure and temperature be non-negative.

Application of the Maxwell equal area rule assigns a unique volume (and hence density) to the fluid for each temperature and pressure, unique that is except at the jump points. Carrying this over to our new coordinates, Fig. 14.4 becomes Fig. 14.5. The change of coordinates does not preserve area, however, and the Maxwell rule has no simple expression in terms of x, a and b. The set of points at which jumps take place is a curve in (a, b)-space, shown heavy in Fig. 14.5, tangent to the a-axis at $(0, 0)$, but not coincident with it.

The fact that van der Waals' equation transforms into the standard equation for the cusp surface is no coincidence, but the algebraic simplicity of the transformation is. The full relation with the 'van der Waals equation plus Maxwell's rule' model of the liquid/gas phase transition is more subtle: we shall come to it in Section 6.

2 Ferromagnetism

Consider now the phenomenon of ferromagnetism. Above a certain temperature T_c a substance may be paramagnetic, below T_c ferromagnetic. We do not wish to go into the physics of these terms. Their essential feature for our present purposes is that ferromagnetic matter is *magnetized* – has an associated magnetic field – even in the absence of an external field. (If all parts of a piece of matter are magnetized parallel to each other, the result is a substantial net magnetic field. The familiar bar magnet is an example of this.) An early approach to this phenomenon was the Weiss model (1907). As with van der Waals we omit the physical arguments (*see* Haug [129]) and look at the equation of state

$$\tanh \eta = \frac{T}{T_c}\eta - \frac{\mu_B H}{\kappa T_c}$$

where H is the external magnetic field and η is a reparametrization of the magnetization M,

$$\tanh \eta = M/M_0,$$

(recall that tanh is an (analytic) diffeomorphism $\mathbb{R} \to \{x \in \mathbb{R} \mid -1 < x < 1)\}$), and M_0, μ_B, κ, T_c are constants. The transition takes place as the temperature T passes the *Curie point* T_c. Let us set

$$\frac{T - T_c}{T_c} = \frac{A}{3}, \qquad -\frac{\mu_B H}{\kappa T_c} = \frac{B}{3},$$

for short, so the equation becomes

$$\tanh \eta = (A+1)\eta + B.$$

Now there is a Taylor series

$$\tanh \eta = \eta - \tfrac{1}{3}\eta^3 + \tfrac{2}{15}\eta^5 + \cdots$$
$$= \eta - \tfrac{1}{3}\eta^3 + t(\eta),$$

say, where $t = O(5)$. The equation of state becomes

$$\eta - \tfrac{1}{3}\eta^3 + t(\eta) = \eta + \frac{A\eta}{3} + \frac{B}{3},$$

or

$$0 = \tfrac{1}{3}\eta^3 - t(\eta) + \frac{A}{3}\eta + \frac{B}{3}.$$

When $A = B = 0$, the right-hand side is 3-determinate by Chapter 4 or Chapter 8. (In fact, it is strongly 3-determinate, by Theorem 8.1 or by inspection of the diffeomorphism constructed in Theorem 4.4.) By a routine use of Theorem 8.7 the unfolding

$$\tfrac{1}{3}\eta^3 - t(\eta) + \frac{A}{3}\eta + \frac{B}{3}$$

is strongly equivalent to its truncation above order 3, so that we have a change of variables

$$(\eta, A, B) \mapsto (x(\eta, A, B), a(A, B), b(A, B))$$

with derivative the identity at $(0,0)$, such that

$$\tfrac{1}{3}\eta^3 - t(\eta) + (A/3)\eta + (B/3) = \tfrac{1}{3}(x(\eta, A, B))^3$$
$$+ \tfrac{1}{3}a(A, B)x(\eta, A, B) + \tfrac{1}{3}b(A, B).$$

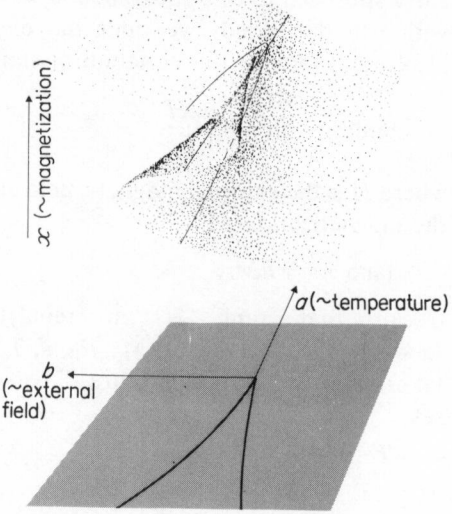

Fig. 14.6

Notice that since we are interested in the *zeros* rather than the *minima* of this function for various A, B, we cannot drop the 'variable constant' discussed in Chapter 8 Section 6 and must include both A and B for transversality. (The algebraic test in Theorem 8.6 is then modified by using E_n^k in place of J_n^k. Once we notice values, x^3 is special in two ways – by having a point of inflexion, and taking value 0 there – and has codimension 2.) This gives the equation of state the form

$$\tfrac{1}{3}(x^3 + ax + b) = 0$$

or

$$x^3 + ax + b = 0,$$

once again. The cusp surface (Fig. 14.6) thus describes the Weiss equation of state exactly, up to a (not fully specified) diffeomorphism with derivative the identity at the Curie point. Any feature of the model invariant under such diffeomorphism (such as the *critical exponents* we consider in Section 8) may be read off with particular ease in this form.

Similar remarks apply to similar theories in which the function $\tanh \eta - \eta$ is replaced by others, such as the Brillouin functions

$$B_J(\eta) = \frac{J+1}{3J}\,\eta - \frac{((J+1)^2 + J^2)(J+1)}{90J^3}\,\eta^3 + \cdots$$

or the Langevin function

$$L(\eta) = \coth \eta - 1/\eta = \eta/3 - \eta^3/45 + \cdots$$

since in each case there is a non-vanishing cubic term.

THERMODYNAMIC POTENTIALS

3 Entropy

For a roomful of air, there are many more states (or a much larger volume in the space of all possible states) for which the air is fairly evenly distributed, than for which it is all in one corner. This is already true for 16 counters on a chess board; it becomes more emphatically true as the numbers involved increase. If individual states are equiprobable, it is thus much more likely that the air will be evenly mixed than that it will be all in one corner. The 'macro-state' of being evenly spread – without regard to *which* evenly spread configuration, or 'microstate', it happens to be in – is much more probable than that of being concentrated in a corner. If we associate to each macrostate μ the number n_μ of microstates that 'belong' to it, we have a measure of the probability of μ. (Really we should speak of a probability *density* on the space of macrostates, with its peak at 'evenly mixed', since the likelihood of being *exactly* evenly mixed is zero.) It is convenient to take log n_μ as the measure of this that we actually use, because if we have two rooms, one with n_1 evenly mixed states for its air, the other with n_2, then the total number of evenly mixed states is $n_1 n_2$. By taking logarithms, we substitute addition for multiplication.

With appropriate constants, this quantity log n_μ is the *entropy S* of the air. Often the probabilities (proportional to e^S) of different macrostates are so sharply peaked at the *most* probable (the ones with maximum entropy) that the corresponding values of micros-copic variables like pressure, density, etc. may be taken to specify exactly the macroscopic state of the system, very soon after any disturbance. (For a full account of the statistical basis of entropy sketched above, including the quantum statistical version, see Landau and Lifshitz [130].)

A purely macroscopic approach to the definition of entropy could be given, systematizing as did Clausius in 1850 the 1824 heat-engine arguments of Carnot; but the concept is much more mysteri-ous that way. (In particular the mathematical treatment is usually perverse; the whole theory cries out to be put in terms of differential forms, a reformulation begun by Jauch [131] shortly before his death.)

Thermodynamics, then, works with the principle that a smooth† function S, entropy, on the space X of macroscopic variables (energy, pressure, magnetization,...) is maximized by the matter we are looking at, subject to whatever constraints apply. (Energy may be fixed by the system being isolated, temperature by connecting it to a large mass whose own temperature cannot be appreciably influenced, and so on.) This principle is then endlessly transformed.

† The assumption that S is *analytic* is often stated, rarely used – except to motivate considering Taylor series, and the polynomials obtained by truncating them.

4 Transforming the Maximum Entropy Principle

It is convenient to illustrate the transformations involved in an abstract example. Consider the two functions on $\mathbb{R}^2$ (using (x, z) as coordinates)

$$p(x, z) = z$$

$$q(x, z) = x^4 + zx^2 + z.$$

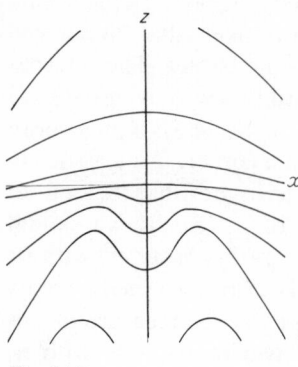

Fig. 14.7

Fig. 14.7 shows the contours of q; those of p are just the z-levels. Neither p nor q has critical points anywhere in the plane, since $\partial p/\partial z = 1$ identically, while $\partial q/\partial z = 1 + x^2$, and neither derivative vanishes. But p restricted to a curve $q(x, z) = $ constant, and q restricted to a curve $p(x, z) = $ constant, do have critical points. Moreover, it is easy to see that these points coincide. (Proof: if $p|_{q^{-1}(c)}$ is degenerate at (x_0, z_0), then $Dp|_{(x_0, z_0)}$ must vanish on all vectors tangent to $q^{-1}(c)$ at (x_0, z_0). But these vectors are precisely those on which $Dq|_{(x_0, z_0)}$ vanishes, since $Dq|_{(x_0, z_0)} \neq 0$. By Chapter 2 Section 3, two linear maps $\mathbb{R}^2 \to \mathbb{R}$ with the same kernel are scalar multiples of each other, so $Dp|_{(x_0, z_0)} = \lambda Dq|_{(x_0, z_0)}$ for some 'Lagrange multiplier' $\lambda \in \mathbb{R}$. Since this condition is symmetric between p and q for $\lambda \neq 0$, it follows that $p|_{q^{-1}(c)}$ is critical at (x_0, z_0) if and only if $q|_{p^{-1}(d)}$ is, where $c = q(x_0, z_0)$, $d = p(x_0, z_0)$.) Moreover it is easy to see in this example that q has a local *minimum* on a curve $p = $ constant exactly where p has a local *maximum* on the corresponding $q = $ constant curve. For a nice treatment of this theory see Spivak [132] Vol. 4, p. 428, where a generalization of this shows that 'maximizing area contained in a given perimeter' (Fig. 14.8(a)) is the same as 'minimizing perimeter containing a given area' (Fig. 14.8(c)): namely, the disc (Fig. 14.8(b)). The infinite-dimensional character of the space of curves involved is not alien to thermodynamics, which becomes similarly infinite-dimensional as soon as any attention is paid to local variations of state (quite apart from involving the microphysics).

Thus, problems of maximization and minimization subject to constraints have many equivalent presentations in terms of extremizing different quantities subject to different constraints. (Sometimes it may be mathematically convenient to solve an extremization problem for given physical constraints by extremizing another quantity subject to constraints which cannot be physically realized.) Hence the entropy maximization principle at the heart of thermodynamics may often usefully be treated by a minimization of

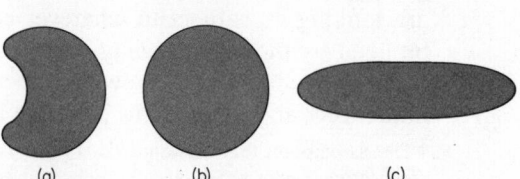

Fig. 14.8   (a)   (b)   (c)

some other function (even perhaps one that, like the energy of an isolated system, cannot *physically* change) over states with the same entropy. If in Fig. 14.7 x is used as a coordinate on curves $p = c$, then we get q as a function of x (varying with c); with x as coordinate on a $q = d$ curve we get p as a function of x (varying with d). The analogues of these functions to be extremized over x that arise in thermodynamics are the entropy (maximized), and a great variety of functions to be minimized (according to constraints) such as the *Gibbs function*, the *Helmholtz potential*, and the *free enthalpy*. We refer to these collectively, as is customary, as *thermodynamic potentials*, and denote a 'general' one by Φ. (Incidentally, the underlying probabilities discussed in Section 3 go in these terms as $e^{-\Phi}$, as opposed to e^{S}, for the probability maxima are associated with *minima* of Φ, *maxima* of S.)

5 Legendre Transformations

Unfortunately, even in the more mathematical thermodynamics texts such as Callen [127], there is confusion between the condition for the above transformations of the extremization problems to work nicely (in the example, that Dp and Dq should not vanish altogether, as distinct from on special curves; the analogue in thermodynamics is satisfied pretty universally), and the much stronger condition for something else. Physicists generally want everything they are looking at to be a function of everything else, a desire they have retained from the mathematics of the nineteenth century. (Mathematicians have moved on from this – the first step was replacing 'multivalued functions' by Riemann surfaces – but it takes physicists almost as long to learn new mathematical viewpoints as it does for the mathematical community to be conscious of new physics. There are honourable exceptions, even now, but when will we see another Euler or Poincaré?) They are of course ahead in this of the numerical sociologists, many of whose computer routines presuppose that everying is a *linear* function of everything else.

Now, in Fig. 14.9(a) we have marked the points for which the equivalent extrema, subject to constraint, occur: Fig. 14.9(b) shows similarly the points in $\mathbb{R}^3$ where

$$P : \mathbb{R}^3 \to \mathbb{R}$$

$$(x, y, z) \mapsto x^4 + zx^2 + yz + z$$

is critical if restricted to curves on which

$$Q : \mathbb{R}^3 \to \mathbb{R}^2$$

$$(x, y, z) \mapsto (y, z)$$

takes a constant value (c_1, c_2), or equivalently where the derivative of Q restricted to a $P = $ constant surface is singular. In the region

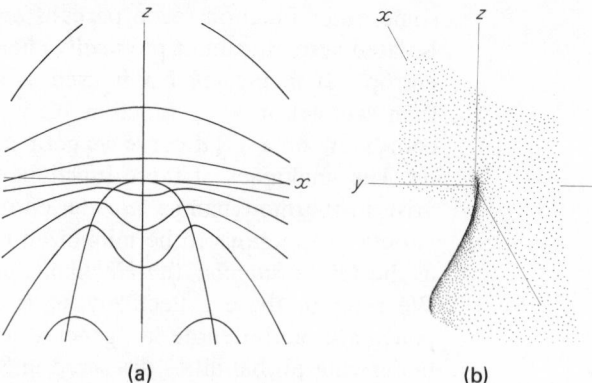

Fig. 14.9

where z is positive, there is a unique value of x which maximizes p or P subject to a fixed value of q or Q; likewise for minimizing q or Q subject to fixed p or P. This encourages the viewpoint that 'x is a function of q or Q' (or p or P), indeed in that region a non-singular one, so that x may be replaced as a coordinate by p, P, q, or part of the range of Q respectively. (This is what the theory of 'Legendre transformations' is all about.) This is visibly not a workable approach around the origin, which corresponds to the point of greatest interest in the thermodynamic context. One of the not altogether minor services that the language of differential topology (including catastrophe theory) can render thermodynamics is to provide a formalism which does not explode at the points of greatest interest. (We have not the space here to develop this formalism to the point of routine calculation, but *see* Dubois and Dufour [132a] and Ekeland [132b].) Of course a physics education involves great experience in calculating successfully despite an exploding formalism, and many physicists will regard the above suggestion as an impertinent piece of needless mystification. The man who has always performed standing up in a hammock may actually find it hard to adjust to a bed.

6 Explicit Potentials

So much for generalities; we come now to particular thermodynamic potentials, functions of the states of a sample of matter, varying with the constants fixed for certain variables. For example, in the case of van der Waals' equation, we are interested in finding the density given the pressure and temperature. The appropriate thermodynamic potential in this case is the Gibbs free energy; conveniently this has been calculated for us in the notation used above by Bell and Lavis [133], who describe the standard thermodynamics involved in some detail. Up to a shear term (which, as they observe, has no effect on the phase transition) it is

$$\Phi(x, p, t) = \frac{16(t+1)}{9} \log\left(\frac{2+2x}{2-x}\right) - \frac{2x(1+p)}{3(1+x)} - 2x.$$

Notice that this is quite distinct from the simplest, algebraic potential

$$\tilde{\Phi}(x, p, t) = \tfrac{1}{4}x^4 + \frac{8t+p}{6}x^2 + \frac{8t-2p}{3}x$$

whose differentiation gives the equation of state

$$x^3 + \frac{8t+p}{3}x + \frac{8t-p}{3} = 0$$

of Section 1, although Φ is constructed exactly to yield this equation. The Maxwell equal area rule does *not* correspond to minimizing $\tilde{\Phi}$, and *does* exactly correspond to choosing the absolute minimum of Φ, as Bell and Lavis show. To celebrate this correspondence, Thom [1] christened 'choose the absolute minimum' the *Maxwell convention*, in contrast to perfect delay.

They conclude that the behaviour 'of a van der Waals fluid (and, incidentally, of a Berthelot or Dieterici fluid) does not correspond to a Riemann–Hugoniot† catastrophe'. This is true, however, only if 'correspond' is taken to mean 'correspond under the simplest algebraic transformation that carries the equation of state to the cusp surface.' It is *false* for a general smooth change of variables. If we expand Φ in a Taylor series to order 5, we obtain

$$x^5\left(\frac{11t-20p}{30} - \frac{3}{10}\right) + x^4\left(\frac{8p-5t}{12} + \frac{1}{4}\right) + x^3\left(\frac{2t-2p}{3}\right)$$

$$+ x^2\left(\frac{2p-2t}{3}\right) + x\left(\frac{8t-2p}{3}\right).$$

With (p, t) as unfolding parameters Φ is thus an unfolding of

$$\Phi_{00} = \frac{x^4}{4} - \frac{3}{10}x^5 + O(6).$$

Now $x^4/4$ is strongly 4-determinate, by the elementary proof of Chapter 4 Section 4. It is easily checked that Φ is a *universal* unfolding of Φ_{00}, and hence that by Theorem 8.6 Φ is equivalent around 0 to the standard unfolding

$$\frac{x^4}{4} + \frac{a}{2}x^2 + bx.$$

But we can say rather more. Define a change of variable by

$$x = y + \tfrac{3}{10}y^2 + \tfrac{2}{15}(t-p). \tag{14.1}$$

† Thom's original name for it. The name 'cusp' is due to Zeeman.

This is a diffeomorphism for $y > -5/3$, with inverse

$$y = x - \tfrac{3}{10}x^2 - \tfrac{2}{15}(t - p) + O(3).$$

Substituting (14.1) in the expansion of Φ, we get

$$-\frac{13}{750}(t - p)y^5 + \left(1 + \frac{62p - 61t}{75}\right)\frac{y^4}{4} + \left(\frac{2t + 7p}{15}\right)y^2 + \left(\frac{8t - 2p}{3}\right)y$$

to fifth order in y, first order in t and p. The 5-jet of the potential Φ_{00} for $t = p = 0$ has the form $y^4/4$. We now use Theorem 8.7: we are in case (a) with $k = 4$, as is quickly checked. Thus the given unfolding is *strongly* equivalent to

$$J^5(\Phi_{00}) + J^2\left(\frac{\partial}{\partial p}\Phi_{p0}\right)p + J^2\left(\frac{\partial}{\partial t}\Phi_{0t}\right)t$$

which in the y-coordinates above is exactly

$$\frac{y^4}{4} + \left(\frac{2t + 7p}{15}\right)y^2 + \left(\frac{8t - 2p}{3}\right)y.$$

This is the standard cusp potential family again; if we put

$$\alpha = \frac{4t + 14p}{15}, \qquad \beta = \frac{8t - 2p}{3}$$

we have the expression

$$\frac{y^4}{4} + \frac{\alpha}{2}y^2 + \beta y.$$

An analysis of the effects of the difference between α and the previous term

$$\tfrac{1}{3}(8t + p)$$

can shed light on what is *not* preserved under strong equivalence (and is recommended as an exercise to readers with an interest in particular properties that might or might not be preserved). One thing that *is* preserved is the direction of the cusp at its point, and hence the direction in which the Maxwell set (or 'curve of first-order phase transition points') leaves $(0, 0, 0)$: by definition, strong equivalence preserves such directions. (Not, however, such things as the *curvature* of the Maxwell set, which involves higher derivatives in p and t; this would require even stronger equivalence, theorems concerning which can be proved by the same methods.) In this case we know this direction already, since it is squeezed between fold curves of the projection of the algebraically simple catastrophe manifold given by the van der Waals equation of state. In other cases we may know it by symmetry. But when a thermodynamic potential is derived from physical reasoning deeper than fitting a heuristic equation of state, its surface of critical points may not be practical to describe in closed form, and such directional information must be found by expansion. Particularly for higher singularities, where there can be several important directions, only the methods we have described provide a systematic route to their certain identification.

To conclude, not merely all the diffeomorphism-invariant information to be obtained from the van der Waals equation of state, but also all that follows from full thermodynamic information about a 'van der Waals fluid', can be found by examining the cusp potential

$$\tfrac{1}{4}x^4 + \frac{a}{2}x^2 + bx.$$

Much information of physical interest is, in fact, diffeomorphism-invariant in this context: even more is invariant under strong equivalence. Similar remarks apply to Berthelot and Dieterici fluids, and to the various models of the ferromagnetic/paramagnetic phase transition discussed in Section 2.

Incidentally, the unfolding and stability problems for a potential Φ and for equations of state ($D\Phi = 0$) are equivalent when we have only one essential variable. But if we have two, then $D\Phi$ becomes essentially an $\mathbb{R}^2$-valued map *and among such has higher codimension than does Φ among $\mathbb{R}$-valued ones*, because it has neighbours that are not gradients at all (*see* the caveat at the end of Chapter 4 Section 9). Thus the stability question may be different for corank ≥ 2, according as in a given problem our equations must, or may not, be equivalent to criticality of some real-valued function. This is quite apart from whether we can explicitly construct such a function. Its theoretical existence, or otherwise, will affect quantitative physical stability questions.

7 The Landau Theory

These examples illustrate the reasons for the enormous similarity of many different models of many different phase transitions. These reasons are the 'typicality' results of Chapter 7 and the determinacy and unfolding results of Chapter 8. They account for the large measure of success enjoyed by the Landau theory, which assumes that the relevant minimized thermodynamic potential Φ is a sufficiently differentiable function of, for example, pressure P, temperature T, and some other quantity η (density or magnetism in the examples above) whose changes describe the changes in phase of the matter. Usually, unless symmetry reasoning compels the contrary, η is taken as 1-dimensional. It is often called an *order parameter* because of examples like magnetization where it measures parallel alignment at different points. The potential is then expanded in the form

$$\Phi(P, T, \eta) = \Phi_0(P, T, \eta) + \alpha(P, T)\eta + A(P, T)\eta^2$$
$$+ B(P, T)\eta^3 + C(P, T)\eta^4 + \cdots.$$

Now, the interesting points are equilibria for things minimizing η, so by choosing the η-origin at one of these points we can suppose without loss of generality that α vanishes there. If A is positive throughout the range we are looking at, the stability of Morse functions (Chapter 4 Section 5) implies that nothing very interesting happens. (Physics texts usually appeal to the weaker result, of the sufficiency of second-order conditions for guaranteeing a minimum or maximum, which suffices for the present discussion.) If A is negative throughout, the matter will not be at the equilibrium being studied, since this is unstable. Thus we are concerned with events at a point where A changes sign: hence where it vanishes. If at this point B is not zero, then Φ has a point of inflexion – again an unstable equilibrium (though less so than a maximum) and the matter will fall down-Φ to some quite different value of η. So for a *continuous* change in state (which is the defining property of a *second-order phase transition*, as distinct from *first-order* ones, like boiling, in which density jumps), we must have $A(P, T) = B(P, T) = 0$. For a minimum at the point of interest, we require $C(P, T) > 0$ at this point (and hence by continuity in a neighbourhood of it).

The above has been a paraphrase of pp. 434–435 of Landau and Lifshitz [130]. It is convenient now to quote directly from p. 436:

'If B does not vanish identically [*for symmetry reasons*] then the transition points are determined by the two equations: $A(P, T) = 0$, $B(P, T) = 0$. Hence in this case the points of continuous phase transition are isolated ones'.

Note the implicit transversality argument in the last sentence. Smooth functions A and B exist whose common zeros are *any closed set whatsoever* in the (P, T)-plane $\mathbb{R}^2$, and this permits some very strange objects (*see* Newman [134]). But *typically* the mapping

$$(A, B) : \mathbb{R}^2 \to \mathbb{R}^2$$

$$(P, T) \mapsto (A(P, T), B(P, T))$$

is transverse to $\{(0, 0)\}$ and hence the zeros are isolated. (And indeed, since the transversality condition here is just that for the Inverse Function Theorem, by the same token A and B may be used, locally, as smooth coordinates in the (P, T)-plane.) This illustrates our general point that transversality arguments are not an unphysical invention of the mathematicians: the sense in which Thom's theorem says 'almost all' is exactly that in which much physical writing says 'all', for the same reasons. Not only is honesty the best policy on ethical grounds, but the systematization of these reasons into transversality theory will be as fruitful in the long term for the sciences as has been the systematization of symmetry reasoning into group theory.

Similar transversality arguments (in the absence of symmetry) show that (α, A) may typically be used as local coordinates in the (P, T)-plane, giving Φ the expression

$$\Phi(\alpha, A, \eta) = \Phi_0(\alpha, A) + \alpha\eta + A\eta^2 + B(\alpha, A)\eta^3 + C(\alpha, A)\eta^4$$
$$+ \cdots.$$

If Φ is smooth, the theorems of Chapter 8 guarantee a smooth change of coordinates producing (up to the shear function Φ_0) the form

$$\tfrac{1}{4}\eta^4 + \frac{a}{2}\eta^2 + b\eta.$$

More 'controls' than just P and T, of course, or symmetries, can lead to higher catastrophes. By taking account of the Splitting Lemma and using further transversality arguments (as in Section 2 of Chapter 7) we may remove the initial hypothesis – explicit or implicit, according to the source – that η is 1-dimensional, although this statement again must be qualified in the presence of symmetry. A row of terms

$$\eta_2^2 + \cdots + \eta_N^2$$

independent of A and B does not influence the physics.

Hence we may explain the capacity of the Landau theory, which involves taking a general Taylor expansion and truncating as seems appropriate (and in one variable intuition is an excellent guide to determinacy!) to yield all the diffeomorphism-invariant information given by models backed by far more physical detail. Hence also the universality of the results.

We have required the absence of symmetry in the above arguments: we shall examine the symmetrical case in Sections 15–20.

FLUCTUATIONS AND CRITICAL EXPONENTS

8 Classical Exponents

The Landau theory finds an archetypal form for the thermodynamic potential by what amounts to a version of catastrophe theory *as such*; that is, the mathematics presented in Chapters 7 and 8. (Without the precise formal machinery whose results are described in these chapters, however, there is room for mathematical error in higher codimensions – with possible physical consequences, see Section 18.) Its other main component, as a physical theory, is one often used also with potentials derived from detailed models: the approximation of the *average* state by the *most probable* one, or equivalently the one that minimizes the appropriate potential. (A refinement treats secondary probability peaks, local potential minima, as *metastable states*, but this does not bear on our present concerns.) We have discussed this approximation in the context of scattering and wave theory (Chapter 12 Section 3); it is usually a good one where the probability has a sharp nondegenerate peak, or equivalently in this context, where the potential has a sharp Morse minimum. Such sharp peaking is common in thermodynamics, as noted in Landau and Lifshitz [130] p. 5:

'If one isolates a small volume in a gas containing, say, only 1/100 of a gram-molecule, then the average relative deviation of the energy of this quantity of matter from its mean value is only $\sim 10^{-11}$. The probability of finding (in a single measurement) a relative deviation, say, of the order of 10^{-6} is given by the fantastically small number $\sim 10^{-3 \times 10^{15}}$.'

Thus this approximation rule may often be used safely.

Consider the archetypal 'cusp catastrophe' thermodynamic potential for a second-order phase transition governed only by two varying constraints (such as P and T) arrived at in the previous section,

$$\tfrac{1}{4}\eta^4 + \frac{a}{2}\eta^2 + b\eta.$$

For this the rule gives both an equation of state

$$\eta^3 + a\eta + b = 0,$$

and a decision between competing 'layers' of the surface so defined. Via the appropriate transformation the rule 'choose the deeper minimum' corresponds exactly to the Maxwell equal area rule, wherever the latter is relevant.

What are the diffeomorphism-invariant predictions of this theory? One is that above the Maxwell set or *coexistence curve*, where the two minima of $\tfrac{1}{4}\eta^4 + \tfrac{1}{2}a\eta^2$ have equal values, we have something like a parabola (Fig. 14.10). More exactly, if we extend the coexistence curve to a smooth curve through the second-order phase transition point (and it is a diffeomorphism-invariant fact that we can) and parametrize this by a, then in the archetypal form the set of jump points is expressed as

$$b = 0, \qquad a + \eta^2 = 0$$

(since $\eta^3 - a\eta = 0$ and $\eta = 0$ is suitable for $a < 0$). Hence by 2-determinacy the corresponding curve, for a *general* potential that satisfies the transversality hypotheses of the Landau theory, must satisfy

$$a - c(\eta^2 + O(3)) = 0, \qquad c > 0.$$

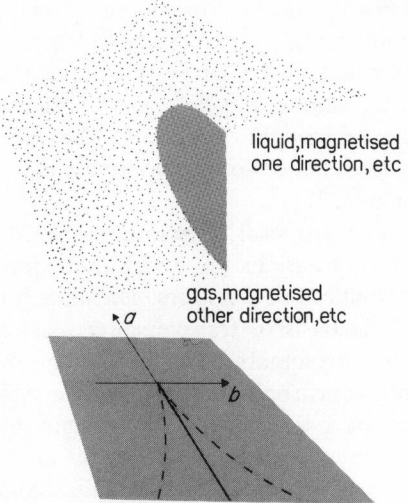

liquid,magnetised
one direction,etc

gas,magnetised
other direction,etc

a

b

Fig. 14.10

(Strictly, we are appealing to the fact that 'diffeomorphism preserves parabolic contact', and the reader who desires greater precision will readily supply it.) In the notation more usual in physics, this says that for all these potentials, and with this approximate hypothesis on the behaviour of the matter governed by them,

$$\eta \sim (-a)^{1/2}.$$

In practice the physically measurable quantity is the height of the jump, but this is clearly related to a around the second order phase transition point by the same *critical exponent*, $\frac{1}{2}$. There are other significant critical exponents associated with the ways physical quantities are interrelated, but this one (universally denoted by β) suffices for our present discussion. We see that the Landau theory gives its value, universally for the wide class of two-variable-constraint phase transitions without special symmetries (indeed for many with them) as $\beta = \frac{1}{2}$.

Inconveniently, in the vast majority of cases, accurate experiment gives values close to $\frac{1}{3}$. The value $\frac{1}{2}$ is not in the range of reasonable experimental error.

9 Topological Tinkering

Many attempts have been made to tinker with the theory and adjust the result, but as we have seen the value $\frac{1}{2}$ is very closely tied to it. Those readers equipped to look at the physical literature will know where to locate examples of such attempts; we mention here only two papers that refer to catastrophe theory. No amount of adjustment of 'conventions', as in Fowler [128], can change β; for any arbitrary choice of coexistence curve gives a curve of 'jump points' whose projection to the (a, η)-plane must lie between that of the fold line (the 'spinodal curve' $a = -3\eta^2$ in the algebraic archetype) and that of the curve forming the edge of the part of the surface lying above the interior of the cusp the curve $4a = -3\eta^2$ in the archetype). Both of these are in parabolic contact with the η-axis, hence so is any curve squeezed between them, analytic at 0 or not.

The proposal of Schulman [135] to allow arbitrary non-smooth transformations (by which the exponents can be altered *ad libitum*) introduces non-smooth *potentials* into the theory. This destroys the whole basis of transversality reasoning, whether implicit as in the usual presentations of the Landau theory, or explicit and formalized into catastrophe theory. Among sufficiently differentiable functions, we have explained why certain phenomena are typical. Among *continuous* functions, measures of some importance say that 'almost all' are *nowhere* differentiable. 'Almost all' functions thus have no derivatives and no Taylor expansions, and the whole procedure cannot start. (A theory using typical smooth potentials in a space of 'truer' physical variables, related to the obvious ones by a non-smooth but *specified* transformation – preferably *derived* from something – would escape this objection and predict specific exponents. But to our knowledge no one has one.) Central to any physical theory is the rule on which transformations are permitted. To bring in more (unless with great caution as a heuristic method, as in Chapter 11), as Schulman [135] suggests, is as obscuring as the failure in Bell and Lavis [133] to consider enough.

10 The Rôle of Fluctuations

The problem, and its resolution, do not in fact lie in the topology. The problem is fundamentally at the statistical level, in the identification of 'most probable' with 'average' state. For Morse extrema the results of this approximation are often splendid, as we saw for optics. In the near vicinity of degenerate singularities, this need not be true. Exact solution of a gradually increasing set of detailed statistical models of organized matter (many of them variations on a theme known as the 'Ising model') since the 1940's, whose critical exponents match experiment rather than the Landau theory, has clarified this point. Without entering into the details of the statistics of microstates, let us consider from a thermodynamic viewpoint what is happening.

As we discussed in Chapter 13, Section 16, a disturbance to a system that tends to minimize an energy function causes it to vibrate, with damping bringing it gently to rest. Near a Morse minimum, the oscillations are well described as being simple harmonic, with amplitude proportional to the square root of the disturbing energy and inversely proportional (as is the period) to the coefficient of the Hessian of Φ in the direction concerned. As the Hessian changes towards degeneracy in this direction, the same disturbing force leads to larger, slower vibrations, which are consequently less damped. Analogous effects occur for electromagnetic behaviour in some crystals approaching phase transition: the vibrations that slow down are there called 'soft modes' and have a large and growing literature. (Associated effects include the speed of light in the crystal going to zero.)

We cannot sensibly treat, say, a fluid near the gas/liquid second-order phase transition point as being subject to damped simple harmonic motion. But the general picture is similar: at a temperature where a best-fit equation of state says the distinction between liquid and gas should not exist, it wanders off from the thermodynamic potential minimum, or entropy/probability maximum, in non-negligible regions for non-negligible periods of time. Small droplets of greater density form (at the expense of reduced density nearby – the fluid moves to both sides of the surface of Fig. 14.4). They vanish again, being not even in equilibrium, much less stable; but they form often and densely enough, and survive long enough, to affect the macroscopic properties of the fluid. Plate 16 shows the phenomenon of *critical opalescence*: light scattered by a fine mist of droplets, inhomogeneities in the fluid, that by strict thermodynamic extremization would not exist.

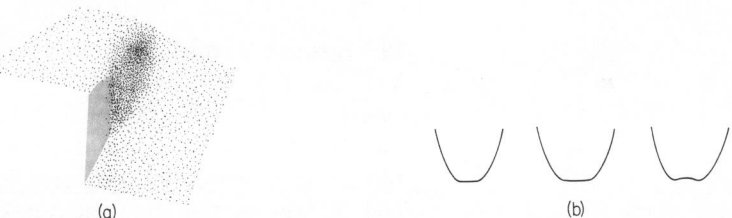

Fig. 14.11. All of the potentials near x^4, of the form $x^4 + ax^2 + bx$ for a,b really small, look like the curves shown in (b) and have an 'almost floor' over which a statistically minimizing system spreads out.

(a)

(b)

Crisp surfaces like Fig. 14.10 should thus be replaced by pictures like Fig. 14.11 with probabilistic cloudiness around the second-order point. In this region the *fluctuations* of the state of the matter must be taken very carefully into account. (The standard Landau theory treats the fluctuations; indeed ν, one of the troublesome 'classical critical exponents' concerns the variation of their size as the critical point is approached; but only *after* making the identification of 'most probable' and 'average'. For an account of the patchwork of approximations 'de façon franchement schizophrénique' needed for the classical fluctuation calculations, *see* Toulouse and Pfeuty [136] p. 48.)

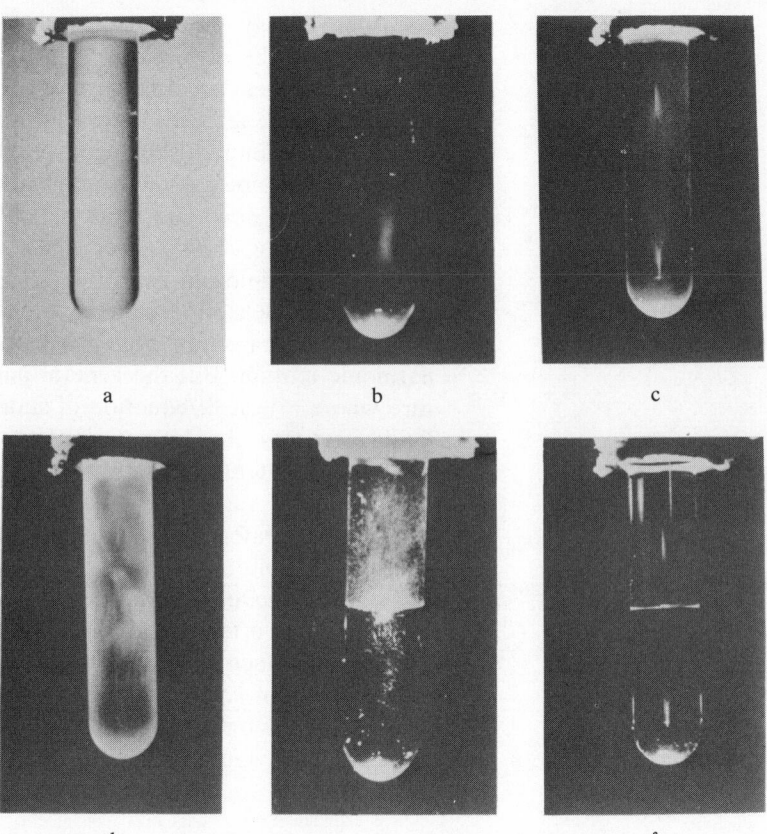

Plate 16. *Critical opalescence: evidence for fluctuations near the critical point (from Stanley [135a]; after Ferrell [225]). (a) T much greater than T_c (b) T a little greater than T_c, (c) T approximately equal to T_c, (d) T a little less than T_c, (e) T less than T_c, (f) T much less than T_c. The liquid is a cyclohexaneaniline mixture.*

11 Spatial Variation

One naturally, asks, seeing Fig. 14.11 and the fading out of the curve of classical 'jump points', how the exponent can be defined at all – let alone carefully measured and found nearer to $\frac{1}{3}$ than to $\frac{1}{2}$. The answer is that one measures in this region not jumps but *correlations*. In the single phase region a higher density at one point tells you nothing, even probabilistically, about the density (or whatever η describes) at another any distance away. Where classically two phases exist, greater instantaneous values of η at separated points are correlated, even if the average over time is near neither surface, because the matter is fluctuating between the two of them. The measurable fact is that its different parts are fluctuating to some extent in step. Exactly in this way in Chapter 11, Section 10, we discussed the measurement of the correlation of the spatial arrangements of dissolved polymer molecules by birefringence effects; the positions themselves are complicated, various, changing over time, and much harder to measure. Analogous measurements are involved in critical exponent determination.

Thus, since spatial variation is intimately involved in what the critical exponents physically refer to, models to predict them cannot treat a single homogeneous density, pressure, magnetization, etc. But exact statistical models are hard to construct (especially for fluids; with crystals the variables may to a good approximation be defined only on a lattice of points) and even harder to solve. The range of exactly solved models is far removed from the tremendous assortment of phase transitions under active study in physics and technology. One needs a 'semi-macroscopic' version of thermodynamics, that does not go right down to the statistical ensemble of microstates, but can take account of spatial variation. The method used, both in the Landau theory and in more refined treatments, is to make the previous single numbers or vectors (such as density or magnetization) into quantities smoothly varying over space. (Much of what we said in Chapter 11 Section 1 on modelling and scale apply here also. Space forbids its repetition here, but the reader should consider it repeated, with emphasis.) This provides us with an infinite-dimensional space of functions, and confronts us with the rigours of functional analysis. For precision we should be more careful in defining the function space and its topology, and probably use the remarkable kinds of space discussed in Section 7 of the previous chapter, but a rigorous treatment of this aspect of the problem has not to our knowledge been given. So we shall follow the path trod many years ago by Dirac, and assume that an infinite-dimensional space is just like a finite-dimensional one, only different.

12 Partition Functions

Corresponding to our general parameter of Sections 3 to 7, living in a finite-dimensional space $E = \mathbb{R}^n$, say, we have now a space Ω of functions $M = (M_1, \ldots, M_n)$ from the set X of spatial positions in our sample to E. Our more general 'thermodynamic potential' is a function $\Phi : \Omega \to \mathbb{R}$, and minimizing Φ corresponds to maximizing a probability density on Ω, given up to a normalizing constant by a *partition function*

$$M \mapsto e^{-\Phi(M)}$$

by analogy with Section 4. But now we are not interested in *most probable M*, but in *average M*:

$$\bar{M} = \int_\Omega M e^{-\Phi(M)},$$

and in similar integrals giving the average correlation and so forth. Note that $\bar{M}$ is a point in Ω, that is, a function $X \to \mathbb{R}$, and could be given more explicitly as

$$\bar{M}(x) = \int_\Omega M(x) e^{-\Phi(M)}.$$

Where Φ is derived as a macroscopic consequence of a *quantum* statistical micro-model, we may consider this integral as describing in some sense a *superposition* of semiclassical states, rather like equation (12.4), though since we are discussing fluctuations rather than oscillations the essentially quantum capacity for interference between states of matter has already been filtered out. (It is interesting to speculate on the possibility of a unified asymptotic analysis, treating phase transitions as caustics in the 'matter-wave' everything is made of.) So what form should Φ take? On this, and the way it varies with temperature, pressure, external magnetic field, etc., depend all the quantities of physical interest.

A great simplification is afforded by the assumption that $\bar{M}(x)$ and the related quantities can be found, not by integrating over the whole of X, but just in some neighbourhood of x; in other words, that the interactions between different points are *local*, or *short range*. For a small neighbourhood U of x in X, then, we suppose that

$$\bar{M}(x) = \int_{\Omega} M(x) e^{-\Phi_U(M|_U)}.$$

(We must use a new functional Φ_U, since the original Φ is defined only on functions with domain all of X.) Now, for all M except a set of infinite codimension (Chapter 8 Section 7), M is k-determined at x for some k. Indeed, a careful analysis of the measure-theoretic aspects of the 'almost all' statements of singularity theory should give an upper bound, K say, to the determinacy at x of the functions that contribute to the integral.† Thus, up to a diffeomorphism of U, if U is sufficiently small then Φ_U depends only on the K-jet

$$j_x^K M = j^K M(x)$$

at x, and the space of these is finite-dimensional. We may take, then,

$$\bar{M}(x) = \int_{E_d^K} M(x) e^{-\phi(j_x^K M)},$$

where d is the dimension of (physical) space, in the notation of Chapter 8. Since 'almost all' functions M are Morse, we may probably take $K = 2$. Most physical calculations, in fact, use $K = 1$.

The next step is to Taylor expand ϕ, and use transversality and determinacy results on this as we did for global thermodynamic potentials in Section 7. (Somewhat more delicate results are needed than those of Chapter 8, since using determinacy arguments on ϕ requires local diffeomorphisms of E_d^K which must be induced by local diffeomorphisms of X. The latter are thus required to adjust M around x, and ϕ around $j_x^K M$, *simultaneously*. A proof akin to that

---

† Of course, if M is constant on X it is finitely determined nowhere, so this argument *requires* the fluctuations that have made the whole discussion necessary.

of the Thom isotopy theorem should work, but it does not seem that one may appeal to any already published result.) The majority of physical treatments, in fact, take

$$\phi(j_x^1 M) = F(M(x)) + K \sum_{i=1}^{n} \sum_{j=1}^{d} \left(\frac{\partial M_i}{\partial x_j}\right)^2$$

where F is a polynomial function of $M(x)$, depending in general upon parameters (explicit, like P, T, H of Sections 1 and 2, or abstract, like A, B, C of the Landau theory, Section 7). This is clearly determinate, transversal, and so on if F is, and exactly the physical (that is, informal transversality and implicit determinacy) reasoning of Section 7 can be and is applied to F.

13 Renormalization Group

Having found the right thing to integrate, what is the next step? As in Chapter 12, the trick is to study its asymptotic behaviour. There, the technique was to consider shorter and shorter wavelengths, so that oscillations become, in the limit, unobservable; here it is to take a larger and larger unit of length (so that *particular* distances are assigned smaller and smaller numbers), with a corresponding effect on the fluctuations. In neither case is the object so much to *find* the limit as to study the way it is approached; backing off from it a little gives an experimentally rewarded picture of what 'really happens'. The scale-changing technique used here is known as the *renormalization group* technique. 'Renormalization' comes from the context (quantum field theory) where it was first applied to resolve problems far more dramatic than exponents of $\frac{1}{3}$ rather than $\frac{1}{2}$; such as infinite predicted rest mass (like the infinite intensities of ray optics). 'Group' comes from the fact that the operations of scale change form in general a semigroup. (For the smooth models described above it *is* actually a group, but applied to a detailed statistical model the change in scale involves forgetting some fine detail, which cannot be recovered by a reverse change.) The language of quantum field theory is like that.

We cannot go into the technical subtleties of renormalization group theory, which are considerable. (For example, the dimension of space becomes a continuous variable ranging from -2 to ∞.) It is clearly desirable to prove that the asymptotic behaviour of these integrals is invariant under the kind of diffeomorphism discussed above, or at least that such predictions as critical exponents are. (If they are not, this is an interesting and somewhat surprising fact in itself.) The whole programme thus outlined is perhaps harder than the analogous theory presented in Chapter 12, which itself is the work of several years by deep mathematicians and physicists, in an area where the dialects of these two tribes are closer than in most. If the above discussion is so much as being more correct than otherwise, in its physics and in the mathematical formulation of the transversality approach, this is our luck more than our breadth of knowledge. But we feel sure that the basic point, which is that implicit catastrophe theoretic arguments are inextricably mixed with all theories of phase transition, will stand. (Even the analytical solutions of exact models require some stability under the interactions varying slightly from those supposed, before they become physically interesting.) They can only benefit from being made explicit and precise.

In analogy to the analysis of how close to a caustic a ray theory may safely be used, for phase transitions there is the *Ginzburg criterion* (*see* Toulouse and Pfeuty [136]) as to how close to a phase transition the Landau theory is applicable, and defining a *critical region* in which it is not. Somewhat surprisingly, this region vanishes for dimensions above 4, making the Landau theory precise – at least, as concerns critical exponents – for a cusp thermodynamic potential family. For a butterfly catastrophe, or *tricritical point*, the critical dimension is 3. Our physical space is thus at the threshold of applicability for this case. The critical dimensions for higher catastrophes (only those where the function being unfolded has a strict minimum are relevant) do not appear to have been computed.

14 Structural Stability of Renormalization

The analysis of the disappearance of the critical region, interestingly, involves a family of differential equations that is *not* structurally stable (in the appropriate Andronov–Pontryagin sense). We have not covered general dynamical systems, but we can follow Toulouse and Pfeuty [136] in discussing this one in descriptive gradient terms. The equations

$$\frac{\mathrm{d}r_0}{\mathrm{d}l} = 2r_0 + bu_0(1 - r_0)$$

$$\frac{\mathrm{d}u_0}{\mathrm{d}l} = (4 - d)u_0 - cu_0^2$$

govern the evolution of the physical parameters r_0 and u_0 as the scale is changed (d is dimension, b and c positive constants). These vector fields, for various d, are not the gradients of any functions, but a dynamical systems theorist would at once transform them into being so. Fig. 14.12 (from pp. 128–129 of Toulouse and Pfeuty [136]) shows them both as vector fields and as slopes of such a function; the 'Gaussian fixed point', when more stable, yields the appropriateness of the Landau theory critical exponents. The saddle

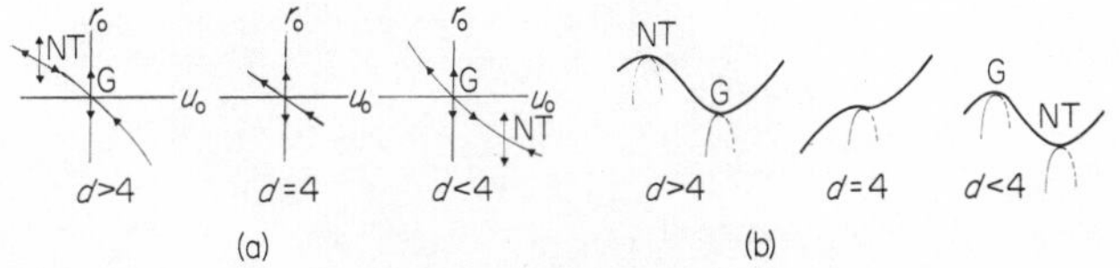

Fig. 14.12. (a) *Exchange of stability between Gaussian fixed
point (G) and non-trivial fixed
point (NT). For $d > 4$, the Gaussian fixed point is more stable
than the non-trivial fixed point.
For $d = 4$, the two fixed points
coincide. For $d < 4$, the stabilities
are interchanged. (b) Representation 'in relief' of the exchange
of stability of Fig. 14.12 (a).
After Toulouse and Pfeuty [136].*

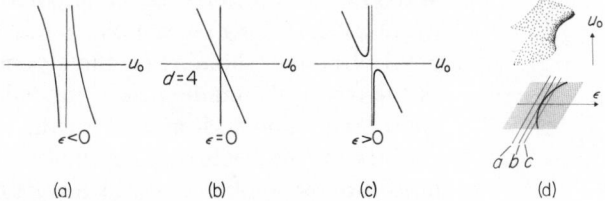

connections of Fig. 14.12(a) (unstable as we saw in Chapter 11 Section 5) are in fact spurious, not corresponding to the equations given. The instability we referred to is that of the one-parameter family, corresponding topologically (as Fig. 14.12(b) suggests) to the gradients of the family

$$x^3 + (4 - d)x^2 - y^2.$$

Aside from the constant-in-d Morse part in the y-direction, this is exactly the non-transverse family discussed in Chapter 8, Section 6. Addition of a small term εx to this family (or a constant ε to the second of the original equations) will remove the bifurcation entirely for $\varepsilon < 0$, and create two separate fold catastrophes for $\varepsilon > 0$. Fig. 14.13 shows the evolution of the fixed points with d for various ε, with the r-coordinate suppressed.

We have been unable to establish whether there is a physical reason for the permanent existence of a fixed point at the origin (*subject to which condition* the family becomes structurally stable). If the addition of a perturbing term εx, changing the picture to Fig. 14.13(a) or (c), cannot be physically excluded, the consequences would seem of physical interest. If not, there is still clearly a rôle for the theory of vector field bifurcations (catastrophe theory's big brother) in the study of mathematical phenomena that arise in this branch of physics, even aside from the aspects discussed above.

Fig. 14.13

THE RÔLE OF SYMMETRY

15 Even Functions

In the Weiss model, described in Section 2, consider the crystal's behaviour in the absence of an external field, that is, with $H = 0$.

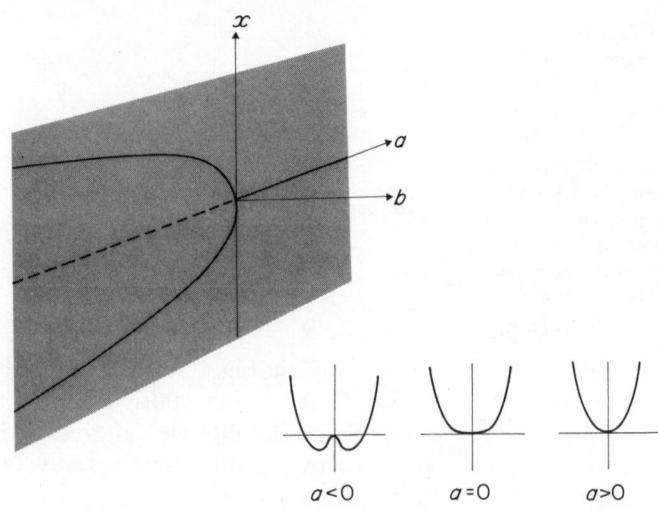

Fig. 14.14

(This is certainly atypical in the 'great outdoors', but often approximated. In the laboratory it can be achieved with great accuracy.) In the coordinates guaranteed by the theorems of Chapter 8, the equation of state takes the form

$$x^3 + ax = 0.$$

The thermodynamic potential can similarly be given the form

$$\tfrac{1}{4}x^4 + \frac{a}{2}x^2$$

(though not every transformation sufficient for the first reduction achieves the second, as we saw for the van der Waals equation). The corresponding $b = 0$ slice of the catastrophe picture of Fig. 14.6, and sample potentials, are shown in Fig. 14.14.

Now the unfolding $\tfrac{1}{4}x^4 + \tfrac{1}{2}ax^2$ is non-transverse, unstable, and atypical. But as an unfolding *among even functions of x*, those with $f(x) = f(-x)$, it is the unique stable single-parameter local family around $x = 0$, up to the usual diffeomorphisms (themselves symmetric under the map $x \mapsto -x$). Since this symmetry can be given strong physical support, holding (in the absence of an external field) as well as the rest of the mathematical modelling involved, this family is the appropriate 'universal model' for the local form of single-parameter families in this context. The whole sequence of discussion in Sections 3 to 14 applies again in this case.

16   The Shapes of Rotating Stars

An interesting example of such an 'even' bifurcation arises in the geometry of stars (Newtonian here, but presumably relativistic

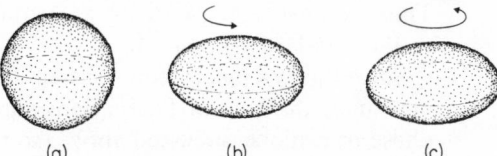

Fig. 14.15   (a)   (b)   (c)

analogues hold). A non-rotating fluid star, for obvious reasons, is spherical (Fig. 14.15(a)). Slow rotations make it an ellipsoid of revolution (exaggerated in Fig. 14.15(b)). Faster rotations produce a general ellipsoid (Fig. 14.15(c)). We may describe these shapes by the *polar* eccentricity e and the *equatorial* eccentricity η. Modulo the usual thermodynamic intricacies as to what may be considered a function of what, Bertin and Radicati [137] give the total energy of the star as

$$E(e, \eta) = \text{gravitational energy} + \text{rotational energy}$$

$$= W_0 \frac{(1-e^2)^{1/2}(1-\eta^2)^{1/6}}{e} \int_0^{\text{arcsine}} \left(1 - \frac{\eta^2}{e^2}\sin x\right)^{-1/2} dx$$

$$+ K_0 \frac{2(1-e^2)^{1/3}(1-\eta^2)^{1/3}}{2-\eta^2},$$

up to a 'shear term' independent of shape; where W_0 and K_0 are constant-absorbing quantities we borrow from their appendix. Treating e as external parameter (other choices, such as angular momentum, could be made) and expanding around 0 in η, we get

$$J^5 E(e, \eta) = \eta^2 \left[\frac{W_0(1-e^2)^{1/3}}{e} \left\{ -\frac{\text{arc sin } e}{6} + \frac{1-(1-e^2)^{1/2}}{2e^2} \right\} \right.$$

$$\left. + \frac{K_0}{6}(1-e^2)^{1/3} \right]$$

$$+ \eta^4 \left[W_0 \frac{(1-e^2)^{1/2}}{e} \left\{ -\frac{5 \text{ arc sin } e}{72} - \frac{1-(1-e^2)^{1/2}}{12e^2} \right. \right.$$

$$\left. \left. + \frac{3}{16e^4} (\text{arc sin } e - e(1-e^2)^{1/2}) \right\} - K_0 \frac{(1-e^2)^{1/3}}{36} \right]$$

$$= t_2(e)\eta^2 + t_4(e)\eta^4,$$

say. (We assume $e > 0$ here.) At first bifurcation, t_2 passes transversely through 0 as e crosses a critical value e_c with $t_4(e_c) > 0$, so by the 'symmetrized' version of Chapter 8, the family is strongly equivalent around $(e_c, 0)$ to

$$(t_4(e_c))\eta^4 + \left((e - e_c) \frac{dt_2}{de}\Big|_{e_c} \right) \eta^2$$

and by a change of origin and scale to the standard form

$$x^4 + ax^2.$$

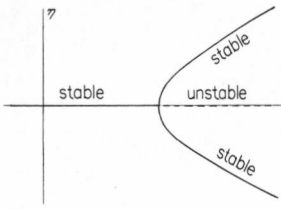

Fig. 14.16

Thus we have Fig. 14.16 for the equilibrium shapes, diffeomorphic to the standard Fig. 14.14.

Since this is a macroscopic system (even more so than the Zeeman machine!) the role of fluctuations should be much less than in the phase transitions discussed above; so the assumption that the system simply minimizes energy should be a good approximation to the 'rotating statics' of the problem. However, the dynamics of the star will be more than usually complicated around the bifurcation point, since the star acquires η as a 'soft mode' of vibration, with infinite period in a linearized treatment. (The larger the vibrations, of course, the more the higher order terms shorten the period.) Whether the effects of being so easily wobbled would show up physically at astronomical distances is a question worth considering; probably, if so, there exist sufficient data to pick out candidates for such effects among nearby stars.

17   Symmetry Breaking

Notice that although the potentials have the specified symmetry throughout, the individual solutions to the minimization problem for $a < 0$ do not. Thus a system which adopts one minimum or the other (Fig. 14.17) 'breaks' the symmetry. (In a context where fluctuations are important, so that the system does not 'adopt a minimum', the corresponding statement is more intricate.) However, the symmetry is not lost to the minimization problem: the *set* of minima still possesses it. We can say that the symmetry of the problem has *fragmented* into a symmetry of the set of solutions, possessed by them only collectively (like a National Park) rather than held by an individual (like the King's New Forest). This is often called 'spontaneous symmetry-breaking' but we think it helpful to reserve 'breaking' as a reference to terms that remove symmetry completely, such as an external magnetic field (in the Weiss model) or an off-axis motion (in the Zeeman machine). We shall call these *symmetry-breaking terms*.

We have assumed throughout this chapter that there are no boundaries (such as $\eta \geq$ some constant) to the range of behaviour allowed. If the Zeeman machine of Chapter 1, Fig. 1.1 had stops to prevent B arising above O, its behaviour around the extreme positions would be very different! Such conditions can arise in the calculation of thermodynamic potentials, and make the *constraint catastrophes* discussed in Chapter 16 Section 7 important. The example from laser physics quoted at the end of that section includes a line of 'second order phase transition' points of the type shown in Fig. 16.9(c) of Chapter 16, where *only* first derivatives, not second or third, vanish. Higher derivative conditions on the jet being unfolded are replaced by a condition of 'location on the boundary'. The existence of a boundary is mathematically more

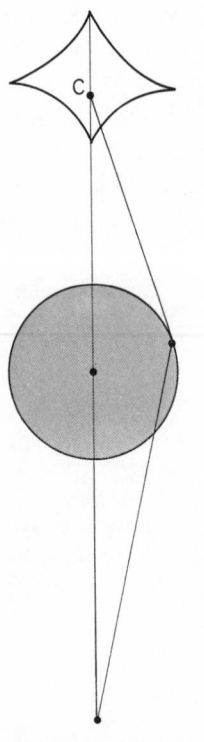

Fig. 14.17. *The Zeeman wheel, held by elastic symmetrically placed at* C, *rests asymmetrically to left or right.*

stable than the presence of symmetry; it cannot be destroyed by a small perturbation. However, the analysis in Thompson [137a] *obtains* the constraint catastrophe formulation by a change of variable that exists by virtue of an assumed circular symmetry, which makes the original problem degenerate unless that symmetry is exact. (As in the last chapter, Section 13, the assumptions made to simplify the sums have complicated the problem!) If higher order terms are added to the Hamiltonian, with only the finite symmetries of the lattice involved, a determinate problem should result: all one can say as yet is that the umbilic catastrophes involved in the effect of symmetry breaking terms will lead to very intricate behaviour in any laser only close to describable by Thompson's model.

However, circular symmetry *can* be physically maintained by the physics to an excellent approximation. The stars in the previous section have it in space before the bifurcation; even afterwards, they have a symmetry that turns them 180° and goes forwards 1/2 a unit of time, stabilized pretty strongly by the physics, and stabilizing in turn the even bifurcation (η and $-\eta$ being rotated versions of the same shape).

18 Tricritical Points

Among even potentials (whether to be treated by the Landau theory, or something deeper), η^4 points are of codimension 1, so that in the (P, T)-plane (for instance) we may expect, and see, whole curves of second order phase transition points. (Just as in the asymmetrical case, two controls gave whole curves of η^3 fold points, and isolated η^4 cusps.) Similarly, at isolated points in the (P, T)-plane, we may expect potentials reducible at the form η^6. Suitable coordinate changes (existing, as before, by the same transversality conditions that implied the points isolated) give the family around such a point the form

$$\eta^6 + A\eta^4 + B\eta^2.$$

The geometry of the set of critical points is shown in Fig. 14.18. (Note that the 'Maxwell set' of Landau theory jump points is only in parabolic contact with the line of second order phase transition points, rather than continuing it smoothly as is sometimes drawn.)

What if symmetry-breaking terms are added to this? For example, if as in the Weiss model η refers to magnetization, what is the effect of an external field H? The usual approach adopted is to add a term strictly linear in H and η, say $hH\eta$, where h is constant; corresponding to the term $-\mu_B H/\kappa T_c$ at the level of the equation of state in Section 2. This may be physically correct in an exact theory, but it is a mathematical error in the setting of transversality arguments, fundamental to the Landau theory and to much of the input of renormalization group methods. For, suppose that a potential in η is

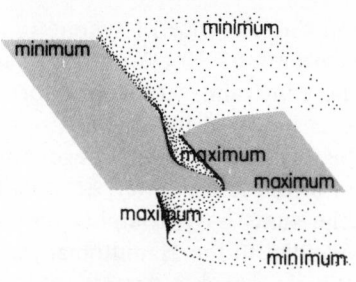

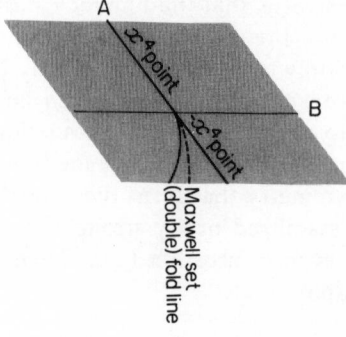

Fig. 14.18

dependent on A, B, H in the particular fashion

$$\Phi(\eta, A, B, H) = \phi(\eta, A, B) + hH\eta$$

exactly, where ϕ is reducible locally to the form $x^6 + ax^4 + bx^2$ by a transformation

$$(\eta, A, B) \mapsto x(\eta, A, B)$$

$$(A, B) \mapsto (a(A, B), b(A, B)).$$

As already indicated, this situation is typical. Furthermore, use of strong 6-determinacy allows us to require that

$$\frac{\partial x}{\partial \eta}(0, 0, 0) = 1,$$

so x and η agree to first order. Expanding η in x (for $A = B = 0$) we get

$$\eta = x + \lambda x^3 + \cdots$$

for some λ. To insist that λ vanish, that is, that x and η agree to third order, would require retaining the 8th order terms of the Taylor expansion of ϕ in η. (To see why, follow the arguments about determinacy in Chapter 8 for 'ultra-strong determinacy', i.e. with this stronger condition on the change of coordinates.) Thus we can truncate at sixth order, as is usual, only at the expense of having the simplest possible algebraic form of ϕ being given by

$$x^6 + ax^4 + bx^2 + hH(x + \lambda x^3 + \cdots)$$

with λ in general non-zero. Application of the unfolding theorems shows this is equivalent to

$$x^6 + ax^4 + bx^2 + hH(x + \lambda x^3),$$

but need not be equivalent to

$$x^6 + ax^4 + bx^2 + hHx,$$

a quite distinct slice of the butterfly catastrophe. (The result is different from that of unfolding x^4, to which x^3 is 'tangent', in the sense discussed in Chapter 8 Section 6.) Linearity in the original coordinates, even if justified on strong physical grounds, does *not* guarantee linearity after the change of coordinates that justifies the dropping of terms of order above 6 in η.

This particular correction in the form of the appropriate 'general unfolding term' does not alter those Landau theory critical exponents that we have yet computed (not all – and we have not begun on the corresponding renormalization group calculations). But it illustrates the subtlety of the points that begin to appear at the verges of the 'non-critical region' where physical intuition may be trusted as a guide to determinacy and transversality.

19 Crystal Symmetries

The complications become very much greater when η has several dimensions and the symmetry is more elaborate. An important case is with η a 3-dimensional vector, and the symmetry that of a crystal. For functions of n variables invariant under the action of a finite group (such as any of the 32 crystallographic point groups) or of a compact group, the results of Chapter 8 go through largely unaltered. They can no longer be stated in such 'cut-down' language, however, and at present we can refer the reader only to the strictly mathematical literature, notably Bierstone [138], Field [78], Poènaru [79–81], Ronga [139], and for a highly applicable treatment Wasserman [81a]. The special case of the crystallographic point groups, with appropriate physical interpretation of the mathematics, will be the subject of a technical report (Ascher, Gay and Poston [140]).

For example, suppose the functions considered of a vector $\eta = (x, y, z)$ satisfy

$$f(-x, y, z) = f(x, y, z) = f\left(\frac{\sqrt{3}}{2} x - \frac{y}{2}, \frac{x}{2} + \frac{\sqrt{3}}{2} y, z\right)$$

(symmetry under the group $3m$ in the 'International Crystallographic' or 'Hermann–Mauguin' notation). Then a typical two-parameter family of such functions will include only functions reducible around $(0, 0, 0)$ to one of the forms:

$$z$$

$$\pm(x^2 + y^2) \pm z^2$$

$$\pm(x^2 + y^2) + z^3$$

$$\pm(x^2 + y^2) \pm z^4$$

$$(3x^2 y - y^3) \pm z^2$$

$$\pm(x^2 + y^2)^2 + \alpha(x^2 + y^2)(3x^2 y - y^3) \pm z^2 \qquad (\alpha \neq 0)$$

$$(3x^2 y - y^3) + (x^2 + y^2)z + \beta z^3 \qquad (\beta \neq 0, \tfrac{9}{2}).$$

Here α and β are diffeomorphism-invariant constants, like the cross-ratio encountered in higher dimensions in the asymmetric case (Chapter 7 Section 6). The first two are stable forms, with no associated bifurcations. The third and fourth give the usual fold and cusp catastrophes with z as essential variable. Those remaining have universal unfoldings (within the class of symmetric functions for this group action)

$$(3x^2 y - y^3) \pm z^2 + t_1(x^2 + y^2)$$

$$(x^2 + y^2)^2 + \alpha(x^2 + y^2)(3x^2 y - y^3) + [t_3(x^2 + y^2)^2]$$

$$+ t_2(3x^2 y - y^3) + t_1(x^2 + y^2)$$

$$(3x^2 y - y^3) + (x^2 + y^2)z + \beta z^3 + [t_3 z^3] + t_2 z^2 + t_1 z.$$

(The square-bracketed terms influence only the differential invariant α or β, rather than 'unfolding' the function topologically.)

20 Spectrum Singularities

Not all of the above polynomials are relevant as thermodynamic potentials (since they lack absolute minima), but there are other possible points of physical interest. For example, at the places where the dispersion relation governing a crystal's mechanical or electronic vibrations gives rise to singularities in the corresponding spectrum, it may locally be treated as the graph of a function from wave vectors to energy values. (For a geometric account, *see* Poston and Budgor [141], though catastrophe-theoretic aspects are not brought out there.) This function inherits the symmetries of the crystal, without any 'strict minimum' requirement, and its bifurcation as external parameters vary – typically reducible to a standard form as above – causes a corresponding bifurcation of the singularity structure of the spectrum.

To sum up: these 'typicality' statements do not mean that for something else to happen is impossible or unphysical. They do mean that something else happening *requires special explanation and justification* as would the discovery of a tricritical point by controlling only a single variable (such as temperature) in either the Landau theory or more sophisticated techniques. The theorems used formalize and deepen the traditional arguments of physics. As Eddington said in a Royal Institution lecture in 1918, 'In two dimensions any two lines are almost bound to meet sooner or later; but in three dimensions, and still more in four dimensions, two lines can and usually do miss one another altogether, and the observation that they do meet is a genuine addition to knowledge.'

15 Laser Physics

This chapter is based entirely on the work of Bob Gilmore (at Tampa, Florida) and Lorenzo Narducci (at Drexel) and we are extremely grateful to them for giving us the free use, not merely of their work, but of a draft account written for us by Gilmore, including all the computations. Our differences from that draft arise only from changes made to improve coherence with the rest of the book.

The work grew out of correspondence between one of us and Gilmore about catastrophe theory in general, and between Gilmore and Narducci about setting recent experimental results in a clear theoretical perspective. It achieves this by using catastrophe theory in an essential way, both to explain the geometry governing laser action and to obtain new quantitative information.

The identification of the laser as a catastrophe is not merely formal. The physical state of a laser is completely specified by an operator, its density operator. We will find that the density operator for a laser factors into two parts, one geometric and one physical. The geometric part is an operator and is completely specified by a point on the associated cusp manifold. The physical part is a number which measures noise. Therefore all physical measurements made on a laser can be expressed in terms of the cusp catastrophe and one other physical parameter. The agreement between theory and experiment (cf. Fig. 15.8) suggests that further improvements in the theory will only be called for (or testable) by major refinements of experiment.

This chapter comprises four parts and an epilogue. The first part contains enough elementary quantum mechanics to provide a language for describing the (classically impossible) laser. The standard rules for calculating with this language require as much explanation as we gave for those of Chapter 8, but we have no space to do so here. The reader unfamiliar with them should skip the computations; we include them however in some detail because they are much simpler than any reference we could give.

The second and third parts contain a derivation of the 'laser equations' and a description of several experiments that have been performed on the laser. In the fourth part the laser equations of motion are studied under an entirely different set of boundary conditions. We find a surprising result that there is a one-to-one correspondence between these two very different physical systems, and that this correspondence is by way of a catastrophe manifold that represents both. The epilogue describes possible ways to exploit such correspondences in future research on critical physical systems.

PRELIMINARIES

1 Atoms

The atoms or molecules in a laser generally have a very complicated energy level structure (Fig. 15.1(a)). The atoms are excited by one means or another from the ground state to a system of highly excited levels (step A). The atoms then decay to some metastable state (step B). They remain in that state until encouraged (stimulated, the s in 'LASER') to emit light radiation by decaying to some well defined lower level (step C). This lower level then decays back to the ground state, and the process starts anew.

This whole process can be very complicated: much more complicated than we wish to study here. For many practical physical purposes it is sufficient to study only the two atomic energy levels which are directly involved in the laser transition (Fig. 15.1(b)). We follow that convenient practice here. In this simplified two-level atomic model (Fig. 15.1(b)), the 'ground' state is not actually the atomic state with least energy.

For our purposes the detailed nature of the mechanisms D-A-B leading from the 'ground' to the 'excited' state are unimportant. It is sufficient to know that we can 'pump' the atoms from the ground to the excited state, from which they can decay, giving up photons to amplify any electromagnetic signal present.

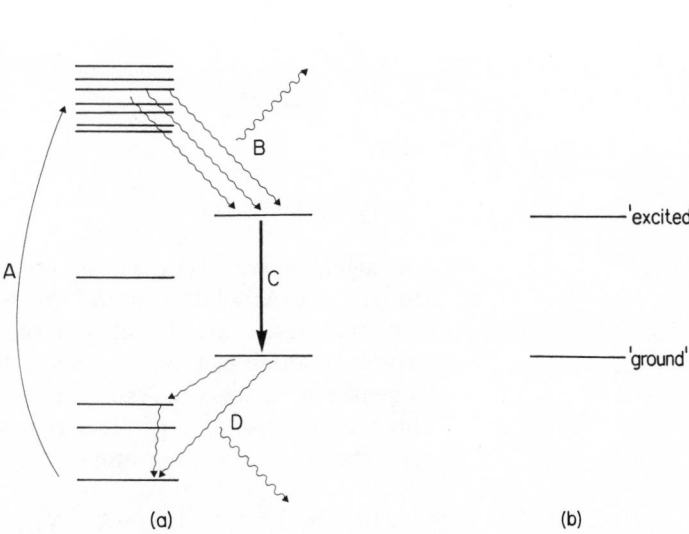

Fig. 15.1   (a)   (b)

These simplified two-state atoms are not 'definitely in' one state or the other unless we take steps to make them so (just as a photon is only definitely polarized a certain way if we have just sent it through a polaroid screen): its state is a *superposition* or linear combination of the two, hence the space of states is the vector space spanned by the 'ground' state $g = \begin{bmatrix} 0 \\ 1 \end{bmatrix}$ and the 'excited' state $e = \begin{bmatrix} 1 \\ 0 \end{bmatrix}$. Since these basic states are something like vibration modes and superpose with a particular phase relationship, the vector space has to be taken as *complex:* since we have a particular basis, we may identify it with the space of ordered pairs of complex numbers, which we denote $\mathbb{C}^2$. To describe motion in this space we shall need to use the various possible operators on the space; or rather, just as in Chapter 8 Section 2 we looked at matrices giving *directions to move* in I_1^4, here we look at similar matrices for $\mathbb{C}^2$. (Technically we are not working in a Lie *group* of invertible operators on $\mathbb{C}^2$ but in its tangent space at the identity transformation – its Lie *algebra.*) What we wrote down in Chapter 8 was actually the identity plus assorted such 'direction matrices' given by various α, β, γ; here we will use the 'direction matrices' alone. The useful ones here turn out to be those spanned by

$$\sigma^+ = \begin{bmatrix} 0 & 1 \\ 0 & 0 \end{bmatrix}, \qquad \sigma^- = \begin{bmatrix} 0 & 0 \\ 1 & 0 \end{bmatrix}, \qquad \sigma^z = \begin{bmatrix} 1 & 0 \\ 0 & -1 \end{bmatrix}. \qquad (15.1)$$

Notice that σ^- takes us in the direction of adding some 'ground' if we have some 'excited', and vice versa for σ^+, while σ^z *increases* 'excited', *decreases* 'ground'. (This is crudely put; but think about what $I + t\sigma$ does for small t, when $\sigma = \sigma^+, \sigma^-, \sigma^z$ in turn.) Important for computations, because they affect the results over time of combining these 'directions' are the *commutation relations*

$$[\sigma^z, \sigma^+] = 2\sigma^+$$
$$[\sigma^z, \sigma^-] = -2\sigma^- \qquad\qquad (15.2)$$
$$[\sigma^+, \sigma^-] = \sigma^z,$$

where

$$[\sigma, \tau] = \sigma\tau - \tau\sigma.$$

Lie algebras are classified (in ways subtly related to the catastrophes, see Arnol'd [40]) and this is actually the simplest; A_1.

A real laser consists of a large number N of atoms. Each two-level atom can be described by a vector in its own two-dimensional space $\mathbb{C}_j^2$. Associated with each atom is a system of three operators $\sigma_j^\pm, \sigma_j^z$. The operators for one atom commute with those for any other atom:

$$[\sigma_j^\#, \sigma_k^\#] = 0 \qquad 1 \le j \ne k \le N; \qquad \# = +, -, z. \qquad (15.3)$$

The Hilbert space H_A for the system of N identical two-level atoms is a tensor product $(\mathbb{C}^2)^{\otimes N}$ of dimension 2^N.

2 Field

The electromagnetic field also has a very complicated structure. In general the electromagnetic field is 'polychromatic', that is, contains components with many different frequencies (sunlight). Even should we look at a monochromatic field (filtered sunlight), it will still contain many different modes. These modes are described by the propagation direction of the photons they contain. Photons with a variety of different momenta $\hbar k_1$, $\hbar k_2$, ... may all have the same frequency $\hbar \omega = \hbar c |k|$ provided the momenta are all the same magnitude $|k_1| = |k_2| = \cdots$. In a laser cavity it is generally possible to arrange mirrors and other devices so that only a single mode of the electromagnetic field is involved in the lasing transition.

We adopt the simplifying assumption that only a single field mode is present (analogous to the convenient two-level atom assumption in Section 1). There may be 0, 1, 2, ... photons present in this mode. Again, the state of the mode may be represented by a vector in a Hilbert space. Basis vectors in this space, called Fock space and denoted H_F, are the vectors $|n\rangle$, $n = 0, 1, 2, \ldots$, corresponding to states with 0, 1, 2, ... photons present. Notice that H_F is infinite-dimensional, since we have put no upper bound on the number of photons permitted. We will use, as we just have, the Dirac notation $|\rangle$ for a vector in H_F, $\langle|$ for a linear function $H_F \to \mathbb{C}$, and the Dirac approach to the difference between finite and infinite dimensions: ignore it unless it bites you. (Here it doesn't.) Again a general state of the field will usually be a *superposition* of basis vectors: we won't generally know the field to be definitely in some state. (*See* Feynman [54] Vol. 3 for a nice discussion of the way, whenever you *can* know 'which of various possibilities', quantum effects disappear. Since laser action *is* a quantum effect, we *don't* want to know always just how many photons there are, if we want a laser!)

Three linear operators on H_F are particularly useful for physical computations. They are the *creation* (a^+) and *annihilation* (a) operators, and their composite the *photon number* $(a^+ a)$, defined by their values on the basis vectors $|n\rangle$:

$$a^+ |n\rangle = \sqrt{(n+1)} |n+1\rangle$$
$$a |n\rangle = \sqrt{n} |n-1\rangle$$

and hence

$$a^+ a |n\rangle = n |n\rangle. \tag{15.4}$$

The names 'creation' and 'annihilation' come from taking definite states towards other definite states with one more or fewer photons; we will explain the name 'photon number' in a moment. These three and the identity I on H_F also generate a Lie algebra $h(4)$, with

$$[a^+a, a^+] = a^+$$
$$[a^+a, a] = -a \qquad\qquad (15.5)$$
$$[a, a^+] = I$$
$$[\text{anything}, I] = 0.$$

3 Interaction

A state of the combined system (atoms + field) is described by a vector $|\varphi\rangle$ in the *tensor product* $H_F \otimes H_A$ of the two spaces. (For an account of '$\otimes$', though not of why it is appropriate here, see for example Dodson and Poston [5].) The time evolution of such states is given by fixing the direction of motion in $H_F \otimes H_A$ at each $|\varphi\rangle$ in a way that (very conveniently) varies linearly with $|\varphi\rangle$. For the historical reason that the equation came before the Dirac notation, this statement (Schrödinger's equation) is usually given without the $|\rangle$ symbols:

$$\mathcal{H}\varphi = i\hbar \frac{\partial \varphi}{\partial t} \qquad\qquad (15.6)$$

where $2\pi\hbar$ is Planck's constant and $\mathcal{H}$ is a linear operator, required to be hermitian for reasons we omit here. In any physical model, of course, $\mathcal{H}$ must be specified fairly explicitly.

Schrödinger's equation for almost anything is too complicated to solve completely: as in Chapter 12 Section 5 we look for solutions that do nothing more than oscillate according to

$$|\varphi_j(t)\rangle = |\varphi_j(0)\rangle \mathrm{e}^{-iE_j t/\hbar}.$$

Inserting this in (15.6) gives an equation for $\varphi_j(0)$; dropping the (0) as is conventional we get the eigenvalue equation

$$\mathcal{H}\varphi_j = E_j\varphi_j. \qquad\qquad (15.7)$$

The number E_j corresponding to each eigenvector $|\varphi_j\rangle$ gives the *frequency* and hence the *energy* of the solution, so that the E_j are the *energy levels* of the system.

So for the laser, or rather for the simplified model for its states described above, we must specify a Hamiltonian $\mathcal{H}$. The one that has been most used is due to Dicke [142]; for an extensive review and bibliography of the mathematical treatment see Haken [143]. The Dicke Hamiltonian is

$$\mathcal{H} = \hbar\omega a^+ a + \sum_{j=1}^{N} \varepsilon \tfrac{1}{2}\sigma_j^z + \sum_{j=1}^{N} (\lambda^* a^+ \sigma_j^- + \lambda a\sigma_j^+). \qquad\qquad (15.8)$$

This operator acts in the Hilbert space $H_F \otimes H_A$. The terms have the following meaning, interpreted in the eigenvalue problem:

(a) The first term measures the energy in the field mode, assuming a single mode is present, with photons of energy $\hbar\omega$. This term acts only on the first factor of a vector like $v \otimes w$ in $H_F \otimes H_A$.

(b) The second term measures the energy in each atomic system. We assume the energies of the excited and ground states as $\varepsilon/2$, $-\varepsilon/2$, respectively. This term acts only on the second factor.

(c) The third term describes photon emission when the jth atom jumps from the excited to the ground state $(a^+\sigma_j^-)$, and excitation of the jth atom from ground to excited state when a photon is absorbed $(a\sigma_j^+)$. The coupling constant λ describes how strongly the atoms and field interact. This term acts on both factors in $H_F \otimes H_A$. It makes the Hamiltonian (15.8) physically interesting and mathematically difficult.

4 Measurement

One more piece of the Dirac notation: if $|\varphi\rangle \in H_F \otimes H_A$, the dual vector $\langle\varphi| : H_F \otimes H_A \to \mathbb{C}$ is defined by taking the value 1 on $|\varphi\rangle$, 0 on everything orthogonal to $|\varphi\rangle$. If φ_j is an eigenvector of $\mathcal{H}$, the energy of the associated oscillating state $|\varphi_j\rangle e^{-iE_jt/\hbar}$ is given by

$$\langle\varphi_j| \mathcal{H} |\varphi_j\rangle = \langle\varphi_j| (E_j |\varphi_j\rangle) = E_j(\langle\varphi_j | \varphi_j\rangle) = E_j. \tag{15.9}$$

The number E_j is the energy we expect to find (and find with certainty if the physical system is exactly in this state). In general, the 'expectation value' for an operator $\mathcal{O}$ for such a state is

$$\langle\varphi_j| \mathcal{O} |\varphi_j\rangle. \tag{15.10}$$

In particular, for the field alone, it is immediate that when the field is exactly in an eigenstate $|n\rangle$ the expected value of a^+a is exactly n, hence the name 'photon number'.

This formalism for computing expectation values of operators is fine provided:

(a) we know the eigenstates $|\psi_j\rangle$;
(b) we know the physical system is in some eigenstate.

However, the more usual circumstance is:

(a′) we cannot compute the eigenstates;
(b′) the physical system is not in any eigenstate.

For a model laser with Hamiltonian (15.8), we face problem (a′). For a real physical system we also face problem (b′). How, then, to proceed? Two types of approaches to these kinds of difficult problem are popular among physicists:

(a″) ignore the problem, in hopes that it will go away (sometimes);

(b″) sneak up on the problem from behind (usual).

The latter approach is somewhat more useful, and is followed here.

If the system *were* in eigenstate $|\psi_j\rangle$ (never mind that we cannot compute $|\psi_j\rangle$), the expectation value of $\mathcal{O}$ would be $\langle j|\mathcal{O}|j\rangle$. If the system is in state $|\psi_i\rangle$ with probability p_i, then the expectation value of $\mathcal{O}$, denoted $\langle\mathcal{O}\rangle$, is

$$\langle\mathcal{O}\rangle = \sum_i p_i\langle\psi_i|\mathcal{O}|\psi_i\rangle. \tag{15.11}$$

This can be written in another way, in terms of something that looks like a projection operator

$$\sum_j p_j\langle\psi_j|\mathcal{O}|\psi_j\rangle = \mathrm{Tr}\left(\sum_j |\psi_j\rangle p_j\langle\psi_j|\right)\mathcal{O}$$
$$= \mathrm{Tr}\rho\mathcal{O}. \tag{15.12}$$

The operator $\rho = \sum_j |\psi_j\rangle p_j\langle\psi_j|$ is called the system *density operator*.

The density operator completely characterizes a physical system.

Problems (a′) and (b′) above are solved (avoided, if you prefer) by working with a density operator. In fact, for many purposes we do not even need ρ, we need only to know some expectation values. For one set of measurements however (represented, as a sample, by Fig. 15.8 below), we will compute ρ explicitly.

Not every operator in infinite dimensions *has* a trace, of course (the identity, I, hasn't), but rigour at this point would require sneaking up from *far* behind. Rigour plus geometric explanation would require starting from the lattice of propositions about the system and going via projection-valued measures to the operators that 'elementary' treatments begin with. We have not space to include Jauch [144] or a volume of Varadarajan [145] here, so we will proceed firmly with the notation of practical physics and Dirac's view of infinite dimensions (compare remarks in Chapter 14, end of Section 11).

THE LASER CATASTROPHE

5 Unfolded Hamiltonian

Rather than analyse a structurally unstable bifurcation problem, as we did repeatedly in Chapter 13, and then add unfolding terms, we will put in the extra term now, since it enters in a slightly complicated way.

Instead of using the simple model Hamiltonian (15.8) to describe a laser, it is more useful to use a slightly more complicated Hamiltonian which takes into account the interaction of the atoms with a classical external electric or magnetic field. To do this, an extra term of the form $\sum_j (\lambda^* \alpha^* \sigma_j^- + \lambda \alpha \sigma_j^+)$ must be included, where α represents the strength of this extra field. The Hamiltonian we use to model a real physical laser is

$$\mathcal{H} = \mathcal{H}(\alpha) + \text{(more stuff)} \tag{15.13a}$$

$$\mathcal{H}(\alpha) = \hbar\omega a^+ a + \sum_{j=1}^{N} \tfrac{1}{2}\varepsilon\sigma_j^z + \sum_{j=1}^{N} \{\lambda^*(a^+ + \alpha^*)\sigma_j^- + \lambda(a + \alpha)\sigma_j^+\}. \tag{15.13b}$$

There are two important reasons for including the extra field term α.

(a) Physically, the external field provides a useful tool for altering experimental conditions.

(b) Mathematically, the result of the analysis becomes structurally stable, giving a universal unfolding of the bifurcation point involved. (It would clearly be desirable that $\mathcal{H}$ itself is in a suitable sense a universal unfolding of the Hamiltonian at the bifurcation point, but no one that we know of yet has the machinery for deciding this.)

The other term 'more stuff' is discussed in the following section.

6 Equations of Motion

The time dependence for an operator expectation value in a pure state $|\psi_j\rangle$ can be computed from the equation

$$\frac{\mathrm{d}}{\mathrm{d}t}\langle\psi_j| \mathcal{O} |\psi_j\rangle = \left(\frac{\mathrm{d}}{\mathrm{d}t}\langle\psi_j|\right)\mathcal{O} |\psi_j\rangle$$

$$+ \langle\psi_j| \mathcal{O}\left(\frac{\mathrm{d}}{\mathrm{d}t}|\psi_j\rangle\right) + \langle\psi_j| \frac{\partial\mathcal{O}}{\partial t} |\psi_j\rangle. \tag{15.14}$$

From Schrödinger's equation (15.6) we know that $i\hbar\dfrac{\partial}{\partial t}|\psi\rangle = \mathcal{H}|\psi\rangle$ and $\langle\psi|\mathcal{H} = -i\hbar\dfrac{\partial}{\partial t}\langle\psi|$. Then (15.14) simplifies to the Heisenberg equation†

$$i\hbar\frac{\mathrm{d}}{\mathrm{d}t}\langle j| \mathcal{O} |j\rangle = \langle j| [\mathcal{O}, \mathcal{H}] |j\rangle + \langle j| i\hbar\frac{\partial\mathcal{O}}{\partial t} |j\rangle. \tag{15.15}$$

If in fact the system is described by a density operator where the

† This material is covered in any good book on elementary quantum mechanics: *see* for instance Dicke and Wittke [146].

probabilities p_j (cf. (15.10)) are independent of time, then (15.15) for pure states $|\psi_j\rangle$ may be multiplied by p_j, and the summation over j carried out to give

$$i\hbar \frac{d}{dt}\langle \mathcal{O}\rangle = \langle[\mathcal{O}, \mathcal{H}]\rangle + \left\langle i\hbar \frac{\partial \mathcal{O}}{\partial t}\right\rangle. \tag{15.16}$$

So to compute the time derivative of the expectation value of a time independent operator $\mathcal{O}$, it is only necessary to compute the expectation of the commutator $[\mathcal{O}, \mathcal{H}]$, since $\partial\mathcal{O}/\partial t = 0$.

When equation (15.16) is used in conjunction with (15.13), it is a simple matter to compute the equations of motion for $\langle a\rangle$, $\langle\sigma_j^-\rangle$, $\langle\sigma_j^z\rangle$. Since $\langle\sigma_j^-\rangle = \langle\sigma_k^-\rangle$ and $\langle\sigma_j^z\rangle = \langle\sigma_k^z\rangle$, all $j, k = 1, 2, \ldots, N$, we drop the subscripts on the atomic operators:

$$i\hbar \frac{d}{dt}\langle a\rangle = \hbar\omega\langle a\rangle + N\lambda^*\langle\sigma^-\rangle + \langle[a, \text{more stuff}]\rangle$$

$$i\hbar \frac{d}{dt}\langle\sigma^-\rangle = \varepsilon\langle\sigma^-\rangle - \lambda\langle(a+\alpha)\sigma^z\rangle + \langle[\sigma^-, \text{more stuff}]\rangle \tag{15.17}$$

$$i\hbar \frac{d}{dt}\langle\sigma^z\rangle = 2\lambda\langle(a+\alpha)\sigma^+\rangle - 2\lambda^*\langle(a^+ + \alpha^*)\sigma^-\rangle$$
$$+ \langle[\sigma^z, \text{more stuff}]\rangle.$$

In deriving (15.17) the commutation relations (15.3) and (15.5) have been used, together with the linearity of the commutator and its differential property $[X, YZ] = [X, Y]Z + Y[X, Z]$.

Equations (15.17) are not really useful as is, because the commutators $\langle[\cdot, \text{more stuff}]\rangle$ do not yet convey any information. While the term 'more stuff' in (15.13) is fairly complicated, its effects on the equations of motion are not. In fact, it has three effects (the first two discussed in Haken [143], the third in Gilmore and Narducci [147]).

(a) *Damping.* If an expectation value, say $\langle a\rangle$, deviates by δa from its equilibrium value $\langle a\rangle_e$, then it decays back to its equilibrium value with a typical exponential decay:

$$\langle a\rangle(t) = \delta a e^{-\gamma_a t} + \langle a\rangle_e.$$

This can be completely taken care of by making substitutions of the form

$$\frac{d}{dt}\langle a\rangle \mapsto \left(\frac{d}{dt} + \gamma_a\right)(\langle a\rangle - \langle a\rangle_e)$$

in (15.17). The equilibrium value of an expectation value is the value that would be obtained in the absence of dissipative losses ('friction').

(b) *Noise.* The extra terms in (15.13) also add a random inhomogeneous driving term to the right-hand side of each of equations (15.17). These noise terms are often neglected in discussions of the laser equations. We do likewise. (As discussed in the previous chapter, Sections 10 ff., casual treatment of fluctuations leads to trouble very close to the critical point: in this context the trouble is known as 'competition between modes'. What we are finding, correctly, is probability *peaks* which may be treated as predictions of possible *averages* when sharp – which they are, except in a neighbourhood of a bifurcation point.)

(c) This extra term also requires that the expectation values $\langle a^+\rangle = \langle a\rangle^*$ are real if α is real. This is very convenient, because the canonical equations (15.22) and (15.43) would not be structurally stable with A, B each one-real-dimensional if $\langle a\rangle$ were complex. In short, (15.17) 'simplify' to

$$i\hbar \left(\frac{\mathrm{d}}{\mathrm{d}t} + \gamma_a\right)(\langle a\rangle - \langle a\rangle_e) = \hbar\omega\langle a\rangle + N\lambda^*\langle\sigma^-\rangle$$

$$i\hbar\left(\frac{\mathrm{d}}{\mathrm{d}t} + \gamma_\sigma\right)(\langle\sigma^-\rangle - \langle\sigma^-\rangle_e) = \varepsilon\langle\sigma^-\rangle - \lambda\langle(a+\alpha)\sigma^z\rangle \quad (15.18)$$

$$i\hbar\left(\frac{\mathrm{d}}{\mathrm{d}t} + \gamma_z\right)(\langle\sigma^z\rangle - \langle\sigma^z\rangle_e) = 2\lambda\langle(a+\alpha)\sigma^+\rangle$$

$$-2\lambda^*\langle(a^+ + \alpha^*)\sigma^-\rangle.$$

7 Mean Field Approximation

Equations (15.18) are still not simple enough. They cannot be solved. The differential equations for single operator expectation values involve also expectation values of operator products such as $\langle a\sigma^z\rangle$ and $\langle a^+\sigma^-\rangle$.

One possibility is to construct the equations of motion for these operators: $\mathrm{d}\langle a^+\sigma^-\rangle/\mathrm{d}t = \dots$, and hope for a system of closed equations. The hope is in vain. Equations of motion for products of two operators involve products of three, and things rapidly get worse.

Another possibility is to make (15.18) a closed system of equations by assuming the expectation values of operator products factor in the sense $\langle a\sigma^+\rangle = \langle a\rangle\langle\sigma^+\rangle$, for example. Then (15.18) become a closed system of nonlinear equations. The nonlinearity leads to very interesting behaviour. In fact, it leads right to the cusp catastrophe.

While the motivation for the factorization is clear (we cannot proceed very well without it), its justification is not. However, it turns out that the relative error made in the computation of any non-zero expectation value using this assumption is of the order $\log N/N$, where N is the number of atoms. In a small laser N might be 10^{12}, so that $\log N/N = 12 \times (2.3)/10^{12} =$ small number. This number should be sufficiently small to please the most ardent δ, ε

mathematician; it causes few problems to the experimental physicist.

The factorization assumption is equivalent to a mean field assumption. This assumption says that to the field system, the atoms look like classical sources, while to the atoms, the field looks classical. Put another way: if we want to consider the field system alone, it is perfectly adequate to replace the atomic operators $\sigma^\pm$, σ^z by their expectation values $\langle\sigma^+\rangle$, $\langle\sigma^z\rangle$, and vice versa. This in turn is equivalent to the assumption that the density operator ρ for the whole interacting field-atom system factors into the direct product of two *reduced* density operators $(\rho = \rho_F \otimes \rho_A)$, one for the field subsystem alone, the other for the atomic subsystem alone. Although we cannot compute ρ for (15.13), we can compute both ρ_F and ρ_A in the mean field approximation. Further, ρ_F (and ρ_A) is an operator which is defined by a point on the cusp manifold.

The factorized equations (15.18) can now be simplified somewhat. This is done by removing the fast time dependence from $\langle a \rangle$, $\langle \sigma^- \rangle$. This is done by making the substitutions $\langle a \rangle(t) = e^{-i\omega t}\langle\tilde{a}\rangle(t)$, $\langle\sigma^-\rangle(t) = e^{-i\varepsilon t/\hbar}\langle\tilde{\sigma}^-\rangle(t)$ and similar substitutions for $\langle a \rangle_e$, $\langle\sigma^-\rangle_e$. Then a little algebra produces a new set of equations with tildes (also $\tilde{\lambda} = \lambda\exp(i(\varepsilon - \hbar\omega)t/\hbar)$). Dropping the tildes for the printer's pleasure, the simplified set of equations is

$$i\hbar\left(\frac{d}{dt} + \gamma_a\right)\langle a \rangle = \hbar\omega\langle a \rangle_e + N\lambda^*\langle\sigma^-\rangle$$

$$i\hbar\left(\frac{d}{dt} + \gamma_\sigma\right)\langle\sigma^-\rangle = \varepsilon\langle\sigma^-\rangle_e - \lambda(\langle a \rangle + \alpha)\langle\sigma^z\rangle \qquad (15.19)$$

$$i\hbar\left(\frac{d}{dt} + \gamma_z\right)(\langle\sigma^z\rangle - \langle\sigma^z\rangle_e) = 2\lambda(\langle a \rangle + \alpha)\langle\sigma^-\rangle^* - 2\lambda^*(\langle a \rangle + \alpha)^*\langle\sigma^-\rangle.$$

In deriving this, we have used the relations $\langle\sigma^-\rangle^* = \langle\sigma^+\rangle$, $\langle a \rangle^* = \langle a^+\rangle$.

8 Boundary Conditions

The equations (15.19) have been studied subject to two quite different sets of physical conditions (Haken [143], Hepp and Lieb [148], Gilmore and Bowden [149]): *equilibrium* and *non-equilibrium* boundary conditions.

The laser model consists of the equations (15.19) together with non-equilibrium assumptions. It is assumed that a pumping mechanism exists to pump the atoms from the ground state into the excited state. The pumping rate affects $\langle\sigma^z\rangle_e$. With no pump $\langle\sigma^z\rangle_e = -1$. With very strong pumping $\langle\sigma^z\rangle_e = +1$. With intermediate pumping rates $\langle\sigma^z\rangle_e$ assumes intermediate values. We therefore take $\langle\sigma^z\rangle_e$ as a convenient *measure* of the pumping rate: it will provide one of our unfolding parameters. In this model it is further assumed (Haken [143]) that $\langle a \rangle_e = \langle\sigma^-\rangle_e = 0$.

The model (15.19) can also be studied subject to thermal equilibrium boundary conditions, with equilibrium temperature T. There is no pumping, all time derivatives vanish, and $\langle\sigma^z\rangle_e$ is determined in a self-consistent way from $\langle a\rangle_e$ and $\langle\sigma^-\rangle_e$. Interesting things happen only if $\langle a\rangle_e \neq 0$, $\langle\sigma^-\rangle_e \neq 0$.

Although the time-dependent equations (15.19) can be studied, it is more useful to look for stationary solutions to these equations. These are obtained by setting all time derivatives equal to zero. Such solutions are denoted by a subscript s. For equilibrium boundary conditions $\langle a\rangle_s = \langle a\rangle_e$, $\langle\sigma^-\rangle_s = \langle\sigma^-\rangle_e$, $\langle\sigma^z\rangle_s = \langle\sigma^z\rangle_e$. For the non-equilibrium case $\langle a\rangle_s \neq \langle a\rangle_e$, $\langle\sigma^-\rangle_s \neq \langle\sigma^-\rangle_e$ even if we do not assume $\langle a\rangle_e = \langle\sigma^-\rangle_e = 0$.

9 Non-equilibrium Stationary Manifold

We now search for stationary solutions to the coupled nonlinear equations (15.19) subject to non-equilibrium (i.e. laser) boundary conditions. These equations simplify (with $\langle a\rangle_e = \langle\sigma^-\rangle_e = 0$) to

$$i\hbar\gamma_a\langle a\rangle_s = N\lambda\langle\sigma^-\rangle_s$$
$$i\hbar\gamma_\sigma\langle\sigma^-\rangle_s = -\lambda(\langle a\rangle_s + \alpha)\langle\sigma^z\rangle_s \qquad (15.20)$$
$$i\hbar\gamma_z(\langle\sigma^z\rangle_s - \langle\sigma^z\rangle_e) = 2\lambda(\langle a\rangle_s + \alpha)(\langle\sigma^-\rangle^* - \langle\sigma^-\rangle).$$

In deriving (15.20) we have chosen α (external field amplitude) real so that $\langle a\rangle$ is real by remark (c) in Section 6, and taken λ real for simplicity. The equation for the stationary values of $\langle a\rangle_s$ is obtained by solving the first of equations (15.20) for $\langle\sigma^-\rangle_s$ and the third for $\langle\sigma^z\rangle_s$ and then plugging both into the second equation. The following equation results after slight tidying:

$$-\gamma_a\gamma_\sigma\langle a\rangle_s = -(\lambda/\hbar)^2(\langle a\rangle_s + \alpha)\left\{N\langle\sigma^z\rangle_e - \frac{4\gamma_a}{\gamma_z}(\langle a\rangle_s + \alpha)\langle a\rangle_s\right\}$$

$$(15.21)$$

This equation is a cubic in the expectation value ('order parameter') $\langle a\rangle_s$ of the field amplitude, so can be put into the canonical form for the cusp catastrophe manifold:

$$X^3 - AX - B = 0$$

where

$$X = \langle a\rangle_s + \tfrac{2}{3}\alpha$$
$$A = \frac{\alpha^2}{3} + \frac{N\gamma_z}{4\gamma_a}\left[\langle\sigma^z\rangle_e - \frac{\gamma_a\gamma_\sigma}{N(\lambda/\hbar)^2}\right] \qquad (15.22)$$
$$B = \tfrac{2}{27}\alpha^3 + \frac{\alpha}{3}\frac{N\gamma_z}{4\gamma_a}\left[\langle\sigma^z\rangle_e - \frac{\gamma_a\gamma_\sigma}{N(\lambda/\hbar)^2}\right] + \frac{\gamma_\sigma\gamma_z}{4(\lambda/\hbar)^2}\alpha.$$

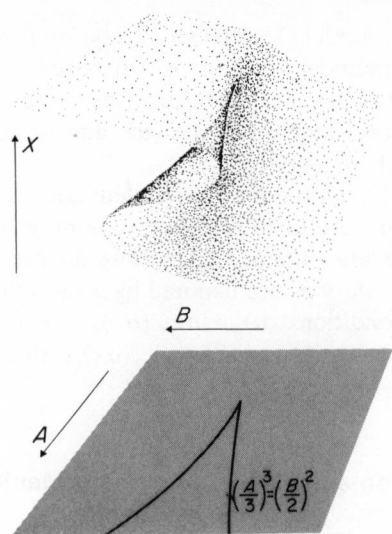

Fig. 15.2

Equations (15.22) identify a point on the cusp manifold (Fig. 15.2) with a laser stationary state with order parameter $\langle a \rangle_s$ determined by (15.22), and order parameter $\langle \sigma^- \rangle_s$ determined by (15.20). Further, the Jacobian of the transformation from α, $\langle \sigma^z \rangle_e$ to A, B is non-singular around $\alpha = 0$, so that we have a universal unfolding of the singular stationary value equation at the control values

$$\alpha = 0,$$

$$\langle \sigma^z \rangle_e = \gamma_a \gamma_\sigma / N(\lambda/\hbar)^2$$

(if this is possible: recall $\langle \sigma^z \rangle_e \leqslant 1$ by definition). Notice that (as in Chapter 14 Section 2) we are unfolding not a 'potential' but an 'equation of state'. Thus the remarks at the end of Section 6 of Chapter 14 mean that for bifurcations of higher corank the existence (or not) of the potential must come into the argument, if not the calculations.

EXPERIMENTS

Comparison of mathematical predictions with the result of experiment is the physicist's time-honoured way of separating the wheat from the chaff. We follow that procedure in this part of the chapter. We describe one experiment in each of the following three sections. The first two are 'rough cut' and 'orthogonal'. They are rough in the sense that they measure only mean or average properties of the field associated with the laser. They are orthogonal in the sense that quite distinct physical controls are varied, and the laser responds quite differently as a function of these changing parameters. The third experiment is a 'second generation' experiment, in that it probes the detailed structure of the electromagnetic field produced by the laser.

10   Laser Transition

If the threshold value $\gamma_a \gamma_\sigma / N(\lambda/\hbar)^2$ is less than 1 and hence physically attainable for $\langle \sigma^z \rangle_e$, we have a standard cusp catastrophe. With the classical external field α exactly zero, we get the usual 'even function' slice (Figs. 6.3, 13.40(a), 14.14, 14.17) with field amplitude $\langle a \rangle_s$ taking off parabolically as the pumping rate increases through the critical point. Since the field *intensity*, and hence the power output of the laser, is proportional to $(\langle a \rangle_s)^2$, this increases linearly to first order with pumping rate, if $\langle \sigma^z \rangle_e$ is indeed a smooth parametrization of the latter. The experimental data fit nicely (Fig. 15.3).

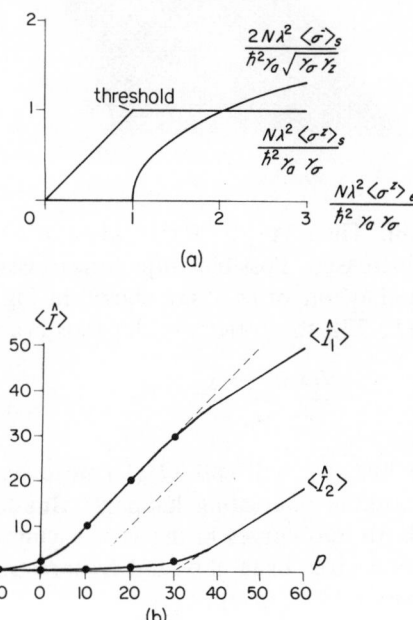

Fig. 15.3. (a) *The amplitude* $\langle a \rangle_s$ *is not easily detectable, since optical devices generally measure intensity. However, the amplitude* $\langle a \rangle_s$ *is related linearly to the atomic polarization* $\langle \sigma^- \rangle_a$ *by equation (15.20). The two curves show* $\langle \sigma^z \rangle_s$ *and* $\langle \sigma^- \rangle_s$ *plotted against equilibrium population inversion* $\langle \sigma^z \rangle_e$. *The axes have been 'stretched' as indicated for convenience of scale (After Graham. See Haken [227], pp. 54–70.). (b) The laser is tuned so that two modes are active at a sufficiently high pump rate p (proportional to A). For p < 0, the laser is below threshold. For 0 < p < 30, the device is above threshold for the first mode. The intensity increases linearly except around the threshold regions for the two modes, where there is some curvature due to fluctuations. Since the two modes compete for the available energy, onset of laser activity in mode 2 suppresses the intensity of mode 1 (break in slope) (After Grossman. See Haken [227], pp. 54–70.).*

As usual, this picture is structurally unstable against symmetry-breaking perturbations, and equally as usual such perturbations are physically realizable – here, by the field α. By changing α, we change the picture in the usual way (Figs. 6.4, 13.40(b)). The contribution of catastrophe theory is to say that pumping rate and α between them capture *all* such effects; any new field term, however exotic (say, density of charmed gremlins) that has a smooth effect on equation (15.21) can be locally rendered impotent by the wand of smooth reparametrization.

11   Optical Bistability

In this class of experiments $\langle \sigma^z \rangle_e$ is held fixed below the threshold for laser activity. This is simply done by not pumping the sample at

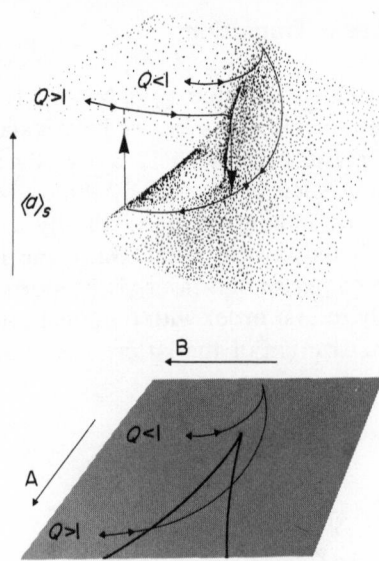

Fig. 15.4

all. Then $A < 0$, $B = 0$ when $\alpha = 0$. The external field is slowly increased. Possible trajectories (assuming delay convention for the behaviour of $\langle a \rangle_s$) are shown in Fig. 15.4. As is easily shown from (15.22), the trajectory depends crucially on the value of

$$\frac{-N(\lambda/\hbar)^2 \langle \sigma^2 \rangle_e}{8\gamma_a \gamma_\sigma}$$

which we will call Q (Gilmore and Narducci [147]). For $Q < 1$, nothing interesting happens. But for $Q > 1$ the trajectory crosses both fold curves in the very asymmetric way shown: delay convention gives bistability and hysteresis. (The histories also diverge for any other convention.)

Optical bistability has been observed in an unpumped sodium vapour confined to a Fabry–Perot cavity at 10^{-4} to 10^{-5} Torr irradiated by an external classical field at a power level $\sim 10\,\mathrm{mW}$ detuned 150 MHz above a $^2S_{1/2} \rightarrow {}^2P_{3/2}$ transition. (Gibbs, McCall and Venkatesan [150]). The data are shown in Fig. 15.5, which plots transmitted power (intensity/second) against incident power. The transmitted amplitude is $\langle a \rangle_s + \alpha$, while the incident amplitude is α. Therefore, Fig. 15.5 can be compared to the predictions made based on the catastrophe by plotting $(X + \alpha/3)^2$ which is proportional to the transmitted power $(\langle a \rangle_s + \alpha)^2$, against α^2, which is proportional to the incident power. This is done in Fig. 15.6.

The detailed form of the curves in Fig. 15.5 makes it appear that behaviour very close to perfect delay is occurring, since the parabolic contact with the vertical of a slice through the fold line is visible in all the pre-jump 'shoulders' where the jump is deep enough to display a vertical part (compare Fig. 15.7). This property

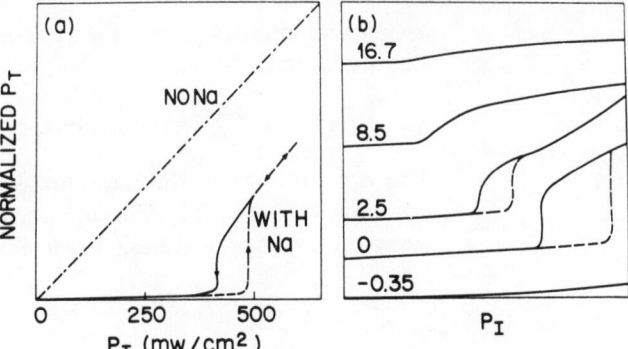

Fig. 15.5. (a) *Optical bistability in sodium vapour. The oscilloscope trace is dashed for increasing output intensities and solid for decreasing. (b) Characteristic curve dependence on Fabry–Perot plate separation. The detuning is given in MHz; the free spectral range was about 1364 MHz.* (Gibbs, McCall and Venkatesan [150].)

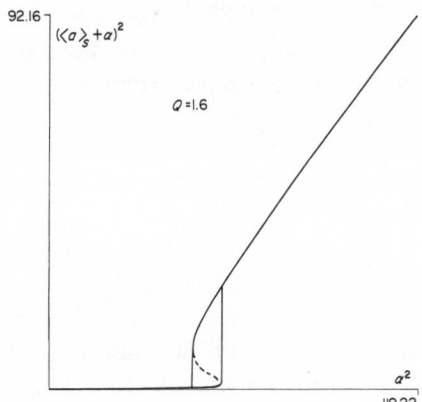

Fig. 15.6. *Catastrophe model predictions for experimental data of Fig. 15.5* (Bonifacio and Lugiato [151], Gilmore and Narducci [147]).

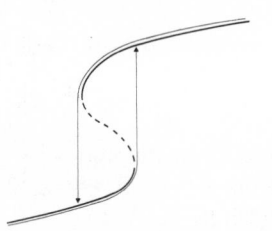

Fig. 15.7

is diffeomorphism-invariant, and is eminently susceptible to numerical test (often even when very little else is, as we illustrate in the next chapter for moving ecological frontiers). It characterizes delay convention completely, since *only* at a fold is the manifold in parabolic contact with the vertical. (At Maxwell jump points, for instance, it is never vertical at all.) It would be interesting to examine the experimental numbers with this question in mind.

12 Photocount Distributions

Photocount experiments count photons (Glauber [152, 153]). If the state of the electromagnetic field is represented by the vector $|\psi\rangle$ in Fock space, then the inner product $\langle m \mid \psi \rangle$ is the probability amplitude for an experiment to record m photons (i.e. that the state contains m photons). The probability $P(m)$ for measuring exactly m photons is the absolute square of this complex number

$$P(m) = |\langle m \mid \psi \rangle|^2 = \langle m \mid \psi \rangle \langle \psi \mid m \rangle. \tag{15.23}$$

If we cannot guarantee that the system is in state $|\psi\rangle$, but can only assign probabilities p_i that the system is in state $|\psi_i\rangle$, then (15.23) is slightly altered:

$$P(m) = \langle m| \left(\sum_i |\psi_i\rangle p_i \langle \psi_i| \right) |m\rangle. \tag{15.24}$$

The operator inside the parenthesis is our old friend, the density operator (15.12). The operator ρ can be expressed in terms of a complete orthonormal basis $|n\rangle$ in Fock space

$$\rho = \sum_{n=0}^{\infty} \sum_{n'=0}^{\infty} |n\rangle \rho_{nn'} \langle n'|. \tag{15.25}$$

Therefore, the probability $P(m)$ which we seek is a particular matrix element of the density operator

$$P(m) = \langle m| \rho |m\rangle = \rho_{mm}. \tag{15.26}$$

The next step in our quest for a photocount probability distribution function is simple arithmetic:

$$\frac{d}{dx} x^n = n x^{n-1}$$

and

$$\frac{d^k}{dx^k} x^n = n(n-1) \cdots (n-k+1) x^{n-k} \to k!\, \delta_{n,k} \quad \text{as} \quad x \to 0. \tag{15.27}$$

Since $a^+ a |n\rangle = n |n\rangle$, it is also possible to write $x^n = \langle n| x^{a^+ a} |n\rangle$. In fact, using the description (15.25) for the density operator, we have

$$\langle x^{a^+ a} \rangle = \text{Tr}\, \rho x^{a^+ a} = \text{Tr} \left(\sum_{n=0}^{\infty} \sum_{n'=0}^{\infty} |n\rangle \rho_{nn'} \langle n'| \right) x^{a^+ a}$$

$$= \sum_{n=0}^{\infty} \sum_{n'=0}^{\infty} \rho_{nn'} \langle n'| x^{a^+ a} |n\rangle$$

$$= \sum_{n=0}^{\infty} \sum_{n'=0}^{\infty} \rho_{nn'} x^n \delta_{nn'} = \sum_{n=0}^{\infty} \rho_{nn} x^n. \tag{15.28}$$

In deriving this, we have used the orthonormality of the states $|n\rangle$: $\langle n' | n \rangle = \delta_{n'n}$ where $\delta_{n'n}$ is the Kronecker delta ($= +1$ if $n' = n$, 0 otherwise). Combining (15.26), (15.27) and (15.28) leads to

$$\frac{1}{k!} \left(\frac{d}{dx} \right)^k \langle x^{a^+ a} \rangle \bigg|_{x=0} = \rho_{kk} = P(k). \tag{15.29}$$

So far we have been quite silent about how to compute the field density operator ρ_F, let alone the expectation $\langle x^{a^+ a} \rangle$. These problems turn out to be delightfully simple too.

Under the mean field approximation introduced in Section 7, the atoms behave like a classical source, as far as the field is concerned. So if we were to write down a Hamiltonian for the field subsystem alone, it would be a linear superposition of the photon number, creation and annihilation operators (cf. (15.8) or (15.13)). For such a Hamiltonian there is a nice theorem (Gilmore and Narducci [147]) which assures us that under suitable conditions (met here) the reduced field density operator has the general form

$$\rho_F \simeq \text{EXP}\,(Ma^+a + Ra^+ + La) \tag{15.30}$$

where $M^* = M$, $R^* = L$, and the squiggle means 'proportional to'. The density operator must be normalized to unity: $\text{Tr}\,\rho = \sum p_i = 1$ (cf. (15.11)). We can work with unnormalized density operators provided expectation values are defined by $\langle \mathcal{O} \rangle = \text{Tr}\,\rho\mathcal{O}/\text{Tr}\,\rho$.

We want to relate the parameters M, R, L to points on the cusp manifold. Before doing so, it is useful here to introduce one additional complication. This involves computing the expectation value of yet another operator. We do this to get a very convenient generating function. The operator is $\text{EXP}\,(ma^+a + ra^+ + la)$:

$$\langle \text{EXP}\,(ma^+a + ra^+ + la) \rangle$$
$$= \frac{\text{Tr}(\text{EXP}\,(Ma^+a + Ra^+ + La)\;\text{EXP}\,(ma^+a + ra^+ + la))}{\text{Tr}\,(\text{EXP}\,(Ma^+a + Ra^+ + La))}. \tag{15.31}$$

The coefficients (M, R, L) and (m, r, l) have nothing to do with each other. The trace in the numerator can be computed (by an argument from a few deep results in Lie group theory (Gilmore [147a]), which space forces us to omit) as

$$\frac{1}{1 - e^{M+m}}\exp\left\{-\left(\frac{LR}{M} + \frac{lr}{m}\right) + \left(\frac{L}{M} - \frac{l}{m}\right)\right.$$
$$\left. \times \left(\frac{R}{M} - \frac{r}{m}\right)\frac{(e^M - 1)(e^m - 1)}{e^{M+m} - 1}\right\}. \tag{15.32}$$

The result is valid for $M + m < 0$. The denominator in (15.31) is a special case of (15.32), and the ratio in (15.31) is easily computed to give

$$\langle \text{EXP}\,(ma^+a + ra^+ + la) \rangle$$
$$= \frac{1 - e^M}{1 - e^{M+m}}\exp\left\{-\frac{lr}{m} + \left(\frac{L}{M} - \frac{l}{m}\right)\left(\frac{R}{M} - \frac{r}{m}\right)\frac{(e^M - 1)(e^m - 1)}{e^{M+m} - 1}\right\}. \tag{15.33}$$

While rather a mess, this expectation value contains everything we need to know:

$$\langle a \rangle = \frac{\partial}{\partial l}\langle \text{EXP}\,(ma^+a + ra^+ + la) \rangle \bigg|_{m=r=l=0} = -R/M \tag{15.34a}$$

$$\langle a^+ \rangle = \frac{\partial}{\partial r}\langle \text{EXP}\,(ma^+a + ra^+ + la) \rangle \bigg|_{m=r=l=0} = -L/M \tag{15.34b}$$

$$\langle a^+a \rangle = \frac{\partial}{\partial m}\langle \text{EXP}\,(ma^+a + ra^+ + la) \rangle \bigg|_{m=r=l=0}$$

$$= \frac{e^M}{1 - e^M} + \left(\frac{-L}{M}\right)\left(\frac{-R}{M}\right). \tag{15.34c}$$

We are now in a position to make contact with the cusp manifold. From (15.34a,b), $R = -M\langle a \rangle$ and $L = -M\langle a^+ \rangle$, so the density operator ρ in (15.30) becomes ($\langle a \rangle = \langle a \rangle^*$)

$$\rho_F \simeq [\rho(\text{geometry})]^{M(\text{physics})}$$

$$\rho(\text{geometry}) \simeq \text{EXP}\{a^+a - \langle a \rangle (a^+ + a)\}$$

$$\langle a \rangle = X - \frac{2\alpha}{3}. \tag{15.35}$$

It is a remarkable occurrence that the field density operator 'factors' into two parts. The operator part depends on the geometry of the cusp catastrophe through the point $X = \langle a \rangle_s + 2\alpha/3$. The number M depends on physics, and is in fact closely related to noise.

For experimental purposes it is convenient to describe a measurement in terms of signal and noise. (For an extensive discussion of how an experimentalist knows what he is measuring is actually what he thinks he is measuring, *see* Arecchi [154].) Such a decomposition is evident in (15.34c) where the signal $\mathcal{S} = |\langle a \rangle|^2$ and the system noise is the correlation function $\langle a^+a \rangle - \langle a^+ \rangle \langle a \rangle$:

$$\mathcal{S} = \left(-\frac{L}{M} \right) \left(-\frac{R}{M} \right) = \left(X - \frac{2\alpha}{3} \right)^2 = \langle a^+ \rangle \langle a \rangle$$

$$\mathcal{N} = \frac{1}{e^{-M} - 1} = \langle a^+a \rangle - \langle a^+ \rangle \langle a \rangle. \tag{15.36}$$

The point X on the cusp manifold is determined (at worst, up to two possible choices – the upper and lower sheets) by the system signal, and M(physics) is determined from system noise by $-M = \log (1 + \mathcal{N}^{-1}) > 0$.

It is now a relatively simple matter (Glauber [152]) to compute the photocount probability distribution function $P(k)$ (15.29). This is first done by setting $r = l = 0$ in (15.33), producing the expectation $\langle (e^m)^{a^+a} \rangle$. If we now set $e^m = x$ we start to get very close to our goal:

$$\langle x^{a^+a} \rangle = \frac{1 - e^M}{1 - xe^M} \exp \left\{ \frac{LR}{M^2} \frac{(1 - e^M)(x - 1)}{1 - xe^M} \right\}. \tag{15.37}$$

This cries out to be expanded in powers of x (consider (15.29)). Moreover, it is vaguely reminiscent (Abramowitz and Stegun [69]) of a Laguerre polynomial generating function

$$\frac{1}{1 - z} \exp \left\{ \frac{-yz}{1 - z} \right\} = \sum_{k=0}^{\infty} L_k(y)z^k, \qquad |z| < 1. \tag{15.38}$$

To make the correspondence between (15.37) and (15.38) less than vague, write

$$\frac{1-e^M}{1-xe^M}=\frac{1-e^M}{1-e^M+e^M(1-x)}=\frac{1}{1+\mathcal{N}(1-x)}=\frac{1}{1+\mathcal{N}}\frac{1}{1-\dfrac{\mathcal{N}}{1+\mathcal{N}}x}$$

$$\frac{LR}{M^2}\frac{(1-e^M)(x-1)}{1-xe^M}=\frac{-\mathscr{S}}{1+\mathcal{N}}\left\{1-\frac{\dfrac{1}{1+\mathcal{N}}x}{1-\dfrac{\mathcal{N}}{1+\mathcal{N}}x}\right\}$$

$$=\frac{\mathscr{S}}{\mathcal{N}(1+\mathcal{N})}\frac{\dfrac{\mathcal{N}}{1+\mathcal{N}}x}{1-\dfrac{\mathcal{N}}{1+\mathcal{N}}x}-\frac{\mathscr{S}}{1+\mathcal{N}}. \tag{15.39}$$

Then we can make the identification

$$-y=\frac{\mathscr{S}}{\mathcal{N}(1+\mathcal{N})},\qquad z=\frac{\mathcal{N}}{1+\mathcal{N}}x,$$

$$\langle x^{a^+a}\rangle=\frac{1}{1+\mathcal{N}}\sum_{k=0}^{\infty}L_k\left[\frac{-\mathscr{S}}{\mathcal{N}(1+\mathcal{N})}\right]\left[\frac{\mathcal{N}x}{1+\mathcal{N}}\right]^k\exp\left[\frac{-\mathscr{S}}{1+\mathcal{N}}\right]. \tag{15.40}$$

The photocount probability distribution function can be written down by comparing (15.40) and (15.29):

$$P(k)=\frac{1}{1+\mathcal{N}}\left[\frac{\mathcal{N}}{1+\mathcal{N}}\right]^k L_k\left[\frac{-\mathscr{S}}{\mathcal{N}(1+\mathcal{N})}\right]\exp\left[\frac{-\mathscr{S}}{1+\mathcal{N}}\right]. \tag{15.41}$$

For experimental arrangements to measure this distribution (assuming the Law of Large Numbers applies) *see* Arecchi [154]. A typical comparison of (15.41) with experiment is shown in Fig. 15.8, from Freed and Haus [155]. Deciding between this and a more refined theory would not be easy.

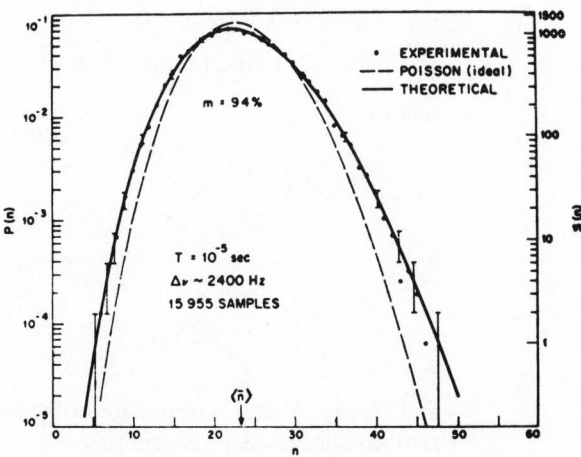

Fig. 15.8. Probability distribution with laser operating above threshold, $T=10^{-5}$ s, $\Delta\nu\sim$ 2400 Hz. (*Freed and Haus* [155].)

ANALYTIC CORRESPONDENCE

In this part of the chapter we reconsider the 'laser' equations of motion (15.19), but this time subject them to a different set of boundary conditions. That is, we search for stationary solutions at thermodynamic equilibrium at temperature T ($\beta = 1/kT$). The results lead again to a cusp catastrophe. Since the details of the treatment with equilibrium boundary conditions are very similar to the treatment of Section 9, the discussion of Sections 12 to 16 is somewhat compressed. The principal result – and surprise – is that the different results predicted by (15.19) under equilibrium and non-equilibrium boundary conditions are related very closely to each other and can be obtained from each other by 'real analytic correspondence' on the cusp manifold. (We might say 'analytic continuation', but this sits uneasily with the usage of mathematicians.)

This displays the omnipresence and unifying power of catastrophe theory in its strongest form: not so much saying 'in certain dimensions everything is a cusp', more enabling us to find a common canonical form for different detailed models so that their correspondence becomes clear, and mapping predictions and experiments for one to predictions and experiments about the other. It is as important and useful to recognize the occurrence of a cusp catastrophe as of a simple harmonic oscillator.

13 Equilibrium Boundary Conditions

Under equilibrium boundary conditions, the steady state and equilibrium expectation values are one and the same, and $\langle \sigma^z \rangle_e$ is determined in a self consistent way. Equations (15.19) reduce (Gilmore and Bowden [156]) to

$$\hbar\omega\langle a\rangle_e + N\lambda^*\langle\sigma^-\rangle_e = 0$$

$$\varepsilon\langle\sigma^-\rangle_e - \lambda(\langle a\rangle_e + \alpha)\langle\sigma^z\rangle_e = 0$$

where

$$\langle\sigma^z\rangle_e = -\left(\frac{\varepsilon}{2\theta}\right)\tanh\beta\theta$$

and

$$\theta^2 = \left(\frac{\varepsilon}{2}\right)^2 + \lambda^2(\langle a\rangle + \alpha)^2. \tag{15.42}$$

We choose λ real without loss of generality, and $\langle a\rangle_e$ is real if α is real, as discussed in Section 6.

14 Equilibrium Manifold

The manifold describing the equilibrium values of the order parameters $\langle a \rangle_e$, $\langle \sigma^- \rangle_e$ may be obtained by eliminating $\langle \sigma^- \rangle_e$, $\langle \sigma^z \rangle_e$ from (15.42)

$$\hbar \omega \langle a \rangle_e - \left(\frac{N\lambda^2}{2\theta} \right)(\langle a \rangle_e + \alpha) \tanh \beta\theta = 0. \tag{15.43}$$

This can be put into the canonical form for the cusp catastrophe by expanding θ and $\tanh \beta\theta$ up to terms in $(\langle a \rangle_e + \alpha)^2$ and applying the theorems of Chapter 8. A little arithmetic shows that around $A = B = 0$, (15.43) is equivalent by the unfolding rules to

$$X^3 - AX - B = 0$$

where

$$X = \langle a \rangle_e + \alpha$$
$$A = -[\varepsilon \hbar \omega - N\lambda^2 \tanh (\beta \varepsilon/2)]/C$$
$$B = \varepsilon \hbar \omega \alpha / C \tag{15.44}$$
$$C = 2N(\lambda^2/\varepsilon)^2 \left[\tanh \left(\frac{\beta \varepsilon}{2} \right) - \left(\frac{\beta \varepsilon}{2} \right) \operatorname{sech}^2 \left(\frac{\beta \varepsilon}{2} \right) \right].$$

The coefficient C is positive for all finite temperatures T, $0 \le T < \infty$.

15 Thermodynamic Phase Transition

In the absence of external field $\alpha = 0$ and $B = 0$. The coefficient A is negative at sufficiently high temperatures, and must remain negative at all temperatures if $N\lambda^2 < \varepsilon \hbar \omega$. However, if $N\lambda^2 > \varepsilon \hbar \omega$, the trajectory in control parameter space passes through the cusp point, the system state passes onto the upper or lower sheet, and the system undergoes a second-order phase transition (Hepp and Lieb [148], Gilmore and Bowden [149]), at the critical temperature T_c determined by

$$\varepsilon \hbar \omega = N\lambda^2 \tanh (\beta_c \varepsilon/2). \tag{15.45}$$

16 Critical Behaviour

The existence of a second order phase transition forces a pleat into the equilibrium surface ('pleat' is what the current French name 'fronce' for the cusp catastrophe means). The unfolding (15.44) is versal, and hence universal: for equilibrium as for non-equilibrium conditions, no additional perturbations that change equation (15.44) smoothly will produce any new surprises for small values. All possible nearby types of critical behaviour are described by the cusp catastrophe.

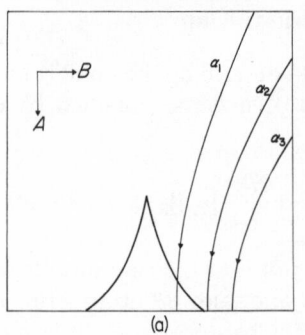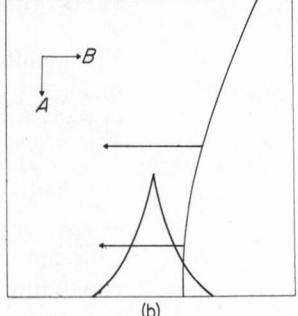

Fig. 15.9

In Fig. 15.9(a) we plot trajectories in control parameter space obtained by holding α fixed at various values and decreasing the temperature. No phase transitions occur, as the system state passes smoothly onto the upper sheet. In Fig. 15.9(b) we illustrate what happens in control parameter space when T is held fixed and α is decreased through zero from its initial positive value, and then made sufficiently negative. Divergence is clearly apparent. There are no phase transitions for $T > T_c$. For $T < T_c$ first-order phase transitions occur. If the delay convention is obeyed, these transitions occur when the trajectory crosses the fold lines. Hysteresis occurs (Gilmore and Narducci [147]).

17 Analytic Correspondence of Experiments

The stationary values of the dynamical equations (15.19) exist in one-to-one correspondence with the points on the cusp manifold, whether these equations are subject to equilibrium or nonequilibrium boundary conditions. Thus, any equilibrium system can be identified with a corresponding non-equilibrium system via the cusp manifold. We can call this a 'real analytic correspondence' because it allows us to map all properties of either (say the equilibrium) into the corresponding properties of the other (nonequilibrium) by a real analytic manifold. To be sure, the curved paths obtained in the non-equilibrium regime by fixing $\langle \sigma^z \rangle_e$ and varying α (Fig. 15.5) are images of paths in the equilibrium regime that are obtained by varying *both* temperature (or β) *and* field (α) simultaneously. The specific mappings are obtained from (15.22) and (15.44). Knowing what the specific mapping is, is less important than knowing that it exists. For example, knowing that a second order phase transition occurs in the non-equilibrium (laser) case guarantees the occurrence of a pleat in the stationary state manifold. This in turn guarantees the existence of first-order phase transitions (jumps). They are lurking there somewhere – it is only a question of finding the appropriate physical controls to produce them. Similarly, knowing the existence of an analytic correspondence from the non-equilibrium to the equilibrium regime guarantees the existence of second- and first-order phase transitions in thermodynamic equilibrium.

To illustrate the importance of this concept consider the photo-count experiments described in Section 12. In the first place, all data were taken for the specific case of zero external field. The predictions (15.41) can immediately be extended to the $\alpha \neq 0$ case through (15.35). If the noise level in the experiment is unchanged, the photocount distributions will have the same shape for all experiments with the same signal $\mathscr{S} = (x - \frac{2}{3}\alpha)^2/M^2$. Therefore, *all physical controls* $\langle \sigma^z \rangle_e$, α *which produce the same value* $\langle a \rangle_s = x - \frac{2}{3}\alpha$ *must yield the same photocount distribution in experiments carried out at constant noise level.*

We now raise the ante even higher by invoking the analytic correspondence concept. In the equilibrium case the reduced field density operator is given by (15.35) with $\langle a \rangle_e$ replaced by $x - \alpha$, and with $-M = 1/kT$. Therefore, the photocount distributions under equilibrium conditions are given by (15.41) with $\mathscr{S} = (x - \alpha)^2$ and $\mathscr{N} = (e^{\beta \hbar \omega} - 1)^{-1}$. Identification of other experimental results for equilibrium and non-equilibrium configurations by way of the cusp manifold is done in exactly the same way.

18 Future Prospects

The universality of the unfoldings (15.22) and (15.43) would suggest that more interesting new effects will be predicted and discovered only when more complicated Hamiltonians are studied. Such might involve three- or four-level systems, or higher symmetries than the simple evenness of the model (15.8) that we have unfolded with a single symmetry-breaking field.

How should such studies be carried out? The work above presents a clear suggestion.

(a) Unfold universally the predictions of the new Hamiltonian.

(b) Explore thoroughly all the critical possibilities under one set of boundary conditions, say in equilibrium.

(c) Carry the information gained to a non-equilibrium regime via the appropriate standard catastrophe model. If the behaviour does *not* correspond in the two regimes (say we get a cusp in one, dual cusp in the other) look for a higher catastrophe between them (like the butterfly by which cusp changes to dual cusp buckling in Chapter 13 Section 8).

16 Biology and Ecology

Thom's prime motivation for studying catastrophes (in the broad sense) was the search for a framework of theory to bind together the enormous quantity of observational data available in biology. It is perhaps ironic that little of the extant data can actually be used to test the theories that have emerged, because the phenomena they predict often require new kinds of data for experimental confirmation! Thom's theories are very broad and philosophical, giving the style of modelling but not the details. The best developed attempt to put Thom's programme into practice is the paper of Zeeman [157] which advances a cusp catastrophe interpretation of the formation of a biochemical frontier and applies it to gastrulation and formation of somites in amphibia, and to the culmination of slime-mould. He also notes that similar frontiers may form between species in ecology.

Our aim here is even more modest. It is to discuss possible catastrophe mechanisms for the formation of frontiers, with emphasis on ecological examples where suitable data exist (though not in abundance) and where the effects are probably simpler; and to explain how our conclusions extend to biochemical frontiers. We begin with an 'economic' model which may shed light on the tendency of bees to occur either as solitary species, or as social species in which the groups that occur are very large. Why is there this noticeable gap? Analysing this by a relatively simple, but not unrealistic model, we are led to a new class of catastrophe: the *constraint catastrophe*, applicable to regions possessing boundaries. The same methods that handle Thom's original seven catastrophes apply to these, though the list is more extensive. We relate these ideas to Zeeman's theory of frontiers, and discuss existing numerical data and experiments that might be performed. Finally we move on to embryology. Zeeman's model of gastrulation and somites requires more embryological background than we could reasonably put into this book, and is in any case very clearly explained in his paper. The reader who desires greater biological detail should consult it, bearing in mind the qualifications that emerge from this chapter.

THE SIZE OF BEE SOCIETIES

1 Bee Economics

There are about 20 000 known species of bee. Of these, all but about 5% are *solitary*, like most insects: they do not live in cooperative colonies. The remainder are *social* species, and form substantial communities. The social bees fall into three main groups: the bumblebees (*Bombinae*), tropical stingless bees (*Melponinae*), and honeybees (*Apinae*). Bumblebee colonies seldom attain enormous size: a colony of 1000 is exceptional. An average colony of *Bombus muscorum* contains about 120 individuals. For tropical stingless bees the size of colony varies from species to species: in some a few dozen, in others several thousands. Honeybees habitually come in groups of thousands, ranging up to 180 000 or more. With rare exceptions (particularly outside the tropics) the distinction is quite sharp: either one sees groups of several thousand individuals, or colonies which are small and primitive if they exist at all.

Why?

Consider an environment producing sufficient nectar to maintain a population density B of bees/km^2. A colony of N bees will be supported in equilibrium by a territory of area N/B, of radius R km, say. Roughly (since territories may overlap) we have

$$N = \pi R^2 B. \tag{16.1}$$

Notice that since $N \geqslant 1$ trivially, this implies

$$R \geqslant R_{min} = \surd(1/\pi B). \tag{16.2}$$

The average time spent travelling to and from the hive on collecting trips (ignoring scouting time) is proportional to the average trip length,

$$2R/3B.$$

Since this is time lost from production, the average productivity of a bee can be expressed as

$$P(R) = C - \frac{A}{B} R = C - \frac{A}{\pi B^{3/2}} \surd N \tag{16.3}$$

where A and C are constants depending on the bees' collection technique, velocity, load capacity and so forth. It is a basic ecological principle that less efficient species among those competing for the same food are squeezed out, so if equation (16.3) were all of bee economics social bees would clearly disappear in favour of some species of solitary bee on the spot.

2 The Advantages of Aggregation

The compensating advantage which permits a choice of large N and R to be competitive is of course division of labour. Computations of the benefits of this go back to the founding father of systematic economics, with Adam Smith's calculation that ten men dividing the labour between them could make 48 000 pins per day, while one could make perhaps 20. Without entering into the details of hive economy, let us suppose that the benefits S of aggregation increase at least linearly with N, perhaps faster, for small N. Many more detailed models could support this. Consequently S increases at least quadratically with R for small R (by equation (16.1)) and we have our first postulate.

(a) The benefit S per bee of aggregation, as a function of R, has a slope dS/dR which is positive and increasing at R_{min}.

Our second postulate is essentially the law of diminishing returns.

(b) The slope dS/dR eventually becomes a decreasing function of R for large R, with

$$\lim_{R \to \infty} \frac{dS}{dR} = 0.$$

This is guaranteed, for instance, if any given task will only subdivide so far. Then the curve S will 'saturate', never rising above some level: doubling the population simply doubles the number of bees doing each subtask, without improving productivity. It will equally follow from the weaker assumption that $S(R)$ increases only logarithmically for large R, or the stronger one that organizational problems, difficulties of waste disposal, etc. produce an actual decrease for large R. (Such points are illustrated by the honeybee's enlarged rectum, compared to other bees, which permits storage of faeces while inside the hive.) Specialization on flower types (Heinrich [157a]) saturates when there are as many specialists as kinds of flower.

It follows from postulates (a) and (b) that dS/dR has at least one maximum at R_i (say), as in Fig. 16.1. We have drawn S as a saturation curve for definiteness, but as noted this is not essential. We have also made the simplifying assumption that dS/dR has *only* one maximum; we will return to the more complicated possibilities later.

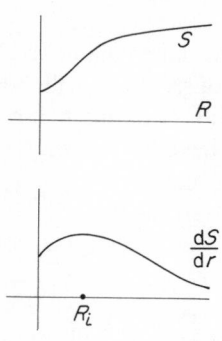

Fig. 16.1

3 Catastrophe Geometry

Now consider the total utility function

$$F(R) = P(R) + S(R)$$

$$= C - \left(\frac{A}{B}\right)R + S(R), \qquad (16.4)$$

from which we have

$$\frac{dF}{dR}(R) = \frac{dS}{dR}(R) - \frac{A}{B}. \qquad (16.5)$$

The graph of dF/dR is thus extremely easy to deduce from that of dS/dR for various A/B (Fig. 16.2); and the elementary fact that where dF/dR increases or decreases through 0, then F has a

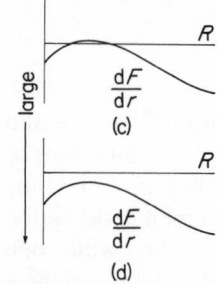

Fig. 16.2

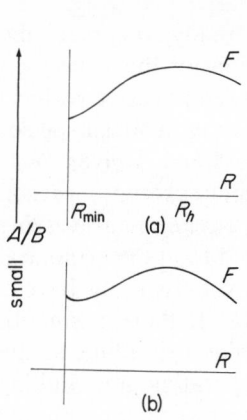

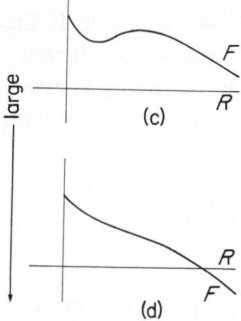

Fig. 16.3

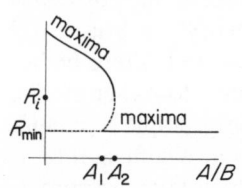

Fig. 16.4

minimum or maximum respectively, gives Fig. 16.3. The constant of integration is unimportant in what follows.

The clear relation with what we did in Section 2 of Chapter 7 leads us to draw the 'graph' of local maxima and minima against A/B, which yields Fig. 16.4, with a fold catastrophe at (A_2, R_i). If B is large enough (i.e. food plentiful enough for a large population density B) to make $A/B < A_1$, then F looks like Fig. 16.3(a). The unique maximum is at a large value R_h for R, which means large colonies. Any species that experiments accidentally with cooperation will rapidly evolve towards colonies of population size $N = \pi R_h^2 B$, since a move in that direction is immediately rewarded by an increase in efficiency and competitive edge, and hence genetically consolidated. ('Rapidly' on an evolutionary timescale, of course; the point is that in an arbitrary century we are unlikely to see a species far from equilibrium.)

An unfavourable environment, supporting a population density B so low that $A/B > A_2$, gives F a unique maximum at R_{min}. In these conditions (unless a change in the bees' 'technology' can suitably change A or the form of F) the social bees cannot efficiently compete with the solitary species, and are wiped out. (Only the bumblebee is found wild in England, and as we saw this tends to form very small colonies, with a wide seasonal population cycle outside our present scope.)

This model is highly simplified, of course. It leaves out special adaptations that permit solitary species to survive anarchistically in the territory of a commune: ingenious ploys like imitating a social species to the point of moving in and sharing its food for no work (a technique not unknown in the genus *Scientist*), and many other subtleties of bee ecology. However, Fig. 16.4 makes clear the implication of postulates (a) and (b) that *sudden* appearance and disappearance of substantial colonies is to be expected (even with only smooth variations of the environment). Moreover, colonies of more than perhaps a few hundred bees (for which equation (16.3) is swamped by random fluctuations in food distribution) and less than a far more substantial population $R_i^2 B$ should be rare, and either growing or shrinking individually, or part of a rapidly evolving species.

4 Variation in Space

Consider an improving environment, initially without social bees and inaccessible to social bees from elsewhere. One would expect the delay convention to operate, with social species not appearing before B has increased to make $A/B < A_1$. (It is unlikely that the bees will develop the substantial genetic modifications necessary for a large society before small movements *towards* it are rewarded.) But once a varied environment has a reservoir of both social and solitary species, variously located but always ready to spread, the Maxwell convention might operate. Competition for the same food favours absolute maxima of efficiency, if representatives of all local maxima are present to compete. This suggests that a social species will be present exactly when and where, for the S and A given by its nature, the carrying capacity B of the environment exceeds a certain value B_c between A/A_2 and A/A_1. In an environment varying smoothly in space, steady in time (often referred to in the ecological literature as an *environmental gradient*) this will typically have a sharp boundary in the form of a curve $B = B_c$. If there is smooth variation in time as well, this curve will vary like a shoreline as the tide rises and falls; creating, destroying, separating and uniting islands of social bee habitation. It need not be associated with any sharp changes in the environment itself, except such as may themselves be caused by the presence or absence of social bees. If the effects of competition are less absolute, permitting local efficiency maxima to survive and giving something closer to the delay convention, the boundary becomes history-dependent, but should remain sharp.

5 Complications

If, while still satisfying postulates (a) and (b), the function S is complicated by the existence of more than one maximum for dS/dR (for instance Fig. 16.5, where S is the sum of several saturation curves, each relating to division of labour on a different task), the catastrophe geometry becomes more complicated. But most of the above analysis still holds.

This use of catastrophe theory is illuminating, if its postulates apply, and illustrates the possible uses of the theory. Such uses will become more subtle and powerful, particularly as and where higher catastrophes are generated by models and related to experimental data. But it is already clear that the comparatively simple dynamics implicit in catastrophe modelling cannot come near a sufficient basis for the whole of mathematical ecology. Indeed, the latter is attracting the research interest of leading experts in dynamical systems theory, precisely because of the many examples of 'chaotic' behaviour (like that to be discussed in Chapter 17 Section 7) displayed

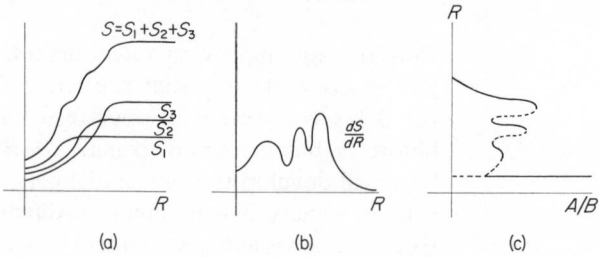

Fig. 16.5

(a)   (b)   (c)

mathematically by plausible ecological models of deceptive apparent simplicity. In this, as in many scientific fields, catastrophe theory is a useful tool but not the 'open sesame' to all the secrets of nature. (With Thom's wider use of the term 'catastrophe' it is an 'open sesame', but not always a useful tool!)

CONSTRAINT CATASTROPHES

6 Boundary Effects

The reader may have noticed that the catastrophe by which local stability of solitariness (with respect to aggregation) changes in Fig. 16.4 is not one of the phenomena studied in Chapters 7 to 9. It is, however, a perfectly stable event of codimension 1. This should be clear from Fig. 16.6, which shows it and its dual. Thus this bifurcation is perfectly stable in a single-parameter family, and yet not in the list given by Thom's theorem. This is not a counterexample to the theorem: like the effects of symmetry (Chapter 7 Section 3, Chapter 14 Sections 15 to 20) it shows the care with which the theorem must be approached. The entire discussion of Chapter 7 rested on the identification of maxima, minima, etc. with points where the derivative vanishes. This is valid inside the domain of a function, but if the domain has a sharp boundary it is false. The relevant condition is now that the first derivative *in directions along the boundary* should vanish (Fig. 16.7). (If the boundary of the domain is zero-dimensional, as in Fig. 16.6, this condition is satisfied automatically.)

Since inequalities are a vital part of the mathematics of the biological and social sciences (a population or chemical concentration cannot be negative, or territorial radius lower than R_{min} in the

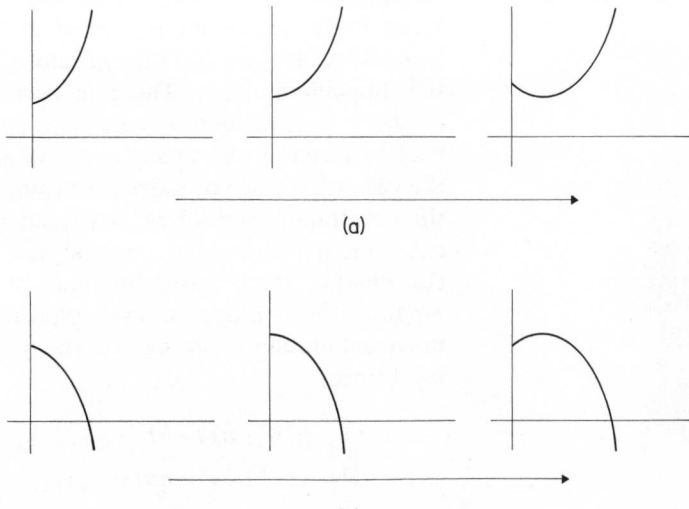

(a)

(b)

Fig. 16.6

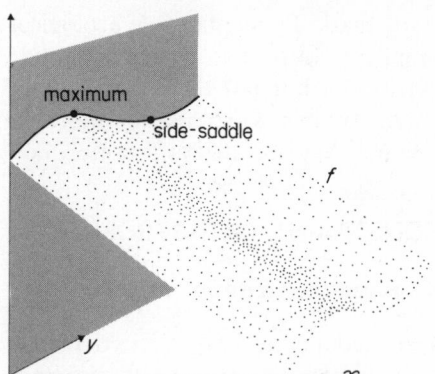

Fig. 16.7

last section; at most N vehicles per hour can use a given stretch of road) the bifurcations of optimization problems in these studies will involve such boundary critical points. (In 'nonlinear programming', as the computational aspect of these problems is called, the extrema are more often than not on boundaries.) The results of Chapters 7 and 8 cannot be applied here, *but the methods can.* In the absence of special conditions such as symmetry or linearity, we may again list the typical phenomena to be expected in r-parameter families for various r. A full treatment will appear in Pitt and Poston [158]; here we illustrate the kinds of *constraint catastrophes* that arise. (Note that the usual catastrophes can still occur in the interior of the domain; indeed most constraint catastrophes have associated families of interior catastrophes of lower codimension.)

7 Classification

Even if the constraint is curved in the original coordinates (for example $x^2 + y^2 + z^2 \leq 1$) it typically gives a smooth boundary, and the Implicit Function Theorem permits us to choose coordinates $(x, y_1, \ldots, y_n)$ around a point of interest, such that the constraint may be given locally simply as $x \geq 0$, as in Fig. 16.8. (We omit here the case where two or more constraints combine to give a corner, or the constraints themselves vary in an r-parameter family and bifurcate, though these can be analysed in the same way.) In such coordinates, stable cases (minima, non-critical points, etc.) apart, r-parameter families for $r \leq 4$ typically can meet functions whose universal unfoldings in this constraint sense can be given the following forms:

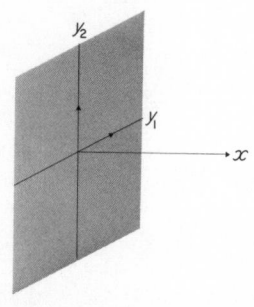

Fig. 16.8

$$r = 1 \quad \pm(x^2 + ax) + M_n(y_1, \ldots, y_n)$$
$$r = 2 \quad \pm(x^3 + bx^2 + ax) + M_n(y_1, \ldots, y_n)$$
$$xy_1 \pm y_1^3 + by_1 + ax + M_{n-1}(y_2, \ldots, y_n)$$

$$r = 3 \quad \pm(x^4 + cx^3 + bx^2 + ax) + M_n(y_1, \ldots, y_n)$$
$$\pm(x^2 + y_1^3 + cxy_1 + by_1 + ax) + M_{n-1}(y_2, \ldots, y_n)$$
$$\pm(xy_1 + y_1^4 + cy_1^2 + by_1 + ax) + M_{n-1}(y_2, \ldots, y_n)$$
$$r = 4 \quad \pm(x^5 + dx^4 + cx^3 + bx^2 + ax) + M_n(y_1, \ldots, y_n)$$
$$xy_1 \pm y_1^5 + dy_1^3 + cy_1^2 + by_1 + ax + M_{n-1}(y_2, \ldots, y_n)$$
$$\pm(x^3 + 3tx^2 y_1 + y_1^3 + [ex^2 y_1] + dxy_1 + cx^2 + by_1 + ax)$$
$$+ M_{n-1}(y_2, \ldots, y_n), \quad t^3 \neq \tfrac{1}{4}$$
$$\pm(x^2 \pm xy_1^2 + tx^4 + [ey_1^4] + dxy_1 + cy_1^2 + by_1 + ax)$$
$$+ M_{n-1}(y_2, \ldots, y_n), \quad t \neq \tfrac{1}{4}$$
$$\pm(x(y_1 + ty_2) + y_1^2 y_2 - y_2^3 + [exy_2] + dy_2^2 + cy_2 + by_1 + ax)$$
$$+ M_{n-2}(y_3, \ldots, y_n), \quad t^2 \neq 1$$

where each M_i is a nondegenerate Morse sum and difference of i squares (compare Chapter 7 Section 7). We do not list the forms

$$x + [\text{something on Thom's list in } (y_1, \ldots, y_n, t_1, \ldots, t_r)]$$

since the only new feature is that 'inwards' is a new sort of inessential variable. The terms in t and e for the case $r = 4$ arise from the same sort of infinite families of types as in Chapter 7 Section 6 and Chapter 14 Section 19; a 4-parameter family can meet one type (for one value of t) which only 5 parameters can meet transversely. Perturb the 4-parameter family a little, and it will include another type, reducible to the same form but not to the same value of t. (In each of the cases listed, the obstruction to equivalence of the types with different values of t can again be geometrically described in terms of the cross-ratio of four lines.) So 'almost everything is stable' only up to $r = 3$, instead of $r = 5$ in the situation of Chapter 7.

Not distinguishing duals, the new list gives 14 catastrophe forms up to t values, in addition to the original 7 (inside or in the boundary), describing everything typically met with up to four external variables. The geometry of the first six of these is shown in Fig. 16.9 (and will be extended to higher cases in Pitt and Poston [158]: space forbids further discussion here). As with interior catastrophes, higher dimensional unfoldings 'organize' lower dimensional ones; for instance (iii) contains a line of (i), a line of (ii), and a line of folds. In (v) and (vi) the x-direction has been suppressed, leaving y as the vertical direction, to make drawing possible. The heavy lines in the (a, b)-plane, in each case, mark the points at which the number of minima changes. Note that although (iii) resembles the cusp (and (vi) the dual cusp) insofar as the region where two minima exist comes to a sharp point, the point is here created by two curves in parabolic contact. Thus for instance a smooth change of variables could make the point convex on one side, concave on the other, whereas the infinite curvature of the single, direction-reversing fold curve of the usual cusp makes biconcavity near the tip a diffeomorphism-invariant property.

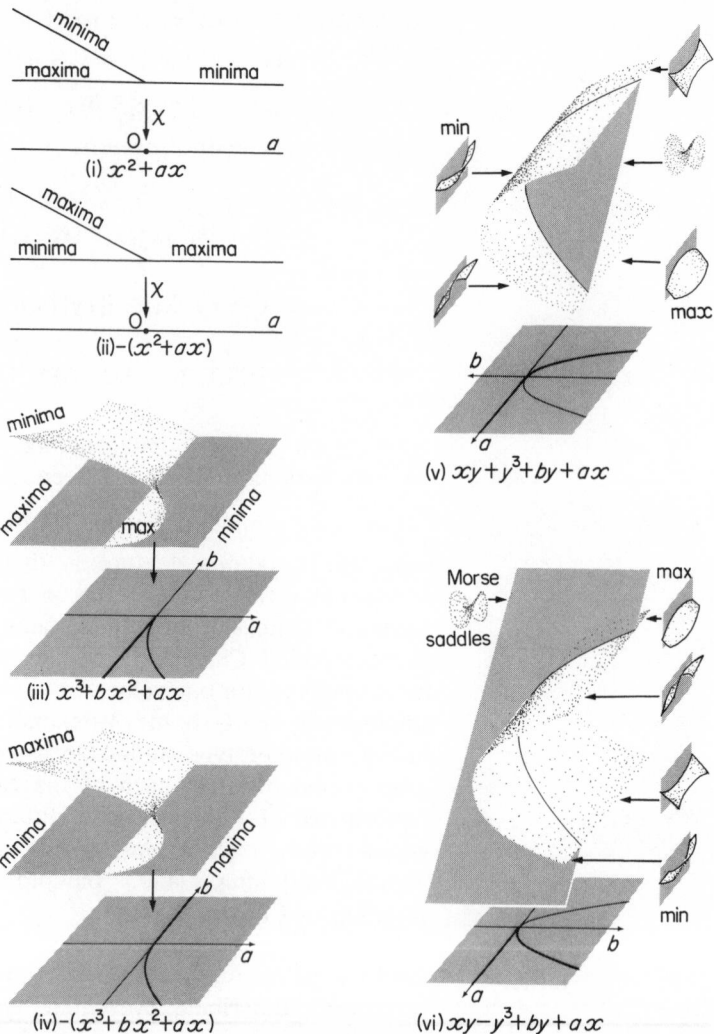

minima
maxima   minima

O↓X
          a
(i) $x^2 + ax$

maxima
minima   maxima

O↓X
          a
(ii) $-(x^2 + ax)$

minima
maxima   minima
max
          b
          a
(iii) $x^3 + bx^2 + ax$

maxima
minima   maxima
          b
          a
(iv) $-(x^3 + bx^2 + ax)$

min   max
b
a
(v) $xy + y^3 + by + ax$

Morse   max
saddles
min
b
a
(vi) $xy - y^3 + by + ax$

Fig. 16.9

This classification is too new for us to offer as many examples and applications as we have given for the interior catastrophes in earlier chapters, but the range of applicability is equally wide. We derived Fig. 16.4 from bee economics, having first thought of it in connection with archaeology (Chapter 17 Section 2 and Renfrew and Poston [159]) and later applied similar arguments to the location and size of shopping centres (Poston and Wilson [160]). More recently, we have found an example in laser physics (see also Chapter 15), involving not only Fig. 16.4 but also Fig. 16.9(iii), of which Fig. 16.4 is the 'front slice'. The explicit model of phonon-assisted transition to the super-radiant (laser action) state given in Thompson [161] is locally reducible by the appropriate determinacy and unfolding theorems for this context to the catastrophe $x^3 + bx^2 + ax$ plus a positive definite 'inessential variable' term y^2.

As noted in Chapter 14 Section 17, there is a close relation between constraint and group-invariant catastrophes; indeed, the problems are often mathematically equivalent. Simpler than Thompson's phonon-assisted laser transition, consider the laser of the previous chapter, with zero external magnetic field α. This reduces to the form $X^4 - AX^2$ (in integrated, 'potential' form) and within the space of even functions, since this is equally the space of functions of X^2, we could change variable to $\eta = X^2$ and get the constraint catastrophe $\eta^2 - A\eta$, $\eta \geqslant 0$. This is exactly what is involved in the change from *amplitude* (which can be negative with opposite phase) to *intensity* (which is essentially positive). Once symmetry is broken, however (by for instance an external field), the transformation confuses rather than assists.

Notice the way in which mathematical stability depends on the stability of the hypotheses; economic constraints, like price $\geqslant 0$, are far more stable than symmetries, like magnetic field $= 0$, which allow transformation to constraint problems. One could transform the economic problem into a symmetric one, break the symmetry, and say 'behold the instability!' – but one could not bring the symmetry-breaking term back through the transformation. Structural stability, we must repeat, is contingent on the formulation of the problem and the class of perturbations judged relevant, as well as the notion of equivalence used.

The same set of standard polynomials, with an interpretation different enough to change the pictures, arises naturally again in Section 14.

TRAVELLING WAVES IN ECOLOGY

8 Choice of Convention and Model

We argued in Section 4 for the relevance of the Maxwell convention to bee socialization dynamics, but these arguments – even if correct for that case – do not apply universally in ecology. They correspond in economic terms to a system (which in its pure form can be found only in the speeches of U.S. Congresspersons) whereby different agents *compete* but do not mutually *interfere*. However, if you can make dingbats slightly cheaper and better than your competitor but he owns all the dingbat shops, he wins. You need the capital to set up your own chain of shops, just to enter the game.

Similarly, if tree species A is better fitted than B for a particular piece of ground G in present conditions, but as a result of history species B is in possession, then B is likely to stay put. Young specimens of A may do better on G than young specimens of B, and likewise for mature specimens; but A can only invade with new plants, by seed, and a mature stand of B can deal very effectively with that. (An extreme example is the yew tree, which not only takes all the sunlight but also poisons the ground below, leaving it bare for some years after the tree is removed.) Thus if we may use catastrophe theory here, the delay convention is suggested. Often something more general than catastrophe theory will be required – explicit ecological models can exhibit bifurcations of other kinds, from the classic differential equation Hopf bifurcation (Hopf [162], Hirsch and Smale [163] Marsden and McCracken [164]) to the onset of 'chaos' – but equally, systems with more elaborate dynamics than a simple minimization or maximization often exhibit behaviour reducible to a catastrophe. Hence we may extend the rule of thumb of all the sciences 'Try a linear model first', when bifurcation phenomena make clear that this won't work, to 'Try a catastrophe model next'. If this gives experimentally verified results, we have captured at least the consequences of the underlying dynamics and made a good step towards analysing the detailed mechanisms involved. We have found, too, a description of the behaviour that has all the robustness guaranteed by structural stability – which is why Thom makes this the cornerstone of his philosophy. If it does not fit experiment, and particularly if phenomena are revealed which definitely cannot be obtained from catastrophe geometry, then we have also learned something important about the dynamics. (Just as the Michelson–Morley experiment showed that the motion of light could not be treated as vibrations in the ether, and prepared the ground for Relativity.) On general grounds, catastrophe theory should often, though not always, be a successful descriptive technique.

9 Frontiers

We can consider the formation, movement, and disappearance of ecological frontiers – sharp differences in the distribution of species – from this point of view. (The possibility of this sort of model was mentioned by Zeeman [157], though more as an illustration of the frontiers that form in the biochemistry of a developing organism.) Such frontiers are common; quoting Connell, Merz and Murdoch [165] p. 103:

'On the other hand the boundaries are often more abrupt than any changes in the physical environment, suggesting either that the physical limiting factor is transient (such as fire) or that biotic agencies may be limiting. Thus a predator may eliminate all individuals outside a safe refuge ... or one species may displace a competitor completely from part of its habitable range'.

More specifically, Connell [166] observes:

'In the course of an investigation of animals of an intertidal rocky shore I noticed that the adults of 2 species of barnacles occupied 2 separate horizontal zones with a small area of overlap, whereas the young of the species from the upper zone were found in much of the lower zone. The upper species, *Chthamalus stellatus*, thus settled but did not survive in the lower zone. It seemed probable that this species was eliminated by the lower one, *Balanus balanoides*, in a struggle for a common requisite which was in short supply'.

The ecological literature contains many such observations for a wide variety of species: instances involving plants are found in Daubenmire [167]; and Ford [168] includes a case occurring for the butterfly *Maniola jurtina*.

For simplicity we shall discuss only one spatial dimension: for example height up a slope, where temperature and drainage vary; or height on the shore for *Chthamalus* versus *Balanus*. The discussion extrapolates to higher spatial dimensions, but the possibilities become more numerous. Suppose environmental conditions vary smoothly in space, and so do the initial biota. (In particular, take land that has been completely cleared by man – or created empty, like slag heaps – so that different invading plant species start with smoothly varying advantages given by the local environment.) How should frontiers develop? And how should they look, once formed?

The simplest model is given by Fig. 16.10. At time 1, across the whole range of the space variable *a*, conditions are much more favourable for plant X, which establishes itself throughout. At this stage, conditions are resistant to invasion by plant Y. But as it matures and increases, X itself changes these conditions, until at time 2 they are favourable at a_c to young members of Y. The typical way to encounter a fold curve (if that is the hignest catastrophe

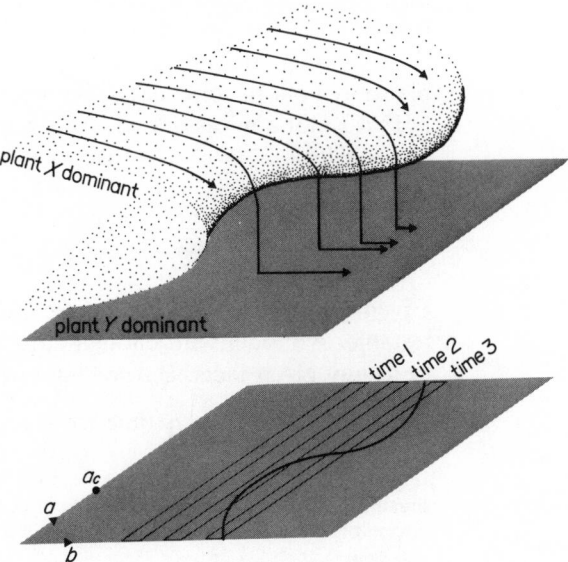

Fig. 16.10

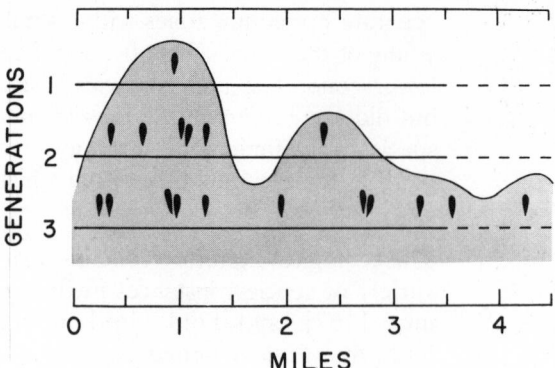

Fig. 16.11. *Distribution of extremely active colonies of* Malacosoma pluviale *along a forested roadside after the first colony was established in the area, with schematic representation of a frontier.* (*Adapted from Wellington* [169].)

present) is parabolically,† as shown: moreover Wassermann [112] has shown that this is structurally stable. It is immensely *un*stable for the whole range to fall over the fold at once; that would require exact homogeneity of soil, drainage, and so forth. Thus small 'spots' of Y appear. This 'spottiness' is certainly what is observed in the successive changes of vegetation in, for instance, land returning to forest; and is observed for many other species (for example the moth *Malacasoma pluviale* studied by Wellington [169]). Beyond this, the model predicts that the initial widening of the segments (or two-dimensional spots in two spatial dimensions) will be explosively fast, following an (asymptotic) square root law. The final disappearance of each spot of X should be similarly dramatic, as should the merging of separate spots of Y. (The observations of moth populations in Wellington [169] p. 444 support this (see Fig. 16.11), but are not extensive enough to provide a really good test.)

Notice that Fig. 16.10 involves implicitly another hypothesis beyond the delay convention: essentially that when one equilibrium disappears, the system moves to another *fast*, like the Zeeman machine. In the present context, this would mean a jump in less than the time interval between species counts in field work. This will plausibly hold in some cases, but even where the other assumptions hold it is not a necessary consequence of them. Where it fails, so do the conclusions. We take this point further in Section 12.

10 Numerical Tests

Given adequate field data, these predictions are quite rigorously testable. A model with enough adjustable parameters can be made to fit any given facts. But a square root law prediction

$$\text{(width of spot)} \sim \alpha(\text{time from first appearance})^{1/2}$$

†Parabolic here does not mean that the curve is *exactly* a parabola, but that the local situation is diffeomorphic to a parabola in contact with its tangent. This implies an (asymptotic) square root law near the point of contact, and is diffeomorphism-invariant.

requires fitting only the one number α, and if false will easily be shown so. (Minor complications arising from the *local* nature of the prediction are dealt with in the usual way.)

Another quantitative test is given by the predicted shape of the frontier. Consider the situation at time 3 in Fig. 16.10, either *en route* to the complete dominance of Y or as a final state. (Grassland with little spinneys, or forest with clearings, can exist as reasonably steady states). The vertical dimension in Fig. 16.10 may reasonably be measured by the number N_X of individuals of species X; we could refine this to consider smooth variations in density of Y when that is dominant, without affecting what follows. The state is as given in Fig. 16.12, with N_X the discontinuous function of position a, marked by a heavy line. The prediction is thus that as we cross from 'no X' to 'X dominant' N_X jumps from zero to C, say, and then increases like $c + \alpha(\text{distance})^{1/2}$ for some α. (Zeeman refers to this effect as the 'ripple before the wave': recall that the jumps are moving outward from each other in this picture, by its context in Fig. 16.10.)

Some data appear to confirm this. Work by Ken Ashton of Auckland University (Ashton [170]), at present in progress, investigates frontiers in forests (clearings, etc.) by the following procedure.

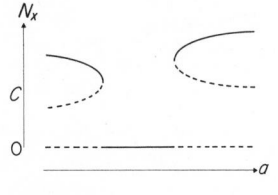

Fig. 16.12

(a) Check for absence of substrate discontinuities (soil, drainage, etc.) and identify all plants in the area down to a specified level.

(b) Map a transect across the frontier region (typically 15 to 30 m long, 2 to 4 m broad). Identify and locate each individual tree in the strip and record its diameter at breast height (dbh).

(c) Test the data for fit against the above theoretical form.

Since the dbh of a tree is closely correlated with its age (*see* Daubenmire [167]) this gives not merely an instantaneous picture, but (allowing for tree deaths) a history of the transect. Though the data are not yet fully analysed, we can give some of the preliminary results, and the general flavour of them.

Fig. 16.13 shows parabolic curves fitted to data for two tree species, *Phyllocladus trichomanoides* and *Leptospermum scoparium*, from a transect in Kauri Park, Auckland. Each curve is numbered with the dbh in cm. The *Leptospermum* frontier has clearly stabilized, whereas the *Phyllocladus* is advancing. The fitted data are in good agreement with the raw data, errors being of the order of 5% or less.

Fig. 16.14 shows a transect taken on Mt Ruapehu in the centre of the North Island of New Zealand. Ruapehu is an active volcano and the vegetation has suffered frequent disturbances from eruptions and man-made fires. Fig. 16.15 shows fitted data for two of the species occurring. There appears to have been a disturbance at the time of germination of trees now 15 cm dbh. After this disturbance,

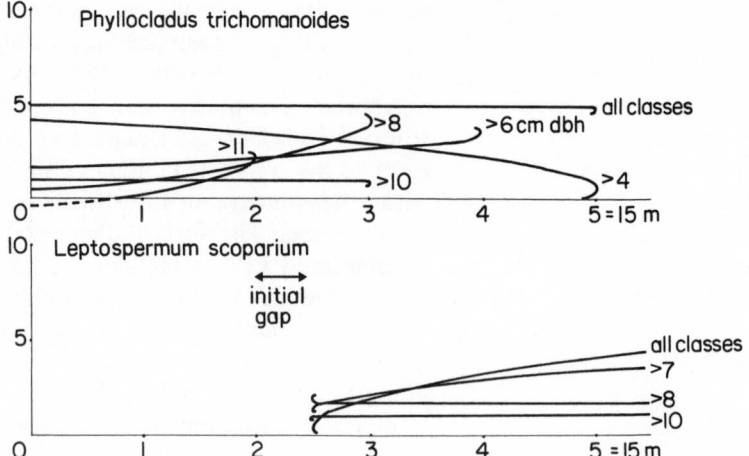

Fig. 16.13. Parabolic curves fitted to preliminary data on distribution of Phyllocladus trichomanoides *and* Leptospermum scoparium *in Kauri Park, Birkdale, Auckland. (Courtesy of Ken Ashton.)*

Phyllocladus has advanced steadily. *Nothofagus* has advanced rapidly and then stabilized.

Even in this preliminary state Ashton's data (of which the above are a small proportion) show some resemblance to the proposed wavefront mechanism, and certainly illustrate the value of fitting parabolic curves to obtain a history of the transect. However, he notes that the basic fold pattern of the waves usually occurs in complex, multiple forms, with successions of waves following each other. This, perhaps, may be traced to environmental disturbances.

Further results are promised, and may be available before this book goes into print. But in any case the final outcome is less important for present purposes than the procedure: our aim is less to proclaim the Triumph of Catastrophe Theory, more to indicate the way that it can be used and quantitatively tested in ecology. If no tests of low-dimensional catastrophe models of this kind show a useful fit, this approach will properly be abandoned. If experience of successful uses of it builds up, however, this will provide a base

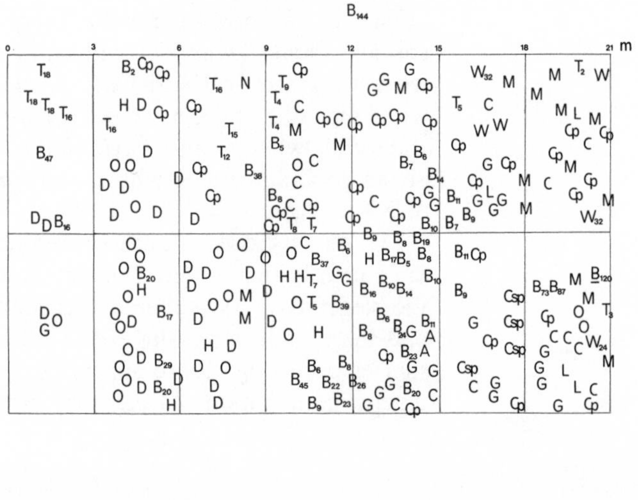

Fig. 16.14. Transect on Ruapehu. Interval 3 m, *width* 2 m × 2.4 m. *Species marked as follows:* (B) Nothofagus solandri *var.* cliffortioides, (T) Phyllocladus alpinus, (D) Dracophyllum filifolium, (O) Olearia nummularifolia, (C) Coprosma foetidissima, (Cp.) C. pseudocuneata, (Csp.) C. spathulata, (M) Myrsine divaricata, (H) Hebe venustula, (N) Nepanax (Pseudopanax) simplex, (L) Leptospermum scoparium, (G) Gleichenia cunninghamii, (A) Astelia sp. B̲ *is a dead tree, and numerals indicate dbh in centimetres (Courtesy of Ken Ashton.)*

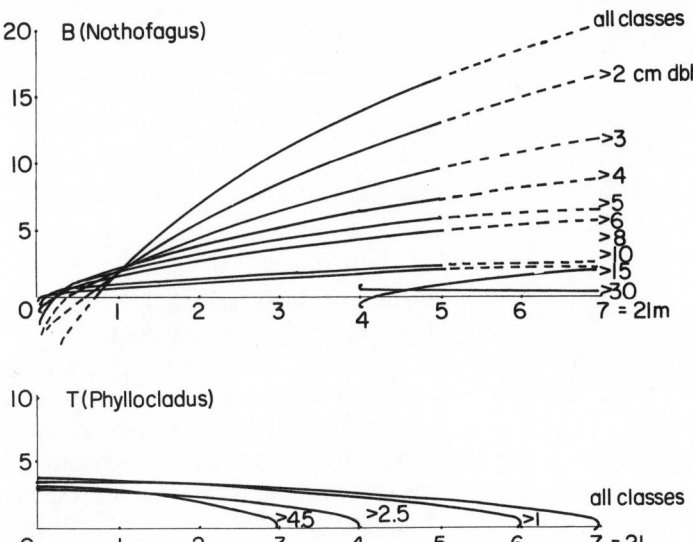

Fig. 16.15. Parabolic curves fitted to data from Fig. 16.14 for Nothofagus *and* Phyllocladus. (*Courtesy of Ken Ashton.*)

upon which a controlled catastrophe theoretic understanding of more complicated phenomena can be built.

11 How Frontiers Stabilize

Still assuming the applicability of catastrophe theory with delay convention and fast jumps, Fig. 16.16 tells the story. Zeeman [157] first pointed out that a frontier of this kind, if it stops moving, should not slow down exponentially, as for instance a highly damped pendulum does in returning to equilibrium; it should slow parabolically,† stopping at a point a_c at a specific time. Later states would look like Fig. 16.16(b). Again the typicality and stability

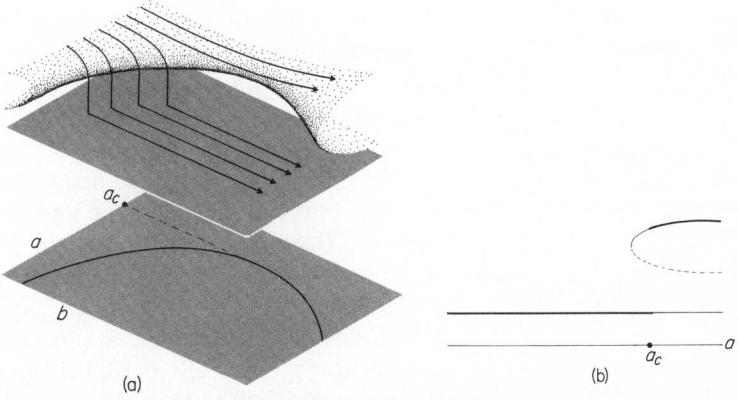

Fig. 16.16

†*See* previous footnote. The parabolic stopping is diffeomorphism-invariant and does not depend – as some authors imagine – on the use of canonical coordinates giving the catastrophe its standard polynomial form. This should be obvious from structural stability of Morse critical points.

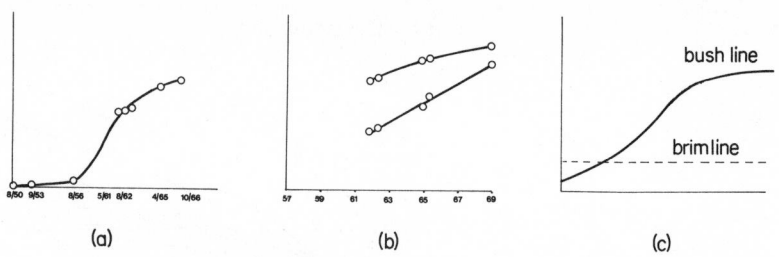

Fig. 16.17

(a) (b) (c)

results of Wassermann [111] substantiate this, given the general hypotheses on the dynamics.

Ashton [170] has evidence for this also; again work is in progress and later results should be more substantial. The Franz Josef glacier in the South Island of New Zealand has been in a state of slow retreat for some time, and various types of plant life have begun to invade the valley as the ice departs: lichens, moss and shrubs. Because the ice is retreating, a walk up the valley is like walking backwards in time at a fixed point in the valley. From photographs of the valley it is possible to plot the appearance and advance of the frontier ('bush line') between shrubs and smaller plants. Fig. 16.17(a) shows the advance of scrub and bush on a moraine on the South side of the valley; Fig. 16.17(b) shows patches of scrub on the North wall; Fig. 16.17(c) shows a tracing (inverted) of the bush line on the North wall taken down valley to where the bush reaches the valley floor. Three of these graphs show slowing which agrees with fitted parabolic curves to within less than 1.4%. Closest fits with exponential slowing are less good. More data, again, are being collected.

12 How Differentiation Begins

Sections 9 and 10 were concerned with the onset of change from the dominance of one species to that of another: this is ecologically common enough. But sometimes an initial continuous variation alters to a discontinuous one, so that we end up with a frontier, but in such a way that at each point in space no sudden jump is seen. How does this fit into our general approach?

The simplest answer that maintains the catastrophe hypotheses that the bifurcations are stable and like those of a smooth gradient dynamic (hypotheses that Zeeman [157] calls repeatability, continuity and homeostasis, the latter being used in the sense of 'tendency towards biochemical equilibrium'), is to drop the hypothesis that the jumps, when equilibria vanish, are fast. This permits things like Fig. 16.18, for the emergence of a patch of Y in the midst of X. Note that at the time of fixing the frontier position (i.e. when all points that are going to have come off the fold) no points have yet moved far from the 'X dominant' state, and those near the fold have moved least; no discontinuity is yet visible. A

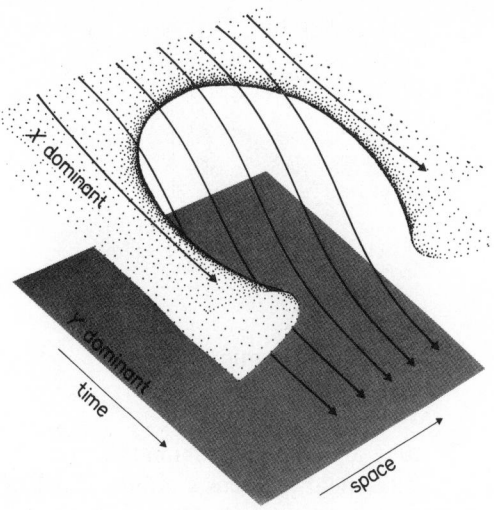

Fig. 16.18

frontier edge begins to appear as the local state of those points in space that have crossed the fold diverges from that of those that have not; evidently it is fixed in space. None of the quantitative deductions in Sections 10 and 11 hold for this situation.

Zeeman's theorem [157] that a frontier must initially move thus rests on at least one more postulate than he mentions, since the requirement that jumps be fast is there only implicit. One might argue against this postulate on the grounds that it requires rapid acceleration near the (equilibrium) fold points, and that this is energetically unreasonable. Such arguments are easily dispelled, as generalities, by experiment: the 'rapid jump followed by quick settling to new equilibrium' hypothesis fits excellently the behaviour of the Zeeman machine. Equally, however, delay convention with *slow* jumps may be exhibited by experimenting with a Zeeman machine submerged in treacle. It is perfectly feasible *a priori* for some biological systems to jump fast, others slowly. (Notice that in continual rapid change – perhaps just moving the equilibria about, not involving catastrophes – the slow seeker after equilibrium will generally be chasing it from so far off that the catastrophe geometry is not useful for describing what happens.)

If this rapid jump postulate is added, then one can deduce that separation into two sheets requires a cusp catastrophe (Fig. 16.19). Commonly, the direction of flow will be far enough from tangent to the cusp direction for a frontier to move visibly in space as successive points meet the fold (as appears to be the case for the bush-line in the Franz Josef glacier, and perhaps the line between *Balanus* and *Chthamalus* in an aging population), and to stabilize away from where it first formed. If the delay and jump conventions were exact hypotheses, then one could say that *typically* this is so, and that nothing else is structurally stable. However, even for an explicit

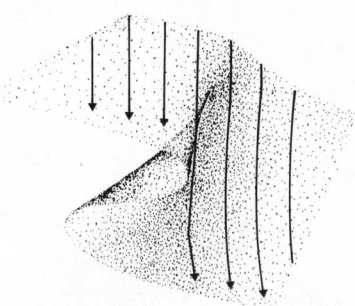

Fig. 16.19

gradient dynamic

$$\mathrm{d}x/\mathrm{d}t = -D_x\phi$$

they are only approximations (see Poston [171]) which become less valid near the cusp point. The lack of a strong attraction to the catastrophe manifold, in a Maxwell convention version, is just what causes the fluctuation and critical exponent difficulties discussed in Chapter 14. Here, it implies that the closer we look around the cusp point, the less a gradient dynamic looks like 'delay and jump'.

The best we can say, then, is that typicality considerations lead us to consider unlikely a phenomenon indistinguishable (at the level of discrimination that makes 'delay and jump' a useful approximation) from Fig. 16.20. It is improbable that we will experimentally 'see' some points passing on one side, some on the other, and none crossing folds. But the probability is neither zero nor easy to quantify.

Taking the delay and jump model as exact, and applying typicality, leads to various quantitative predictions (Zeeman [157]) such as parabolic growth of the height of the jump, frontier starting with non-zero speed and infinite† acceleration due to the cusp's infinite curvature. But, in the light of the above discussion, experimental tests present difficulties. Zeeman's theory of parabolic stopping, discussed in the previous section, is much more robust.

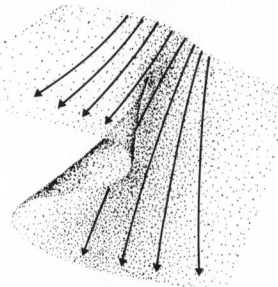

Fig. 16.20

---

†Infinite, but decreasing so fast as to make Corollary 1 of Zeeman [157] §9 – initial movement at constant speed – a reasonable approximation.

EMBRYOLOGY

13 Cell Differentiation

Consider an embryo: first a fertilized ovum, then a blastula – a round blob of identical cells. The cells are identical not merely to the eye, but in potential: in many species a blastula divided at random into two equal pieces will produce two complete individuals with no missing parts. But the descendants of these cells are individually identifiable as liver cells, nerve cells, skin cells, etc., in very specific geometric relationships. The central problem of embryology is to explain how this emergence of patterned difference, this *morphogenesis*, is brought about.

Now, although the plants and other biota considered in previous sections consist of millions of cells, an ecology is in many ways less complicated than an embryo. We did not need the chemistry or aerodynamics of bees to arrive at equation (16.3) of Section 1, and many ecological analyses can work in terms of a small number of interacting species, with a fairly simple model of the behaviour of each. (Analogously, the structural engineer is rarely concerned with the quantum solid state physics of steel and concrete; bulk properties like elastic moduli suffice for his/her analyses.) But a cell contains many thousand distinct compounds, none without importance, generally associated with particular sites in the cell (nucleus, ribosomes, etc.). How should we set about modelling this?

The central approach of recent decades, with great achievements to its credit, has been the *genetic switching model* (in which genes for producing different proteins are either on or off, and interact like the switches in a digital computer.) We defer consideration of this until the next section, and look now at smooth models.

The cell may be treated as a semi-permeable bag of chemicals, with exchange of molecules taking place between it and its neighbours and reactions described by a differential equation going on inside. The first significant approach to a differentiation model in these terms was made by Turing [172], who was also the first real theoretician of the digital computer. For equations showing how diffusion can itself produce spatial variation of species distributions in a *homogeneous* environment (not varying, even smoothly), see Hadeler, an der Heiden and Rothe [173]. Turing's paper generated a substantial literature (though much smaller than that on abstract 'Turing machines') which received new life when approached with modern dynamical systems methods by Smale [174, 175]. Plainly, the dynamics of a cell cannot be treated for all purposes as a differential equation in the average chemical concentrations, since not merely are the chemicals localized in space, but most cells contain immense quantities of folded membrane to complicate the reactions. Collapsing an evolutionary argument to teleology, such a complicated structure would not be there if it did not make a substantial difference to the cell dynamics.

However, a smooth model with reactions at each point (replacing the finite cell structure with a continuum) and diffusion between them has its interest, both for biology, and for chemicals in solution (for which the model is much less approximate). Much work has been done on such *reaction-diffusion equations,* particularly since the spectacular pattern formations of the Zhabotinskii reaction became known. (For a study with photographs, *see* Winfree [176, 177]; for mathematical literature Howard [178].) The work of Kopell and Howard [179] in particular makes clear that a naive approach based mainly on the point dynamics – whether gradient or more general – is inadequate; but, as for caustics, catastrophe geometry appears to play a rôle: *see* for instance Guckenheimer [180]. In the mathematically more elementary approach of the Brussels school (*see* Prigogine [181] for example), the majority of the bifurcation diagrams seem to be special, unstable sections of cuspoid catastrophes; which suggests that a synthesis with catastrophe theory should be possible – and useful – once a certain amount of work has been done to tidy the mathematical notation.

The elementary catastrophe theory approach of treating the cells as a continuum of differential equations, with diffusion rather serving to ensure continuity of the equations' parameters in space than entering actively into the dynamics, is thus in various ways false in principle. But, as with the equations of fluid mechanics and the catastrophe theoretic approach to stream functions, whose fundamental dubiety and great experimental success are described in Chapter 11, this is no reason not to use it. It should be used as a predictor of what we may usefully look for, rather than of what we will necessarily see. At present, only experiment serves to distinguish those aspects of cellular growth and differentiation which it describes effectively from those that it cannot. The development of more sophisticated theories will include criteria for its safe use, in the same way that we have a criterion for how close to a caustic a ray approximation may be used (Chapter 12 Section 8), or the Ginzburg criterion for the Landau theory to be applicable (Chapter 14 Section 13). In physics ray theory and Landau theory are used up to the limits of their strict applicability – indeed often beyond – for their great convenience. Spacecraft orbits are not computed relativistically.

Biology is likely to have a similar permanent place for elementary catastrophe theory. The hypothesis that the bifurcations of the local differential equations are gradient-like (i.e. that they may be reduced to bifurcations of suitably constructed Liapunov functions, *not* that the dynamic need be gradient) is the most restrictive. But many conclusions reached in the first instance by thought about gradient-like behaviour hold more widely. For example, Fig. 16.21 shows a stable codimension 1 phenomenon that cannot be reduced to a fold.

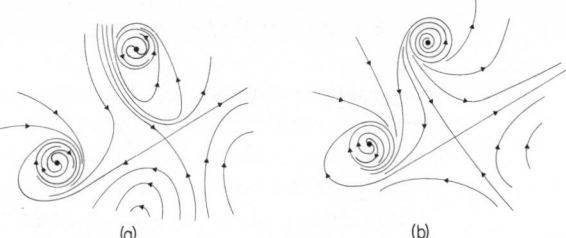

Fig. 16.21

(a)   (b)

In (a) there are two attracting cycles, representing two oscillatory states possible for a cell. In the change to (b), the upper one disappears by a non-local interaction with the saddle, and any cell that has been going round it moves across to the other cycle – commonly, in explicit models, quite fast. (If the dynamics then returns to (a), the cell will not notice: we would expect to model the observed behaviour quite well with delay convention, though the usual caveats apply.) The set of such jump points is typically smooth and of codimension 1, in the same way that fold lines and surfaces are; and the same reasoning that yields Zeeman's law of parabolic stabilization of frontiers applies here too. As this illustrates, the mathematical toolkit needed by the biologist is less the sort of list of theorems that we gave in Chapter 8 than an education in dynamical systems theory generally. The inclusion of a suitably oriented mathematician in a biological team may be advocated; and since the recent overproduction of qualified Ph.D.'s there is no shortage of supply.

These remarks of course depend on the proposition that such mathematics is indeed usefully applicable to embryology. (We would claim that for ecology this is established.) The chief exhibit here is Zeeman and Cooke's theory of gastrulation (Zeeman [157, 182], Cooke and Zeeman [183]), but this has a background of detailed embryology which is much harder to summarize for the non-specialist than even fluid dynamics, the application to which we have devoted most space for introductory treatment (Chapter 11). We simply cannot do it justice in the space available here, and refer the reader to the original sources, conveniently collected together in Zeeman [48]. The model has not yet gained universal acceptance, and even if it does will doubtless be modified *en route*, but it has certainly been suggesting successful experiments. It has been interesting, in our conversations with biologists previously unimpressed by catastrophe theory, to meet reactions that here at least was a theory that could be taken seriously and judged *as biology*, independently of the mathematical reasoning – some rigorous, some by analogy – that lay behind it. This in itself is an important point to have reached, since the profundities of Thom's book are not in the same way communicable. Zeeman's work, which he describes as an attempt to put into practice the programme of that book, has required some years of mutual effort in developing a common language with Cooke and other biologists.

14 Switching Catastrophes

We mentioned above, but did not enlarge upon, the important gene switching model of cell dynamics. In some circles this paradigm has become so dominant that the working of the cell is seen almost literally as the operation of a large computer program. In this picture differentiable dynamics, from Turing onwards, has no rôle to play. A finite automation operates by combinatorial laws, abjuring the infinitesimal calculus and all its works. Although the program in Appendix I is for use in the calculus, nobody would use the calculus to analyse the workings of the program. From this point of view, molecular biology is central, as the key to locating all the switching rules and reading the Great Program of Life, morphogenesis included.

There is much to be said for this approach. It has yielded not merely abstract, but medical, information; 'deciphering' various diseases, clarifying the rôle of antigens in immunology, and so forth. But the implication that we will understand the system once the switching relationships become known gives one pause: apart from the common difficulty in understanding a program written by someone else, there are deep theorems to consider. For a general abstract Turing machine, the only way *even in principle* to learn such elementary facts about its behaviour as whether it stops in a finite time is ... run it and see! (There are other types of abstract automaton under study now, including some with built-in oracles, but a miasma of formal undecidability pervades the whole subject.) The sense in which one can understand such a machine's behaviour is certainly weaker than that in which we would like to understand morphogenesis. This of course does not imply that a cell cannot accurately be described by a finite automaton model.

However, it is hard to accept that gene switching rules can be altogether adequate as a model. Certainly the cell includes chemicals in various concentrations, whose reactions everyone agrees to model by differential equations, and these reactions often constitute among other things the function of the cell in the larger organism. (Removing waste, storing fat ... the whole chemical metabolism.) Such reactions are greatly affected by conditions external to the cell, since these conditions are what they evolved to deal with. The controlling switching system must receive feedback from the dynamics, since it cannot know *a priori* what enzyme production will be needed at various times. The connection between switches, most commonly modelled by a gene turning on or off according to which side of a threshold value X_c a particular concentration X lies, is thus responsive to the dynamics. At this point the habitual implicit hypothesis that X is a *function* (perhaps nonlinear) of various $Y_1, \ldots, Y_N$; waste product concentrations, temperature, compounds diffusing from other cells, creeps in.

This may oversimplify the rôle of chemical reaction dynamics quite as much as an entirely differential model underrates that of the switches. Experimentally, important genes do switch on and off and it is not satisfactory to suppose that the jumps of a catastrophe model alone suffice to describe this (precisely because such models predict things like parabolically increasing size of jump: the difference between on and off is not something that *can* grow parabolically.) We must allow the 'threshold value X_c of X' switch model, but we hope that the reader who has reached this chapter via the previous ones will find a simply functional relationship $X(Y_1, \ldots, Y_N)$ implausibly special. If the dynamics *do* give a unique value of X for each combination of the influencing factors $Y_1, \ldots, Y_N$, one can show by generalizing Chapter 6 to more general dynamics that this will be stable under small perturbations; so we can't say that typically it *won't* happen (using 'typical' in the sense of Chapter 7) but it is certainly highly special. We have explored the ubiquity of multiple equilibria, and bifurcations of the pattern of equilibria, in far simpler systems: why should the complexity of cell dynamics, in this one respect, be so simple?

Let us take the simplest dynamics beyond a trivial functional relationship $X = X(Y_1, \ldots, Y_N)$: a gradient-like bifurcation geometry with 'delay and fast jump' convention closely approximating the behaviour. The first thing that we find is, once more, Zeeman's law of parabolic stopping. The frontier in Fig. 16.22 (between those cells that have, and those that have not, changed X through the critical value X_c and flipped their genetic switches) moves exactly like the frontier in Fig. 16.16. In many catastrophe models we can insert such a threshold level for the jumps to pass through, and get a stable model with similar features. The changed model will not have *identical* features, since when the observable is a discrete on/off or its consequences we change such predictions as the parabolic local geometry of Fig. 16.12. Ashton's observations described in Section 10 clearly occur in a context where switching is *not* operating. In forest ecology there is no obvious reason why it should, though ecological discussion in 'limiting factor' terms involves precisely this assumption (together with that of the functional

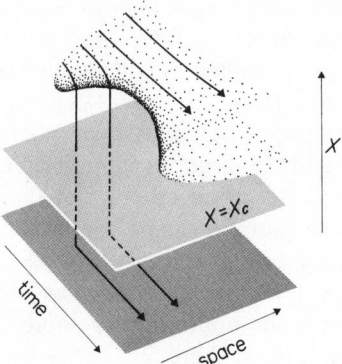

Fig. 16.22

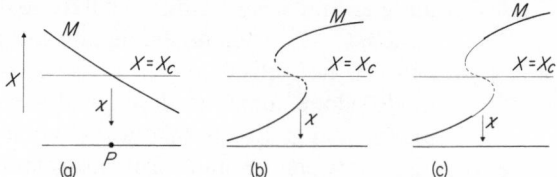

Fig. 16.23

relationship questioned above). The quantitative geometric implications of these models are quite distinct enough for experimental programmes like Ashton's to decide between them in individual cases.

Behaviour newer to catastrophe theory appears when the catastrophe surface and the threshold level meet. In Fig. 16.23(a) movement in the control space through the point P brings immediate, *reversible* sudden change (more familiar as a result of the Maxwell convention, in catastrophe theory; the standard model for gene-switchers), while Fig. 16.23(b) shows hysteresis. If the 'control' variable shown is physical space there will be a frontier in Fig. 16.23(a) which will move by cells crossing the switch threshold under smooth variation of the catastrophe manifold M with time; but delay convention in Fig. 16.23(b) will typically produce a set of states like the heavy curve in Fig. 16.23(c), with a discontinuity whose position is not affected by small changes in M.

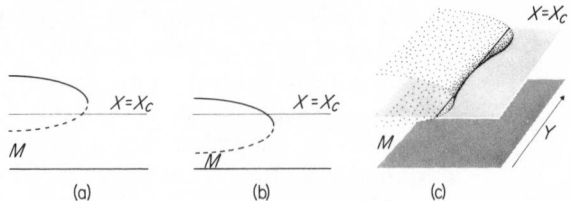

Fig. 16.24

The transition between different geometries like Figs 16.24(a) and (b) as another control Y varies is typically as shown in Fig. 16.24(c). The local geometries typically to be encountered with up to four control variables are described by the list of polynomials in Section 7, since exactly the same determinacy and unfolding algebra applies whether $X = X_c$ is later interpreted as a boundary defining a constraint, or a threshold. The associated behaviour, and the geometry of the bifurcation sets, is however quite different for the two interpretations: note the differences between Figs 16.24(c) and 16.9(iii). Both types of geometry will be analysed in Pitt and Poston [158].

The integration of such 'switching catastrophe' geometry into detailed theories of particular phenomena in morphogenesis will involve quite as arduous interdisciplinary effort as did Zeeman and Cooke's gastrulation model, but the attempt seems worth making. Notice that here again we are talking about the application, not of Thom's *list*, but of Thomist *methods*. The only mathematics that can safely be appealed to without any understanding of the reasoning behind it is the multiplication table, and such blind appeal leads to inefficiency even there.

17 The Problems of Social Modelling

Attempts to apply catastrophe theory in the social sciences *face* the general problems of methodology in those sciences, rather than automatically *solving* them. Instead of attempting to survey the great variety of suggestions now in the literature (for many of them see Zeeman [48]) we discuss with examples the ways catastrophe theory may or may not be useful in particular contexts, and how such uses may be tested. The issues involved have been greatly clarified by the recent explosion in well founded physical applications, which we have illustrated in previous chapters; some critics in the social context have made objections in general terms for which we can now give counterexamples in successful physics. This leaves us free to examine problems – and there are many – that are more peculiar to the social sciences.

We conclude with an examination of a simple model illustrating that while the geometry we have treated in this book may have a useful future in social modelling, the hypotheses that make it applicable are far from universal truths. Conclusions drawn from them may be useful (we consider that they often will be) but they cannot be Theorems about Society.

1 Identification of Variables

Most of the variables now basic to physics took centuries to find. 'Velocity' was perhaps understood by Eudoxus (cf. Zeeman [184]) but did not become part of the common scientific understanding for nearly two millenia. 'Energy', 'momentum', 'temperature', 'probability' (as a *number*), 'charge' have all come, since then, from the patient work of men of genius. (The process still goes on, as in the present search for 'charm' among quarks, and the problem of recognizing it if it occurs.) These physical quantities were not obvious, and only their immense coherence and success has enabled the world view implicit in them to become dominant in the sciences. The world looks quite different through them than it did to the mediaeval scholar.

A similar shift in quantities by which to think has been taking place in the social sciences. A mediaeval student of the *just price* concerned with the evil of money lent at *usury*, cannot compare quantities with a theorist of the *market price* concerned with too high or low an *interest rate*. We would quote Blake, 'a fool sees not the same tree that a wise man sees', but for the suggestion that one of these economists is a fool: the crucial point is the difference. The modern analysis has not invalidated the older: indeed, the recent histories of food in Poland and petroleum in the U.S.A. illustrate that a government disrespectful of the popular view of the just price has as much trouble as those that neglect consideration of the market price.

Self-proclaimed 'scientific' and 'objective' economic theories notwithstanding, ideology/world view/paradigm is crucial to the choice of a model. This is not less true in physics: Einstein could not accept 'God playing dice' in quantum mechanics, and Michelson (of the Michelson–Morley experiment) died in 1933 without ever accepting Relativity. Both of these questions may be seen as ones of what variables 'really' *describe* the system: the details of how they interrelate are almost secondary. We cannot have an intelligent argument over what Hamiltonian to model the laser by (Chapter 15) until we have agreed that 'states' live in a complex Hilbert space on which 'observables' are operators. No such general agreement exists in any social science, except within competing schools; consequently, intelligent debate is usually *within* schools. Argument between, say, Marxist and Monetarist economists is more often a slanging match. (This has an unfortunate corollary. A scholar observing correctly that debate within his own group is constructive and intelligent, and that between it and another group is the reverse, is apt to conclude that it is *his school* that is intelligent and the other not.)

Any social model, verbal or mathematical, thus involves an implicit ideology. Presentations of mathematical modelling of most persuasions attempt to mask this fact, as a wander through the volumes of (for example) *Econometrica* will confirm: the desire to mask it is itself as ideological as the detailed connection with the thoughts of Chairman Mao that Chinese papers in high energy physics sometimes seek to establish.

None of this means that science cannot be very effectively carried on, any more than the question of the ultimate 'truth' of the space $H^2 \cap H_0^1$ as a description of strut positions in Chapter 13 Section 7 prevents the success of elasticity theory. Rather, it means that criticism of any mathematical model, in catastrophe terms or not, should distinguish between objections to the ideology or to the mathematics, to a greater extent than has hitherto occurred. We may illustrate this conveniently with one of the more tentative models, a geometrical view of censorship (Isnard and Zeeman [185]). We will not discuss its mathematical virtues or vices; we are concerned here with implicit world view. The model suggests that the response of those in a position to censor art is describable by a cusp picture (Fig. 17.1) where A is 'aesthetic value' and B 'erotic content'. The detailed presentation shows it to be at least a reasonable geometrization of the liberal establishment view of how censorship decisions (by the community or by authorities) are made. A radical critique might be as follows.

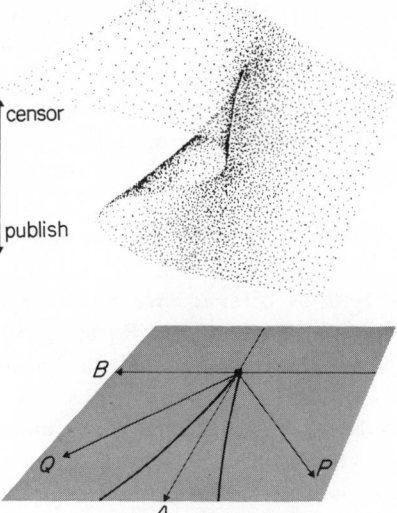

Fig. 17.1

'Judgement of aesthetic value is contingent on response to sexual content, so the variables cannot be separated. (Indeed, if erotic content is not an aesthetic value itself, it is surely the reverse of anaesthetic.) More importantly, those whose sexual anxieties make them into censors respond not directly to erotic content but to sexual threat, and what they find threatening is the suggestion of lowered defences in peaceful copulation. Violent sex is common in *permitted* books and films, particularly in the laudable context of war: what made the old blue movies and comics totally unacceptable was their non-violence. People hardly ever got hurt in them (cf. Anon [186] for a wide sample). The distinction between *Straw Dogs* and most "hardcore porn" remains the savagery of the former. A better description, if you insist on a catastrophe picture would be to use axes P and Q in Fig. 17.1 as "violence" and "sexuality" respectively'.

We do not seek to establish either this view of sexual politics or the alternative model suggested within it. Rather, our point is that between those who express in laws the idea that erotic content requires 'redeeming' by literary value, and those who consider censorship an act of *people* who need redemption, there is a great gulf fixed, which has nothing to do with mathematics. If both sides turned numerical, they would perceive and measure such different quantities that their hypotheses would not be testable against each other's data.

The rôle of mathematics is not to mediate between models, but to clarify individual theories and reveal more of their implications than meets the eye, so that these implications may be compared with observation. We have shown in various chapters how effectively catastrophe theory can obtain new predictions from much tested physical hypotheses. We may class optics, fluid mechanics, elasticity, etc. as 'models that imply catastrophes'. We outline an example of the same kind, starting from widely assumed general *economic* hypotheses, in the next section, before turning to the subtleties of using catastrophes *as* models.

2   The Archaeology of Sudden Change

For reasons of space we must treat this theory briefly: far more detail is in Renfrew and Poston [159]. A somewhat similar analysis applies to the size and distribution of shopping centres (Poston and Wilson [160]).

Both modern and ancient settlement patterns in agricultural areas show great variety between extreme *dispersal* of the population into individual scattered farmhouses, and *nucleation* into towns of up to 5000 people (though without necessarily a 'central' rôle, of serving people who do not live in the town). Moreover, the archaeological record shows sudden alterations between these patterns. In the archaeological literature of all parts of the world, these changes are assumed to be strong evidence (in the absence of signs of natural disaster) for invasion by a new 'people', laying waste towns or building them.

They are not.

The economics of division of labour for people are much as for bees (discussed in Chapter 16 Section 2). The economic disadvantages of travelling long distances to work in the fields correspond to those of long trips in search of nectar. If we assume that people remain for significant periods only in arrangements that locally maximize economic advantage – the hypothesis of 'rational economic man' – the geometry of Chapter 16 Section 3 applies. (Indeed, the arguments and discussion were first worked out for this case rather than that one.) Consequently, gradual smooth changes in soil fertility or population, or *slow* external influence through diffusion of technology, or agricultural methods and crops, force an 'advantage-maximizing' population to make sudden changes. (In Renfrew and Poston [159] the discussion is extended to the advantages of defensibility and 'non-economic' ones of easier social or religious gatherings, and movement between plain and hill sites.) The hypotheses analogous to (a) and (b) of Chapter 16 Section 2 are easily, and probably often, satisfied.

Such sudden changes, therefore, under very general assumptions should fairly commonly occur without sudden outside causes. Reasonable economic hypotheses *imply* the occurrence of such jumps, and jumps are not therefore evidence of invasion.

Much work is clearly needed to quantify this analysis for particular cases, but here it is the qualitative conclusion that is most relevant to previous work. Just as a large body of experimental data on electric field variations in growing plants was cancelled by an experiment showing that the apparatus behaved the same way if you left out the plant, much archaeological argument evaporates upon a proof that you still get sudden change if you leave out the invaders. It is not always clear from the texts, which concentrate on the currently *un*defeated views, how much the progress of every science depends on such clearing actions.

3 Catastrophes as Models

The archaeological model above does not *require* catastrophe theory (in the sense that the analysis of the flows in the six-roll mill of Chapter 11 does) any more than the van der Waals cusp (Chapter 14) did. It can be derived and explained, and will stand or fall, without it. Catastrophe theory merely made it a natural way to think, and provides evidence that the two catastrophes involved (the fold and a constraint catastrophe) are structurally stable, and typical phenomena in the dimensions of the model. It suggests that many catastrophes will be found in mathematical economics, once that subject abandons its concern with ever subtler hypotheses that imply unique optima to which the 'invisible hand of the market' guides us all, and faces the multiple optima to be found in the marketplace.

How much, however, can we deduce directly from catastrophe theory? Given a particular view of what kinds of variable describe the system in general (or, say, stuck with the views of the 1950, 1960 and 1970 census planners, since we can't go back to those years with different questions) how much can we say about typical geometries to be expected in the data? The idea is to assume that the state of the system is usually near a point on a catastrophe manifold, and argue from typicality.

Voices are sometimes raised against this procedure, on the grounds that the force (potential, Liapunov function, . . .) holding us to the catastrophe manifold is unspecified; the same voices, *mutatis mutandis*, that objected to the unexplained force acting at a distance in Newton's model of gravity. It is quite legitimate to say, with Newton, '*hypotheses non fingo*' and proceed to test the model. One may still hope for an improved theory that explains the forces in more detail, as General Relativity does for gravitation, but it is pointless to make it a *requirement*. (No one asks for the potential which holds social systems to the planes and hyperplanes so diligently sought by current numerical sociology. We did not need to discuss the dynamic for the brilliantly successful application to polymer flow in Chapter 11.) To test the model we must, as Newton did, obtain predictions by correct geometry within it, and compare them with observation. What testable predictions can we obtain from the general assumption here?

The answer depends, rather subtly, on the dimensions involved. If we can reasonably identify r external variables (such as position in space and time), *where r is fairly small*, and postulate N internal variables (no size restriction) over which the extremizing is done, we are well set up. (We *must* be able to assume that the system is usually in a state which is locally extreme in *all N* directions.) Typically, barring symmetries etc., the N variables will behave locally according to an appropriate list of catastrophes met with up to r controls. (If some of the N internal variables are limited by inequalities, this list must include the constraint catastrophes of Chapter 16 Sections 6, 7.) Moreover for a catastrophe of corank m, almost any m of the N variables serve for a good description. This is because the Splitting Lemma (Chapter 6) guarantees that the behaviour lies entirely in some m-dimensional manifold C in $\mathbb{R}^N$, which moves smoothly with external parameter change. For instance if $m = 1$, then C is a curve: as Fig. 17.2 illustrates, either x or y gives a perfectly good parametrization of C. Replacing (x, y) by $(x_1, \ldots, x_N)$, typically any x_i will do as well. All of this is genuinely independent of N, which can be any finite number, however large.

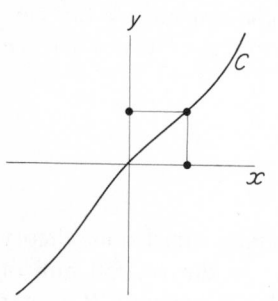

Fig. 17.2

This is the setting of the ecological discussion of Chapter 16 Sections 8 to 12, which appears experimentally promising as we go to press. The dimension r can be reasonably taken as 2 – one spatial, one temporal – for the questions concerned, and data collected so far sit well with the typicality predictions for $r = 2$. The unformalized typicality arguments basic to both Landau and 'renormalized' phase transition theories, as discussed in Chapter 14 Sections 7 and 12, are certainly extremely successful.

However, what if r, as distinct from N, is large? In macroeconomics it assuredly is: for instance world economics is affected by harvests, and thence by weather, in many different places. Weather is not yet fixed by utility maximization, and for economics must therefore be treated as an external variable. (In meteorology it is internal, by definition, but the dynamics are certainly more general than gradient systems and probably 'chaotic' in the sense illustrated in Section 7.) The reader can doubtless multiply such external factors for any given macroeconomic problem faster than (s)he can write them down. We can illustrate the resulting difficulties excellently with the fold catastrophe, so let us assume for a moment that nothing more complicated happens.

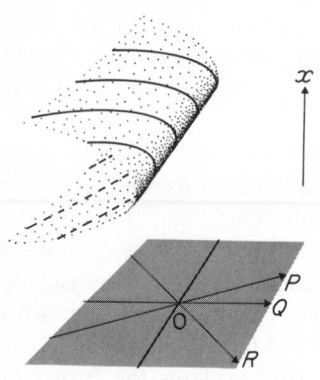

Fig. 17.3

In Fig. 17.3 we show a 2-dimensional control governing x via catastrophe dynamics. Supposing that at point O there is a potential

$$x^3 + (\text{Morse in rest of behaviour})$$

then typically (by transversality and the theorems of Chapter 8) there will be a fold *curve* through O. Any of the axes marked P, Q, R is transversal to it, so any of P, Q, R will serve as the single unfolding variable that a fold point needs. However . . .

Fig. 17.2 showed that almost any internal variable would do to label behaviour confined to C; we could use either x or y *and forget*

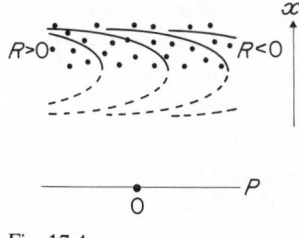

Fig. 17.4

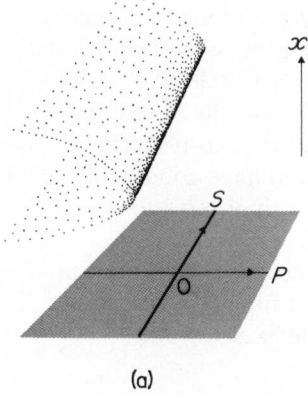

(a)

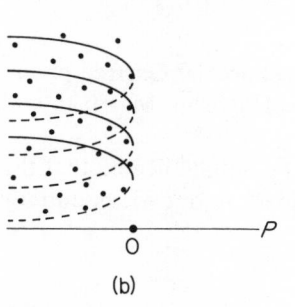

(b)

Fig. 17.5

the other. More technically, almost any projection onto *m* behaviour variables preserves the structure of a corank *m* catastrophe. For unfolding variables, deep theorems show the *existence* of a projection onto a few unfolding variables, which preserves the catastrophe structure; but this time almost all other projections obliterate it. Fig. 17.4 shows the result of treating *P* as control and simply forgetting the fact that *R* may vary. The $R = $ constant curves of Fig. 17.3 project as shown to Fig. 17.4. The dots, which represent 'perfect' data lying exactly on the catastrophe surface, display no fold geometry in Fig. 17.4. To display the fold nicely we need to find the right projection to the *P*-axis, parallel to the fold line: this will very rarely consist of just dropping variables. It will not in general even be linear; but just to specify its linear part requires $r - 1$ numbers. Projection to the two variables needed for a cusp requires $2(r - 2)$. With 100 control variables in the data, this means some 200 numbers to estimate. If one important control, like *R*, in which substantial variation happens, is missing from the 100, the result can still be like Fig. 17.4.

Furthermore, recall that the projection involves a control-dependent change of behaviour parametrization *x*. If control *S* is projectable to *P* by simply forgetting it, but the fold is not horizontal over the *S*-axis (Fig. 17.5(a)), then forgetting *S* can spread the picture vertically as in Fig. 17.5(b).

A combination of unmeasured parameters like *R* and *S*, therefore, can spread exact fold data evenly over the (P, x)-plane!

(Of course, if we can restrict ourselves to a truly one-dimensional *slice* of control space, moving only on the *P*-axis, these problems do not arise. The success of the physical sciences has depended on just this tactic, of holding more and more variables fixed until repeatable results are obtained. But this is not always possible in the social sciences.)

To some extent, then, the theorem that all but one control around a fold can be made 'dummy' is, though true, a snare and a delusion: we need a lot of data about the other controls *in order* to discard them.

In the problems of the social sciences, where many of the controls are unknown and 'uncontrollable', the graphs formed by experimental data may thus fail to show up perfectly good catastrophes because the projection involved is not suitable. No amount of transforming the data will help: only a better grasp of the effect of the extra controls, or a more carefully designed experiment to keep more fixed. This is a very serious problem and cannot be dismissed by waving the magical wand of catastrophe theory. Conversely, however: if the data form good curves or surfaces, then the choice of control variables must have been very good indeed.

The work of Cover [187], discussed usefully in Bremermann [188] pp. 17–23, is very relevant to this question. It shows that if we attempt to separate a set of *N* points in $\mathbb{R}^n$ into two classes by using a polynomial depending on M adjustable parameters, one class corresponding to the polynomial taking positive value, the other negative; then for M large the probability of success drops suddenly and sharply from near certainty to near impossibility as N becomes larger than $2M$. In other words, if a model classifies more than twice as many points as it has adjustable parameters, it is unusually good in a strong sense. When we are using, not an arbitrary polynomial, but one of a small list of standard types, this sense is strengthened. More work in this direction might yield useful tests of goodness of fit for catastrophe models.

As in physical cases (e.g. Chapter 11 Section 5), it is crucial to distinguish between an exact or explicit set of equations and an exact theory. Many workers on physically bifurcating phenomena have extended their efforts to the social sciences. An excellent example is Weidlich [188a], who sets up detailed hypotheses leading to a form of the Fokker–Planck equation of statistical physics; elaborate, detailed, careful mathematics. By the methods of Chapter 15 his conclusions reduce to a cusp catastrophe picture, which is more clearly explanatory and contains all his testable predictions (which are no more quantitative, without values for unknown parameters, than the censorship models of Section 1). Thus simply assuming a cusp model is not experimentally different and is far more robust. The cusp is insensitive to small perturbations, while small changes completely alter the appearance (and sometimes the substance) of statistical equations. The stability of Weidlich's model (which he does not discuss) is given only by catastrophe theory.

Just as at least ten distinct statistical processes give a Poisson distribution (so statistical measurements cannot tell us which process is operating), many processes give – as Thom's theorem predicts – a cusp catastrophe. The hypotheses of a Poisson or cusp distribution are at least testable (when we are fortunate and have good data), and it is no less scientific to investigate their fit with observation than to fit planetary motion against ellipses.

In the following three sections we discuss three particular models which may be described as 'catastrophe-fitting', to illustrate the variety of possible approaches and problems involved.

4 Prison Riots

In this section we examine a study of disturbances at Gartree prison during 1972, carried out by Zeeman, Hall, Harrison, Marriage and Shapland [189].

Preliminary statistical factor analysis of the data indicated that the factors influencing disorder may be divided into two essentially independent groups:

tension (frustration, distress)

alienation (division, lack of communication, polarization).

Further, increased tension tends to increase the disorder, while increased alienation tends to lead to more sudden and violent outbreaks of disorder. This suggests a cusp catastrophe, with tension as normal factor and alienation as splitting factor. Further hypotheses are added to fix the dynamical behaviour of the system, in the form of *feedback flow* on the catastrophe manifold, indicated by the arrows in Fig. 17.6. (The use of feedback flows is a favourite device with Zeeman, but it is beyond the present scope to say much about it. For mathematical references *see* Takens [190, 190a].) The system is assumed to 'home' vertically onto the catastrophe manifold (the 'fast flow') and then follow the feedback ('slow flow'). By examining Fig. 17.6 it will be seen that for low levels of alienation the system tends to a stable position of neutral disorder, but for higher levels of alienation it oscillates back and forth within the bifurcation set of the cusp catastrophe, jumping alternately from the top sheet to the lower one.

This behaviour will be partially masked by random noise, but the oscillatory nature of the behaviour should nonetheless be observable.

More statistical factor analysis gives as measure of the tension the sum of the weekly standardized values of men reporting sick, men reporting sick at work, governor's applications and welfare visits. The alienation is measured by the sum of the numbers of inmates in the punishment wing, and requesting segregation. To reveal trends the data are smoothed, and the result is Fig. 17.7. This indicates the time path of the controls, plotted week by week. Serious incidents are plotted as circles; solid circles indicate new forms of mass protest. These are assessed for seriousness on a scale from 0 to 10.

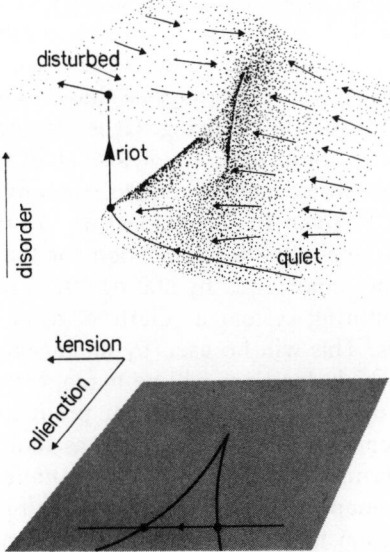

Fig. 17.6

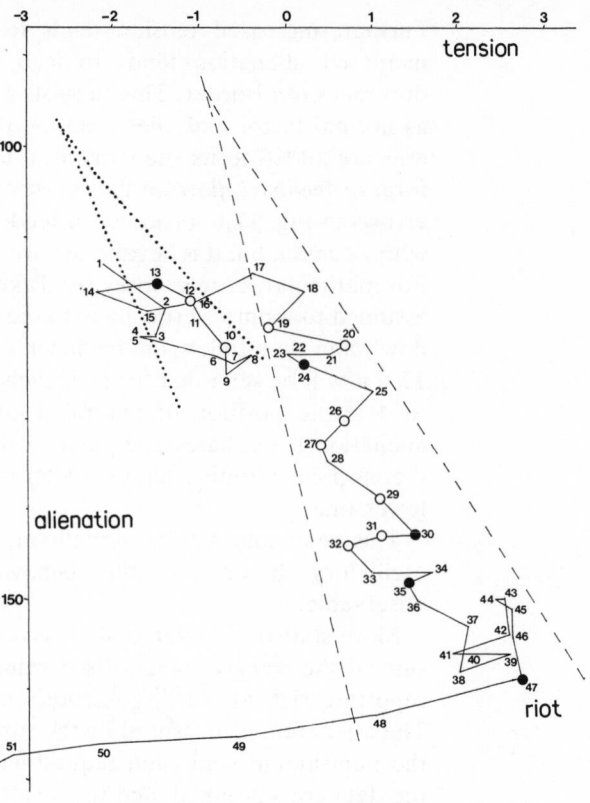

Fig. 17.7

(The assessment was by an independent panel of seven assessors, and their figures showed high correlation between each other.)

The data exhibit clear oscillations. However, to obtain good-looking cusps, it seems necessary to draw two different cusps; one (dotted) for the first 17 weeks, another (dashed) for the rest. This must be considered a defect of the model, but need not be serious: Zeeman *et al.* [189] suggest that it may be due to the 'personality' of the institution changing. (It is also possible that the cusp could be drawn with a curve, to enclose all of the oscillations: the model as it stands would permit that to some extent.) A closer analysis of the data is given by Zeeman *et al.* [189], together with some conclusions, including a justification for the policy of 'playing it cool'.

The paper ends by stating: 'It is hoped to institute an on-going monitoring system at Gartree, using better measures of the variables. This will be used by management as part of the information on which decisions will be made. Only in this way do we think that the model can be tested for predictive accuracy.' This monitoring system (not hidden cameras, as ludicrously asserted by *Newsweek* magazine (Panati [191]) but continued collection of similar data to that mentioned) has been in operation for some time. On the one occasion that the track crossed the cusp line, there was a riot.

5 Bistability of Perception

There is some evidence in favour of the occurrence of catastrophes in perception, in connection with ambiguous or 'multistable' figures. In the array of pictures shown in Fig. 17.8, the fourth figure from the left in the top row was shown by Fisher [192, 193] to be perceived with equal probability as the face of a man or the figure of a girl. Attneave [194] reported that if the figure is included in a sequence (the top row of Fig. 17.8) then the perception of the middle figure is biased according to the direction in which the sequence is viewed: towards the man by viewing from the left to the right, and towards the girl by viewing in the opposite direction.

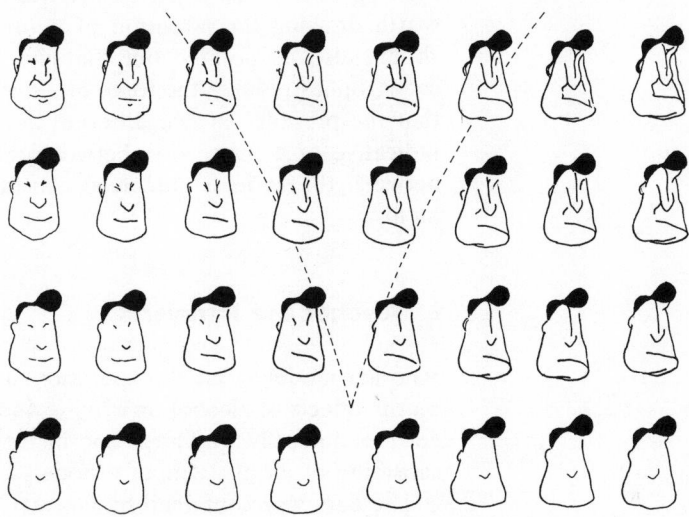

Fig. 17.8

These observations suggest a one-dimensional section of a cusp catastrophe (Fig. 17.9) with delay convention. Fig. 17.8 is an attempt to enlarge the sequence into a full two-dimensional array with a cusp catastrophe appearing, by introducing a factor corresponding to the detail in the picture. A cusp bifurcation set, based on subjective evidence, is sketched. It should be possible to extend Attneave's experiments to test whether a cusp catastrophe is indeed present, but no data exist at the time of writing. Since Attneave reported repeatable behaviour for the top row, the problems considered in Section 3 should not render work on the whole array impractical.

We have learned just as we go to press of a considerable body of careful experimental work on this kind of perceptual model by Inouye and others (*see* Inouye [199a], Brown, Inouye, Williams and Borrus [199b]).

Notice that while Fig. 17.9 need not be a slice of a cusp (it could equally be the $Q = 0$ slice of Fig. 17.10) the cusp is *very* common: we have given explicit examples from naval architecture to quantum

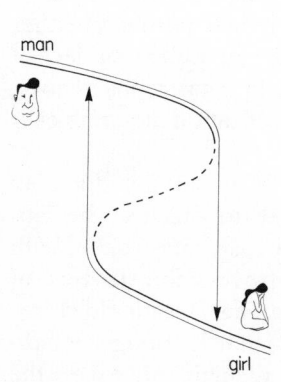

Fig. 17.9

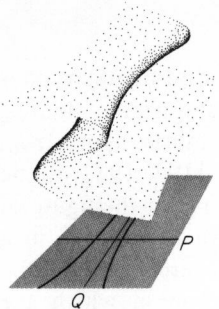

Fig. 17.10

optics. It is therefore a reasonable practice, whenever investigating a bistable phenomenon which displays hysteresis, to look for another control dimension that provides a smooth 'way round the back' of a cusp that organizes the two fold jump points. Poston [171] describes new aspects found in this way of a soap film catastrophe machine due to Courant [195].

This model may have some bearing on Zeeman's theories of brain behaviour (Zeeman [196–198]) in which he considers the brain as a large number of coupled nonlinear oscillators, for which he shows that elementary catastrophe bifurcations *may* (but do *not* always) occur. Such 'dynamical systems' models of the brain go back at least to Ashby [199], and since catastrophe theory is a part of dynamical systems theory and bears on the questions Ashby raises, it may be worth drawing the attention of neurologists to this connection. It should also be pointed out that the 'typicality' of the elementary catastrophes reduces the force of an argument sometimes advanced: that the presence in two different areas of the same catastrophe is indicative of a connection between them. Evidence of connections between the variables involved can of course strengthen the argument.

6 Alcohol and Introverts

Another model based on experimental data is that of Zeeman [200], on the effects of alcohol on driving ability. It fits the data fairly well, yet paradoxically we find it one of the most unconvincing and least satisfying of all of Zeeman's models.

The data were obtained by Drew, Colquhoun and Long [201]. A subject was placed in a driving simulator and asked to drive at normal speed, this being recorded. He was then given a quantity of alcohol and the experiment was repeated. The variation in speed from normal was plotted against a psychological measure of introversion/extroversion (the 'Bernreuter scale'), with the results of Fig. 17.11. The response is marked: extroverts tend to drive at normal speed even after imbibing alcohol, whereas introverts either drive too fast or too slow. Such scales, of course, are the subject of controversy quite independently of any use in catastrophe theory, but we will not discuss their merits here. Identification of variables is not easier in psychology than elsewhere.

Zeeman, referring to Fig. 17.11, asks: 'Why do the dots fall roughly into the shape of a cusp?' His model (to which, to be fair, we have not the space to do full justice) is a cusp catastrophe, with control variables the measure of introversion/extroversion and the driving speed (in Zeeman's convenient terminology respectively the 'splitting' and 'normal' factors, that is, along the cusp axis and transverse to it). As behaviour variable he takes the subject's *estimate* of his speed. This leads to the graph of Fig. 17.12.

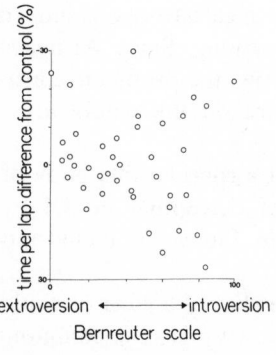

Fig. 17.11

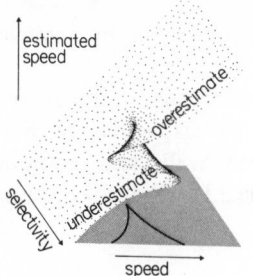

Fig. 17.12

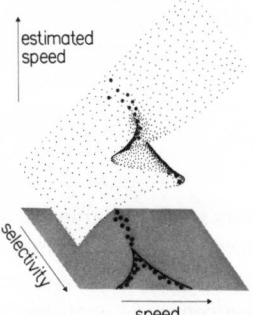

Fig. 17.13

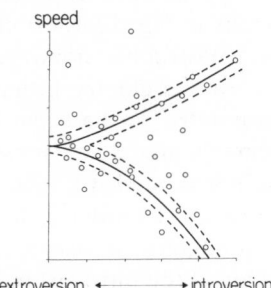

Fig. 17.14

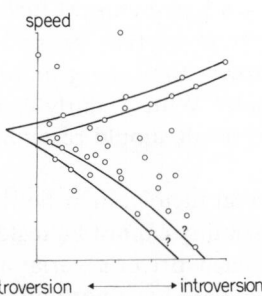

Fig. 17.15

'Extroverts' are able to drive at what they estimate to be their normal speed, and do so. 'Introverts', on the other hand, are denied this possibility because the central sheet of surface is inaccessible. Zeeman postulates that they do the best they can, and hence 'hover on the borderline of catastrophe'. Thus the subjects' behaviour is distributed over the catastrophe manifold as in Fig. 17.13; the projection into control space gives points clustering around the cusped curve of the bifurcation set. Fig. 17.14 shows a cusp fitted to the data of Fig. 17.11. With the small 'standard deviation' indicated by dotted lines, 50% of the dots lie within this deviation from the cusp. Zeeman adds 'trebling this deviation will include over 80% of the individuals'.

The first objection is that this is not as good a fit as it may seem. It is actually quite hard to decide on a satisfactory test for goodness of fit of curved graphs; but what is clear is that the quoted remark on trebling the deviation has little force: not only do 80% of the dots come within the region around the curve, but that region fills approximately half of the whole graph. Given *any* distribution of 40 dots other than a random 'plum pudding' it ought to be possible to fit a curve such that the region around it having half the area of the graph contains most of the points!

The next objection is that the model is very artificial. The state variable – estimated speed – is not one for which data exist. Then it is unclear why an introvert's estimate of his speed should be so consistently in error, denying him precisely that which he is seeking. It is more reasonable that he should do his best, and perch near the fold curve; but one would expect him to oscillate between over- and under-estimating his speed. If, as Zeeman suggests, he later learns to stick to a fixed estimate with considerable precision, why can he not do so at his *normal* speed?

Further, this assumption affects our judgement of the fit of the data; for by the theory, the points should cluster *within* the bifurcation set, more as in Fig. 17.15. And now we need a greater 'standard deviation' to include the same number of points.

But these are mere quibbles. The real objection is that this whole mechanism is an *ad hoc* construction, dictated by the initial assumption that *the graph is a cusp*. Despite the date of publication, this model goes back to very early days, when cusp catastrophe modelling was thought to mean 'look for a cusp'. Having accepted Fig. 17.11 as a cusped curve, one is looking willy-nilly at a bifurcation set; this lives in a control space; and the observed speed of the subject perforce is a *control* variable. This is a somewhat perverse assumption: the individual's response is most naturally a behaviour variable, and until this possibility is ruled out it is unreasonable to assume otherwise. Having taken the speed as a control variable, two problems arise. The first is to find the state variable; the second to explain why everything hovers over the top of the bifurcation set, with the rest of the catastrophe manifold being virtually devoid of population. The use of 'estimated speed' is designed to achieve solutions to exactly these problems.

This is not to say that Zeeman's model is wrong: just that it seems unduly complicated. Two simpler alternative models, which we shall describe below, seem to fit the data just as well, and without considerable refinement we cannot hope to decide which (if any) of the three is correct.

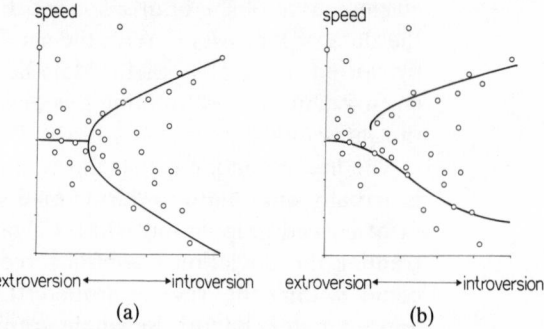

Fig. 17.16

Suppose we do the obvious, and take speed as state variable. It is still reasonable to use introversion/extroversion as a control variable – indeed as a splitting factor, since that is its effect. As yet we have no normal factor, so Fig. 17.11 is a *projection*, viewed from the side, of some catastrophe manifold. *If* as a first approximation we ignore the second control variable altogether then we *might* expect to see a *section* of a cusp catastrophe, with the data clustered around a curve of the usual 'pitchfork' shape, as in Fig. 17.16(a) or (b). The most significant feature of the cusp catastrophe, sideways on, is its parabolic 'hole' of unstable states: one could try fitting parabolae through, or inside, the given data points. (To be consistent with one remark made concerning Zeeman's model, we must insist that the data are outside the parabola, not around it.)

Now it is clear that the fit here is of the same sort of order as for Fig. 17.14 (and it would be hard to devise an effective measure of goodness of fit to compare them.) The branching at the introvert but not at the extrovert end is entirely natural (especially if, as does Zeeman, we ignore the group of three reckless individuals at the top left). However, the graph is not a *section* of a catastrophe, but a *projection*, and we have no right to dismiss a missing control factor as if it never existed (as we saw in Section 3). If we take as model Fig. 17.17(a), the normal factor being unknown, then the graph of Fig. 17.11 will be this, viewed from the side. With a fairly even distribution of points, within fixed limits, the result should resemble Fig. 17.17(b). Which it does.

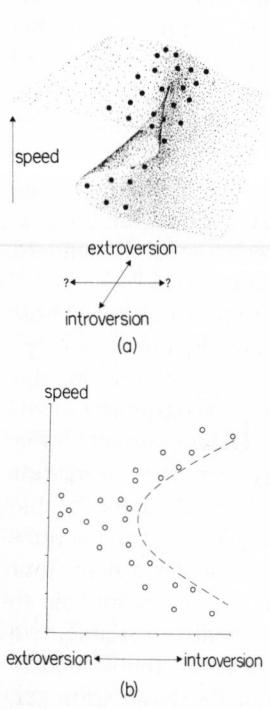

Fig. 17.17

A plausible candidate for the missing normal factor might be the driver's normal speed: unfortunately this possibility cannot be tested from the given data, since they give only each driver's variation from his own normal. It is quite likely that the data have normalized away the normal factor!

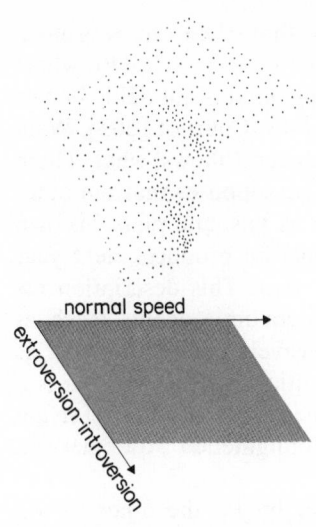

normal speed

extroversion-introversion

Fig. 17.18

Pursuing this line of thought: it may be that the observed divergence is an artefact of this normalization, and there is no catastrophe at all: the apparent 'hole' in the plotted points may just be a region where the points are thinly spread, which could occur for a single-valued (non-catastrophic) graph with steep slopes in the corresponding places, as in Fig. 17.18.

This raises more questions than it answers, but it does illustrate how crucially a catastrophe model depends on the initial assumptions made.

7 Beyond Elementary Catastrophe Theory

Not all dynamic phenomena are driven by the maximizing or minimizing of potential functions, with consequent catastrophe geometry as these functions vary. Many that are not can be related to catastrophe theory less directly; by the construction of suitable Liapunov functions, as for the Duffing equation (Holmes and Rand [202]); by variational principles as for optics (Chapter 12); by stream functions (Chapter 11); and so on. Catastrophe theory (in the sense we have used the phrase) is not universally applicable, but the frequent statement 'it applies only to gradient systems' is simply a misunderstanding brought about by the limited range of early examples.

Conversely, however, we do not wish to leave the reader with the impression that where optimization *is* the name of the game, catastrophe theory can be uncritically applied. It is worth illustrating this with an example where two agents are trying to optimize a situation, each from his/her own point of view.

If you start producing books, when no one else is, you will not sell many. There will be no book-buying habit among the public, and no distribution industry to take and display your products to Hull and Halifax. On the other hand, if other people are producing books in huge numbers, yours will be invisible among so many, and again you will sell rather few. Your sales will be best when other output exists but is moderate. The most profitable number to print, of course, is exactly what you can sell.

This description is oversimplified, of course, but since we shall be illustrating complicated behaviour following from simple assumptions this does not spoil the argument: realistically complicating the assumptions will further complicate the behaviour. (One must distinguish two cases. In the disorganized complexity of the aggregates of 'simple' systems studied in physics, like bulk matter, even such complication as the components actually display will often statistically simplify for large enough samples. But in the highly structured, organized complexity of an economy or a brain, complications at each level are often compounded – rather than averaging out – at the next level up.) We further simplify, in fact, to the case of

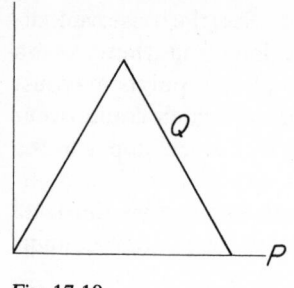

Fig. 17.19

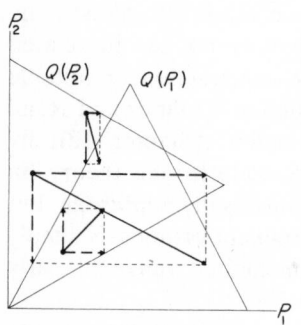

Fig. 17.20

just two identical producers, and suppose that when one is producing P items, the optimum production for the other is $Q(P)$, where the graph of Q is shown in Fig. 17.19. This kind of model is known as a *Cournot duopoly* and goes back to 1843 (Cournot [203], Wald [204]). Further, suppose that each producer follows what Thom [205] calls a '*stratégie myope*': each year, he supposes that the other producer will produce the same next year as this, and plans his own production for next year accordingly. Thus he produces next year what would have been his optimum this year. This description fits depressingly well the strategy of many economic agents (such as farmers) whose behaviour has been intensively studied, as well as the public economic thinking of many politicians. (The myopia is a convenience for quick description, *not* a necessary condition for the wild behaviour below. Models with more 'enlightened' strategies can also bite.)

Denote the production of one producer by P_1, the other by P_2. Typical changes of the pair (P_1, P_2) from one year to the next are shown in Fig. 17.20; the broken arrows are the moves that the producers are trying for (optimizing according to Q, on the assumption that the other's production does not change). The solid arrows show the resulting combined change.

One might hope that this set-up would settle to a steady production rate for each, or at least a regular cycle. But a typical sequence of moves is shown in Fig. 17.21, in which no obvious pattern emerges. We observe what has become known in mathematical ecology, from a paper by Li and Yorke [206], as *chaos*. The lack of obvious pattern has been refined, for many examples of this kind, to mathematical proof that no known statistical techniques can distinguish the behaviour of systems that are known beforehand to be 'deterministic' from purely random ones. (*See* Guckenheimer [207] and Guckenheimer, Oster and Ipaktchi [208] for a general discussion.) All characteristics of 'statistical' behaviour, like the Law of

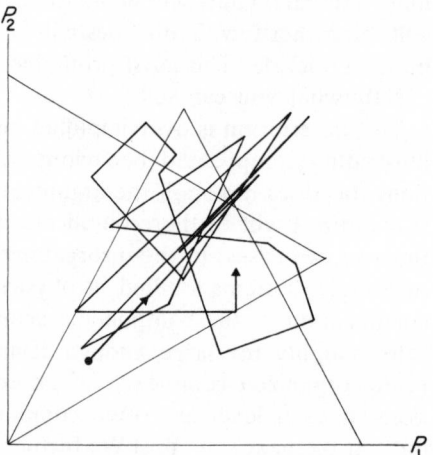

Fig. 17.21

Large Numbers, can be obeyed perfectly by simple *deterministic* models of this kind. (Thus the behaviour of stock market prices, which fits excellently a 'random walk' model in which previous movements have no predictive value for the future, could equally well be a deterministic system displaying chaos.) For the case we are considering, Rand [209] has shown directly that regular cycles of all periods, $1, 2, \ldots, 2001, \ldots$ years exist, but are all *unstable*: deviate from one by a hairsbreadth and, rather than continuing to follow it approximately, the system moves away from it rapidly and the cycle becomes useless, even as an approximation. Approximate knowledge of the present situation, however accurate, is useless for long-term prediction: arbitrarily close points have futures which diverge utterly. The *typical* behaviour is indeed the type of chaos shown in Fig. 17.21.

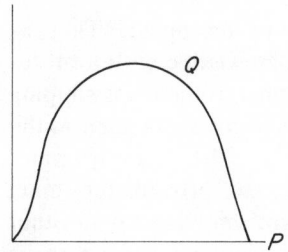

Fig. 17.22

This exotic behaviour is not a consequence of special features of the function Q, such as being linear up and linear down, with a sharp peak. Rand [209] shows that the behaviour with a smooth Q as in Fig. 17.22 is even harder to describe, and that its complexity is perfectly stable under small perturbations of Q. (Two types of stability occur, and must not be confused: the drastic instabilities within a system with given Q persist, *stably*, as Q changes slightly.)

More complicated behaviour still can be expected from models of many economic structures elaborate enough even to approach realism. It has recently been shown by Smale [210] that the systems studied in ecology and in the General Equilibrium Theory of mathematical economics are not a special, tame class, as was widely supposed. Absolutely all of the chaotic behaviour, 'strange attractors' (Guckenheimer, Oster and Ipaktchi [208]) 'omega-explosions' and other freakish familiars of the wizards of modern dynamical systems theory can stably occur in this class of models. (Thom's more general use of the term 'catastrophe' takes these phenomena into account – but then a general theory is lacking, so the practical advantage is small: the philosophical gain may be greater.) When this is realized widely enough for computer-model economists not to treat such phenomena as program-bugs, and not to steer their parameters or their model away from realism to eliminate them, explicit examples with detailed models 'drawn from life' will become common too. Unless this is too unpopular with the people who pay for large computer studies in economics.

Thus, the fact that a system is full of subunits *aiming* at maximizing a simple function, or family of functions, does not bring it within the scope of the mathematics treated in this book unless, most of the time, such optimal behaviour is actually achieved. Often it will not be, and more powerful tools are needed in the analysis, even for theoretical purposes. Adequate mathematics for *planning* in the presence of such phenomena is a still far distant goal.

18 Catastrophe Theory: Whither Away?

1 The Present State

Catastrophe theory is already beginning to disappear. That is, 'catastrophe theory' as a cohesive body of knowledge with a mutually acquainted group of experts working on its problems, is slipping into the past, as its techniques become more firmly embedded in the consciousness of the scientific community. The 'strictly pure' mathematicians among those who found and proved the main theorems of the subject have remained 'pure', moving on to other areas, near or distant, where mathematical problems remain to be solved. Identifiable areas for future work which will be required for applications are many: they include 'equivariant' catastrophe theory (catastrophes with symmetry), the relation between differential and topological equivalence, the use of infinite-dimensional state spaces, and numerous variants of the elementary catastrophes such as the time catastrophes of Wassermann [112] and the constraint catastrophes mentioned in Chapter 16. These are only those most closely connected with the elementary catastrophes; and the whole field of dynamical systems and bifurcation theory will receive increasing attention. The potential scope for alliances between the powerful topological methods of mathematics, and powerful numerical methods (such as the finite element method) used by practical research workers, is enormous.

Those who remain attached to catastrophe theory itself are working on the relation between the content of the theorems and the nature of the world. Alone, or increasingly in collaboration with experts in *things* (waves, bribes, embryos ...) they are *applying* mathematics. (We prefer not to say that they are doing 'applied mathematics' since, by an unfortunate quirk of British academic history, far too much of what goes by that name is neither mathematics of any conceptual integrity, nor is it applied to anything of practical significance, or in areas where it might be proved wrong by experiment. This is not our opinion alone: it is stated at length in the Royal Society report *Postgraduate Training in the UK 4: Applied Mathematics.*) The second half of this book gives some indication of the spread of these applications: the bibliography gives far more. Even to compile such a bibliography is now quite hard, with the literature scattered from the physical chemistry to the town planning journals. In a few years time it will be not only impossible, but a pointless exercise. For a while after Newton and Leibniz one could usefully list all works on the differential calculus; for a while longer one could list all the original and the main expository writings; but who would attempt to do so today? The ideas and techniques of the calculus have become absorbed into the background of almost every scientist worthy of that description. The same goes for probability theory, variational methods, the theory of infinite-dimensional vector spaces, and all of the other branches of mathematics that make up the toolkit of the working scientist.

Catastrophe theory, while in a sense going beyond the calculus (by virtue of its topological flavour) is nevertheless best viewed as a new development *within* the calculus. In our view the oft-repeated assertion that 'one does not need to understand the mathematics behind catastrophe theory in order to be able to apply it' is misleading. Unless one has *some* feeling for the mathematics (and the mathematical chapters of this book are an absolute minimum) one's applications of the theory are open to serious errors. (The same goes, in applications, for the subject to which the mathematics is applied: hence catastrophe theory is increasingly becoming a team sport.)

One of the subject's beauties, to a mathematician, is the way that it has both brought to light and explained the recurrence of certain mathematical structes in field after field of the sciences. But this does not and will not interest most scientists, who are usually content to call a mathematical result after whoever *in their field* first noticed it. Its presence under other names in other fields, whether discovered earlier or later, has little influence. And shared structures cause even less excitement than shared theorems. Thus the feature of catastrophe theory of perhaps the greatest interest to the non-specialist – its broad unifying power – is of far less interest to the specialist. The cusp, for example – the simplest 'interesting' catastrophe – has innumerable names in innumerable fields, and is as omnipresent as we saw in Chapter 7 that it must be. In each long-established field it has been more or less completely analysed, almost always adequately for the purpose at hand. Precisely because it is one of the simplest catastrophes, barehanded analysis like that of Chapter 5 is enough to answer any questions about its geometry. Such analyses have been performed over and over and over again, to the complete satisfaction of those involved. On the other hand, such facts as the structural stability of the cusp can *not* be established by elementary methods. But few people are excited by answers to questions – however relevant to their work – that they have not asked. What most scientists say they want is numbers.

Catastrophe theory, like the calculus generally, *gives* numbers. It also gives answers to topological questions, and since it was invented and first preached by topologists, it was these answers that were stressed. Hence the myth, soon circulating and still widespread, that it is 'purely qualitative'. Once understood by scientists who thought in physical numbers, fed it physical numbers, and asked it numerical questions, the theory has begun to produce numerical answers. These answers, it must be said, are based on mathematical descriptions currently accepted by scientists; an observation sometimes interpreted (e.g. Croll [116]) as 'catastrophe theory provides no new information about anything'. This is just the general point that mathematics as a whole is 'tautological', needing hypotheses to obtain conclusions. No mathematical theorem ever gives 'information' in the sense of Information Theory – information that what *need* not be true *is* true in some instance – only experiment can do that. The 'information' in a scientific theory lies in its hypotheses about what is, and these contain all its predictions. As has been observed many times over the centuries in rebuttal of 'no content' objections to mathematical reformulations, the problem is *getting the information out.*† And here catastrophe theory is a powerful new weapon – as the latter half of this book demonstrates. It is not a *replacement* for existing methods, but an *addition* to them. We have given many examples of this, including many numerical answers to numerical questions. We hope that this book will be a tool in the production of many more, as well as suggesting new kinds of question which to us seem just as physical (or chemical, or biological, or . . .).

2   The Future

In the short term, the currently 'hard' sciences – physics, chemistry, engineering – are likely to benefit most from catastrophe theory, which will serve, in conjunction with current techniques and theories, as a tool for discovering more consequences of mathematical models whose hypotheses are already reasonably established but whose implications are still being explored. Chapters 10 to 15 are examples. ('To him that hath, shall be given.') The *stably* nonlinear behaviour of catastrophe phenomena, which no reparametrization or small adjustment can change, points up the absurdity of the simplifying assumptions like linearity or convexity used in most socio-mathematical modelling . . . but often (Chapter 17 Section 3) increases the data requirements beyond even the presently unattainable needs of the current improbably simple models. ('From him that hath not, even that which he hath shall be taken away'.)

In the long term, the balance will probably be different. The field of biology presents the same difficulties of organized complexity as do the social sciences, but hard data and repeatable experiments are available. The real triumphs of molecular biology are slowly ceasing to obscure the fact that knowing the axioms of even a simple system does not imply an understanding of its behaviour. (Biochemistry is

---

† One is reminded of the king in the fairy story, complaining about the cloak of invisibility provided by the wizard: 'I still keep bumping into things!'

of course one of the 'hard' sciences mentioned above, and since so many reactions are governed by free energy minimization the subject will be enriched by the new calculation techniques and geometry, as soon as the relevant pieces of mathematical and chemical knowledge are joined in the right individuals or teams. But this is not our point here.) Statistical mechanics, for example, does not attempt to analyse the bulk behaviour of a disorganized mass of simple molecules by including in the model the detailed known properties of (for example) carbon dioxide. Instead it considers 'hard spheres' and the like. Living systems involve *organized* complexity, and many kinds of molecule, whose individual complexity is crucial to the subtle behaviour. Can anyone expect to be able to deduce, from only a full molecular description of a baby, that the life of Man is threescore years and ten? That wisdom teeth will come in a few decades? Science must work with many descriptions at many levels, with the larger scale model (as discussed for fluid dynamics in Chapter 11 Section 1) sometimes explicitly contradicting what is known about smaller scale phenomena. In the opposite direction, most quantum mechanical calculations concerning solids or liquids make assumptions about simultaneity believed false by those who make them. Relativistic quantum chemistry is not in prospect.

Biology is likely to be the testing ground, over the next few decades, of mathematical ways of modelling the aggregated behaviour of systems with underlying organized complexity. In this effort, catastrophe theory will play a major rôle; yet we can be sure in advance that it will not be the whole story: neither even will dynamical systems theory, of which it is a part. But the bifurcation of real-valued functions does describe the simplest kinds of essentially nonlinear behaviour. As Chapters 10 and onwards show, many phenomena reduce either rigorously or heuristically to such descriptions, *and certainly no further* without loss of useful understanding. The insights of the arguments sketched in Chapter 7, and a wealth of experience with examples, insist that the nature of the elementary catastrophes will be essential knowledge in the mathematization of biology.

Moreover, this knowledge will not be lacking. If catastrophe theory were relevant *solely* to biology, then a failure to apply it gloriously within a few years could lose face and opportunity, and it might be forgotten. But increasingly in the hard sciences catastrophe theory will be assimilated into the generally known calculus. The scientist will use the general determinacy rules as routinely as (s)he now uses the special case that a positive definite Hessian implies a minimum. As this complex of results and viewpoints becomes as firmly embedded in the scientific mentality as the habit of writing down differential equations, there will be no shortage of expertise in the many versions of what 'catastrophe theory' evolves into. Quite apart from the final success or failure of the current promising efforts, such as the work of Zeeman [157] and Cooke in embryology, biology will no more 'forget' catastrophe theory than it will the calculus.

More generally, we may predict that in the future the problems posed by biology, as yet struggling towards precision of statement, will affect the mainstream of mathematics, as have those posed by physics in the past. As methods relevant to organized complexity develop in this laboratory science, the social sciences will benefit in proportion. The new concepts – fusing with, changing, and adding to present understanding – may allow the definition and measurement of quantities more central to the health of the body politick than a 'standard of living' that includes useless packaging discarded, or a 'gross national product' that includes machines whose productivity is defined in megadeaths. In the absence of a communicable wisdom as to what we are, all predictions for the future of our scientific culture (if not of the race) reduce to negation. If *any* mathematical methods can aid the growth of such wisdom, then catastrophe theory will be part of them.

Appendix 1 Computer Program for Determinacy and Unfoldings

By D. R. Olsen, S. R. Carter
and A. Rockwood

For the reader wishing to apply catastrophe theory to complicated situations in the hard sciences, and with access to a good computer, we include an ALGOL program that will perform the more essential computations of Chapter 8: strong determinacy, local determinacy, codimension, and test for transversality. (Ordinary determinacy, using Theorem 8.4, presents more algebraic intricacy, but an economical routine is promised before our publication date.) For algebraic convenience the 'constant' polynomial 1 is included in the computations, as discussed in Chapter 8 Section 6, so that for the sense we have mainly used the reader should reduce the computed codimension by 1 and ignore the constant term.

Most scientifically-arising functions and unfoldings are not, of course, polynomials, as we have seen; this routine requires that it be supplied with a k-jet. Formal differentiation is a well developed subject, however, so that there is no point in our including a routine here. The reader wishing to test functions and unfoldings expressed directly in terms of sin, tanh, etc. should run the program below in tandem with one for symbolic differentiation, such as the rather sophisticated one developed at MIT, to which the following references apply:

> ACM *Proceedings* of the Second Symposium on Symbolic and Algebraic Manipulation, Los Angeles, California, March 1971.
> *ACM Communications* **14,** No. 8, 1971.
> *ACM Journal* **18,** No. 4, 1971.
> 'MACSYMA – the fifth year', *SIGSAM Bulletin* August 1974.

Elementary non-polynomial functions will be accepted by a version now in preparation.

The program will test for strong and local k-determinacy for specified k (and ask for another k if the answer is 'no'). When determinacy has been established, it will test any unfolding offered (in polynomial form) for transversality, and give the codimension. If the unfolding offered is not transversal, the program will provide a list of polynomial unfolding terms which complete it. (Thus in the sample output given at the end, where no unfolding terms are offered, it simply lists a choice of cobasis to give a universal unfolding.)

Earlier versions of this program, which did less, were quite costly, but this problem has now been overcome. The example given comes from a run of ten, which as a whole cost $1.21.

```
!*******************************************************************
!*                                                               *!
!*        THIS IS A PROGRAM TO CHECK POLYNOMIALS IN ANY NUMBER OF  *!
!*        VARIABLES FOR STRONG OR LOCAL K-DETERMINACY FOR ANY GIVEN *!
!*        VALUE OF K.                                              *!
!*                NOTE:    THE PROGRAM WILL HANDLE ANY VALUE OF K IN *!
!*                         ANY NUMBER OF VARIABLES BUT THE CORE     *!
!*                         REQUIREMENTS ARE EXCESSIVE FOR SOME MACHINES *!
!*                         IF THE NUMBER OF VARIABLES OR THE VALUE OF *!
!*                         K BECOMES LARGE. (NEW PROGRAMMING TECHNIQUES *!
!*                         BEING DEVELOPED AT BRIGHAM YOUNG UNIVERSITY *!
!*                         WILL SOLVE THIS PROBLEM)                *!
!*        POLYNOMIALS ARE ENTERED AND REPRESENTED AS A LIST OF TERMS *!
!*                WITH EACH TERM BEING REPRESENTED BY A INTEGER     *!
!*                COEFFICIENT AND A LIST OF INTEGER EXPONENTS       *!
!*                                                                 *!
!*        AUTHORS:        DAN R. OLSEN JR,  STEPHAN R. CARTER,      *!
!*                        ALYN ROCKWOOD                             *!
!*        CONSULTING:     HELAMAN R. R.P.FERGUSON                   *!
!*        SPONSORS:       BRIGHAM YOUNG UNIVERSITY                  *!
!*                                COMPUTER SCIENCE DEPARTMENT       *!
!*                        EYRING RESEARCH INSTITUTE                 *!
!*                                                                 *!
!*******************************************************************!
BEGIN
EXTERNAL PROCEDURE SORT,UNFOLD,ORDINARY;

EXTERNAL BOOLEAN PROCEDURE KDET;
EXTERNAL PROCEDURE PARTIAL;
INTEGER N.VARIABLES, N.TERMS,CMND;
BOOLEAN STRONG;
INTEGER K;
WRITE("WELCOME TO THE BLESSED WORLD OF CATASTROPHE THEORY[CC]");
WRITE("[T]NUMBER OF VARIABLES =[SB]");
READ(N.VARIABLES);
WRITE("[T]NUMBER OF TERMS IN THE POLYNOMIAL =[SB]");
READ(N.TERMS);
        BEGIN ;SET UP STORAGE FOR FUNCTION AND GET FUNCTION;
        INTEGER  ARRAY TERMS[1:N.VARIABLES,1:N.TERMS],
                DEGREE[1:N.TERMS],
                PART[1:N.VARIABLES,1:N.VARIABLES,1:N.TERMS],
                COEF[1:N.TERMS],PCOEF[1:N.VARIABLES,1:N.TERMS];
        INTEGER I,J,K,L,M,N;
        WRITE("IT IS NOW TIME TO INPUT THE POLYNOMIAL [C]");
        WRITE("EACH OF THE[S]");
        PRINT(N.TERMS,3,0);
        WRITE(" TERMS OF THE POLYNOMIAL");
        WRITE("IS INPUT [C] USING THIS FORMAT[B]");
        WRITE("[CT]C ");
        FOR I + 1 UNTIL N.VARIABLES DO WRITE("D ");
        WRITE("[2CB]");
        WRITE("WHERE C IS THE COEFFICIENT AND D IS THE DEGREE[CB]");
        WRITE("PLEASE INPUT THE TERMS [CC]");
        FOR I + 1 UNTIL N.TERMS DO
                BEGIN
                WRITE("TERM "); PRINT(I,3,0); WRITE("[SB]");
                READ(COEF[I]);
                DEGREE[I] + 0;
                FOR J + 1 UNTIL N.VARIABLES DO
                        BEGIN
                        READ(TERMS[J,I]);
                        DEGREE[I] + DEGREE[I] + TERMS[J,I];
                        END;
                END;
        SORT(TERMS,N.VARIABLES,N.TERMS,COEF,DEGREE);
        PARTIAL(TERMS,COEF,PART,PCOEF,N.VARIABLES,N.TERMS);
        WRITE("WHAT K DO YOU WANT TO TRY? (0 TO STOP) [B]");
        READ(K);
        WHILE K#0 DO
                BEGIN
                IF KDET(PART,PCOEF,N.VARIABLES,N.TERMS,K,STRONG) THEN
```

```
                            BEGIN
                            IF STRONG THEN
                                    BEGIN
                                    WRITE("*** STRONGLY DETERMINED ***[CCB]");
                                    UNFOLD(PART,PCOEF,N.VARIABLES,N.TERMS,K);
                                    END
                            ELSE
                                    BEGIN
                                    WRITE("*** LOCALLY DETERMINED ***[CCB]");
                                    WRITE("DO YOU WANT TO TRY ORD");
                                    WRITE("INARY DETERMINACY? [CB]");
                                    WRITE("1 = YES, 0=NO[CB]");
                                    READ(CMND);
                                    IF CMND=1 THEN ORDINARY(K);
                                    K:=K+1;
                                    UNFOLD(PART,PCOEF,N.VARIABLES,N.TERMS,K);
                                    END;
                            END
                    ELSE
                            WRITE("FAILED - - TRY A LARGER K[CCB]");
                    WRITE("WHAT K DO YOU WANT TO TRY? ");
                    WRITE("(0 TO STOP) [B]");
                    READ(K);
                    END;
            END;
    END
    !COMPUTES THE MATRIX SIZE, ALLOCATES THE MATRIX AND CALLS SOLVED TO
            CHECK FOR K-DETERMINACY;
    BOOLEAN PROCEDURE KDET(PART,PCOEF,N.VAR,N.TERMS,K,STRONG);
    VALUE N.VAR,N.TERMS,K;
    INTEGER ARRAY PART;
    INTEGER ARRAY PCOEF;
    INTEGER N.VAR,N.TERMS,K;
    BOOLEAN STRONG;
    BEGIN
    EXTERNAL BOOLEAN PROCEDURE SOLVED;
    EXTERNAL INTEGER PROCEDURE KTERM;
    INTEGER COL,ROW,I,J,DEG,MINDEG,K1,START;
    MINDEG:=K+1;
    KDET:=TRUE;
    FOR I:=1 UNTIL N.VAR DO   !LOOP THRU PARTIALS;
            BEGIN
            DEG:=0;
            FOR J:=1 UNTIL N.VAR DO   !LOOP THRU EXPONENTS;
                    DEG:=DEG+PART[I,J,1];
            IF DEG<MINDEG THEN
                    MINDEG:=DEG;
            END;
    IF K<=MINDEG THEN
            BEGIN
            WRITE("ERROR *** K TO SMALL[CCB]");
            KDET:=FALSE;
            GOTO OUTR;
            END;
    ROW:=N.VAR*KTERM((K-MINDEG+1),N.VAR);
    COL:=KTERM((K+1),N.VAR)-KTERM(MINDEG-1,N.VAR);
    WRITE("THE MATRIX HAS BEEN ALLOCATED AT "); PRINT(ROW); WRITE(" BY ");
    PRINT(COL); WRITE("[CB]");
            BEGIN
            INTEGER ARRAY N.MATRIX[1:ROW,1:COL],D.MATRIX[1:ROW,1:COL];
            INTEGER I,J;
            KDET:=SOLVED(N.MATRIX,D.MATRIX,ROW,COL,K,MINDEG,N.TERMS,PART,
                    PCOEF,N.VAR,STRONG);
            END;
    OUTR: END
    !SETS UP MATRIX AND CALLS FOR GAUSSIAN REDUCTION TO CHECK TO SEE IF
    THE POLYNOMIAL IS DETERMINED AT K;
    BOOLEAN PROCEDURE SOLVED(N.MATRIX,D.MATRIX,ROW,COL,K,MINDEG,N.TERMS,PART,
```

```
                PCOEF,N,VAR,STRONG);
INTEGER ARRAY N.MATRIX,D.MATRIX,PART,PCOEF;
INTEGER ROW,COL,K,MINDEG,N.VAR,N.TERMS;
BOULEAN STRONG;


BEGIN
EXTERNAL PROCEDURE FILLMATRIX;
EXTERNAL BOOLEAN PROCEDURE SINGULAR;
INTEGER ARRAY POINTER[1:COL],TERM.LIST[1:COL,1:N.VAR];
INTEGER EACH,K1,TERM.PNT,K1.START,MONPNT,PARTITION,K1DEG;
FILLMATRIX(N.MATRIX,D.MATRIX,PART,PCOEF,N.VAR,N.TERMS,K,K1.START,MINDEG,
        COL,ROW,PARTITION,TERM.LIST,K1DEG);
MONPNT_1;
STRONG_TRUE;
FOR TERM.PNT_1 UNTIL COL DO
        POINTER[TERM.PNT]_TERM.PNT;
FOR TERM.PNT_1 UNTIL COL DO
        BEGIN
        N.MATRIX[ROW,TERM.PNT]_0;
        D.MATRIX[ROW,TERM.PNT]_1;
        END;
SOLVED _ SINGULAR(N.MATRIX,D.MATRIX,POINTER,ROW,COL,PARTITION,STRONG,
                TERM.LIST,K1DEG,N.VAR);
END
|THIS PROCEDURE PERFORMS A GAUSSIAN REDUCTION OF THE MATRIX;
|IF PARTITION IS EQUAL TO ZERO THE ROUTINE CHECKS UNFOLDINGS
        OTHERWISE IT CHECKS FOR K-DETERMINACY;
|ALL ARITHEMATIC IS RATIONAL WITH NUMERATOR AND DENOMINATOR STORED
        IN IDENTICAL ARRAYS;
BOULEAN PROCEDURE SINGULAR(N.MATRIX,D.MATRIX,POINTER,ROW,COL,PARTITION,STRONG,TERM.LIST,
INTEGER ARRAY N.MATRIX,D.MATRIX,TERM.LIST;
INTEGER ARRAY POINTER;
INTEGER ROW,COL,PARTITION,K1.DEG,N.VAR;
BOULEAN STRONG;


BEGIN
BOULEAN SING.FLG;
INTEGER ROWI,COLI,COLSRCH,COLPNT,TEMP,K,T,UNFLDCNT;
BOULEAN FOUND,USED;

|THIS PROCEDURE REDUCES FRACTIONS TO PREVENT OVERFLOW;
PRUCEDURE REDUCE(N,D);
INTEGER N,D;
        BEGIN
        INTEGER N.DIV,D.DIV,I,INC.END,DA,NA;
        IF N=0 THEN GOTO END.REDUCE;
        NA_N;
        IF NA<0 THEN NA_-NA;
        DA_D;
        IF DA<0 THEN DA_-DA;
        INC.END_DA-NA;
        IF INC.END<0 THEN INC.END_-INC.END;
        IF INC.END>NA THEN INC.END_NA;
        IF INC.END>DA THEN INC.END_DA;
        IF N=D THEN
                BEGIN
                N_1;
                D_1;
                END
        ELSE
                BEGIN
                FOR I_2 UNTIL INC.END DO
                        BEGIN
                        N.DIV_N/I;
                        D.DIV_D/I;
                        WHILE ((N.DIV*I=N)AND((D.DIV*I)=D)) DO
                                BEGIN
                                N_N.DIV; D_D.DIV;
```

```
                                    N.DIV.N/I;
                                    D.DIV.D/I;
                                    END;
                        END;
                END;
        END.REDUCE; END REDUCE;


COLPNT.1;
UNFLOCNT.0;
FOR ROWI.1 UNTIL ROW DO
        BEGIN
        IF COL<COLPNT THEN
                BEGIN
                ROWI.ROW;
                END
        ELSE
BEGIN
USED.FALSE;
FOR COLSRCH.COLPNT UNTIL COL DO
        BEGIN
        IF N.MATRIX[ROWI,POINTER[COLSRCH]]#0 THEN
                BEGIN
                TEMP.POINTER[COLPNT];
                POINTER[COLPNT].POINTER[COLSRCH];
                POINTER[COLSRCH].TEMP;
                COLSRCH.COL+3;
                USED.TRUE;
                END;
        IF COLSRCH#COL+3 THEN
                BEGIN
                INTEGER N,D,NUM,DENUM,R;
                COLI.POINTER[COLPNT];
                IF N.MATRIX[ROWI,COLI]#D.MATRIX[ROWI,COLI] THEN
                        BEGIN
                        NUM.N.MATRIX[ROWI,COLI];
                        DENUM.D.MATRIX[ROWI,COLI];
                        FOR R.ROWI+1 UNTIL ROW DO
                                BEGIN
                                N.N.MATRIX[R,COLI];
                                IF N#0 THEN
                                        BEGIN
                                        N.N*DENUM;
                                        D.D.MATRIX[ROW,COLI];
                                        D.D*NUM;
                                        REDUCE(N,D);
                                        N.MATRIX[R,COLI].N;
                                        D.MATRIX[R,COLI].D;
                                        END;
                                END;
                        END;
                IF ROWI > PARTITION THEN
                        BEGIN
                        IF N.MATRIX[ROWI,COLI]#0 AND
                            COLI >= K1DEG THEN STRONG.FALSE;
                        END;
                N.MATRIX[ROWI,COLI].1;    |SET PIVOT TO 1;
                D.MATRIX[ROWI,COLI].1;
                COLPNT.COLPNT+1;
                FOR COLSRCH.COLPNT UNTIL COL DO
                        BEGIN
                        INTEGER MUL.N,MUL.D,NN,DN,DA,NA,
                                DI,NI,COLA,R;
                        COLI.POINTER[COLSRCH];
                        COLA.POINTER[COLPNT-1];
                        MUL.N.N.MATRIX[ROWI,COLI];
                        IF MUL.N#0 THEN
                                BEGIN
                                MUL.D.D.MATRIX[ROWI,COLI];
```

```
                                        D.MATRIX[ROWI,COLI]_1;
                                        N.MATRIX[ROWI,COLI]_0;
                                        FOR R_(ROWI+1) UNTIL ROW DO
                                                BEGIN
                                                NA_N.MATRIX[R,COLA];
                                                IF NA#0 THEN
                                                    BEGIN
                                                    DA_D.MATRIX[R,COLA];
                                                    DI_D.MATRIX[R,COLI];
                                                    NI_N.MATRIX[R,COLI];
                                                    DN_DI*MUL.D*DA;
                                                    NN_NI*MUL.D*DA-
                                                        NA*MUL.N*DI;
                                                    REDUCE(NN,DN);
                                                    D.MATRIX[R,COLI]_DN;
                                                    N.MATRIX[R,COLI]_NN;
                                                    END;
                                                END;
                                END;
                        END;
                END;
                IF PARTITION <= 0 AND NOT USED THEN
                        BEGIN
                        T_ROW-ROWI+1;
                        IF T<=K1DEG THEN
                                BEGIN
                                WRITE("UNFOLDING TERM ");
                                PRINT(T,3);
                                WRITE(" IS REDUNDANT [CB]");
                                UNFLDCNT_UNFLDCNT+1;
                                END;
                        END;
                END;
        SING.FLG_TRUE;
        IF PARTITION<=0   THEN
                BEGIN
                WRITE("[CC]CODIMENSION = ");
                PRINT(COL-COLPN1+1+K1DEG-UNFLDCNT,3);
                WRITE("[CB]");
                WRITE("ADDITIONAL UNFOLDING TERMS ARE[CCB]");
                FOR COLSRCH _ COLPNT UNTIL COL DO
                        BEGIN
                        COLI _ POINTER[COLSRCH];
                        FOR ROW1 _ 1 UNTIL N.VAR DO
                                PRINT(TERM.LIST[COLI,ROW1],3);
                        WRITE("[C]");
                        END;
                WRITE("[CB]");
                END
        ELSE
                BEGIN
                FOR COLSRCH_COLPNT UNTIL COL DO
                        BEGIN
                        IF POINTER[COLSRCH]>=K1DEG THEN
                                BEGIN
                                SING.FLG_FALSE;
                                COLSRCH_COL+1;
                                END;
                        END;
                END;
        SINGULAR_SING.FLG;
        END
        ;GENERATES THE SET OF ALL POLYNOMIALS WHICH EQUAL K-JETS OF THE
                PARTIALS MULTIPLIED BY ALL MONOMIALS OF DEGREE D SUCH THAT
                0<=D<=K+1        AND STORES IN THE MATRIX WITH ROWS REPRESENTING
```

```
            POLYNOMIALS AND COLUMNS REPRESENTING TERMS WITH THE COEFICIENTS
            OF THE TERMS AS ELEMENTS;
PROCEDURE FILLMATRIX(N.MATRIX,D.MATRIX,PART,PCOEF,N.VAR,N.TERM,K,K1.START,
            MINDEG,COL,ROW,PARTITION,TERM.LIST,K1DEG);
INTEGER ARRAY N.MATRIX,D.MATRIX,PART,PCOEF,TERM.LIST;
INTEGER N.VAR,K,K1.START,COL,ROW,N.TERM,MINDEG,PARTITION,K1DEG;

BEGIN
EXTERNAL BOOLEAN PROCEDURE NEXTMON;
INTEGER P.PNT,KLV,TERM.PNT,D,D.SAVE,ROW.P,J,I,T,L,M,X;
INTEGER ARRAY MONOM[1:N.VAR],DEG.LIST[MINDEG:(K+1)],TERM[1:N.VAR];
BOOLEAN F,Z;
            FOR I.1 UNTIL ROW DO        ;ZERO OUT MATRIX;
                        BEGIN
                        FOR J.1 UNTIL COL DO
                                    BEGIN
                                    N.MATRIX[I,J].0;
                                    D.MATRIX[I,J].1;
                                    END;
                        END;
TERM.PNT.1;
FOR KLV.MINDEG UNTIL (K+1) DO
            BEGIN
            DEG.LIST[KLV].TERM.PNT;
            K1DEG.TERM.PNT;
            FOR I.1 UNTIL N.VAR DO
                        MONOM[I].0;
            WHILE NEXTMON(MONOM,N.VAR,KLV) DO
                        BEGIN
                        FOR I.1 UNTIL N.VAR DO
                                    BEGIN
                                    TERM.LIST[TERM.PNT,I].MONOM[I];
                                    END;
                        TERM.PNT.TERM.PNT+1;
                        END;
            END;
ROW.P.0;
FOR KLV.K+1-MINDEG STEP -1 UNTIL 0 DO
            BEGIN
            Z.FALSE;
            FOR I.1 UNTIL N.VAR DO
                        MONOM[I].0;
            IF KLV=0 THEN
                        Z.TRUE
            ELSE
                        PARTITION.ROW.P;            ;PREVIOUS ROW.P AS PARTITION POINTER;
            WHILE NEXTMON(MONOM,N.VAR,KLV) OR Z DO
                        BEGIN
                        Z.FALSE;
                        FOR J.1 UNTIL N.VAR DO   ;LOOP THRU PARTIALS;
                                    BEGIN
                                    D.0;
                                    ROW.P.ROW.P+1;
                                    FOR T.1 UNTIL N.TERM DO
                                                BEGIN
                                                D.0;
                                                FOR L.1 UNTIL N.VAR DO
                                                            BEGIN
                                                            TERM[L].MONOM[L]+PART[J,L,T];
                                                            D.D+TERM[L];
                                                            END;
                                                IF D<=K+1 THEN
                                                            BEGIN
                                                            D.DEG.LIST[D];
                                                            F.FALSE;
                                                            WHILE NOT F DO
                                                                        BEGIN
```

```
                                                F.TRUE;
                                                FOR M.1 UNTIL N.VAR DO
                                                        BEGIN
                                                        IF TERM.LIST[D,M]#
                                                            TERM[M] THEN
                                                                BEGIN
                                                                F.FALSE;
                                                                M.N.VAR;
                                                                END;
                                                        END;
                                                IF NOT F THEN D.D+1;
                                                END;
                                        N.MATRIX[ROW,P,D].PCOEF[J,T];
                                        END;
                                END;
                        END;
                END;
        END;
K1.START.DEG.LIST[K+1];
END
{SORTS TERMS OF A POLYNOMIAL INTO ASCENDING DEGREE ORDER}
PROCEDURE SORT (A,V,T,C,D); {A=ARRAY V=VARIABLES T=TERMS C=COEF ARRAY D=DEGREE ARRAY}
INTEGER ARRAY A,D,C; VALUE V,T; INTEGER V,T;
BEGIN
INTEGER I,TEMPI,M,N,L; REAL TEMP;
FOR I.1 UNTIL (T-1) DO
        BEGIN
        IF D[I] > D[I+1] THEN
                BEGIN
                FOR N.1 UNTIL  V DO
                        BEGIN
                        TEMPI.A[N,I]; {SWAP}
                        A[N,I].A[N,I+1];
                        A[N,I+1].TEMPI;
                        END;
                TEMP.C[I];
                C[I].C[I+1];
                C[I+1].TEMP;
                TEMP.D[I];
                D[I].D[I+1];
                D[I+1].TEMP;
                I.I-2;
                IF I <= -1 THEN I.1;
                END;
        END;
END
{COMPUTES THE NUMBER OF TERMS OF DEGREE <= K IN V VARIABLES}
INTEGER PROCEDURE KTERM (K,V);
{COMPUTE (K+V) THINGS TAKEN V AT A TIME;
VALUE K,V;
INTEGER V,K;
BEGIN
INTEGER N,TOP,BOTTOM,I;
N.V+K;   TOP.1;   BOTTOM.1;
IF V<K THEN
        BEGIN
        FOR I . V+1 UNTIL N DO TOP.TOP*I;
        FOR I . 1 UNTIL K DO BOTTOM.BOTTOM*I;
        END
ELSE
        BEGIN
        FOR I . K + 1 UNTIL N DO TOP.TOP*I;
        FOR I . 1 UNTIL V DO BOTTOM.BOTTOM*I;
        END;
KTERM.TOP/BOTTOM;
END
```

```
|TAKES ALL PARTIAL DERIVITIVES OF A POLYNOMIAL|
PROCEDURE PARTIAL(TERMS,COEF,PART,PCOEF,V,T);
        VALUE V,T;
INTEGER ARRAY COEF,PCOEF;
INTEGER ARRAY TERMS,PART;
INTEGER V,T;
BEGIN
INTEGER I,N,M;
FOR I_1 UNTIL T DO        |LOOP THRU TERMS|
        BEGIN
        FOR N_1 UNTIL V DO                |LOOP THRU PARTIALS|
                BEGIN
                FOR M_1 UNTIL V DO                |TAKE PARTIAL OF EACH TERM|
                        BEGIN
                        IF N=M THEN
                                BEGIN
                                PCOEF[N,I]_COEF[I]*TERMS[M,I];
                                IF PCOEF[N,I]=0 THEN
                                        PART[N,M,I]_999999
                                ELSE
                                        PART[N,M,I]_TERMS[M,I]-1;
                                END
                        ELSE
                                BEGIN
                                PART[N,M,I]_TERMS[M,I];
                                END;
                        END;
                END;
        END;
END
|GENERATES SUCCESSIVE MONOMIALS OF DEGREE K|
BOOLEAN PROCEDURE NEXTMON(MONOM,N,VAR,K);
VALUE N,VAR,K;
INTEGER ARRAY MONOM;
INTEGER N,VAR,K;
BEGIN
        INTEGER SUM,I;
        IF K=0 THEN
                BEGIN
                NEXTMON_FALSE;
                GOTO RETURN;
                END;
        NEXTMON_TRUE;
LOOP1:  MONOM[N,VAR]_MONOM[N,VAR]+1;
LOOP2:  SUM_0;
        FOR I_1 UNTIL N,VAR DO
                SUM_SUM+MONOM[I];
        IF SUM>K THEN
                BEGIN
                I_N,VAR;
                WHILE MONOM[I]=0 DO I_I-1;
                IF I=1 THEN
                        BEGIN
                        NEXTMON_FALSE;
                        GOTO RETURN;
                        END;
                MONOM[I]_0;
                MONOM[I-1]_MONOM[I-1]+1;
                GOTO LOOP2;
                END;
        IF SUM=K THEN GOTO RETURN;
        GOTO LOOP1;
RETURN: END
|COMPUTES MATRIX SIZE NEEDED TO COMPUTE UNFOLDING, ALLOCATES MATRIX
        AND CALLS ROUTINE TO CHECK UNFOLDINGS|
PROCEDURE UNFOLD(PART,PCOEF,N,VAR,N,TERMS,K);
INTEGER ARRAY PART,PCOEF;
INTEGER  N,VAR,N,TERMS,K;
```

```
BEGIN
EXTERNAL INTEGER PROCEDURE KTERM;
INTEGER UNFOLDING.TERMS,COL,ROW,C,R;
EXTERNAL PROCEDURE VERSAL;
UNFOLDING.TERMS:=0;
WHILE UNFOLDING.TERMS=0 DO
        BEGIN
        WRITE("HOW MANY UNFOLDING TERMS DO YOU HAVE? [B]");
        READ (UNFOLDING.TERMS);
        COL_KTERM(K,N,VAR);
        ROW_N.VAR*COL+UNFOLDING.TERMS;
                BEGIN
                INTEGER ARRAY NUMER[1:ROW,1:COL],DENOM[1:ROW,1:COL];
                FOR R:=1 UNTIL ROW DO
                        BEGIN
                        FOR C := 1 UNTIL COL DO
                                BEGIN
                                NUMER[ROW,COL]:=0;
                                DENOM[ROW,COL]:=1;
                                END;
                        END;
                VERSAL(NUMER,DENOM,K,N,VAR,N.TERMS,PART,PCOEF,ROW,COL,
                        UNFOLDING.TERMS);
                END;
        WRITE("ENTER 0 FOR ANOTHER UNFOLDING [B]");
        READ(UNFOLDING.TERMS);
        END;
END;
|FILLS MATRIX, ENTERS USER UNFOLDING TERMS AND CALLS SINGULAR TO
        COMPUTE UNFOLDING;
PROCEDURE VERSAL(NUMER,DENOM,K,N.VAR,N.TERMS,PART,PCOEF,ROW,COL,UNFLD.TRMS);
INTEGER ARRAY NUMER,DENOM,PART,PCOEF;
INTEGER K,N.VAR,N.TERMS,ROW,COL,UNFLD.TRMS;
BEGIN
EXTERNAL PROCEDURE FILLMATRIX;
EXTERNAL BOOLEAN PROCEDURE SINGULAR;
EXTERNAL INTEGER PROCEDURE MAP;
INTEGER ARRAY TERM.LIST[1:COL,1:N.VAR],POINT[1:COL],MONOM[1:N.VAR];
INTEGER I,L,J,TERMS,COEF;
BOOLEAN DUMMY,DUM2;
J_0;
I_0;
L_0;
FILLMATRIX(NUMER,DENOM,PART,PCOEF,N.VAR,N.TERMS,(K-1),I,0,COL,ROW,J,
        TERM.LIST,L);
FOR I_1 UNTIL UNFLD.TRMS DO
        BEGIN
        WRITE("HOW MANY TERMS IN UNFOLDING TERM");PRINT(I);WRITE(" ? [B]");
        READ (TERMS);
        FOR J_1 UNTIL TERMS DO
                BEGIN
                WRITE("[C]COEF = [B]");
                READ(COEF);
                WRITE("EXPONENTS = [B]");
                FOR L_1 UNTIL N.VAR DO READ(MONOM[L]);
                NUMER[(ROW-I+1),MAP(MONOM,N.VAR,TERM.LIST)]_COEF;
                END;
        END;
FOR I_1 UNTIL COL DO POINT[I]_I;
J_0;
DUMMY_SINGULAR(NUMER,DENOM,POINT,ROW,COL,-1,DUM2,TERM.LIST,UNFLD.TRMS,N.VAR);
END;
|SEARCHS FOR THE MATRIX COLUMN CORRESPONDING TO THE TERM IN MONOM;
INTEGER PROCEDURE MAP(MONOM,N.VAR,TERM.LIST);
INTEGER ARRAY MONOM,TERM.LIST;
INTEGER N.VAR;
BEGIN
INTEGER MP,I,INC;
```

```
BOOLEAN FOUND;
MP←0;
I←1;
WHILE MP≠0 DO
        BEGIN
        FOUND←TRUE;
        FOR INC←1 UNTIL N.VAR DO
                BEGIN
                IF MONOM[INC]≠TERM.LIST[I,INC] THEN
                        BEGIN
                        FOUND←FALSE;
                        INC←N.VAR+1;
                        END;
                END;
        IF FOUND THEN
                MP←I;
        I←I+1;
        END;
MAP←MP;
END;
```

Sample output

```
.RUN KDET

WELCOME TO THE BLESSED WORLD OF CATASTROPHE THEORY

        NUMBER OF VARIABLES = 3
        NUMBER OF TERMS IN THE POLYNOMIAL = 4
IT IS NOW TIME TO INPUT THE POLYNOMIAL
EACH OF THE    4 TERMS OF THE POLYNOMIALIS INPUT
 USING THIS FORMAT
        C D D D

WHERE C IS THE COEFFICIENT AND D IS THE DEGREE
PLEASE INPUT THE TERMS              means

TERM    1 1 2 0 1          1x²z
TERM    2 1 0 3 0                + 1.y³
TERM    3 1 0 0 4                    + 1.z⁴
TERM    4 1 0 1 3                      + 1.yz³
WHAT K DO YOU WANT TO TRY? (0 TO STOP) 4
THE MATRIX HAS BEEN ALLOCATED AT            60 BY            52
*** STRONGLY DETERMINED ***

HOW MANY UNFOLDING TERMS DO YOU HAVE? 0


CODIMENSION =   10        (=9 if constants are discounted)
ADDITIONAL UNFOLDING TERMS ARE

    0   0   2          z²
    0   1   1          yz
    0   0   1          z
    1   0   0          x
    1   1   0          xy
    2   0   0          x2
    0   0   3          z³
    0   1   0          y
    0   3   0          y³
    0   0   0          constant term 1

ENTER 0 FOR ANOTHER UNFOLDING 1
WHAT K DO YOU WANT TO TRY? (0 TO STOP) 0


END OF EXECUTION.
```

Appendix 2 Catastrophes in Numerical Analysis

As well as *requiring* new numerical techniques, catastrophe theory may also help to understand old ones – indeed the two aspects are part of the same thing. We can use cuspoid catastrophes to investigate instability phenomena in the solutions of polynomial equations. Acton [211] p. 201 cites the polynomial (due to Wilkinson [212])

$$f(x) = \prod_{r=1}^{20} (x+r) = x^{20} + 210x^{19} + \ldots + 20!$$

whose zeros are $-20, -19, \ldots, -1$. Adding 2^{-23} to the coefficient of x^{19} leads to the distribution of roots in the complex plane shown in Fig. A2.1. Serious changes have occurred in the roots from -9 to -20, ten becoming complex with relatively large imaginary parts. It is well known that instability of this kind occurs for *degenerate* polynomials, with repeated roots. But for $f(x)$ Acton [211] remarks: 'the roots could scarcely be more isolated'. Where does this instability arise? To quote Acton [211] again:

'This example is horrifying indeed. For if we have actually seen one tiger, is not the jungle immediately filled with tigers, and who knows where the next one lurks? Wilkinson shows that the difficulty is caused by the regular spacing of the roots . . .'

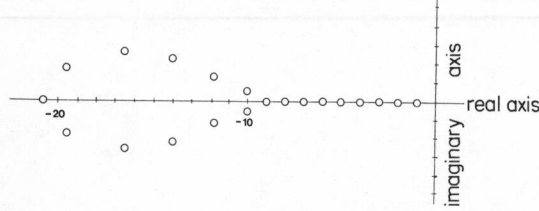

Cuspoid geometry helps to tame the tigers. For varying coefficients $a_{r-1}, \ldots, a_0$ the solutions of the polynomial equation

$$x^r + a_{r-1}x^{r-1} + \cdots + a_0 = 0$$

define the catastrophe manifold of a cuspoid catastrophe of type A_r, with control variables $a_{r-1}, \ldots, a_0$ (of which a_{r-1} may be disconnected by the Tschirnhaus transformation $x \mapsto x + a_{r-1}/r$) and x as behaviour variable. The above f, having all 20 roots real, lives

Fig. A2.2

over that part of the birfurcation set for which all roots are real – the innermost 'pocket' of the cuspoid. Fig. A2.2, drawn from Woodcock and Poston [20], illustrates for A_7 the way that this pocket is typically very small; other pictures from the same source show that it is also fragile, and tends to disappear altogether as control variables change. Thus a small perturbation to a coefficient may perturb f out of this region altogether, with some real roots going complex. When two real roots meet and become complex, at least one of them must move more than half the original distance between them, so it is to be expected that changes of the order of magnitude of the spacing of the roots may occur. The region with all roots real becomes rapidly smaller as r increases, so higher degree polynomials are especially susceptible – if they have all their roots real. But the same smallness means that few have. (Similar observations apply to the higher catastrophes in several essential variables.)

Alternatively, we may note that the zeros of f by no means 'could scarcely be more isolated'. In comparison to the sizes of the coefficients they are extremely closely bunched: all twenty roots jammed into an interval of length 19, compared to coefficients rising to $20! \sim 2.4 \times 10^{18}$. Thus, seen on a suitable scale, the graph of $f(x)$ resembles Fig. A2.3(a), with a very flat bottom, and is nearly as degenerate as the polynomial x^{20}, whose degeneracy is extreme. Quite small perturbations (especially to coefficients of higher degree terms) can lift one end of the graph, as in Fig. A2.3(b), stripping away real roots, and consequently moving them an appreciable distance into the complex plane.

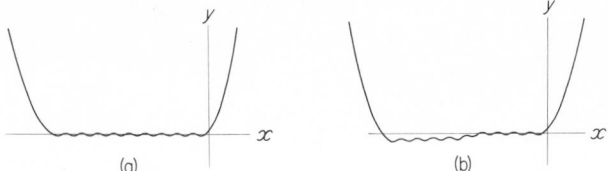

Fig. A2.3

Thus it is not the *regular* spacing of the roots, but the *bunching*, that causes trouble. Any doubting numerical analyst should try the following experiment: scatter 20 points θ_i randomly between -1 and -20. Then the polynomial

$$g(x) = \prod_{i=1}^{20} (x + \theta_i)$$

should exhibit quite as drastic sensitivity as does f. It is not the tiger's fearful symmetry, but his predilection for prey that gathers in herds, that we must beware. (All the polynomials encountered in passing from f to that of Fig. A2.1 are just as unstable as f, but far less regular in their roots. And while all roots being real tends to lead to bunching, it is not a necessary feature, for the same reason.) Symmetry *is* fearful, however, not in producing tigers of this kind, but in making actual degeneracy commoner than it might be, as we saw in Chapter 13 Section 13.

By analysing cuspoid geometry, it should be possible to obtain quantitative estimates as to how small 'small' is – clearly the important thing – or what we mean by 'near' degeneracy. But even without this we have a much better grip on the danger spots. The jungle is not full of tigers. They inhabit only narrow regions around the bifurcation set of the relevant cuspoid, and can be detected by the relatively close bunching of zeros (a feature which persists under perturbation, so may *stably* be used as test). But once we have found a tiger, the best we can do is to handle him more cautiously by using higher accuracy in the computations. If we have found him in a *designed* system, the same high accuracy must be used in building to the design if the computations are to be relevant to it. Numerical instability is either an indicator of what Acton [211] aptly calls 'perverse formulations' or of structural instability of the system. Either a change in description of the problem, or a careful imperfection sensitivity analysis (cf. Chapter 13) is absolutely required. A blind rewriting of the unstable program with all variables taken to thirty decimal places is an expensive recipe for – quite possibly – physical disaster.

Guide to the Literature

The literature of catastrophe theory is varied and widely scattered, and the background demanded to read it varies both in subject and in level. We append a brief guide to some of the more prominent works.

1. *The voice of the Master.* In the beginning was Thom [1]. It is highly compressed, extremely demanding of the reader, and could do with a guide of its own. The exposition is much illuminated by a familiarity with 1960's dynamical systems theory (as can be gained by reading the survey paper of Smale [175]), analogies with which form a continual undercurrent and a source of ideas. The book is often provocative: it must be approached with an open mind and each word weighed carefully. Even the most frivolous statements may conceal gold. The more other works on the subject you have read, the more you will get out of Thom.

2. *General introduction.* There are many of these, but a proportion fail to make clear the restrictions that apply to Thom's classification theorem, and are over-optimistic about its universality. For clarity of exposition Zeeman [213] is excellent (the preprint version [44] has more material and is in some ways better). Chillingworth [214] and Stewart [215, 216] are also recommended. At a higher mathematical level there is the very readable survey of Arnol'd [59].

3. *The mathematics.* This amounts to the greatest single portion of the literature, but is mostly in the form of the original papers and accessible only to people with a good grounding in differential topology and other areas of mathematics. Poston and Stewart [25] provide some motivation without proofs, but since this was a forerunner to the present book Chapters 7 and 8 of this one are more appropriate as a reference. The Research Note, however, is cheaper. The most readable expositions of the classification theorem are probably Bröcker and Lander [9] and Lu [28]. Trotman and Zeeman [22] includes a study of stratifications, used to obtain the global typicality of catastrophe forms in codimension up to 5; Bröcker and Lander omit this material. Another treatment is Wassermann [36]. For the *topological* (rather than the differentiable) theory see Gibson, Wirthmüller, du Plessis and Looijenga [27]. The real hard core occurs in the papers of Mather and Arnol'd listed below, of which the latter publications are of more interest to people who wish to *apply* the theory. Deakin [218] uses a different approach, via formal power series, which provides useful motivation (especially for people who prefer to think algebraically rather than geometrically). For general dynamical systems theory consult Chillingworth [52] for motivation and background, Hirsch and Smale [163] for a graduate level treatment. Background on manifolds, and calculus of several variables, is to be found in Dodson and Poston [5] and Spivak [8]. For pictures of catastrophes see Thom [42], Bröcker and Lander [9], Woodcock and Poston [20] and Zeeman [48].

4. *Applications to physics.* Each applications chapter of the present volume lists suitable references for the material therein. For ships, the only reference yet is Zeeman [48]. For optics, the papers of Berry [61–64, 75] and co-workers are highly recommended, plus Arnol'd [59, 74, 217] and Chazarain [67]. Duistermaat [41] is almost impenetrable. Elastic structures are covered well in Thompson and Hunt [105, 110] and Sewell [96]. For fluid flow, the papers of Mackley and co-workers should be consulted, together with the references therein. Golubitsky [86] and Guckenheimer [219] deal with shockwaves, an interesting subject which we have been forced by space to omit. Holmes and Rand [202] is a good introduction to the use of catastrophe theory in nonlinear differential equations.

5. *Applications to biology.* Thom [42] is stimulating but hard to follow, Thom [220] is more readable and a good introduction. The archetype biological model of Thomist proportions is Zeeman [157], which is full of interesting ideas. (Do not take the word 'theorem' too seriously here.) Several papers in Waddington [221] are relevant. The available literature should grow rapidly in this area.

6. *Applications to the behavioural sciences.* Many of the main ideas are well discussed in Isnard and Zeeman [185]. Zeeman, Hall, Harrison, Marriage and Shapland [189] is of interest for the experimental data included. Zeeman's collected papers [48] cover the ground fairly thoroughly. The best source for anorexia at the time of writing is the draft (Zeeman [44]) of the *Scientific American* article [213]. Renfrew and Poston [159] underlines the need to use Thom's methods, rather than just his theorems. Poston [222] discusses what these methods mean in judging 'simplicity'. The September 1978 *Behavioral Science* will be a special issue on catastrophe theory.

7. *Philosophy and criticism.* Thom's interest is more philosophical than the present volume, see for example Thom's interviews (Fogel, Hue and Thom [223], Walgate [224], Dickson and Thom [225]); and of course Thom [1]. Much of the criticism of catastrophe theory to date has been purely verbal: for a rare excursion into print see Croll [116] who, despite a few misunderstandings, makes some good points; or Sussmann and Zahler [1a]. The latter has enjoyed a certain notoriety, but its usefulness is seriously marred by repeated errors.

8. *Other sources.* The British Broadcasting Corporation's *Horizon* programme devoted 30 minutes to catastrophe theory (and the programme is on record with the BBC). A. E. R. Woodcock (Thompson Biology Laboratory, Williams College, Mass.) has made a film of catastrophe geometry, as has Nelson Max (Brown University, Math. Dept., Providence R. I.).

Bibliography of Catastrophe Theory

All the items listed in this section either make explicit reference to catastrophe theory or to very closely associated areas of mathematics, or are 'implicit' applications of catastrophe theory to the sciences. The coverage of the first category is as complete as we have been able to make it, though doubtless not exhaustive; for the second category we have been more selective, listing only papers especially relevant to the material discussed. (The inclusion of an item in this list does not imply that its content is without error: a small proportion of the items listed in fact contain serious mistakes. Rather than prejudice the issue, we leave it to the reader to form his own opinions.)

Abraham, R. Predictions for the future of differential equations. In *Symposium on Differential Equations and Dynamical Systems, Warwick 1968–69*. Lecture Notes in Mathematics 206 (D. R. J. Chillingworth, ed.). Springer, Berlin and New York, 1971. pp. 163–166.

Abraham, R. Introduction to morphology. In *Quatrième Rencontre entre Mathématiques et Physique* 1972, Vol. 4, fasc. 1. Mathematics Department, University of Lyons, 1972. Tome 9, pp. 38–114.

Abraham, R. and Robbin, J. W. *Transversal Mappings and Flows*. Benjamin, New York, 1967.

Amson, J. C. Equilibrium and catastrophic modes of urban growth. In *London Papers in Regional Science 4* (E. L. Cripps, ed.). Pion, London, 1973. pp. 108–128.

Amson, J. C. Catastrophe theory: a contribution to the study of urban systems? *Environ. Planning B* **2,** 177–221, 1975.

Anon. How catastrophe may teach us all the wrong lessons. *Times Higher Educational Supplement,* 5th December 1975. p. 12.

Anon. The magnificent seven. *Manifold* **14,** 6–13, 1973.

Anon. Catastrophes in action. *Manifold* **14,** 26–31, 1973.

Anon. Bibliography. *Manifold* **14,** 36–37, 1973.

Antonelli, P. Transporting a pure mathematician into theoretical biology. *Proceedings of the Conference on Mathematics, Statistics, and the Environment.* Ottawa, 1974.

Arnol'd, V. I. Singularities of smooth maps. *Uspehi Mat. Nauk* **23,** 3–44, 1968; *Russian Math. Surveys* **23,** 1–43, 1968.

Arnol'd, V. I. On braids of algebraic functions and the cohomology of swallowtails. *Uspehi Mat. Nauk* **23,** 247–248, 1968.

Arnol'd, V. I. On matrices depending on parameters. *Uspehi Mat. Nauk* **26,** 101–114, 1971; *Russian Math. Surveys* **26,** 29–43, 1971.

Arnol'd, V. I. Lectures on bifurcations and versal families. *Uspehi Mat. Nauk* **27,** 119–184, 1972; *Russian Math. Surveys* **27,** 54–123, 1972.

Arnol'd, V. I. Integrals of rapidly oscillating functions and singularities of projections of Lagrangian manifolds, *Funkcional Anal. i Priložen* **6,** 61–62 1972; *Functional Anal. Appl.* **6,** 222–224, 1972.

Arnol'd, V. I. Normal forms for functions near degenerate critical points, the Weyl groups of A_k, D_k, and E_k, and Lagrangian singularities. *Funkcional. Anal. i Priložen.* **6,** 3–25, 1972; *Functional Anal. Appl.* **6,** 254–272, 1972.

Arnol'd, V. I. Classification of unimodal critical points of functions. *Funkcional. Anal. i Priložen.* **7,** 75–76, 1973; *Functional Anal. Appl.* **7,** 230–231, 1973.

Arnol'd, V. I. Remarks on the stationary phase method and Coxeter numbers. *Russian Math. Surveys* **28,** 19–48, 1973.

Arnol'd, V. I. Normal forms for functions in the neighbourhood of degenerate critical points. *Uspehi Mat. Nauk* **29,** 11–49, 1974; *Russian Math. Surveys* **29,** 10–50, 1974.

Arnol'd, V. I. Critical points of smooth functions, *Proc. Internat. Congr. Math. Vancouver* 1974. pp. 19–39.

Arnol'd, V. I. Critical points of smooth functions and their normal forms. *Uspehi Mat. Nauk* **30,** 3–65, 1975; *Russian Math. Surveys* **30,** 1–75, 1975.

Arnol'd, V. I. Classification of bimodal critical points of functions. *Funkcional. Anal. i Priložen.* **9,** 49–50, 1975; *Functional Anal. Appl.* **9,** 43–44, 1975.

Arnol'd, V. I. Local normal forms of functions. *Invent. Math.* **35,** 87–109, 1976.

Arnol'd, V. I. Wave front evolution and equivariant Morse lemma. *Commun. Pure Appl. Math.* **29,** 557–582, 1976.

Ascher, E. and Poston, T. Catastrophe theory in scientific research. *Research Futures* **2,** 15–18, 1976, Battelle Memorial Institute, Battelle, Ohio.

Ascher, E., Gay, D. A. and Poston, T. Equivariant bifurcation of thermodynamic potentials in crystallography. To appear.

Ashton, K. Catastrophe theory: applications in biology. Duplicated seminar notes, University of Auckland, 1976.

Atkinson, G. Catastrophe theory in geography – a new look at some old problems. Mimeographed notes, 1976.

Baas, N. A. On the models of Thom in biology and morphogenesis. Preprint, University of Virginia, 1972.

Banchoff, T. F. Polyhedral catastrophe theory I. Maps of the line to the line. In *Dynamical Systems. Proceedings of the Symposium at Salvador, Brazil, 1971* (M. M. Peixoto, ed.). Academic Press, New York and London, 1973. pp. 7–22.

Banchoff, T. and Strauss, C. A reinvestigation of the centro-surface of the ellipsoid. To appear in Francis, G. K. (ed.). *Graphic Techniques in Geometry and Topology,* Proc. of Special Session, Amer. Math. Soc. Evanston, Illinois, April 1977.

Battro, A. M. Morphogénèse des Limnées, adaptation vitale et théorie des catastrophes. *Bull. de Psychologie, Hommage à Jean Piaget, Paris* 1977.

Battro, A. M. Réflexions sur une psychologie écologique expérimentale. Le probleme de l'échelle de l'environment. *L'Année Psychologique,* 1977.

Battro, A. M. Stabilité structurelle et psychogénèse. To appear.

Battro, A. M. Le geometría de la inestabilidad y la teoría de las catástrofes, *Criterio (Buenos Aires)* no. 1722, 463–468, 1975.

Beer, M. Endliche Bestimmtheit und universelle Entfaltungen von Keimen mit Gruppenoperation. Diplomarbeit, University of Regensburg, 1976.

Bell, G. M. and Lavis, D. A. Thermodynamic phase changes and catastrophe theory. To appear.

Benguigui, L. and Schulman, L. S. Topological classification of phase transitions. *Phys. Lett. A* **45,** 315–316, 1973.

Berry, M. V. *The Diffraction of Light by Ultrasound.* Academic Press, New York and London, 1966.

Berry, M. V. Attenuation and focussing of electromagnetic surface waves rounding gentle bends. *J. Phys. A* **8,** 1952–1971, 1975.

Berry, M. V. Cusped rainbows and incoherence effects in the rippling-mirror model for particle scattering from surfaces. *J. Phys. A* **8,** 566–584, 1975.

Berry, M. V. Catastrophes and semi-classical mechanics. In *Rencontre de Cargèse sur les Singularités et leurs Applications* (F. Pham, ed.). Institut d'Études Scientifiques de Cargèse, Publ. Math. Dept., Univ. Nice, 1975. pp. 133–136.

Berry, M. V. Waves and Thom's theorem. *Adv. Phys.* **25,** 1–25, 1976.

Berry, M. V. Semi-classical mechanics in phase space. *Phil. Trans. Roy. Soc.* To appear.

Berry, M. V. Focusing and twinkling: critical exponents from catastrophes in non-Gaussian random short waves. Preprint, Univ. of Bristol, 1977.

Berry, M. V. and Hannay, J. H. Umbilic points on Gaussian random surfaces. Preprint, Univ. of Bristol, 1977.

Berry, M. V. and Mackley, M. R. The six-roll mill: unfolding an unstable persistently extensional flow. *Phil. Trans. Roy. Soc.*, in press.

Berry, M. V. and Mount, K. E. Semiclassical approximations in wave mechanics. *Rep. Prog. Phys.* **35,** 315–397, 1972.

Berry, M. V. and Nye, J. F. Fine structure in caustic junctions. *Nature, Lond.* **267,** 34–36, 1977.

Bierstone, E. Local properties of smooth maps equivariant with respect to finite group actions. *J. Diff. Geom.* **10,** 523–540, 1975.

Boardman, J. M. Singularities of differentiable mappings. *Publ. Math. IHES* **33,** 21–57, 1967.

Bochnak, J. and Kuo, T.-C. Different realizations of a non-sufficient jet. *Indag. Math.* **34,** 24–31, 1972.

Bröcker, Th. Differentierbäre Abbildungen. Lecture notes, University of Regensburg, 1973.

Bröcker, Th. and Lander, L. *Differentiable Germs and Catastrophes.* London Mathematical Society Lecture Notes 17. Cambridge University Press, Cambridge, 1975.

Brown, B. L., Inouye, D., Williams, R. and Borrus, K. A catastrophe theory account of dichotic listening. Preprint, Dept. of Psychology, Stanford University, 1976.

Callahan, J. J. Singularities of plane maps. *Amer. Math. Monthly* **81,** 211–240, 1974.

Callahan, J. J. Singularities of plane maps II. Sketching catastrophes. Preprint, University of Warwick, 1976; *Amer. Math. Monthly* in press.

Callahan, J. J. The geometry of $E_6 = x^3 + y^4$, *anorexia nervosa*, and the method of tableaus for visualizing five dimensional objects. To appear in Francis, G. K. (ed.), *Graphic Techniques in Geometry and Topology*, Proc. of Special Session, Amer. Math. Soc. Evanston, Illinois, April 1977.

Carpenter, G. A. Travelling wave solutions of nerve impulse equations. Thesis, University of Wisconsin, 1974.

de Carvalho, M. S. B. Liapunov functions for diffeomorphisms. Thesis, University of Warwick, 1973.

Casti, J. and Swain, H. Catastrophe theory and urban processes. Research Memorandum RM-75-14, Laxenburg, 1975.

Chapple, G. Catastrophe theory. *New Zealand Listener* **82,** no. 1915, 16–17, 1976.

Chaudhuri, D. Rai and Jones, R. B. Unstable singularities in the σ model. *J. Phys. A* **9,** 1349–1357, 1976.

Chazarain, J. Solutions asymptotiques et caustiques. In *Rencontre de Cargèse sur les Singularités et leurs Applications* (F. Pham, ed.). Institut d'Études Scientifiques de Cargèse, Publ. Math. Dept., Univ. Nice, 1975. pp. 43–78.

Chenciner, A. Travaux de Thom et Mather sur la stabilité topologique. *Sem. Bourbaki* no. 424, 1972–73.

Chenciner, A. Singularités des applications différentiables et catastrophes élémentaires. In *Rencontre de Cargèse sur les Singularités et leurs Applications* (F. Pham, ed.). Institut d'Études Scientifiques de Cargèse, Publ. Math. Dept., Univ. Nice, 1975. pp. 1–5.

Chillingworth, D. R. J. Elementary catastrophe theory. *Bull. Inst. Math. Appl.* **11,** 155–159, 1975.

Chillingworth, D. R. J. The catastrophe of a buckling beam. In *Dynamical Systems –*

Warwick 1974. Lecture Notes in Mathematics 468 (A. Manning, ed.). Springer, Berlin and New York, 1975. pp. 86–91.

Chillingworth, D. R. J. *Differential Topology with a View to Applications.* Research notes in mathematics 9. Pitman Publishing, London, 1976.

Chillingworth, D. R. J. (ed.). *Catastrophe Theory in Infinite Dimensions.* To appear.

Chillingworth, D. R. J. and Furness, P. Reversals of the earth's magnetic field. In *Dynamical Systems – Warwick 1974.* Lecture Notes in Mathematics 468 (A. Manning, ed.). Springer, Berlin and New York, 1975. pp. 91–98.

Chilver, H. Wider implications of catastrophe theory. *Nature, Lond.* **254,** 381, 1975.

Chow, S.-N., Hale, J. K. and Mallet-Paret, J. Applications of generic bifurcation I. *Arch. Rat. Mech. Anal.* **59,** 159–188, 1975.

Chow, S.-N., Hale, J. K. and Mallet-Paret, J. Applications of generic bifurcation II. *Arch. Rat. Mech. Anal.* **62,** 209–236, 1976.

Connor, J. N. L. Multidimensional canonical integrals for the asymptotic evaluation of the *S*-matrix in semiclassical collision theory, *Faraday Discuss. Chem. Soc.* **55,** 51–58, 1973.

Connor, J. N. L. Semiclassical theory of molecular collisions: three nearly coincident classical trajectories. *Molec. Phys.* **26,** 1217–1231, 1973.

Connor, J. N. L. Evaluation of multidimensional canonical integrals in semiclassical collision theory. *Molec. Phys.* **26,** 1371–1377, 1973.

Connor, J. N. L. Semiclassical theory of molecular collisions: real and complex valued classical trajectories for collinear atom Morse oscillator collisions. *Molec. Phys.* **28,** 1569–1578, 1974.

Connor, J. N. L. Catastrophes and molecular collisions. *Molec. Phys.* **31,** 33–55, 1976.

Cooke, J. and Zeeman, E. C. A clock and wavefront model for control of the number of repeated structures during animal morphogenesis. *J. Theoretical Biology* **58,** 455–476, 1976.

Courant, R. Soap film experiments with minimal surfaces. *Amer. Math. Monthly* **47,** 168–174, 1940.

Croll, J. Is catastrophe theory dangerous? *New Scientist,* 17 June 1976. pp. 630–632.

Damon, J. Comparing topological and C^∞ stability. In *Rencontre de Cargèse sur les Singularités et leurs Applications* (F. Pham, ed.). Institut d'Études Scientifiques de Cargèse, Publ. Math. Dept., Univ. Nice, 1975. pp. 137–142.

Deakin, M. A. B. The formal power series approach to elementary catastrophe theory. Preprint, Monash University, 1976.

Delany, S. R. *Triton.* Bantam Books, New York, 1975.

Dendrinos, D. S. Mode choice, transport pricing and urban form. Mimeographed, 1975.

Dickson, D. and Thom, R. Was Newton's apple a cusp or a swallowtail? *Times Higher Education Supplement,* 5th December 1975. p. 13.

Dodson, C. T. J. and Dodson, M. M. Simple non-linear systems and the cusp catastrophe. Preprint, York University, 1974.

Dodson, M. M. Evolution and the fold catastrophe. In *Rencontre de Cargèse sur les Singularités et leurs Applications* (F. Pham, ed.). Institut d'Études Scientifiques de Cargèse, Publ. Math. Dept., Univ. Nice, 1975. pp. 126–127.

Dodson, M. M. Darwin's law of natural selection and Thom's theory of catastrophes. To appear in *Math. Biosci.*

Dodson, M. M. Quantum evolution and the fold catastrophe. To appear.

Dodson, M. M. and Hallam, A. Allopatric speciation and the fold catastrophe. To appear.

Dubois, J.-G. and Dufour, J.-P. La théorie des catastrophes I. La machine à catastrophes. *Ann. Inst. Henri Poincaré* **20,** 113–134, 1974.

Dubois, J.-G. and Dufour, J.-P. La théorie des catastrophes II. Dynamiques gradientes à une variable d'état. *Ann. Inst. Henri Poincaré* **20**, 135–151, 1974.

Dubois, J.-G. and Dufour, J.-P. La théorie des catastrophes V. Transformées de Legendre et thermodynamique. Preprint, Dept. de Math., Univ. de Québec, Montréal, 1977.

Dubois, J.-G., Dufour, J.-P. and Stanek, O. La théorie des catastrophes III. Caustiques de l'optique géometrique. *Ann. Inst. Henri Poincaré* **24**, 243–260, 1976.

Dubois, J.-G., Dufour, J.-P. and Stanek, O. La théorie des catastrophes IV. Deploiements universels et leurs catastrophes. *Ann. Inst. Henri Poincaré* **24**, 261–300, 1976.

Duistermaat, J. J. Oscillatory integrals, Lagrange immersions, and unfolding of singularities. *Commun. Pure Appl. Math.* **27**, 207–281, 1974.

Eisenbud, D. and Levine H. The topological degree of a finite C^∞ map germ. In *Structural Stability, the Theory of Catastrophes, and Applications in the Sciences*. Lecture Notes in Mathematics 525 (P. J. Hilton, ed.). Springer, Berlin and New York, 1976. pp. 90–98.

Ekeland, I. Duality in nonconvex optimization and calculus of variations. Technical Summary Report 1675, Math. Res. Center, Univ. Wisconsin, Madison 1976.

Erber, T. and Latal, H. G. A state-area principle for (magnetic) condensation processes, *Bull. Acad. Royale Belgique (Sci.)* **9**, 1019–1042, 1967.

Erber, T. and Sklar, A. Macroscopic irreversibility as a manifestation of micro-instability. In *Modern Developments in Thermodynamics* (B. Gal-Or, ed). Israel University Press, Jerusalem; Wiley, New York, 1974. pp. 281–301.

Erber, T., Latal, H. G. and Harmon, B. N. The origin of hysteresis in simple magnetic systems. *Adv. Chem. Phys.* **20**, 71–133, 1971.

Fankhauser, H. R. Katastrophentheorie – Ergänzungen. *Acta Phys. Austriaca* **40**, 377–380, 1974.

Fankhauser, H. R. Phasenübergänge als Katastrophen – Ein Beispiel. *Helv. Phys. Acta* **47**, 486–490, 1974.

Ferguson, H. R. P. Preliminary to catastrophe theory in the behavioural sciences: how to make cusp proverbs. *Proceedings, Symposium on the Behavioural Sciences, Brigham Young University 1976* (A. Bergin, ed.), BYU Press, Salt Lake City, to appear.

Ferguson, J. A. Investment decisions and sudden changes in transport. *Surveyor* **9**, 10–11, 1976.

Field, M. Transversality in *G*-manifolds. Preprint, University of Warwick, 1976.

Fogel, J.-F., Hue, J.-L. and Thom, R. La planète de l'oncle Thom, *Le Sauvage*, January 1977. pp. 74–80.

Förster, W. Katastrophentheorie. *Acta Phys. Austriaca* **39**, 201–211, 1974.

Fowler, D. H. The Riemann–Hugoniot catastrophe and van der Waals' equation. In *Towards a Theoretical Biology* (C. H. Waddington, ed.). Edinburgh University Press, Edinburgh, Vol. 4, 1972. pp. 1–7.

Fowler, D. H. *See* Thom, R. *Structural Stability and Morphogenesis.*

Francis, G. K. (ed.). *Graphic Techniques in Geometry and Topology*, Proc. of Special Session, Amer. Math. Soc. Evanston, Illinois, April 1977.

Francis, G. K. From Riemann surfaces to catastrophe machines. To appear in Francis, G. K. (ed.). *Graphic Techniques in Geometry and Topology*, Proc. of Special Session, Amer. Math. Soc. Evanston, Illinois, April 1977.

Fukutome, H. Theory of the unrestricted Hartree–Fock equation and its solutions IV. *Progr. Theor. Phys.* **53**, 1320–1336. 1975.

Furutani, N. A new approach to traffic behaviour. Preprint, University of Tokyo, 1974.

Gaffney, Properties of finitely determined map germs. Thesis, Brandeis Univ., June 1975.

Gaffney, T. On the order of determination of a

finitely determined germ. *Invent. Math.* **37,** 83–92, 1976.

Gibson, C. G. *Singular points of smooth mappings: a geometric introduction.* In preparation.

Gibson, C. G. Wirthmüller, K., du Plessis, A. A. and Looijenga, E. *Topological Stability of Smooth Mappings.* Lecture Notes in Mathematics 552. Springer, Berlin and New York, 1977.

Giimore, R. Structural stability of the phase transition in Dicke-like models. *J. Math. Phys. A* **18,** 17–22, 1977.

Godwin, A. N. Elementary catastrophes. Thesis, University of Warwick, 1971.

Godwin, A. N. Three dimensional pictures for Thom's parabolic umbilic. *Publ. Math. IHES* **40,** 117–138, 1971.

Godwin, A. N. Methods for Maxwell sets of cuspoid catastrophes. Preprint, Lanchester Polytechnic, Rugby, 1974.

Godwin, A. N. Topological bifurcation for the double cusp polynomial. *Proc. Cambridge Philos. Soc.* **77,** 293–312, 1975.

Golubitsky, M. Contact equivalence for Lagrangian submanifolds. In *Dynamical Systems – Warwick 1974.* Lecture Notes in Mathematics 468 (A. Manning, ed.). Springer, Berlin and New York, 1975. pp. 71–73.

Golubitsky, M. Regularity and stability of shock waves for a single conservation law. In *Rencontre de Cargèse sur les Singularités et leurs Applications* (F. Pham, ed.). Institut d'Études Scientifiques de Cargèse, Publ. Math. Dept., Univ. Nice, 1975. pp. 84–88.

Golubitsky, M. An introduction to catastrophe theory and its applications. Lecture notes, Queens College, New York, 1976.

Golubitsky, M. and Guillemin, V. *Stable Mappings and their Singularities.* Graduate Texts in Mathematics 14. Springer, Berlin and New York, 1973.

Goodwin, B. Review of Thom, R. *Stabilité Structurelle et Morphogénèse. Nature, Lond.* **242,** 207–208, 1973.

Graham, R. Phase-transition-like phenomena in lasers and nonlinear optics. In *Synergetics* (H. Haken, ed.). Teubner, Stuttgart, 1973. pp. 71–86.

Gray, A. A proof of the polynomial division theorems via smoothing operators. Preprint, La Trobe University, 1976.

Gromoll, D. and Meyer, W. On differentiable functions with isolated critical points. *Topology* **8,** 361–370, 1969.

Grossman, S. Fluctuations near phase transitions in restricted geometries. In *Synergetics* (H. Haken, ed.). Teubner, Stuttgart, 1973. pp. 54–70.

Guckenheimer, J. Caustics, In *Proceedings of the UNESCO Summer School, Trieste 1972.* International Atomic Energy Authority, Vienna, pp. 281–289.

Guckenheimer, J. Review of Thom, R. *Stabilité Structurelle et Morphogénèse. Bull. Amer. Math. Soc.* **79,** 878–890, 1973.

Guckenheimer, J. Bifurcation and catastrophe. In *Dynamical Systems. Proceedings of the Symposium at Salvador, Brazil, 1971* (M. M. Peixoto, ed.). Academic Press, New York and London, 1973. pp. 95–110.

Guckenheimer, J. Catastrophes and partial differential equations. *Ann. Inst. Fourier* **23,** 31–59, 1973.

Guckenheimer, J. Solving a single conservation law. In *Dynamical Systems – Warwick 1974.* Lecture Notes in Mathematics 468 (A. Manning, ed.). Springer, Berlin and New York, 1975. pp. 108–134.

Guckenheimer, J. Caustics and non-degenerate Hamiltonians. *Topology* **13,** 127–133, 1974.

Guckenheimer, J. Constant velocity waves in oscillating chemical reactions. In *Structural Stability, the Theory of Catastrophes, and Applications in the Sciences.* Lecture Notes in Mathematics 525 (P. J. Hilton, ed.). Springer, Berlin and New York, 1976. pp. 99–103.

Guckenheimer, J. Shocks and rarefactions in two space dimensions. To appear in *Arch. Rat. Mech. Anal.*

Guckenheimer, J. Isochrons and phaseless sets. To appear in *J. Math. Biol.*

Guckenheimer, J. On the bifurcation of maps of the interval. To appear.

Guseĭn-Zade, S. M. Intersection matrices of some singularities of functions of two variables. *Funkcional. Anal. i Priložen.* **8,** 15, 1974; *Functional Anal. Appl.* **8,** 10–13, 1974.

Guseĭn-Zade, S. M. Dynkin diagrams for singularities of functions of two variables. *Funkcional. Anal. i Priložen.* **8,** 23–30, 1974; *Functional Anal. Appl.* **8,** 295–300, 1974.

Güttinger, W. Catastrophe geometry in physics and biology. *Physics and Mathematics of the Nervous System,* Lecture Notes in Biomathematics 4 Springer, Berlin and New York, 1974, pp. 2–30.

Hahn, H. Geometrical aspects of the pseudo steady state hypothesis in enzyme reactions. *Physics and Mathematics of the Nervous System,* Lecture Notes in Biomathematics 4 Springer, Berlin and New York, 1974. pp. 528–545.

Haken, H. (ed.), *Synergetics.* Teubner, Stuttgart, 1973.

Harrison, P. J. and Zeeman, E. C. Applications of catastrophe theory to macroeconomics. Symposium on Applied Global Analysis, Utrecht, 1973.

Heatley, B. Local stability properties equivalent to catastrophe theory. Thesis, University of Warwick, 1974.

Hilbert, D. Über die Singularitäten der Diskriminantenflache. *Math. Ann.* **30** (1887) 437–441; Gesammelte Abhandlungen, Vol. 2, 117–120. Springer, Berlin, 1933.

Hilton, P. J. (ed.) *Structural Stability, the Theory of Catastrophes, and Applications in the Sciences.* Lecture Notes in Mathematics 525. Springer, Berlin and New York, 1976.

Hilton, P. J. Structural stability, catastrophe theory, and their applications to the sciences and engineering. *Research Futures 1.* Battelle Memorial Institute, Ohio, 1976.

Hilton, P. J. Unfolding of singularities. *Colloquium on functional analysis,* Campinas, Brazil, 1974.

Holford, R. L. Modifications to ray theory near cusped caustics. Preprint, Bell Telephone Laboratories, 1972.

Holmes, P. J. and Rand, D. A. The bifurcations of Duffing's equation: an application of catastrophe theory. *J. Sound Vib.* **44,** 237–253, 1976.

Hughes, A. An application of catastrophe theory, *Math. Gazette* **61,** 1–20, 1977.

Inoue, M. Catastrophes and fluctuations of polarization in anisotropic dielectrics. *J. Chem. Phys.* **68,** 3351–3354, 1976.

Inouye, D. The dynamic microstructure of evaluative processes: structural stability models of judgement and intentional action. To appear.

Isnard, C. A. and Zeeman, E. C. Some models from catastrophe theory in the social sciences. In *Use of Models in the Social Sciences* (L. Collins, ed.). Tavistock, London, 1976. pp. 44–100.

James, I. M. Singularities and catastrophes: a sketch. Duplicated notes, Summer Research Institute, Australian Mathematical Society, Monash University, 1974.

Jänich, K. Caustics and catastrophes. In *Dynamical Systems – Warwick 1974.* Lecture Notes in Mathematics 468 (A. Manning, ed.). Springer, Berlin and New York, 1975. pp. 100–101.

Jänich, K. Caustics and catastrophes. *Math. Ann.* **209,** 161–180, 1974.

Källman, C. G. Lee–Wick states as an example in Thom's catastrophe theory. *Phys. Lett.* A **56,** 70, 1976.

Kilmister, C. W. The concept of catastrophe (review of Thom, R. *Stabilité Structurelle et Morphogénèse*). *Times Higher Educational Supplement,* 30th November 1973. p. 15.

Kilmister, C. W. Population in cities. *Math. Gazette* **60,** 11–24, 1976.

King, H. C. Real analytic germs and their varieties at isolated singularities. *Invent. Math.* **37,** 193–200, 1976.

Klahr, D. and Wallace, J. G. *Cognitive*

Development – an Information-processing View. Lawrence Erlbaum Associates, Hillsdale, N.J., 1976; Wiley, New York and London 1976.

Komorowski, J. On Thom's idea concerning Guggenheim's one-third law in phase transitions. Preprint, University of Warwick, 1977.

Kozak, J. J. and Benham, C. J. Denaturation; an example of a catastrophe. *Proc. Nat. Acad. Sci. U.S.A.* **71,** 1977–1981, 1974.

Kuiper, N. H. C^1-equivalence of functions near isolated critical points. *Symposium on infinite-dimensional topology* (R. D. Anderson, ed.). Annals of Mathematical Studies 69. Princeton University, 1972.

Kuo, T.-C. On C^0-sufficiency of jets. *Topology* **8,** 167–171, 1969.

Kuo, T.-C. A complete determination of C^0-sufficiency in J^r (2,1). *Invent. Math.* **8,** 225–235, 1969.

Kuo, T.-C. The jet space $J^r(n, 1)$. In *Proceedings of the Liverpool Singularities Symposium.* Lecture Notes in Mathematics 192 (C. T. C. Wall, ed.). Springer, Berlin and New York, 1971. pp. 169–177.

Kuo, T.-C. Characterizations of v-sufficiency of jets. *Topology* **11,** 115–131, 1972.

Lacher, R. C., McArthur, R. and Buzyna, G. Catastrophic changes in circulation flow patterns. Preprint, Florida State University, 1977.

Lassalle, M. G. Une demonstration du théorème de division pour les fonctions différentiables. *Topology* **12,** 41–62, 1973.

Lassalle, M. G. Déploiement universel d'une application de codimension finie. *Ann. Scient. École Norm. Super.* **7,** 219–234, 1974.

Latour, F. Stabilité des champs d'applications différentiables; généralisation d'un théorème de Mather. *C.R. Acad. Sci. Paris* **268,** 1331, 1969.

Lax, P. D. Asymptotic solutions of oscillatory initial value problems. *Duke Math. J.* **24,** 627–646, 1957.

Levine, H. I. Singularities of differentiable mappings. Notes of lectures by R. Thom, Bonn, 1959; also in *Proceedings of the Liverpool Singularities Symposium.* Lecture Notes in Mathematics 192 (C. T. C. Wall, ed.). Springer, Berlin and New York, 1971. pp. 1–89.

Ljaško, O. V. Decomposition spaces of singularities of functions. *Funkcional. Anal. i Priložen.* **10,** 49–56, 1976.

Looijenga, E. Structural stability of families of C^∞-functions and the canonical stratification of $C^\infty(N)$. Preprint, IHES 1974.

Looijenga, E. On the semi-universal deformations of Arnol'd's unimodular singularities. Preprint, University of Liverpool, 1975.

Looijenga, E. On the semi-universal deformation of a simple-elliptic singularity Part I. Unimodularity. Topology **16,** 257–262, 1977.

Looijenga, E. On the semi-universal deformation of a simple-elliptic singularity Part II. Geometry of the discriminant locus. Preprint, University of Nijmegen, 1976.

Lu, Y.-C. Sufficiency of jets in $J^r(2, 1)$ via decomposition. *Invent. Math.* **10,** 119–127, 1970.

Lu, Y.-C. *Singularity Theory and an Introduction to Catastrophe Theory.* Springer, Berlin and New York, 1976.

Lu, Y.-C. and Chang, S.H. On C^0-sufficiency of complex jets. *Can. J. Math.* **25,** 874–880, 1973.

Ludwig, D. Uniform asymptotic expansion at a caustic. *Commun. Pure Appl. Math.* **19,** 215–250, 1966.

Magnus R. J. On universal unfoldings of certain real functions on a Banach space. *Mathematics Report* 100, Battelle, Geneva, 1976. *Math. Proc. Cambridge Philos. Soc.* **81,** 91–95, 1977.

Magnus, R. J. Determinacy in a class of germs on a reflexive Banach space. *Mathematics Report* 103, Battelle, Geneva, 1976. To appear in *Math. Proc. Cambridge Philos. Soc.*

Magnus, R. J. On the orbits of a Lie group action. *Mathematics Report* 105, Battelle, Geneva, 1976.

Magnus, R. J. Universal unfoldings in Banach spaces: reduction and stability. *Mathematics Report* 107, Battelle, Geneva, 1977. To appear in Chillingworth, D. R. J. (ed.) *Catastrophe Theory in Infinite Dimensions.*

Magnus, R. and Poston, T. On the full unfolding of the von Kármán equation at a double eigenvalue. *Mathematics Report* 109, Battelle, Geneva, 1977. To appear in Chillingworth, D. R. J. (ed.) *Catastrophe Theory in Infinite Dimensions.*

Magnus, R. and Poston, T. A strictly infinite-dimensional 'fold catastrophe'. *Mathematics Report* 110, Battelle, Geneva, 1977.

Malgrange, B. The preparation theorem for differentiable functions. In *Differential Analysis, Bombay Colloquium.* Oxford University Press, Oxford and New York, 1964. pp. 203–208.

Malgrange, B. *Ideals of Differentiable Functions.* Oxford University Press, Oxford and New York, 1966.

Manning, A. (ed.) *Dynamical Systems – Warwick 1974.* Lecture Notes in Mathematics 468. Springer, Berlin and New York, 1975.

Markus, L. *Lectures in Differentiable Dynamics.* CBMS Regional Conference Series in Mathematics 3. American Mathematical Society, Providence, R.I., 1971.

Markus, L. Dynamical systems – five years after. In *Dynamical Systems – Warwick 1974.* Lecture Notes in Mathematics 468 (A. Manning, ed.). Springer, Berlin and New York, 1975. pp. 354–365.

Marsden, J. E. *Applications of Global Analysis in Mathematical Physics.* Lecture Notes 2. Publish or Perish, Boston, 1974.

Marsden, J. E. and McCracken, M. *The Hopf Bifurcation and its Applications.* Applied Mathematics Series 19. Springer, Berlin and New York, 1976.

Martinet, J. *Singularités des Fonctions et Applications Différentiables.* Lecture Notes. PUC, Rio de Janeiro, 1974.

Martinet, J. Déformations verselles des fonctions numériques. Catastrophes élémentaires de R. Thom. In *Rencontre de Cargèse sur les Singularités et leurs Applications* (F. Pham, ed.). Institut d'Études Scientifiques de Cargèse, Publ. Math. Dept., Univ. Nice, 1975. pp. 6–19.

Martinet, J. Déploiements versels des applications différentiables et classification des applications stables. In Burlet, O. and Ronga, F. (eds.). *Singularités d'Applications Différentiables, Plans-sur-Bex 1975,* Lecture Notes in Mathematics 535, Springer, Berlin and New York, 1976. pp. 1–44.

Martinet, J. Déploiements stables des germes de type fini, et determination finie des applications différentiables. Preprint, Math. Dept. Univ. Strasbourg, 1976.

Mather, J. Stability of C^∞-mappings I. The division theorem. *Ann. Math.* **87,** 89–104, 1968.

Mather, J. Stability of C^∞-mappings II. Infinitesimal stability implies stability. *Ann. Math.* **89,** 254–291, 1969.

Mather, J. Stability of C^∞-mappings III. Finitely determined map germs. *Publ. Math. IHES* **35,** 127–156, 1968.

Mather, J. Stability of C^∞-mappings IV. Classification of stable germs by R-algebras. *Publ. Math. IHES* **37,** 223–248, 1969.

Mather J. Stability of C^∞-mappings V. Transversality. *Adv. Math.* **4,** 301–336, 1970.

Mather, J. Stability of C^∞-mappings VI. The nice dimensions. In *Proceedings of the Liverpool Singularities Symposium.* Lecture Notes in Mathematics 192 (C. T. C. Wall, ed.). Springer, Berlin and New York, 1971. pp. 207–253.

Mather, J. Right equivalence. Preprint, University of Warwick, 1969.

Mather, J. Notes on topological stability. Preprint, Harvard University, 1970.

Mather, J. On Nirenberg's proof of Malgrange's preparation theorem. In *Proceedings of the Liverpool Singularities Symposium.* Lecture Notes in Mathematics 192 (C. T. C. Wall, ed.). Springer, Berlin and New York, 1971. pp. 116–120.

Mather, J. Stratifications and mappings. In *Dynamical Systems. Proceedings of the*

Symposium at Salvador, Brazil, 1971 (M. M. Peixoto, ed.). Academic Press, London and New York, 1973. pp. 195–232.

Mather, J. How to stratify mappings and jet spaces. In Burlet, O. and Ronga, F. (eds.). *Singularités d'Applications Différentiables, Plans-sur-Bex 1975*, Lecture Notes in Mathematics 535, Springer, Berlin and New York, 1976. pp. 128–176.

Max, N. New films of the butterfly catastrophe, and critical points of functions from the plane. To appear in Francis, G. K. (ed.). *Graphic Techniques in Geometry and Topology*, Proc. of Special Session, Amer. Math. Soc. Evanston, Illinois, April 1977.

Mees, A. I. The revival of cities in medieval Europe: an application of catastrophe theory. *Regional Sci. Urban Econ.* **5,** 403–425, 1975.

Michor, P. Classification of elementary catastrophes of codimension ≤6. Universität Linz Institutsbericht 51, 1976.

Michor, P. The preparation theorem on Banach spaces. To appear in Chillingworth, D. R. J. (ed.). *Catastrophe Theory in Infinite Dimensions.*

Milnor, J. *Morse Theory.* Annals of Mathematical Studies 51, Princeton University, 1963.

Mitchison, G. Topological models in biology: an art or a science? MRC Molecular Biology Unit, Cambridge, 1973.

Monson, S. R. An APL implementation of Kushnirenko's Theorem to find the Milnor number of a polynomial. Preprint, Math. Dept, Brigham Young University, 1977.

Morin, B. Calcul Jacobien, *Ann. Sci. École Norm. Super.* **8,** 1–98, 1975.

Morse, M. The critical points of a function of n variables. *Trans. Amer. Math. Soc.* **33,** 72–91, 1931.

Nicolis, G. and Auchmuty, J. F. G. Dissipative structures, catastrophes, and pattern formation: a bifurcation analysis. *Proc. Nat. Acad. Sci. U.S.A.* **71,** 2748–2751, 1974.

Nirenberg, L. A proof of the Malgrange preparation theorem. In *Proceedings of the Liverpool Singularities Symposium.* Lecture Notes in Mathematics 192 (C. T. C. Wall, ed.). Springer, Berlin and New York, 1971. pp. 97–105.

Noguchi, H. and Zeeman, E. C. *Applied Catastrophe Theory.* Bluebacks, Kodansha, Tokyo, 1974. (In Japanese.)

Olenick, R. and Erber, T. A λ transition of two magnetic dipoles. *Amer. J. Phys.* **42,** 338–339, 1974.

Palamodov, V. P. The multiplicity of a holomorphic mapping. *Funkcional. Anal. i Priložen.* **1,** 54–65, 1967; *Functional Anal. Appl.* **1,** 218–226, 1967.

Panati, C. Catastrophe theory. *Newsweek*, 19th January 1976. pp. 46–47.

Pattee, H. H. Discrete and continuous processes in computers and brains. *Physics and Mathematics of the Nervous System*, Lecture Notes in Biomathematics 4, Springer, Berlin 1974, pp. 128–149.

Pearcey, T. The structure of an electromagnetic field in the neighbourhood of a cusp of a caustic. *Philos. Mag.* **37,** 311–317, 1946.

Peixoto, M. M. (ed.) *Dynamical systems. Proceedings of the Symposium at Salvador, Brazil, 1971.* Academic Press, New York and London, 1973.

Pham, F. *Introduction à l'Étude Topologique des Singularités de Landau.* Gauthier-Villars, Paris, 1967.

Pham, F. Remarque sur l'équisingularité universelle. Preprint, University of Nice, 1970.

Pham, F. Classification des singularités. CNRS preprint Vol. 13, Strasbourg, 1971.

Pham, F. (ed.) *Rencontre de Cargèse sur les Singularités et leurs Applications.* Institut d'Études Scientifiques de Cargèse, Publ. Math. Dept. Univ. of Nice, Sept. 1975.

Pham, F. Caustics and microfunctions. RCP-25 **23,** 91–104, 1976. IRMA, CNRS Strasbourg.

Pitt, D. H. and Poston, T. Generic buoyancy and metacentric loci. To appear.

Pitt, D. H. and Poston, T. Determinacy and unfoldings in the presence of a boundary. To appear.

Poènaru, V. *Analyse Différentielle*. Lecture Notes in Mathematics 371. Springer, Berlin and New York, 1974.

Poènaru, V. The Maslov index for Lagrangian manifolds. In *Dynamical Systems – Warwick 1974*, Lecture Notes in Mathematics 468 (A. Manning, ed.). Springer, Berlin and New York, 1975. pp. 70–71.

Poènaru, V. Zakalyukin's proof of the (uni)versal unfolding theorem. In *Dynamical Systems – Warwick 1974*. Lecture Notes in Mathematics 468 (A. Manning, ed.) Springer, Berlin and New York, 1975. pp. 85–86.

Poènaru, V. Théorie des invariantes C^∞: stabilité structurelle équivariante I and II. Preprints, Orsay, 1975.

Poènaru, V. *Singularités C^∞ en Présence de Symétrie*. Lecture Notes in Mathematics 510. Springer, Berlin and New York, 1976.

Porteous, I. R. Geometric differentiation – a Thomist view of differential geometry. In *Proceedings of the Liverpool Singularities Symposium II*. Lecture Notes in Mathematics 209 (C. T. C. Wall, ed.) Springer, Berlin and New York, 1971. pp. 121–127.

Porteous, I. R. The normal singularities of a submanifold. *J. Diff. Geom.* **5,** 543–564, 1971.

Porteous, I. R. Nobel prizes for catastrophes. *Manifold* **15,** 34–36, 1974.

Poston, T. Do-it-yourself catastrophe machine. *Manifold* **14,** 40, 1973.

Poston, T. Various catastrophe machines. In *Structural Stability, the Theory of Catastrophes, and Applications in the Sciences*. Lecture Notes in Mathematics 525 (P. J. Hilton, ed.). Springer, Berlin and New York, 1976. pp. 111–126.

Poston, T. The computational rules of catastrophe theory. In *Overlapping Tendencies in Operations Research, Systems Theory, and Cybernetics* (E. Billeter, M. Cuenod and S. Klaczko, eds). Birkhäuser, 1976. pp. 461–469.

Poston, T. The elements of catastrophe theory *or* the honing of Occam's razor. To appear in Cooke, K. L. and Renfrew, A. C. (eds.). *Transformations*: *Mathematical Approaches to Culture Change*, Academic Press, 1978.

Poston, T. and Stewart, I. N. *Taylor Expansions and Catastrophes*. Research Notes in Mathematics 7. Pitman Publishing, London, 1976.

Poston, T. and Stewart, I. N. The geometry of binary quartic forms II. Foliation by cross-ratio. Preprint, University of Warwick, 1977.

Poston, T. and Wilson, A. G. Facility size vs. distance travelled: urban services and the fold catastrophe. *Environ. Planning A* **9,** 681–686, 1977.

Poston, T. and Woodcock, A. E. R. On Zeeman's catastrophe machine. *Proc. Cambridge Philos. Soc.* **74,** 217–226, 1973.

Poston, T., Stewart, I. N. and Woodcock, A. E. R. *The geometry of the higher catastrophes*. In preparation.

Put, M. van der. Some properties of the ring of germs of C^∞ functions. *Compositio Math.* **34,** 1977.

Rand, D. Arnol'd's classification of simple singularities of smooth functions. Duplicated notes. Math. Inst. Univ. Warwick, April 1977.

Renfrew, A. C. and Poston, T. Discontinuities in the endogenous change of settlement pattern. To appear in Cooke, K. L. and Renfrew, A. C. (eds.). *Transformations*: *Mathematical Approaches to Culture Change*, Academic Press, 1978.

Rockwood, A. The canonical strip. Preprint, Math. Dept. Brigham Young Univ. Provo, Utah, 1976.

Rockwood, A. and Burton, R. An inexpensive technique for displaying algebraically defined surfaces. To appear in Francis, G. K. (ed.). *Graphic Techniques in Geometry and Topology*, Proc. of Special Session, Amer. Math. Soc. Evanston, Illinois, April 1977.

Ronga, F. Stabilité locale des applications équivariantes. Preprint, Mathematics Department, University of Geneva, 1976.

Ronga, F. and Cerveau, D. Applications topologiquement stables, Séminaire de Topologie, Dijon 1975.

Rössler, O. E. Adequate locomotion strategies for an abstract organism in an abstract environment – a relational approach to brain function. *Physics and Mathematics of the Nervous System*, Lecture Notes in Biomathematics 4, Springer, Berlin and New York, 1974. pp. 342–369.

Rössler, O. E. A synthetic approach to enzyme kinetics. *Physics and Mathematics of the Nervous System*, Lecture Notes in Biomathematics 4, Springer, Berlin and New York, pp. 546–582.

Rozestraten, R. J. A., Battro, A. M. and Dos Santos Andrade, A. A visual catastrophe: the reversal of the Oppel–Kundt illusion in the open field. *Abstract guide of the XXIst International Congress of Psychology*, Paris, July 1976. p. 316.

Ruelle, D. and Takens, F. On the nature of turbulence. *Commun. Math. Phys.* **20,** 167–192, 1971.

Schulman, L. S. Phase transitions as catastrophes. In *Symposium on Differential Equations and Dynamical Systems, Warwick 1968–69.* Lecture Notes in Mathematics 206 (D. R. J. Chillingworth, ed.). Springer, Berlin and New York, 1971. pp. 98–100.

Schulman, L. S. Tricritical points and type three phase transitions. *Phys. Rev. Ser. B* **7,** 1960–1967, 1973.

Schulman, L. S. Stable generation of simple forms. *J. Theor. Biol.* **57,** 453–468, 1976.

Schulman, L. S. and Revzen, M. Phase transitions as catastrophes. *Collect. Phenom.* **1,** 43–47, 1972.

Schwarz, G. Smooth functions invariant under the action of a compact Lie group. *Topology* **14,** 63–68, 1975.

Sergeraert, F. La stratification naturelle de $C^\infty(M)$. Thesis, Orsay, 1971.

Setterstrom, R. D. The double Zeeman catastrophe machine. Preprint, Math. Dept. Brigham Young Univ. Provo, Utah, 1976.

Sewell, M. J. On the connexion between stability and the shape of the equilibrium surface. *J. Mech. Phys. Solids* **14,** 203–230, 1966.

Sewell, M. J. Kitchen catastrophe. *Math. Gazette* **59,** 246–249, 1975.

Sewell, M. J. Some mechanical examples of catastrophe theory. *Bull. Inst. Math. Applic.* **12,** 163–172, 1976.

Sewell, M. J. Elementary catastrophe theory. In *Proceedings of the International Conference on Problem Analysis in Science and Engineering, Waterloo University 1975,* to appear.

Sewell, M. J. Review of Thom, R. *Stabilité Structurelle et Morphogénèse. Math. Gazette* in press.

Sewell, M. J. On Legendre transformations and elementary catastrophes. *Technical Summary Report,* 1707, Math. Res. Center, University of Wisconsin, Madison, 1976.

Sewell, M. J. Elastic and plastic bifurcation theory. Preprint, University of Reading, 1976.

Sewell, M. J. Some global equilibrium surfaces. *Technical Summary Report,* 1714, Math. Res. Center, Univ. of Wisconsin, Madison, 1977.

Sewell, M. J. A survey of plastic buckling. In *Study No. 6 (Stability),* Solid Mechanics Division, University of Waterloo, Ontario, Canada, 1972. Chap. 5.

Sewell, M. J. Degenerate duality, catastrophes and saddle functionals. Preprint, Univ. of Reading, 1977.

Siersma, D. Singularities of C^∞ functions of right-codimension smaller or equal than eight. *Indag. Math.* **25,** 31–37, 1973.

Siersma, D. Classification and deformation of singularities. Thesis, Amsterdam, 1974.

Smale, S. On gradient dynamical systems. *Ann. Math.* **74,** 199–206, 1961.

Smale, S. Differentiable dynamical systems. *Bull. Amer. Math. Soc.* **73,** 747–817, 1967.

Smale, S. Global analysis and economics I. Pareto optimum and a generalisation of

Morse theory. In *Dynamical Systems. Proceedings of the Symposium at Salvador, Brazil, 1971* (M. M. Peixoto, ed.). Academic Press, New York and London, 1973. pp. 531–544.

Smale, S. Global analysis and economics II. Extension of a theorem of Debreu. *J. Math. Econ.* **1,** 1–14, 1974.

Smale, S. Pareto optima and price equilibria. To appear.

Smith, T. R. Continuous and discontinuous response to smoothly decreasing effective distance: an analysis with special reference to 'overbanking' in the 1920's. To appear in *Environ. Planning A* **9.**

Stanley, H. E. *Introduction to Phase Transitions and Critical Phenomena.* Oxford University Press, London and New York, 1971.

Starobin, L. Our changing evolution: strategies for 1980. *General Systems* **21,** 3–46, 1976.

Stefan, P. A remark on right k-determinacy. Preprint, Bangor University, 1974.

Stewart, I. N. *Concepts of Modern Mathematics.* Penguin, Harmondsworth, Middlesex, 1974.

Stewart, I. N. The seven elementary catastrophes. *New Scientist* **68,** 447–454, 1975.

Stewart, I. N. The geometry of binary quartic forms part I. Preprint, University of Warwick, 1976.

Stewart, I. N. Catastrophe theory. *Math. Chronicle,* **5,** 140–165, 1977.

Stewart, I. N. Catastrophe Theory. In *Encyclopaedia Britannica,* Special Supplement to the Yearbook 1977. To appear.

Sussmann, H. J. Catastrophe theory. *Synthèse* **31,** 229–270, 1975.

Sussmann, H. J. Catastrophe theory – a preliminary critical study. To appear in *Biennial Meeting of the Philosophy of Science Association, Chicago, October 1976.*

Sussmann, H. J. and Zahler, R. S. Catastrophe theory as applied to the social and biological sciences: a critique. To appear in *Synthèse.*

Takens, F. A note on sufficiency of jets. *Invent. Math.* **13,** 225–231, 1971.

Takens, F. Singularities of functions and vector fields. *Nieuw. Arch. Wisk.* **20,** 107–130, 1972.

Takens, F. *Introduction to global analysis.* Mathematics Institute, Utrecht University, 1973.

Takens, F. Constrained differential equations. In *Dynamical Systems – Warwick 1974.* Lecture Notes in Mathematics 468 (A. Manning, ed.) Springer, Berlin and New York, 1975. pp. 80–82.

Takens, F. Constrained equations: a study of implicit differential equations and their discontinuous solutions. In *Structural Stability, the Theory of Catastrophes, and the Application in the Sciences.* Lecture Notes in Mathematics 525 (P. J. Hilton, ed.). Springer, Berlin and New York, 1976. pp. 143–234.

Takens, F. Implicit differential equations: some open problems. In *Singularités d'Applications Différentiables, Plans-sur-Bex,* 1975. (Burlet, O. and Ronga, F. eds.), Lecture Notes in Mathematics 535, Springer, Berlin and New York, 1976. pp. 237–253.

Tall, D. O. A long term learning schema for calculus and analysis. *Math. Educ. Teachers* **2,** No. 2, 1975.

Tall, D. O. Conflicts and catastrophes in the learning of mathematics, *Math. Educ. Teachers* **3,** No. 2, 1976.

Teissier, B. Sur la version catastrophique de la règle des phases de Gibbs et l'invariant δ des singularités d'hypersurfaces. In *Rencontre de Cargèse sur les Singularités et leurs Applications* (F. Pham, ed.). Institut d'Études Scientifiques de Cargèse, Publ. Math. Dept., Univ. Nice, 1975. pp. 105–113.

Thom, R. Une lemme sur les applications différentiables. *Bol. Soc. Mat. Mexicana* **1,** 59–71, 1956.

Thom, R. Les singularités des applications différentiables. *Ann. Inst. Fourier* **6,** 43–87, 1956.

Thom, R. La stabilité topologique des applications polynomiales. *L'Enseignement Math.* **8,** 24–33, 1962.

Thom, R. Sur la théorie des enveloppes. *J. Math. Pures Appl.* **41,** 177–192, 1962.

Thom, R. Local properties of differentiable mappings. In *Differential Analysis, Bombay Colloquium.* Oxford University Press, Oxford and New York, 1964.

Thom, R. L'équivalence d'une fonction différentiable et d'un polynome. *Topology* **3,** 297–307, 1965.

Thom, R. On some ideals of differentiable functions. *J. Math. Soc. Japan* **19,** 255–259, 1967.

Thom, R. Topologie et signification. *L'Age Sci.* **4,** 219–242, 1968.

Thom, R. Comments on C. H. Waddington: the basic ideas of biology. In *Towards a Theoretical Biology* (C. H. Waddington, ed.). Edinburgh University Press, Edinburgh, Vol. 1, 1968. pp. 32–41.

Thom, R. Une théorie dynamique de la morphogénèse. In *Towards a Theoretical Biology* (C. H. Waddington, ed.). Edinburgh University Press, Edinburgh, Vol. 1, 1968. pp. 152–179.

Thom, R. A mathematical approach to morphogenesis: archetypal morphologies. In *Heterospecific Genome Interaction.* Wistar Institute Symposium Monograph 9. Wistar Institute Press, Tel Aviv, 1969.

Thom, R. Topological models in biology. *Topology* **8,** 313–335, 1969; also in *Towards a Theoretical Biology* (C. H. Waddington, ed.). Edinburgh University Press, Edinburgh, Vol. 3, 1970. pp. 89–116.

Thom, R. Ensembles et morphismes stratifiés. *Bull. Amer. Math. Soc.* **75,** 240–284, 1969.

Thom, R. Sur les varietés d'ordre fini. In *Global Analysis (Papers in Honour of K. Kodaira).* Tokyo, 1969. pp. 397–401.

Thom, R. The bifurcation subset of a space of maps. In *Manifolds, Amsterdam 1970* (N. H. Kuiper, ed.). Lecture Notes in Mathematics 197. Springer, Berlin and New York, 1971. pp. 202–208.

Thom, R. Topologie et linguistique. In *Essays on Topology and Related Topics (Dedicated to G. de Rham)* (A. Haefliger and R. Narasimhan, eds). Springer, Berlin and New York, 1970. pp. 226–248.

Thom, R. Les symmetries brisées en physique macroscopique et la mécanique quantique. CNRS, RCP-25 10, 1970.

Thom, R. Singularities of differentiable mappings. *See* Levine, H. I.

Thom, R. Stratified sets and morphisms: local models. In *Proceedings of the Liverpool Singularities Symposium.* Lecture Notes in Mathematics 192 (C. T. C. Wall, ed.). Springer, Berlin and New York, 1971. pp. 153–164.

Thom, R. Sur le cut-locus d'un varieté plongée. *J. Diff. Geom.* **6,** 577–586, 1972.

Thom, R. Structuralism and biology. In *Towards a Theoretical Biology* (C. H. Waddington, ed.). Edinburgh University Press, Edinburgh, Vol. 4, 1972. pp. 68–82.

Thom, R. *Stabilité Structurelle et Morphogénèse.* Benjamin, New York, 1972. Translated as *Structural Stability and Morphogenesis* (see below).

Thom, R. A global dynamical scheme for vertebrate embryology. AAAS 1971, some mathematical questions in biology 4. In *Lectures on Mathematics in the Life Sciences 5.* American Mathematical Society, Providence R.I. 1973. pp. 3–45.

Thom, R. Phase transitions as catastrophes. Conference on Statistical Mechanics, Chicago, 1971.

Thom, R. On singularities of foliations. International Conference on Manifolds, Tokyo University, 1973.

Thom, R. Langage et catastrophes: éléments pour une sémantique topologique. In *Dynamical Systems. Proceedings of the Symposium at Salvador, Brazil, 1971.* (M. M. Peixoto, ed.). Academic Press, London and New York, 1973. pp. 619–654

Thom, R. De l'icône au symbole: esquisse d'une theorie du symbolisme. *Cahiers Internat. Symbolisme* **22–23,** 85–106, 1973.

Thom, R. Sur la typologie des langues naturelle: essai d'interpretation psycholinguistique. In *Formal Analysis of Natural Languages* Éditions Moutin, Paris, 1973.

Thom, R. *Modèles mathématiques de la morphogénèse.* Editions 10–18, UGE, Paris, 1974.

Thom, R. La théorie des catastrophes: état present et perspectives. *Manifold* **14,** 16–23, 1973; also in *Dynamical Systems – Warwick 1974.* Lecture Notes in Mathematics 468 (A. Manning, ed.) Springer, Berlin and New York, 1975. pp. 366–372.

Thom, R. La linguistique, discipline morphologique exemplaire. *Critique* No. 322, 235–245, 1974.

Thom, R. *Structural Stability and Morphogenesis* (translated D. H. Fowler). Benjamin-Addison Wesley, New York, 1975. Translation of Thom, R. *Stabilité Structurelle et Morphogénèse* with additional material.

Thom, R. Gradients in biology and mathematics, and their competition. AAAS 1974, some mathematical questions in biology VII. In *Lectures on Mathematics in the Life Sciences 6,* American Mathematical Society, Providence, R.I., 1975.

Thom, R. D'un modèle de la science à une science des modèles. To appear.

Thom, R. Answer to Christopher Zeeman's reply. In *Dynamical Systems – Warwick 1974.* Lecture Notes in Mathematics 468 (A. Manning, ed.). Springer, Berlin and New York, 1975. pp. 384–389.

Thom, R. Introduction à la dynamique qualitative. *Astérisque* **31,** 3–13, 1976.

Thom, R. Catastrophes et équations quasi-linéaires. In *Rencontre de Cargèse sur les Singularités et leurs Applications* (F. Pham, ed.). Institut d'Études Scientifiques de Cargèse, Publ. Math. Dept., Univ. Nice, 1975. pp. 89–90.

Thom, R. The two-fold way of catastrophe theory. In *Structural Stability, the Theory of Catastrophes and Applications in the Sciences.* Lecture Notes in Mathematics 525 (P. J. Hilton, ed.). Springer, Berlin and New York, 1976. pp. 235–252.

Thom, R. Structural stability, catastrophe theory, and applied mathematics. *SIAM Review* **19,** 189–201, 1977.

Thom, R. and Sebastiani, M. Un résultat sur la monodromie. *Invent. Math.* **13,** 90–96, 1971.

Thom, R. and Zeeman, E. C. Catastrophe theory: its present state and future perspectives. In *Dynamical Systems – Warwick 1974.* Lecture Notes in Mathematics 468 (A. Manning, ed.). Springer, Berlin and London, 1975. pp. 366–389.

Thompson, J. M. T. Instabilities, bifurcations, and catastrophes. *Phys. Lett. A* **51,** 201–203, 1975.

Thompson, J. M. T. Catastrophe theory in elasticity and cosmology. In *Rencontre de Cargèse sur les Singularités et leurs Applications* (F. Pham, ed.). Institut d'Études Scientifiques de Cargèse, Publ. Math. Dept., Univ. Nice, 1975. pp. 100–104.

Thompson, J. M. T. Designing against catastrophe. 3rd International Congress on Cybernetics and Systems, Bucharest, 1975.

Thompson, J. M. T. Experiments in catastrophe. *Nature, Lond.* **254,** 392–395, 1975.

Thompson, J. M. T. Catastrophe theory and its role in applied mechanics. In *14th International Congress on Theoretical Applications of Mathematics, Delft, 1976.* North Holland, Amsterdam. To appear.

Thompson, J. M. T. and Gaspar, Z. A buckling model for the set of umbilic catastrophes. Preprint, Engineering Department, University College London, 1977.

Thompson, J. M. T. and Hunt, G. W. *A General Theory of Elastic Stability.* Wiley, London and New York, 1973.

Thompson, J. M. T. and Hunt, G. W. Dangers of structural optimization. *Engng Optimization* **1,** 99, 1974.

Thompson, J. M. T. and Hunt, G. W. Towards a unified bifurcation theory. *J. Appl. Math. Phys. (ZAMP)* **26,** 581–604, 1975.

Thompson, J. M. T. and Hunt, G. W. A bifurcation theory for the instabilities of optimization and design. In *Mathematical Methods in the Social Sciences* (D. Berlinski, ed.). *Synthèse,* to appear.

Thompson, J. M. T. and Shorrock, P. A. Bifurcational instability of an atomic lattice. *J. Mech. Phys. Solids* **23,** 21–37, 1975.

Thompson, J. M. T. and Shorrock, P. A. Hyperbolic umbilic catastrophe in crystal fracture. *Nature, Lond.* **260,** 598–599, 1976.

Thompson, J. M. T., Tulk, J. D. and Walker, A. C. An experimental study of imperfection-sensitivity in the interactive buckling of stiffened plates. In *Buckling of Structures, IUTAM Symposium, Cambridge Mass., 1974* (B. Budiansky, ed.). Springer, Berlin and New York, 1976. pp. 149–159.

Thompson, M. Class, caste, the curriculum cycle and the cusp catastrophe. In *Rubbish Theory*. Paladin, London, to appear.

Thompson, M. The geometry of confidence: an analysis of the Enga *te* and Hagen *moka;* a complex system of ceremonial pig-giving in the New Guinea highlands. Preprint, Portsmouth Polytechnic, 1973; also in *Rubbish Theory*. Paladin, London, to appear.

Tougeron, J.-C. *Idéaux de Fonctions Différentiables*. Springer, Berlin and New York, 1972.

Trinkaus, H. and Prepper, F. On the analysis of diffraction catastrophes. *J. Phys. A Math. Gen.* **10,** 1977.

Trotman, D. J. A. and Zeeman, E. C. Classification of elementary catastrophes of codimension $\leqslant 5$. In *Structural Stability, the Theory of Catastrophes, and Applications in the Sciences*. Lecture Notes in Mathematics 525 (P. J. Hilton, ed.). Springer, Berlin and New York, 1976. pp. 263–327.

Ursell, F. Integrals with a large parameter: several nearly coincident saddle-points. *Proc. Cambridge Philos. Soc.* **72,** 49–65, 1972.

Varchenko, A. N. Newton polyhedra and estimates for oscillatory integrals, *Funkcional. Anal. i Priložen.* **10,** 13–38, 1976.

Varchenko, A. N. Zeta-function of monodromy and Newton's diagram. *Invent. Math.* **37,** 253–262, 1976.

Waddington, C. H. (ed.) *Towards a Theoretical Biology*, 4 vols. Edinburgh University Press, Edinburgh, 1968–1972.

Waddington, C. H. A catastrophe theory of evolution. *Ann. N.Y. Acad. Sci.* **231,** 32–42, 1974.

Walgate, R. René Thom clears up catastrophes. *New Scientist* **68,** 578, 1975.

Walker, W. The analysis of sudden reversals of predator–prey data. Preprint, Auckland University, 1976.

Wall, C. T. C. (ed.) *Proceedings of the Liverpool Singularities Symposium*. Lecture Notes in Mathematics 192. Springer, Berlin and New York, 1971.

Wall, C. T. C. (ed.) *Proceedings of the Liverpool Singularities Symposium II*. Lecture Notes in Mathematics 209. Springer, Berlin and New York, 1971.

Wall, C. T. C. Introduction to the preparation theorem. In *Proceedings of the Liverpool Singularities Symposium*. Lecture Notes in Mathematics 192 (C. T. C. Wall, ed.) Springer, Berlin and New York, 1971. pp. 90–96.

Wall, C. T. C. Stratified sets: a survey. In *Proceedings of the Liverpool Singularities Symposium*. Lecture Notes in Mathematics 192 (C. T. C. Wall, ed.). Springer, Berlin and New York, 1971. pp. 133–140.

Wall, C. T. C. Lectures on C^{∞}-stability and classification. In *Proceedings of the Liverpool Singularities Symposium*. Lecture Notes in Mathematics 192 (C. T. C. Wall, ed.) Springer, Berlin and New York, 1971. pp. 178–206.

Wall, C. T. C. Regular stratifications. In *Dynamical Systems – Warwick 1974*. Lecture Notes in Mathematics 468 (A. Manning, ed.) Springer, Berlin and New York, 1975. pp. 332–344.

Wassermann, G. *Stability of Unfoldings*. Lecture Notes in Mathematics 393. Springer, Berlin and New York, 1974.

Wassermann, G. (r, s)-stability of unfoldings, Preprint, University of Regensburg, 1974.

Wasserman, G. stability of caustics. In *Rencontre de Cargèse, sur les singularités et leurs Applications* (F. Pham, ed.). Institut d'Études Scientifiques de Cargèse, Publ. Math. Dept., Univ. Nice, 1975. pp. 128–132.

Wassermann, G. (r, s)-stable unfoldings and catastrophe theory. In *Structural Stability, the Theory of Catastrophes, and Applications in the Sciences.* Lecture Notes in Mathematics 525 (P. J. Hilton, ed.). Springer, Berlin and New York, 1976. pp. 253–262.

Wassermann, G. Stability of unfoldings in space and time. *Acta. math.* **135**, 57–128, 1975.

Wassermann, G. Classification of singularities with compact abelian symmetry. *Regensburger Mathematische schriften* 1, Department of Mathematics, University of Regensburg, 1977.

Whitney, H. The general type of singularity of a set of $2n-1$ smooth functions of n variables. *Duke Math. J.* **45**, 220–283, 1944.

Whitney, H. The singularities of smooth n-manifolds into $(2n-1)$-space, *Ann. Math.* **62**, 247–293, 1955.

Whitney, H. Mappings of the plane into the plane. *Ann. Math.* **62**, 374–470, 1955.

Whitney, H. Elementary structure of real algebraic varieties. *Ann. Math.* **66**, 545–556, 1957.

Wilson, A. G. Catastrophe theory and urban modelling: an application to modal choice. *Environ. Planning A* **8**, 351–356, 1976.

Wilson, A. G. Nonlinear and dynamic models in geography: towards a research agenda. Working Paper 160, School of Geography, University of Leeds, 1976.

Wilson, A. G. Towards models of the evolution and genesis of urban structure. Working Paper 166, School of Geography, University of Leeds, 1976.

Wilson, A. G. Equilibrium and transport system dynamics. Working Paper 171, School of Geography, University of Leeds, 1976.

Wilson, F. W. Smoothing derivatives of functions with applications. *Trans. Amer. Math. Soc.* **139**, 413–428, 1969.

Woodcock, A. E. R. Discussion paper: cellular differentiation and catastrophe theory. *Ann. N.Y. Acad. Sci.* **231**, 60–76, 1974.

Woodcock, A. E. R. Embryology, differentiation, and catastrophe theory. *Manifold* **15**, 17–33, 1974.

Woodcock, A. E. R. The development of biological form: toward an understanding? Preprint, Williams College, Mass., 1976.

Woodcock, A. E. R. and Poston, T. *A Geometrical Study of the Elementary Catastrophes.* Lecture Notes in Mathematics 373. Springer, Berlin and New York, 1974.

Woodcock, A. E. R. and Poston, T. A higher catastrophe machine. *Proc. Cambridge Philos. Soc.* **79**, 343–350, 1976.

Zakalyukin, V. M. A versality theorem. *Funkcional. Anal. i Priložen.* **7**, 28–31, 1973.

Zakalyukin, V. M. On Lagrange and Legendre singularities. *Funkcional. Anal i Priložen.* **10**, 26–36, 1976.

Zeeman, E. C. Breaking of waves. In *Symposium on Differential Equations and Dynamical Systems, Warwick 1968–69.* Lecture Notes in Mathematics 206 (D. R. J. Chillingworth, ed.). Springer, Berlin and New York, 1971. pp. 272–281.

Zeeman, E. C. The geometry of catastrophe. *Times Literary Supplement*, 10th December 1971. pp. 1556–1557.

Zeeman, E. C. Differential equations for the heartbeat and nerve impulse. In *Towards a Theoretical Biology* (C. H. Waddington, ed.), Edinburgh University Press, Edinburgh, 1968–72. Vol. 4, pp. 8–67; also in *Dynamical Systems. Proceedings of the Symposium at Salvador, Brazil, 1971* (M. M. Peixoto, ed.). Academic Press, New York and London, 1971. pp. 683–741.

Zeeman, E. C. C^0-density of stable diffeomorphisms and flows. *Proceedings of the Symposium on Dynamical Systems, Southampton University, 1972.*

Zeeman, E. C. An essay on dynamical systems. *Report to the SRC on the 1968–1971 Programme of Differential Equations at University of Warwick, 1972.*

Zeeman, E. C. A catastrophe machine. In *Towards a Theoretical Biology* (C. H. Waddington, ed.). Edinburgh University Press, Edinburgh, Vol. 4, 1972. pp. 276–282.

Zeeman, E. C. Catastrophe theory in brain modelling. *Int. J. Neurosci.* **6,** 39–41, 1973.

Zeeman, E. C. Applications of catastrophe theory. *Manifolds,* Tokyo, 1973.

Zeeman, E. C. On the unstable behaviour of stock exchanges, *J. Math. Econ.* **1,** 39–49, 1974.

Zeeman, E. C. Catastrophe theory: a reply to Thom. *Manifold* **15,** 4–15, 1974; also in *Dynamical Systems – Warwick 1974.* Lecture Notes in Mathematics 468 (A. Manning, ed.) Springer, Berlin and New York, 1975. pp. 373–383.

Zeeman, E. C. Research ancient and modern. *Bull. Inst. Math. Appl.* **10,** 272–281, 1974.

Zeeman, E. C. Primary and secondary waves in developmental biology. AAAS 1974. Some mathematical questions in biology VIII. In *Lectures on Mathematics in the Life Sciences 7.* American Mathematical Society, Providence, R.I., 1974. pp. 69–161.

Zeeman, E. C. Levels of structure in catastrophe theory. *Proceedings of the International Congress of Mathematics, Vancouver,* 1974. pp. 533–546.

Zeeman, E. C. Differentiation and pattern formation. Appendix to Cooke, J. Some current theories of the emergence and regulation of spatial organisation in early animal development. *A. Rev. Biophys. Bioengng* **4,** 210–215, 1975.

Zeeman, E. C. Catastrophe theory in biology. In *Dynamical Systems – Warwick 1974.* Lecture Notes in Mathematics 468 (A. Manning, ed.). Springer, Berlin and New York, 1975. pp. 101–105.

Zeeman, E. C. Catastrophe theory. Preprint, University of Warwick, 1975.

Zeeman, E. C. Catastrophe theory. *Scient. Am.* **234,** 65–83, 1976.

Zeeman, E. C. A mathematical model for conflicting judgements caused by stress, applied to possible misestimations of speed caused by alcohol. *Br. J. Math. Statist. Psych.* **29,** 19–31, 1976.

Zeeman, E. C. The umbilic bracelet and the double cusp catastrophe. In *Structural Stability, the Theory of Catastrophes, and Applications in the Sciences.* Lecture Notes in Mathematics 525 (P. J. Hilton, ed.). Springer, Berlin and New York, 1976. pp. 328–366.

Zeeman, E. C. Prison disturbances. In *Structural Stability, the Theory of Catastrophes, and Applications in the Sciences.* Lecture Notes in Mathematics 525 (P. J. Hilton, ed.). Springer, Berlin and New York, 1976. pp. 402–406.

Zeeman, E. C. Gastrulation and formation of somites in amphibia and birds. In *Structural Stability, the Theory of Catastrophes, and Applications in the Sciences.* Lecture Notes in Mathematics 525 (P. J. Hilton, ed.). Springer, Berlin and New York, 1976. pp. 396–401.

Zeeman, E. C. Euler buckling. In *Structural Stability, the Theory of Catastrophes, and Applications in the Sciences.* Lecture Notes in Mathematics 525 (P. J. Hilton, ed.). Springer, Berlin and New York, 1976. pp. 373–395.

Zeeman, E. C. Brain modelling. In *Structural Stability, the Theory of Catastrophes, and Applications in the Sciences.* Lecture Notes in Mathematics 525 (P. J. Hilton, ed.). Springer, Berlin and New York, 1976. pp. 367–372.

Zeeman, E. C. Catastrophe theory. *Proc. Roy. Instn.* in press.

Zeeman, E. C. Applications de la théorie des catastrophes a l'étude du comportement humain. To appear.

Zeeman, E. C. Duffing's equation in brain modelling. To appear in *Bull. Inst. Math. Appl.*

Zeeman, E. C. *Catastrophe Theory: Selected*

Papers (*1972–1977*). Addison-Wesley Reading, Mass., 1977.

Zeeman, E. C. A catastrophe model for the stability of ships. Preprint, University of Warwick 1977, to appear in *Proc. Esc.* *Lat.-Am. Math.* **3,** (1976), IMPA, Rio de Janeiro, Brazil.

Zeeman, E. C., Hall, C., Harrison, P. J., Marriage, H. and Shapland, P. A model for institutional disturbances. *Br. J. Math. Statist. Psych.* **29,** 66–80, 1976.

References

1. Thom, R. *Stabilité Structurelle et Morphogénèse.* Benjamin, New York, 1972.

1a. Sussman, H. J. and Zahler, R. S. Catastrophe theory as applied to the social and biological sciences: a critique. To appear in *Synthèse.*

2. Zeeman, E. C. A catastrophe machine. In *Towards a Theoretical Biology* (C. H. Waddington, ed.). Edinburgh University Press, Edinburgh, 1968–72. Vol. 4, pp. 276–282.

3. Poston, T. and Woodcock, A. E. R. On Zeeman's catastrophe machine. *Proc. Cambridge Philos. Soc.* **74,** 217–266, 1973.

4. Dubois, J.-G. and Dufour, J.-P. La théorie des catastrophes I. La machine à catastrophes. *Ann. Inst. Henri Poincaré* **20,** 135–151, 1974.

5. Dodson, C. T. J. and Poston, T. *Tensor Geometry.* Pitman Publishing, London, 1977

6. Stewart, I. N. and Tall, D. O. *The Foundations of Mathematics.* Oxford University Press, London and New York, 1977.

7. Zeeman, E. C. The umbilic bracelet and the double cusp catastrophe. In *Structural Stability, the Theory of Catastrophes, and Applications in the Sciences,* Lecture Notes in Mathematics 525. (P. J. Hilton, ed.). Springer, Berlin and New York, 1976. pp. 328–366.

8. Spivak, M. *Calculus on Manifolds.* Benjamin, New York 1965.

9. Bröcker, Th. and Lander, L. *Differentiable Germs and Catastrophes.* London Mathematical Society Lecture Notes 17. Cambridge University Press, London, 1975.

10. Mather, J. Stability of C^∞-mappings I. The division theorem. *Ann. Math.* **87,** 89–104, 1968.

11. Mather, J. Stability of C^∞-mappings II. Infinitesimal stability implies stability. *Ann. Math.* **89,** 254–291, 1969.

12. Mather, J. Stability of C^∞-mappings III. Finitely determined map germs. *Publ. Math. IHES* **35,** 127–156, 1968.

13. Mather, J. Stability of C^∞-mappings IV. Classification of stable germs by R-algebras. *Publ. Math. IHES* **37,** 223–248, 1969.

14. Mather, J. Stability of C^∞-mappings V. Transversality. *Adv. Math.* **4,** 301–336, 1970.

15. Mather J. Stability of C^∞-mappings VI. The nice dimensions. In *Proceedings of the Liverpool Singularities Symposium.* Lecture Notes in Mathematics 192 (C. T. C. Wall, ed.) Springer, Berlin and New York, 1971. pp. 207–253.

16. Milnor, J. Morse theory. Annals of Mathematics Studies 51. Princeton University, 1963.

17. Abraham, R. and Robbin, J. W. *Transversal Mappings and Flows.* Benjamin, New York, 1967.

18. Golubitsky, M. and Guillemin, V. *Stable Mappings and their Singularities.* Graduate Texts in Mathematics 14. Springer, Berlin and New York, 1973.

19. Salmon, G. *Lessons Introductory to the Modern Higher Algebra.* Hodges, Figgis and Co., Dublin, 1885.

20. Woodcock, A. E. R. and Poston, T. *A Geometrical Study of the Elementary Catastrophes.* Lecture Notes in Mathematics 373. Springer, Berlin and New York, 1974.

21. Poston, T. Various catastrophe machines. In *Structural Stability, the Theory of Catastrophes, and Applications in the Sciences*. Lecture Notes in Mathematics 525 (P. J. Hilton, ed.). Springer, Berlin and New York, 1976. pp. 111–126.

22. Trotman, D. J. A. and Zeeman, E. C. Classification of elementary catastrophes of codimension ≤ 5. In *Structural Stability, the Theory of Catastrophes, and Applications in the Sciences*. Lecture Notes in Mathematics 525 (P. J. Hilton, ed.). Springer, Berlin and New York, 1976. pp. 263–327.

23. Thompson, J. M. T. and Hunt, G. W. A bifurcation theory for the instabilities of optimization and design. In *Mathematical Methods in the Social Sciences* (D. Berlinski, ed.). *Synthèse*, to appear.

24. Fischer, A. E. and Marsden, J. E. Three lectures on the dynamics of general relativity. Duplicated notes, University of Paris VI, 1975.

25. Poston, T. and Stewart, I. N. *Taylor Expansions and Catastrophes*. Research Notes in Mathematics 7. Pitman Publishing, London, 1976.

26. Arnol'd, V. I. Normal forms for functions near degenerate critical points, the Weyl groups of A_k, D_k, and E_k, and Lagrangian singularities. *Funkcional Anal. i Priložen* **6**, 3–25, 1972; *Functional Anal. Appl.* **6**, 254–272, 1972.

26a. Rand, D. Arnol'd's classification of simple singularities of smooth functions. Duplicated notes, Math. Inst., Univ. Warwick, April 1977.

27. Gibson, C. G., Wirthmüller, K., du Plessis, A. A. and Looijenga, E. *Topological Stability of Smooth Mappings*. Lecture Notes in Mathematics 552. Springer, Berlin and New York, 1977.

28. Lu, Y.-C. *Singularity Theory and an Introduction to Catastrophe Theory*. Springer, Berlin and New York, 1976.

29. Levine, H. I. Singularities of differentiable mappings. Notes of lectures by R. Thom. Bonn, 1959. Also in *Proceedings of the Liverpool Singularities Symposium*. Lecture Notes in Mathematics 192. (C. T. C. Wall, ed.). Springer, Berlin and New York, 1971.

30. Gibson, C. G. *Singular points of smooth mappings: a geometric introduction*. In preparation.

30a. Magnus, R. J. On the orbits of a Lie group action. *Mathematics Report* 105, Battelle, Geneva, 1976.

31. Martinet, J. *Singularités des Fonctions et Applications Différentiables*. Lecture Notes. PUC, Rio de Janeiro, 1974.

32. Siersma, D. Singularities of C^∞ functions of right codimension smaller or equal than eight. *Indag. Math.* **25**, 31–37, 1973.

33. Siersma, D. Classification and deformation of singularities. Thesis, Amsterdam, 1974.

34. Martinet, J. Déploiements versels des applications différentiables et classification des applications stables. In Burlet, O. and Ronga, F. (eds.), *Singularités d'Applications Différentiables, Plans-sur-Bex 1975*, Lecture Notes in Mathematics 535, Springer, Berlin and New York, 1976. pp. 1–44.

35. Martinet, J. Déploiements stables des germes de type fini, et détermination finie des applications différentiables. Preprint, Math. Dept. Univ. Strasbourg, 1976.

36. Wasserman, G. *Stability of Unfoldings*. Lecture Notes in Mathematics 393. Springer, Berlin and New York, 1974.

37. Stefan, P. A remark on right k-determinacy. Preprint, Bangor University, 1974.

38. Palamodov, V. P. The multiplicity of a holomorphic mapping. *Funkcional Anal. i Priložen.* **1**, 54–65, 1967; *Functional Anal. Appl.* **1**, 218–226, 1967.

39. Poston, T., Stewart, I. N. and Woodcock, A. E. R. *The Geometry of the Higher Catastrophes*. In preparation.

40. Arnol'd V. I. Critical points of smooth functions and their normal forms. *Uspehi Mat. Nauk* **30,** 3–65, 1975; *Russian Math. Surveys* **30,** 1–75, 1975.

41. Duistermaat, J. J. Oscillatory integrals, Lagrange immersions, and unfolding of singularities. *Commun. Pure Appl. Math.* **27,** 207–281, 1974.

42. Thom, R. *Structural Stability and Morphogenesis.* Benjamin–Addison Wesley, New York, 1975. Translation of ref. 1 with additional material; translated by D. H. Fowler.

43. Callahan, J. J. Singularities of plane maps II. Sketching catastrophes. Preprint, University of Warwick, 1976; *Amer. Math. Monthly,* in press.

43a. Callahan, J. J. The geometry of $E_6 = x^3 + y^4$ and *anorexia nervosa.* To appear in Francis, G. K. (ed.), *Graphic Techniques in Geometry and Topology,* Proc. of Special Session, Amer. Math. Soc. Evanston, Illinois, April 1977.

44. Zeeman, E. C. Catastrophe theory. Preprint, University of Warwick, 1975. (Preprint version of: Catastrophe theory. *Scient. Am.* **234,** 65–83, 1976.)

45. Chenciner, A. Travaux de Thom et Mather sur la stabilité topologique. *Sem. Bourbaki,* No. 424, 1972–3.

46. Godwin, A. N. Three dimensional pictures for Thom's parabolic umbilic. *Publ. Math. IHES* **40,** 117–138, 1971.

47. Godwin, A. N. Elementary catastrophes. Thesis, University of Warwick, 1971.

47a. Poston, T. and Woodcock, A. E. R. On Zeeman's catastrophe machine. *Proc. Cambridge Philos. Soc.* **74,** 217–226, 1973.

47b. Rockwood, A. and Burton, R. An inexpensive technique for displaying algebraically defined surfaces. To appear in Francis, G. K. (ed.). *Graphic Techniques in Geometry and Topology,* Proc. of Special Session, Amer. Math. Soc. Evanston, Illinois, April 1977.

48. Zeeman, E. C. *Catastrophe Theory: Selected Papers (1972–1977).* Addison-Wesley, Reading, Mass., 1977.

49. Pitt, D. H. and Poston, T. Generic buoyancy and metacentric loci. To appear.

50. Robb, A. M. *Theory of Naval Architecture.* Griffin, London, 1952.

51. Cayley, A. On the centro-surface of an ellipsoid. *Trans. Cambridge Philos. Soc.* **12,** 319–365, 1873; Collected works **520,** 316–365.

51a. Banchoff, T. and Strauss, C. A reinvestigation of the centro-surface of the ellipsoid. To appear in Francis, G. K. (ed.). *Graphic Techniques in Geometry and Topology,* Proc. of Special Session, Amer. Math. Soc. Evanston, Illinois, April 1977.

52. Chillingworth, D. R. J. *Differential Topology with a View to Applications.* Research Notes in Mathematics 9. Pitman Publishing, London, 1976.

53. Berry, M. V. and Mackley, M. R. The six-roll mill: unfolding an unstable persistently extensional flow. *Phil. Trans. Roy. Soc.,* in press.

54. Feynman, R. P., Leighton, R. B. and Sands, M. *The Feynman Lectures in Physics,* 3 Volumes. Addison Wesley, Reading, Mass., 1963.

55. Crowley, D. G. Frank, F. C., Mackley, M. R. and Stephenson, R. G. Localized flow birefringence of polyethylene oxide solutions in a four roll mill. *J. Polymer Sci., Polymer Phys.* **14,** 1111–1119, 1976.

56. Frank, F. C. and Mackley, M. R. Localized flow birefringence of polyethylene oxide solutions in a two roll mill. *J. Polymer Sci., Polymer Phys.* **14,** 1121–1131, 1976.

57. Truesdell, C. and Noll, W. The nonlinear field theories of mechanics. In *Handbuch der Physik,* III/3 1–579 (S. Flugge, ed.). Springer, Berlin, 1965.

58. Mackley, M. R. and Keller, A. Flow induced polymer chain extension and its relation to fibrous crystallization. *Phil. Trans. Roy. Soc.* **278,** 29, 1975.

59. Arnol'd V. I. Critical points of smooth functions. *Proc. Internat. Congr. Math. Vancouver* 1974, pp. 19–39.

60. Cayley, A. A memoir upon caustics. *Phil. Trans. Roy. Soc.* **147,** 273–312, 1857; Collected works **145,** 336–380.

61. Berry, M. V. Cusped rainbows and incoherence effects in the rippling-mirror model for particle scattering from surfaces. *J. Phys.* A **8,** 566–584, 1975.

62. Berry, M. V. Catastrophes and semiclassical mechanics. In *Rencontre de Cargèse sur les Singularités et leurs Applications* (F. Pham, ed.). Institut d'Études Scientifiques de Cargèse, Publ. Math. Dept., Univ. Nice, 1975. pp. 133–136.

63. Berry, M. V. Attenuation and focussing of electromagnetic surface waves rounding gentle bends. *J. Phys.* A **8,** 1952–1971, 1975.

64. Berry, M. V. Waves and Thom's theorem. *Adv. Phys.* **25,** 1–25, 1976.

65. Berry, M. V. and Nye, J. F. Fine structure in caustic junctions. *Nature, Lond.* **267,** 34–36, 1977.

66. Maslov, F. P. *Perturbation Theory and Asymptotic Methods.* Moscow, 1965; French translation Dunod, Paris, 1972.

67. Chazarain, J. Solutions asymptotiques et caustiques. In *Rencontre de Cargèse sur les Singularités et leurs Applications* (F. Pham, ed.). Institut d'Études Scientifiques de Cargèse, Publ. Math. Dept., Univ. Nice, 1975. pp. 43–78.

68. Airy, G. B. On the intensity of light in the neighbourhood of a caustic. *Trans. Cambridge Philos. Soc.* **6,** 379–403, 1838.

69. Abramowitz, H. and Stegun, A. *Handbook of Mathematical Functions.* U.S. National Bureau of Standards, Washington, 1964.

70. Holford, R. L. Modifications to ray theory near cusped caustics. Preprint, Bell Telephone Laboratories, 1972.

71. Ludwig, D. Uniform asymptotic expansion at a caustic. *Commun. Pure Appl. Math.* **19,** 215–250, 1966.

72. Pearcey, T. The structure of an electromagnetic field in the neighbourhood of the cusp of a caustic. *Philos. Mag.* **37,** 311–317, 1946.

73. Connor, J. N. L. Semiclassical theory of molecular collisions: three nearly coincident classical trajectories. *Molec. Phys.* **26,** 1217–1231, 1973.

74. Arnol'd V. I. Remarks on the stationary phase method and Coxeter numbers. *Russian Math. Surveys* **28,** 19–48, 1973.

75. Berry, M. V. *The Diffraction of Light by Ultrasound.* Academic Press, London and New York, 1966.

76. Bryant, H. C. and Jarmie, N. The glory. *Scient. Am.* **231,** 60–71, 1974.

77. Berry, M. V. and Mount, K. E. Semiclassical approximations in wave mechanics. *Rep. Progr. Phys.* **35,** 315–397, 1972.

77a. Khare, V. and Nussenzveig, H. M. Theory of the Glory. *Phys. Rev. Lett.* **38,** 1279–1282, 1977.

78. Field, M. Transversality in *G*-manifolds. Preprint, University of Warwick, 1976.

79. Poènaru, V. *Analyse Différentielle.* Lecture Notes in Mathematics 371. Springer, Berlin and New York, 1974.

80. Poènaru, V. Théorie des invariantes C^∞: stabilité structurelle équivariante I and II. Preprints, Orsay, 1975.

81. Poènaru, V. *Singularités C^∞ en Présence de Symétrie.* Lecture Notes in Mathematics 510. Springer, Berlin and New York, 1976.

81a. Wasserman, G. Classification of singularities with compact abelian symmetry. *Regensburger Mathematische Schriften* 1, Department of Mathematics, University of Regensburg, 1977.

82. Connor, J. N. L. Catastrophes and molecular collisions. *Molec. Phys.* **31,** 33–55, 1976.

83. Fraser, A. B. and Mach, W. H. Mirages. *Scient. Am.* **234,** 102–111, 1976.

84. Thom, R. The two-fold way of catastrophe theory. In *Structural Stability, the Theory of Catastrophes, and Applications*

in the Sciences. Lecture Notes in Mathematics 525 (P. J. Hilton, ed.). Springer, Berlin and New York, 1976. pp. 235–252.

85. Guckenheimer, J. Solving a single conservation law. In *Dynamical Systems – Warwick 1974.* Lecture Notes in Mathematics 468. (A. Manning, ed.). Springer, Berlin and New York, 1975. pp. 108–134.

86. Golubitsky, M. Regularity and stability of shock waves for a single conservation law. In *Rencontre de Cargèse sur les Singularités et leurs Applications* (F. Pham, ed.). Institut d'Études Scientifiques de Cargèse, Publ. Math. Dept., Univ. Nice, 1975. pp. 84–88.

87. Davis, S. S. A preliminary investigation of sonic boom waveforms near focusing ray systems. In *Third Conference on Sonic Boom Research, NASA SP-255,* 1971. pp. 133–146.

88. Obermeier, F. Das Verhalten eines Überschallknalles in der Umgebung einer Kaustik. Max-Planck-Institut für Strömungsforschung (Gottingen), Bericht 28/1976.

89. Sturtevant, B. and Kulkarny, V. A. The focusing of weak shock waves. *J. Fluid Mech.* **73,** 651–671, 1976.

90. Wanner, J.-C. L., Vallée, J., Vivier, C. and Théry, C. Theoretical and experimental studies of the focus of sonic Booms. *J. Acoust. Soc. Am.* **52,** 13–32, 1972.

90a. Berry, M. V. Focusing and twinkling: critical exponents from catastrophes in non-Gaussian random short waves. Preprint, Univ. Bristol, 1977.

91. Arnol'd, V. I. Wave front evolution and equivariant Morse lemma. *Commun. Pure Appl. Math.* **29,** 557–582, 1976.

92. Lumley, J. L. and Panofsky, H. A. *The Structure of Atmospheric Turbulence.* Interscience, New York, 1964.

92a. Berry, M. V. and Hannay, J. H. Umbilic points on Gaussian random surfaces. Preprint, Univ. Bristol, 1977.

93. Mallory, J. K. Abnormal waves on the South-East coast of S. Africa. *Internat. Hydrographic Rev.* **51,** 99–129, 1974.

94. Smith, R. Giant waves. *J. Fluid Mech.* **77,** 417–432, 1976.

95. Euler, L. *Methodus Inveniendi Lineas Curvas Maximi Minimive Proprietate Gaudentes* (*Appendix, De Curvis Elasticis*). Marcum Michaelum Bousquet, Lausanne and Geneva, 1744.

96. Sewell, M. J. Elastic and plastic bifurcation theory. Preprint, University of Reading, 1976.

97. Sewell, M. J. A survey of plastic buckling. In *Study No. 6* (*Stability*). Solid Mechanics Division, University of Waterloo, Ontario, Canada, 1972. Chapter 5.

98. Butterworth, J. W. Frame instability. In *Structural Instability* (W. J. Supple, ed.). IPC Science and Technology Press, Guildford, 1973. pp. 54–63

99. Bishop Berkeley. The analyst: a discourse addressed to an infidel mathematician, 1734. Excerpted in Newman, J. R., *The World of Mathematics.* Simon and Schuster, New York, 1956.

100. Gromoll, D. and Meyer, W. On differentiable functions with isolated critical points. *Topology* **8,** 361–370, 1969.

101. Magnus, R. J. Universal unfoldings in Banach spaces: reduction and stability. *Mathematics Report* 107, Battelle, Geneva, 1977. To appear in Chillingworth, D. R. J. (ed.) *Catastrophe Theory in Infinite Dimensions.*

102. Magnus, R. J. On universal unfoldings of certain real functions of a Banach space. *Mathematics Report* 100, Battelle, Geneva, 1976. *Math. Proc. Cambridge Philos. Soc.* **81,** 91–95, 1977.

103. Magnus, R. J. Determinacy in a class of germs on a reflexive Banach space. *Mathematics Report* 103, Battelle, Geneva, 1976. To appear in *Math. Proc. Cambridge Philos. Soc.*

104. Magnus, R. J. and Poston, T. A strictly infinite-dimensional 'fold catastrophe'.

Mathematics Report 110, Battelle, Geneva, 1977.

105. Thompson, J. M. T. and Hunt, G. W. *A General Theory of Elastic Stability.* Wiley, New York and London, 1973.

106. Dieudonné, J. *Treatise on Analysis*, Volume 1. Academic Press, New York and London, 1970. Also *Foundations of Modern Analysis.* Academic Press, New York and London, 1960.

107. Chillingworth, D. R. J. The catastrophe of a buckling beam. In *Dynamical Systems – Warwick 1974.* Lecture Notes in Mathematics 468 (A. Manning, ed.). Springer, Berlin and New York, 1975. pp. 86–91.

108. Ball, J. M. Initial boundary value problems for an extensible beam. *J. Math. Anal. Appl.* **42,** 61–90, 1973.

109. Keller, J. B. and Antman, S. (eds) *Bifurcation Theory and Non-linear Eigenvalue Problems.* Benjamin, New York, 1969.

110. Thompson, J. M. T. and Hunt, G. W. Towards a unified bifurcation theory. *J. Appl. Math. Phys. (ZAMP)* **26,** 581–604, 1975.

111. Wasserman, G. (r, s)-stability of unfoldings and catastrophe theory, In *Structural Stability, the Theory of Catastrophes, and Applications in the Sciences.* Lecture Notes in Mathematics 525 (P. J. Hilton, ed.). Springer, Berlin and New York, 1976. pp. 253–262.

112. Wasserman, G. Stability of unfoldings in space and time. *Acta Math.* **135,** 57–128, 1975.

113. Zeeman, E. C. Euler buckling. In *Structural Stability, the Theory of Catastrophes, and Applications in the Sciences.* Lecture Notes in Mathematics 525 (P. J. Hilton, ed.). Springer, Berlin and New York, 1976. pp. 373–395.

114. Koiter, W. T. On the stability of elastic equilibrium. Dissertation, Delft, 1945; NASA Technical Translation F10 (1967) 833.

115. Sewell, M. J. On the connexion between stability and the shape of the equilibrium surface. *J. Mech. Phys. Solids* **14,** 203–230, 1966.

116. Croll, J. Is catastrophe theory dangerous? *New Scientist* 17 June 1976, 630–632.

117. Roorda, J. Stability of structures with small imperfections. *J. Engng Mech. Div. Am. Soc. Civ. Engrs* **91,** 87–106, 1965.

118. Wasserman, G. (r, s)-stability of unfoldings. Preprint, University of Regensburg, 1974.

118a. Porteous, I. R. The normal singularities of a submanifold. *J. Diff. Geom.* **5,** 543. 564, 1971.

119. Thompson, J. M. T. and Gaspar, Z. A buckling model for the set of umbilic catastrophes. Preprint, Engineering Department, University College London, 1977.

120. Thompson, J. M. T. and Hunt, G. W. Dangers of structural optimization. *Engng Optimization* **1,** 99–110, 1974.

121. Bauer, L. and Reiss, E. Nonlinear buckling of rectangular plates. *SIAM J.* **13,** 603–626, 1965.

122. Supple, W. J. Post-buckling behaviour of thin plates. In *Structural Instability* (W. J. Supple, ed.). IPC Science and Technology Press, Guildford, 1973.

123. Chow, S.-N., Hale, J. K. and Mallet-Paret, J. Applications of generic bifurcation I. *Arch. Rat. Mech. Anal.* **59,** 159–188, 1975.

124. Chow, S.-N., Hale, J. K. and Mallet-Paret, J. Applications of generic bifurcation II. *Arch. Rat. Mech. Anal.* **62,** 209–236, 1976.

125. Magnus, R. J. and Poston, T. On the full unfolding of the von Kármán equation at a double eigenvalue. *Mathematics Report* 109, Battelle, Geneva, 1977. To appear in Chillingworth, D. R. J. (ed.) *Catastrophe Theory in Infinite Dimensions.*

126. Truesdell, C. *Rational Thermodynamics.* McGraw-Hill, New York 1969.

127. Callen, H. B. *Thermodynamics.* Wiley, New York and London, 1960.

128. Fowler, D. H. The Riemann–Hugoniot catastrophe and van der Waals' equation. In *Towards a Theoretical Biology* (C. H. Waddington, ed.). Edinburgh University Press, Edinburgh, 1968–1972. Vol. 4, pp. 1–7.

129. Haug, A. *Theoretical Solid State Physics,* Vol. 1. Pergamon, Oxford and New York, 1972.

130. Landau, L. D. and Lifshitz, E. M. *Statistical Physics,* 2nd English edition (translated by Peierls and Peierls). Pergamon, Oxford and New York, 1959.

131. Jauch, J.-M. Thermodynamics and differential forms. Preprint MS-R-7217, Mathematics Department, University of Denver.

132. Spivak, M. *A Comprehensive Introduction to Differential Geometry,* 5 Vols. Publish or Perish, Boston, 1970 and 1975.

132a. Dubois, J.-G. and Dufour, J.-P. La théorie des catastrophes V: transformées de Legendre et thermodynamique. Preprint, Dept. de Math., Univ. de Québec, Montréal, 1977.

132b. Ekeland, I. Duality in nonconvex optimization and calculus of variations. *Technical Summary Report* 1675, Math. Res. Center, Univ. Wisconsin, Madison, 1976.

133. Bell, G. M. and Lavis, D. A. Thermodynamic phase changes and catastrophe theory. To appear.

134. Newman, M. H. A. *Elements of the Topology of Plane Sets of Points.* Cambridge University Press, Cambridge, 1961.

135. Schulman, L. S. Tricritical points and type three phase transitions. *Phys. Rev. Ser. B* 7, 1960–1967, 1973.

135a. Stanley, H. E. *Introduction to Phase Transitions and Critical Phenomena.* Oxford University Press, London and New York, 1971.

136. Toulouse, G. and Pfeuty, P. *Introduction au Groupe de Renormalization et à ses Applications.* Presses Universitaires de Grenoble, Grenoble, 1975.

137. Bertin, G. and Radicati, L. A. The bifurcation from the MacLaurin to the Jacobi sequence as a second-order phase transition. *Astrophys. J.* **206,** 815–821, 1976.

137a. Thompson, J. M. T. Instabilities, bifurcations, and catastrophes. *Phys. Lett. A* **51,** 201–203, 1975.

138. Bierstone, E. Local properties of smooth maps equivariant with respect to finite group actions. *J. Diff. Geom.* **10,** 523–540, 1975.

139. Ronga, F. Stabilité locale des applications équivariantes. Preprint, Mathematics Department, University of Geneva, 1976.

140. Ascher, E., Gay, D. A. and Poston, T. Equivariant bifurcation of thermodynamic potentials in crystallography. To appear.

141. Poston, T. and Budgor, A. B. A geometrical approach to calculating the energy and frequency spectra of crystals. *J. Comp. Phys.* **19,** 1–28, 1975.

142. Dicke, R. H. Coherence in spontaneous radiation processes. *Phys. Rev.* **93,** 99–110, 1954.

143. Haken H. Cooperative phenomena in systems far from thermal equilibrium and in nonphysical systems. *Revs Mod. Phys.* **47,** 67–121, 1975.

144. Jauch, J.-M. *Foundations of Modern Quantum Mechanics.* Addison Wesley, Reading, Mass., 1968.

145. Varadarajan, V. S. *Geometry of Quantum Theory,* 2 Volumes. Van Nostrand, New York, 1968.

146. Dicke, R. H. and Wittke, J. P. *Introduction to Quantum Mechanics.* Addison Wesley, Reading, Mass., 1960.

147. Gilmore, R. and Narducci, L. M. Relation between equilibrium and non-equilibrium critical properties of the Dicke model. To appear.

147a. Gilmore, R. Baker–Campbell–Hausdroff formulas. *J. Math. Phys.* **15,** 2090–2092, 1974.

148. Hepp, K. and Lieb, E. H. On the super-radiant phase transition for molecules in a quantized radiation field: the Dicke maser model. *Ann. Phys. N. Y.* **76,** 360–404, 1973.

149. Gilmore, R. and Bowden, C. M. Coupled order parameter treatment of the Dicke model. *Phys. Rev. Ser. A*, **13,** 1898–1907, 1976.

150. Gibbs, H. M., McCall, S. L. and Venkatesan, T. N. C. Differential gain and bistability using a sodium filled Fabry-Perot interferometer. *Phys. Rev. Lett.* **36,** 1135–1138, 1976.

151. Bonifacio, R. and Lugiato, L. A. Cooperative effects and bistability for resonance fluorescence. *Optics Commun.* in press.

152. Glauber, R. J. (ed.) *Quantum Optics, Proceedings of the International School of Physics 'Enrico Fermi' Rendiconti 42.* Academic Press, New York and London, 1969.

153. Glauber, R. J. Coherence and quantum detection. In *Quantum Optics. Proceedings of the International School of Physics 'Enrico Fermi' Rendiconti 42.* (R. J. Glauber, ed.). Academic Press, New York and London, 1969. pp. 15–56.

154. Arecchi, F. T. Photocount distributions and field statistics. In *Quantum Optics. Proceedings of the International School of Physics 'Enrico Fermi' Rendiconti 42* (R. J. Glauber, ed.). Academic Press, New York and London, 1969. pp. 57–110.

155. Freed, C. and Haus, H. A. Photoelectron statistics produced by a laser operating below and above the threshold of oscillation. *IEEE J. Quantum Electron.* **2,** 190–195, 1966.

156. Gilmore, R. and Bowden, C. M. Bifurcation properties of Dicke Hamiltonians. *J. Math. Phys.* **17,** 1617–1625, 1976.

157. Zeeman, E. C. Primary and secondary waves in developmental biology, AAAS 1974, *Some Mathematical questions in biology* VIII. In *Lectures on Mathematics in the Life Sciences* 7. American Mathematical Society, Providence, R. I., 1974. pp. 69–161.

157a. Heinrich, B. Bumblebee foraging and the economics of sociality. *American Scientist* **64,** 384–395, 1977.

158. Pitt, D. H. and Poston, T. Determinacy and unfoldings in the presence of a boundary. To appear.

159. Renfrew, A. C. and Poston, T. Discontinuities in the endogenous change of settlement pattern. To appear in Cooke, K. L. and Renfrew, C. A. (eds.) *Transformations: Mathematical Approaches to Culture Change*, Academic Press, New York and London, 1978.

160. Poston, T. and Wilson, A. G. Facility size vs. distance travelled: urban services and the fold catastrophe. *Environ. Planning A*, **9,** 681–686, 1977.

161. Thompson, B. V. The phase transition in a modified Dicke model. *J. Phys. A* **8,** 126–132, 1975.

162. Hopf, E. Abzweigung einer periodischen Lösung von einer stationaren Lösung einer Differentialsystems. *Ber. Verh. Sachs, Akad. Wiss. Leipzig. Math. Phys.* **95,** 3–22, 1943.

163. Hirsch, M. W. and Smale, S. *Differential Equations, Dynamical Systems, and Linear Algebra.* Academic Press, New York and London, 1974.

164. Marsden, J. E. and McCracken, M. *The Hopf Bifurcation and its Applications.* Applied Mathematics Series 19. Springer, Berlin and New York, 1976.

165. Connell, J. H., Mertz, D. B. and Murdoch, W. M. *Readings in Ecology and Ecological Genetics.* Harper and Row, New York, 1970.

166. Connell, J. H. The influence of interspecific competition and other factors on the distribution of the barnacle *Chthamalus stellatus. Ecology* **42,** 710–723, 1961.

167. Daubenmire, R. *Plant Communities.* Harper and Row, New York, 1968.

168. Ford, E. B. Evolution studied by observation and experiment. In *Readings in Genetics and Evolution*. Oxford University Press, Oxford and New York, 1973.

169. Wellington, W. G. Qualitative changes in populations in unstable environments. *Can. Entomol.* **96**, 436–451, 1964.

170. Ashton, K. Private communication. Data available from Department of Mathematics, University of Auckland, New Zealand.

171. Poston, T. Various catastrophe machines. In *Structural Stability, the Theory of Catastrophes, and Applications in the Sciences*. Lecture Notes in Mathematics 525 (P. J. Hilton, ed.). Springer, Berlin and New York, 1976. pp. 111–126.

172. Turing, A. M. The chemical theory of morphogenesis. *Phil. Trans. Roy. Soc. B* **237**, 32, 1952.

173. Hadeler, K. P., an der Heiden, U. and Rothe, F. Nonhomogeneous spatial distributions of populations. *J. Math. Biol.* **1**, 165–176, 1974.

174. Smale, S. On gradient dynamical systems. *Ann. Math.* **74**, 199–206, 1961.

175. Smale, S. Differentiable dynamical systems. *Bull. Amer. Math. Soc.* **73**, 747–817, 1976.

176. Winfree, A. T. Spatial and temporal organization in the Zhabotinsky reaction. Aahron Katchalsky Memorial Symposium, Berkeley, 1973.

177. Winfree, A. T. Rotating chemical reactions. *Scient. Am.* **230**, 82–95, 1974.

178. Howard, L. N. Bifurcations in reaction-diffusion problems. *Adv. Math.* **16**, 246–258, 1975.

179. Kopell, N. and Howard, L. N. Slowly varying waves. *Studies in Appl. Math.* **LVI** no. 2, 95–146, 1977.

180. Guckenheimer, J. Constant velocity waves in oscillating chemical reactions. In *Structural Stability, the Theory of Catastrophes, and Applications in the Sciences*. Lecture Notes in Mathematics 525 (P. J. Hilton, ed.). Springer, Berlin and New York, 1976. pp. 99–103.

181. Prigogine, I. *Introduction to Thermodynamics of Irreversible Processes*. Wiley, New York and London, 1967.

182. Zeeman, E. C. Differentiation and pattern formation. Appendix to Cooke, J. Some current theories of the emergence and regulation of spatial organization in early animal development. *Ann. Rev. Biophys. Bioengng*, **4**, 210–215. 1975.

183. Cooke, J. and Zeeman, E. C. A clock and wavefront model for control of the number of repeated structures during animal morphogenesis. *J. Theoretical Biology* **58**, 455–476, 1976.

184. Zeeman, E. C. Research ancient and modern. *Bull. Inst. Math. Appl.* **10**, 272–281. 1974.

185. Isnard, C. A. and Zeeman, E. C. Some models from catastrophe theory in the social sciences. In *Use of Models in the Social Sciences* (L. Collins, ed.). Tavistock, London, 1976. pp. 44–100.

186. Anon. Dirty Comics, a History of the Eight Pagers. Uitgeverij Tong-Dordrecht.

187. Cover, T. M. Geometrical and statistical properties of systems of linear inequalities with applications in pattern recognition. *IEEE Trans Electron. Compon.* **14**, 326–334, 1965.

188. Bremermann, H. Pattern recognition by deformable prototypes. In *Structural Stability, the Theory of Catastrophes, and Applications in the Sciences*. Lecture Notes in Mathematics 525 (P. J. Hilton, ed.). Springer, Berlin and New York, 1976. pp. 15–57.

188a. Weidlich, W. Dynamics of interacting social groups. In *Cooperative Effects* (H. Haken, ed.), pp. 269–282.

189. Zeeman, E. C., Hall, C., Harrison, P. J., Marriage, H. and Shapland, P. A model for institutional disturbances. *Br. J. Math. Statist. Psych.* **29**, 66–80, 1976.

190. Takens, F. Constrained equations: a study of implicit differential equations and their discontinuous solutions. In P.

J. Hilton (ed.) *Structural Stability, the theory of Catastrophes, and their Applications in the Sciences,* Lecture Notes in Mathematics 525, Springer, Berlin and New York, 1976. pp. 143–234.

190a. Takens, F. Implicit differential equations: some open problems. In Burlet, O. and Ronga, F. (eds.) *Singularités d'Applications Différentiables, Plans-sur-Bex 1975,* Lecture Notes in Mathematics 535, Springer, Berlin and New York, 1976. pp. 237–253.

191. Panati, C. Catastrophe theory. *Newsweek* 19 January 1976, 46–47.

192. Fisher, G. H. Preparation of ambiguous stimulus materials. *Percept. Psychophys.* **2,** 421–422, 1967.

193. Fisher, G. H. Ambiguity of form: old and new. *Percept. Psychophys.* **4,** 189–192, 1968.

194. Attneave, F. Multistability in perception. *Scient. Am.* **225,** 62–71, 1971.

195. Courant, R. Soap film experiments with minimal surfaces. *Amer. Math. Monthly* **47,** 168–174, 1940.

196. Zeeman, E. C. Catastrophe theory in brain modelling. *Int. J. Neurosci.* **6,** 39–41, 1973.

197. Zeeman, E. C. Brain modelling. In *Structural Stability, the Theory of Catastrophes, and Applications in the Sciences.* Lecture Notes in Mathematics 525. (P. J. Hilton, ed.). Springer, Berlin and New York, 1976. pp. 367–372.

198. Zeeman, E. C. Duffing's equation in brain modelling. To appear in *Bull. Inst. Math. Appl.*

199. Ashby, W. R. *Design for a Brain.* Chapman and Hall, London, 1952.

199a. Inouye, D. The dynamic microstructure of evaluative processes: structural stability models of judgement and intentional action. To appear.

199b. Brown, B. L., Inouye, D., Williams, R. and Borrus, K. A catastrophe theory account of dichotic listening. Preprint, Dept. of Psychology, Stanford University, 1976.

200. Zeeman, E. C. A mathematical model for conflicting judgements caused by stress, applied to possible misestimations of speed caused by alcohol. *Br. J. Math. Statist. Psych.* **29,** 19–31, 1976.

201. Drew, G. C., Colquhoun, W. P. and Long, H. A. Effect of small doses of alcohol on a skill resembling driving. Medical Research Council Memorandum 38, 1959.

202. Holmes. P. J. and Rand, D. A. The bifurcations of Duffing's equation: an application of catastrophe theory. *J. Sound Vibr.* **44,** 237–253, 1976.

203. Cournot, A. Exposition de la Théorie des Chances et des Probabilités, 1843. Translated by D. Irwin, Homewood, Illinois as *The Mathematical Principles of the Theory of Wealth.*

204. Wald, A. On some systems of equations of mathematical economics. *Econometrica* **19,** 368–408, 1957.

205. Thom, R. Théorie des jeux dans les variétés. In *Rencontre de Cargèse sur les Singularités et leurs Applications* (F. Pham. ed.). Institut d'Études Scientifiques de Cargèse, Publ. Math. Inst., Univ. Nice, 1975. pp. 20–26.

206. Li, T.-Y. and Yorke, J. A. Period three implies chaos. *Amer. Math. Monthly* **82,** 985–992, 1975.

207. Guckenheimer, J. On the bifurcation of maps of the interval. To appear.

208. Guckenheimer, J., Oster, G. and Ipaktchi, A. The dynamics of density-dependent population modes. To appear.

209. Rand, D. A. Exotic phenomena in simple games. Preprint, University of Warwick, 1977.

210. Smale, S. On the differential equations of species in competition, *J. Math. Biol.* **3,** 5-7, 1976.

211. Acton, F. S. *Numerical Methods that Work.* Harper and Row, New York, 1970.

212. Wilkinson, J. H. The evaluation of the zeros of ill-conditioned polynomials, part 1. *Numerische Math.* **1,** 150–166, 1959.

213. Zeeman, E. C. Catastrophe theory. *Scient. Am.* **234,** 65–83, 1976.

214. Chillingworth, D. R. J. Elementary catastrophe theory. *Bull. Inst. Math. Applic.* **11,** 155–159, 1975.

215. Stewart, I. N. The seven elementary catastrophes. *New Scientist* **68,** 447–454, 1975.

216. Stewart, I. N. Catastrophe theory. *Math. Chronicle.* in press.

217. Arnol'd, V. I. Lectures on bifurcations and versal families. *Uspehi Mat. Nauk* **27,** 119–184, 1972; *Russian Math. Surveys* **27,** 54–123, 1972.

218. Deakin, M. A. B. The formal power series approach to elementary catastrophe theory. Preprint, Monash University, 1976.

219. Guckenheimer, J. Shocks and rarefactions in two space dimensions. *Arch. Rat. Mech. Anal.* in press.

220. Thom, R. *Modèles mathématiques de la morphogénèse,* Éditions 10–18. UGE Paris, 1974.

221. Waddington, C. H. (ed.) *Towards a Theoretical Biology,* 4 Vols. Edinburgh University Press, Edinburgh, 1968–1972.

222. Poston, T. The elements of catastrophe theory *or* the honing of Occam's razor. To appear in Cooke, K.L. and Renfrew, A. C. (eds.), *Transformations: Mathematical Approaches to Culture Change,* Academic Press, 1978.

223. Fogel, J.-F., Hue, J.-L. and Thom, R. La planète de l'oncle Thom. *Le Sauvage,* January 1977. pp. 74–80.

224. Walgate, R. René Thom clears up catastrophes. *New Scientist* **68,** 578, 1975.

225. Dickson, D. and Thom, R. Was Newton's apple a cusp or a swallowtail? *Times Higher Education Supplement,* 5th December 1975. p. 13.

226. Ferrell, R. A. Fluctuations and Superconductors (W. S. Govee and F. Chilton, eds.). Stanford Research Inst., 1968.

227. Haken, H. (ed.). *Synergetics,* Teubner, Stuttgart, 1973.

Index